CAMBRIDGE LIBRARY COLLECTION

Books of enduring scholarly value

Life Sciences

Until the nineteenth century, the various subjects now known as the life sciences were regarded either as arcane studies which had little impact on ordinary daily life, or as a genteel hobby for the leisured classes. The increasing academic rigour and systematisation brought to the study of botany, zoology and other disciplines, and their adoption in university curricula, are reflected in the books reissued in this series.

The Vegetation of New Zealand

When botanist Leonard Cockayne (1855–1934) first received an invitation from the German publisher Engelmann to write an account of the botany of New Zealand, much of it was still unknown. He spent the period from 1904 to 1913 immersed in fieldwork, and his first edition was not published until 1921. In this 1928 second edition Cockayne extensively updates the text, adding the results of further research from the intervening years. This work gives detailed descriptions of New Zealand's plant life, but Cockayne also considers the history of botanical study of the islands, from Captain Cook's voyages in the eighteenth century onwards, and includes the arrival of colonial plant collectors and an overview of important publications by New Zealand botanists. The descriptions of vegetation cover the sea coast, the lowlands, mountains, and outlying islands, and there are extensive photographs, offering a comprehensive guide to New Zealand's botany.

Cambridge University Press has long been a pioneer in the reissuing of out-of-print titles from its own backlist, producing digital reprints of books that are still sought after by scholars and students but could not be reprinted economically using traditional technology. The Cambridge Library Collection extends this activity to a wider range of books which are still of importance to researchers and professionals, either for the source material they contain, or as landmarks in the history of their academic discipline.

Drawing from the world-renowned collections in the Cambridge University Library, and guided by the advice of experts in each subject area, Cambridge University Press is using state-of-the-art scanning machines in its own Printing House to capture the content of each book selected for inclusion. The files are processed to give a consistently clear, crisp image, and the books finished to the high quality standard for which the Press is recognised around the world. The latest print-on-demand technology ensures that the books will remain available indefinitely, and that orders for single or multiple copies can quickly be supplied.

The Cambridge Library Collection will bring back to life books of enduring scholarly value (including out-of-copyright works originally issued by other publishers) across a wide range of disciplines in the humanities and social sciences and in science and technology.

The Vegetation
of New Zealand

Leonard Cockayne

CAMBRIDGE UNIVERSITY PRESS

Cambridge, New York, Melbourne, Madrid, Cape Town,
Singapore, São Paolo, Delhi, Tokyo, Mexico City

Published in the United States of America by Cambridge University Press, New York

www.cambridge.org
Information on this title: www.cambridge.org/9781108032384

© in this compilation Cambridge University Press 2011

This edition first published 1928
This digitally printed version 2011

ISBN 978-1-108-03238-4 Paperback

Die
Vegetation der Erde

Sammlung
pflanzengeographischer Monographien

herausgegeben von

A. Engler
em. Professor der Botanik

und

O. Drude
em. Professor der Botanik

XIV.

The Vegetation of New Zealand

By

L. Cockayne
Ph. D., F. R. S., F. L. S., F. N. Z. Inst.

Second Edition

With a Frontispiece, 106 Figures on 87 Plates and 3 Maps

Leipzig

Verlag von Wilhelm Engelmann

1928

The
Vegetation of New Zealand

By

Dr. L. Cockayne

F. R. S., F. N. Z. Inst.

Hon. Botanist New Zealand State Forest Service

With a Frontispiece, 106 Figures on 87 Plates and 3 Maps

2nd edition, almost entirely rewritten, thoroughly
revised, and enlarged

Leipzig

Verlag von Wilhelm Engelmann

1928

House in Dunedin (South Otago District), the residence of Mr. W. A. Thomson,
built in the middle of the last century of tree-fern trunks from the original forest,
but now surrounded by a dense artificial plantation of various species of *Nothofagus*.
Photo. Bathgate.

Preface.

Early in the year 1904 I had the honour to receive a letter from Prof. Dr. A. ENGLER inviting me to contribute to the comprehensive "Vegetation der Erde" a volume written in English dealing with the plant-geography of New Zealand. At the same time Prof. ENGLER sent me a synopsis of the proposed work, which he had prepared, and on that, with certain modifications, this book is based.

At that time many wide areas in New Zealand were botanically unexplored, and a great part of the remainder was imperfectly known, consequently it was essential for me to acquire a first-hand knowledge of at least typical examples of the vegetation of each botanical district. This preliminary work was steadily carried on year by year, but it was not completed until June, 1913, at which period the actual writing of the book was begun. By the end of March, 1914, the work was completed, and soon afterwards the manuscript was forwarded to Berlin.

It is no easy matter, even when in close touch with the publisher, to see a scientific work through the press. How much then are the difficulties increased when half the circumference of the Globe separates publisher and author. Nor are these difficulties lessened when the copy for the printer to deal with is in a foreign tongue. But consider the infinitely greater difficulties which arose through the long years of the gigantic world-struggle and the subsequent time of reconstruction!

Much of the book only came into my hands in the form of paged proofs, so little more than verbal corrections could be made. Fortunately this was not the case from page 209 onwards, for of this portion galley proofs were available. These reached me in June, 1920, rather more than six years after the manuscript had left New Zealand, and, thanks to the publisher, I was permitted to make certain important alterations designed to bring the latter part of the work up to the present-day state of knowledge of the vegetation and flora of New Zealand. Not that the other part of the book is greatly deficient in this respect, since much of the theme is the *primitive* plant-covering.

With regard to the classification of the vegetation, with some modifications Warming's system of 1909 is followed. But the system adopted for such a book as this is not of great moment, since, above all else, the aim should be to present as vivid and accurate a picture as possible of the actual vegetation of the country. This surely is the first step in a plant-geographical description of any country. And it is the more necessary in a region, such as New Zealand, possessing a truly virgin vegetation which is rapidly becoming modified, or even destroyed.

As for the biology of the plants the somewhat novel method is adopted of giving detailed statistics regarding the growth-forms of the species and of certain of their vegetative parts. This procedure should be useful for comparative purposes both in the region dealt with and elsewhere. Obviously in deciding certain points, such as relative size or texture of leaves, the personal equation comes in, but where many species are concerned this should not affect the general result.

No attempt is made at completeness. On the contrary, owing partly to the limited space available and partly to the great variety of New Zealand plant communities, the matter is greatly condensed. Many species, especially those which are rare, are not mentioned; it is, after all, the common ones which are of prime importance.

The reader not acquainted with the New Zealand flora has been kept in mind. The leading physiognomic plants are treated at considerable length for each section of the vegetation, while the growth-forms of many species are described when they first appear in the text. Vernacular names are specially avoided.

Since 1914 I have ceased calling the tussock formations of New Zealand "steppe" because they do not fit into the usual plant-geographical conception of that term. I have therefore substituted "tussock-grassland" as a self explanatory name for a distinct type of vegetation. This term appears once or twice in the latter part of the book, but generally "steppe" remains. It must be remembered, then, that the latter is ecologically distinct from true steppe, indeed it has much wider physiological capabilities and can maintain itself intact under a surprising variety of conditions.

The meaning of "epharmonic", as used in the biological chapters, must be explained, since its significance according to my usage is somewhat different to Vesque's definition of the term. In this book, and in my other publications, by "epharmonic variation" is meant a change in its form, or physiological behaviour, beneficial to an organism evoked by the operation of some environmental stimulus. Such a change may be called an epharmonic adaptation as distinguished from such adaptations as cannot be traced to any direct action of the environment.

Apart from those mentioned in the text with regard to photographs, or special information, which they have generously supplied, many botanists and others have given valuable and much-appreciated assistance with regard to this book. To name all would extend this preface far beyond its alloted space, to give merely a partial list would be invidious. Therefore, I thank most sincerely one and all.

I must, however, express my gratitude to Prof. Dr. L. Diels who has devoted much valuable time to seeing the book through the press. Nor can I neglect thanking Prof. Dr. A. Engler for having allowed me the great privilege of contributing to this famous series of monographs of which he and Prof. Dr. O. Drude are the distinguished Editors.

Ngaio, Wellington, New Zealand,
January 15th, 1921.   **L. Cockayne.**

Preface to the Second Edition.

Owing to the first edition of this book being disposed of within one year of its appearance and to the many favourable reviews in scientific journals Geh. Ober-Regierungsrat Prof. Dr. A. ENGLER and Herr WILHELM ENGELMANN invited me to prepare a second edition. This unexpected but truly welcome invitation I most willingly and gratefully accepted in the hope that the far wider knowledge of the vegetation which I had acquired since 1913 would ensure the production of a work more worthy of its great subject. Be this as it may, this edition is practically a new book, by far the larger part having been rewritten and the remainder thoroughly revised. That the making of a book so greatly changed was possible is due in large measure to my having botanically re-explored much of the Region, thanks especially to the carrying-out of plant-ecological studies at all altitudes of the tussock-grassland and forest areas for the Department of Agriculture and the State Forest Service respectively. Not only were localities visited with which I was more or less familiar but many others were examined the vegetation of which was unknown. In addition I have received important communications concerning areas previously unbotanized from Drs. H. H. ALLAN and W. MCKAY and Messrs. J. A. MACPHERSON, W. R. B. OLIVER, G. SIMPSON and J. SCOTT THOMSON (working together) and W. A. THOMSON. Also during the period which has elapsed since the manuscript of the first edition was sent to the publisher (beginning of 1914 to end of 1927), as may be seen in Part I, Chapter II, much has been published in various branches of New Zealand botany and of this a good deal finds a place in the book. Here it is not feasible to mention all those who have contributed information or material including photographs but, unless by some oversight, they are cited in the body of the book, but to one and all I express my gratitude. In addition, attention must be called to my great indebtedness to my friend Dr. H. H. ALLAN to whom I sent for criticism all the rough manuscript as soon as it was prepared and whose advice was invaluable. Furthermore it is my pleasing duty to thank most sincerely my old friend, Prof. Dr. L. DIELS, Director of the Botanical Garden and Museum of Berlin and Honorary Member of the New Zealand Institute, who is correcting the proofs of this edition — a most generous action — and so saving the waste of time entailed in their coming out to me and travelling twice round the world.

Ngaio, Wellington, New Zealand,
January 13th, 1928.   L. Cockayne.

List of Plates and Figures.

Contents.

Part I.
Introduction.

XVI
Contents.

Part II.
The Vegetation of Primitive and Semi-primitive New Zealand.

Section I.
The Vegetation of the Sea-coast.

Pages

Section II.

The Vegetation of the Lowlands and Lower Hills.

Section III.
The Vegetation of the High Mountains.

Part IV.
The Flora of New Zealand, its Distribution and Composition.

The following abbreviations are used for the Botanical Districts and Subdistricts.

NA. = The North Auckland.
SA. = The South Auckland.
K. = The Kaipara.
W. = The Waikato.
T. = The Thames.
EC. = The East Cape.
VP. = The Volcanic Plateau.
EW. = The Egmont-Wanganui.
RC. = The Ruahine-Cook.
SN. = The Sounds-Nelson.
NE. = The North-eastern.
NW. = The North-western.
W. = The Western.
E. = The Eastern.
NO. = The North Otago.
SO. = The South Otago.
F. = The Fiord.

Part I.
Introduction.

Chapter I.
Preliminary Remarks.

It is not generally understood that the New Zealand Botanical Region possesses certain fundamental characteristics absent elsewhere. In the first place, the two main islands are together far and away greater in area than other masses of land equally remote from the nearest continent. Nor is this extreme isolation a thing of to-day but it dates far back into geological time, so far indeed that a flora has developed — considering only the angiosperms — with four-fifths of its species endemic, while many bear virtually no relationship to others elsewhere. The flora, too, falls naturally into the few clear-cut elements dealt with in Part IV, Chapter II, so the phytogeographer generally need be in little doubt regarding the supposed origins and relationships of the species with which he has to deal.

Perhaps even more striking than the systematic position of the plants is their distributional ecology, sociology and life-forms, all three bound up, in large measure, with the peculiar climatic conditions and geomorphological features of the islands, though neither the effect of the complete isolation of the region nor its geological history must be overlooked. Can any other part of the globe *equal in area* boast a vegetation which has to cope with such a variety of circumstances? Consider the rainfall of more than 6000 mm. in some parts and less then 350 mm. in others, the violent rainless hot winds and the piercing subantarctic gales with a cold downpour, the warm sheltered frostless valleys, the alpine heights with perpetual snowfields and huge glaciers descending to extremely low altitudes, or the lesser heights where snow lies all the winter and far into the summer. Then there are the conditions supporting luxuriant subtropical rain-forest in many parts and subantarctic *Nothofagus* forest in others; there is the vegetation which can tolerate the neighbourhood of boiling springs and fumaroles, or soils containing an excess of magnesia or of salt; there are enormous unstable stony-debris fields in the drier mountains and the scoria slopes of the volcanoes,

both with their highly-specialized plant-life; finally there is the gradual latitudinal change in the flora with species suddenly giving out and others
coming in.

Perhaps the most striking feature of New Zealand phytogeography,
dealt with more fully further on, is the vegetation having gained its structure unexposed to the attacks of grazing and browsing animals, the moa
(*Dinornis*) excepted. Nor did the aborigines cause any change except to a
comparatively limited extent in the neighbourhood of their villages. The
absence of this grazing-mammal factor stands out clearly in the light of
what has happened since the white man occupied the land. Besides the
main islands there are others adjacent thereto as also isolated groups far
away northwards, southwards and eastwards. These far-distant groups of
quite small islands have each their special floras containing not only a true
New Zealand element but marked by a high degree of endemism. Nor is
this all, their vegetation is of a more or less peculiar character and the
most southerly of the southern groups — the Macquarie Islands —, though
only as far distant from the equator as north Yorkshire, possesses a flora
of only 34 species of vascular plants!

The New Zealand flora, existing as it does under such diverse conditions, obviously must be made up of many life-forms, yet, in harmony
with the mildness of the climate, there are certain wide-spread, general
features. Thus, for the most part, the plants are woody or semi-woody;
they can endure no high degree of frost, in fact many (probably nearly all)
are close to their frost-resisting capacity; a large percentage are evergreen;
ferns are extremely abundant; bryophytes, often of great size, are a most
frequent feature. Thus there is a strong hygrophytic element which is
largely of palaeotropic origin. On the other hand, there are many life-forms
more or less restricted to definite habitats, many of which forms are so
strongly xerophytic that, as DIELS was the first to point out, they do not
seem suited to their present-day conditions — a supposition strongly supported by the fact that these xerophytes grow in company with typical
mesophytes. Then there is that peculiar biological group consisting of plants
which persist in a juvenile form distinct from that of the adult for many
years — in some cases 50 years and more — and many are xerophytes
at one stage and mesophytes at another. Finally, it must be noted that
wild hybrids are extremely common and that such do not consist of one
or two individuals but frequently of great polymorphic swarms.

At the present time the plant-covering of New Zealand differs greatly
from what it was in the comparatively recent pre-European days. Then,
as already noted, almost all the vegetation was primeval. But now by far
the greater part of the lowland belt bears a stamp of a European character.
All the same, thanks to the rugged, mountainous nature of much of the land,
and to the fact that many scenic reserves and large national parks have

been established, there are still numerous areas — some of great extent —
which are not only virgin, or nearly so, but which represent almost all the
primitive types of vegetation. Nevertheless, these splendid remnants of the
primeval world bid fair, except in inaccessible situations, to be seriously
modified. This deplorable state of affairs is the result not so much of
destruction for the rightful purposes of settlement, but rather for there
having been turned loose in the forests and on the pastoral lands animals
inimical to vegetation (deer of various kinds, rabbits) simply for purposes
of so-called "sport". The deer, assisted in some places by cattle which
have become wild, are doing excessive damage over wide areas to forest
undergrowth and frequently forbidding all natural regeneration. Some of
the virgin plant communities described in the first edition of this work have
been altered greatly or even destroyed. Vegetation, which came into being
in the absence of grazing and browsing mammals, is ill-equipped to withstand their onslaught. Ecologically, it is all-important to study what happens,
but the effect is heart-rending.

Far behind the grazing animal as a modifying and destroying element
are the introduced plants. These in their hundreds of species and vast
numbers of individuals have spread through the length and breadth of the
region; but, when they come in contact with the truly virgin vegetation,
they — one and all — come to a halt. Nevertheless, once there is an
open place, especially where the soil has been disturbed, exotic species
more or less readily gain a footing. So, too, indigenous species may contend
for the mastery in the new habitats. Thus there is every transition from
primitive to artificial vegetation and on this the following primary classification used in this book depends.

The Vegetation is made up of plant-communities which, if actually
virgin are (1) **Primitive,** but if not changed so fundamentally as to have
lost their primeval stamp they are (2) **Modified,** while all other communities
which differ greatly from the primitive and have been made directly or
indirectly by man's action are called (3) **Induced.** Induced vegetation consists of the three following classes: (1) **Indigenous-induced** when the
dominant member or members are indigenous species which have come into
the community by man's indirect action, e. g. *Pteridium* heath, *Danthonia*
grassland; (2) **Exotic-induced** when the dominant member or members are
exotic species which have come into the community by man's indirect action,
e. g. *Ulex* thicket, *Plantago Coronopus* salt-meadow; and (3) **Artificial-
induced** where the community has been directly made by man by ploughing
and sowing or other means, e. g. *Eucalyptus* plantation, wheat field. A fairly
detailed account of these non-primitive communities is given in Part III,
Chapter II, but here they are defined since in Part II not only the actual
primitive communities are dealt with but also those which are slightly modified. This treatment differs from that of the first edition since it not only

1*

attempts to depict the plant-covering of primitive New Zealand but it seeks
to show more clearly the early stages in the development of the new vegetation by keeping the descriptions of both classes side by side. An account
of this evolution of a vegetation new to the globe seems particularly important since there can be very little really primitive vegetation in Europe.
But in New Zealand, there can be estimated to some extent the reaction
of the primitive vegetation to the new factors — grazing and browsing animals
and exotic plants, the latter mostly of life-forms absent in the New Zealand flora.

The foregoing classification is not nearly so simple to put into practice
as it seems. This may best be seen from a few examples. Who can say,
for instance, how far much of the *Pteridium* heath is truly primitive? Yet
who in the pre-ecological days of New Zealand would have questioned its
apparently virgin character. So, too, with certain areas of tussock-grassland
the question arises, have they replaced forest? No association bears the
stamp of one truly primitive more than that of Jack's Pass near Hanmer (NE.),
made up almost entirely of mountain plants; but the remains of burnt trees
and small patches of forest show that it is altogether an indigenous-induced
community and not a primitive one slightly modified. Near Clinton (SO.)
there is a remnant of forest undergrowth of a primitive stamp consisting
of *Fuchsia excorticata*, yet it owes its existence to the other trees having
been removed for firewood, the *Fuchsia*, thanks to its bad reputation for
burning, having been left. In the Urewera Country (EC.) there are considerable stands of the last-mentioned species, but such are quite primitive
and represent a stage of forest development or retrogression.

In dealing with the groups of vegetation in Part II, it has seemed advisable to write sometimes in the present and sometimes in the past tense,
using the former when the community has been studied recently and the
latter when this took place some years ago, and there is reason to believe
that since then it has been more or less modified or even destroyed. As
an example, a certain piece of forest of the Chatham Islands visited by me
in 1901 had stock turned into it soon after my visit, so it is now greatly
altered.

In Part II, when dealing whith the communities, the associations are
arranged rather into groups of such than into definite associations, for the
latter, like the contents of species, are generally an expression of their
describer's opinion rather than distinct entities. To arrange the entire vegetation of New Zealand according to the natural small groups of which it
consists is a task of great magnitude requiring many workers, working
according to a definite system.

The primary aim of this book is to present an accurate picture of the
whole plant-covering of the New Zealand Botanical Region. In order to do
this some uniform system of classification is necessary, but its nature seems

to me to be a matter of comparatively small importance, the essential points being that it be simple, readily understood, and used as consistently as the complexity of the vegetation will permit. Such classification is here based, not on habitat or succession, but on the actual combinations of the plants themselves — though their dynamic relations are frequently discussed. Thus forest is a distinct community made up of trees growing closely, but it flourishes on almost every kind of soil no matter what the chemical and physical constitution may be and it is the climax of different successions originating on habitats distinct from one another, e. g. swamp, bog, unstable dune, dune-hollow, deep pumice, stony river-bed, glacial deposits and soil made from the weathering of most classes of rock.

Habitat — using this term in a restricted sense without reference to climate — is, however, by no means ignored in my classification. On the contrary, in order to obtain a clear picture of the vegetation, as well as for convenience of study, habitats of a specially distinct character are made use of for the primary name of the vegetation, or groups of communities, which constitute their covering, notwithstanding these groups may belong to different formations. Thus it will be seen that dune, rock, stony-debris and other striking habitats are used as descriptive titles for the classes of vegetation which they support, though the latter is to be distinguished in the long run by its ecological and floristic characters.

Having arranged the vegetation into its primary divisions according to its indigenous or exotic composition, there comes the much-debated question of its ecological nomenclature. Here WARMING's system of 1909 is, in the main, followed. The communities are divided into the plant-formation, plant-association, subassociation and plant-colony, and there are also to be distinguished successions and temporary associations; these terms may be respectively defined or explained as follows:

The **plant-formation** is an assemblage of plants of definite life-forms, one form or more than one — frequently many — being represented, and for the formation as a whole there is a representative flora. Thus, though primarily an ecological conception, it is also floristic and the latter character is sometimes of considerable importance. Particular plant-formations need not be restricted to a particular phytogeographic region, they may be world-wide, and such may be disignated *major formations* and their regional divisions *minor formations,* but for convenience the word "minor" is omitted in this book. The **plant-association** is a portion of a plant-formation distinguished by a definite floristic composition. **A subassociation** is a community within an association distinguished by its slightly different floristic composition, especially the dominance of some species other than that considered dominant for the association as a whole. Thus the taraire and the kauri subassociations occur in the kauri-broad-leaved tree association. The **colony** is a more or less pure group of one or more species

in the association or subassociation, as a tree-fern colony in a forest association or a *Ranunculus Lyallii* colony in a herb-field association. **Dominance,**
mentioned above, is not a matter of the largest number of any species
present, but of that species which *physiognomically* dominates, e. g. in
indigenous-induced steppe the commonest plant is the tiny *Poa maniototo,*
but the flat cushions of *Raoulia lutescens* dominate (Fig. 59).

All associations are in process of change, but where there is apparent
stability the community may be called a **climax association,** and a **temporary association** where more or less rapid change is in progress. **Succession** depends upon the alteration brought about in its habitat by the
association thereon — both of plants and animals — which leads eventually to a new association being established, as where the *Raoulia* association of stable stony river-bed, through colonization of the raoulia mats
or cushions by other plants, adds humus to the substratum and so paves
the way for *Festuca* tussock-grassland. This particular kind of change is
biological succession. Such may also come about through the "struggle" &c.
of plant with plant, as when the epiphytic *Metrosideros robusta* kills its
host, *Dacrydium cupressinum,* and its life-form, now that of a tree,
replaces its former host, or where the *Metrosideros* itself is killed by the
weight of epiphytic *Asteliae* its life-form encourages. Also the presence
of gracing animals, or a plant-disease, readily brings about succession.
Geomorphological change leads to a **spurious succession** in which plants
may play no part, and a new habitat arises ready for a new association
or it may be a new formation. An interesting example of the latter is
where the wind transforms a dune into a sand-hollow or sand-plain, the
original sandgrass association having nothing to do with the new vegetation
of the latter. Gradual geomorphological change, however, may go hand
in hand with biological action as in the case of river-bed, cited above,
where, but for the raising of the bed above the stream's reach there would
be no succession.

Should succession lead to the incoming of life-forms different from
those of the preceding association, such a succession is not here considered
part of the original formation, though the habitat may not be greatly
changed; thus the vegetation of unstable dunes belongs to a different formation from that of the subsequent shrubland and this, again, from that
of the final dune-forest.

Before closing this chapter a brief explanation must be given of the
taxonomic terms and conceptions used throughout this book which are those
put forth by H. H. ALLAN and myself in recent publications. The underlying conceptions are, (1) that the term "species" can quite well be defined,
(2) that the units on which it is based are virtually invariable, and (3)
that the content of these is not a matter of opinion but of fact.

The terms used by ALLAN and myself and their meaning are set forth in the following statement taken from one of our joint papers (1926: 12), the only alteration being the bold type. "As our fundamental field-unit we here take with LOTSY[1]) (1916, p. 27), the **jordanon,** which may be defined as "a group of externally alike individuals which breed true when bred among themselves." A jordanon that is not closely related to any other jordanon is easily recognizable in the field as an "invariable species," and such we term a **simple species,** e. g. *Hebe cupressoides.* More often groups of closely-related jordanons are met with, and such groups we term **compound species,** as the more familiar terms "aggregate species and collective species" have now a wider and vaguer connotation. Jordanons that are sufficiently distinct to allow of effective diagnosis we call **varieties.** It should be emphasized that as so defined the variety is of as much importance as the species. Compound species to which in the past taxonomy has attached "intermediates" we term **linneons.** LOTSY (*loc. cit.,* p. 27) uses the term linneon "to replace the species in the Linnean sense, and to designate a group of individuals which resemble one another more than they do any other individuals." As modified by us the term linneon is used to include not only groups of allied jordanons but of the hybrids between them as well. In its widest sense it includes two or more closely related species, their jordanons, and hybrids of all categories. This is what in practice the so-called "Linnean species" have become, coupled, not seldom, with what we later define as "epharmones." The intermediates we consider to be **hybrids.** Unfixed forms, due to habitat conditions, which change when these are sufficiently modified we call **epharmones.** Of course, strictly speaking, the "normal form" is as much an epharmone as any other, and sometimes it would be difficult to say which form should be classed as the "normal" one, but in practice this seldom occasions any difficulty...... We use the term **form** where it is convenient to waive the exact status of an individual or group, or where that is unknown. With BAILEY (1924, p. 25) we use the term **cultigen** for a form of unknown origin found only in cultivation."

Finally, it may be pointed out that the account in this book of the vegetation is generally based upon my personal examination — frequently too rapid — of such areas as I have been able to visit. Such are in all the botanical districts, except the Kermadec and the Macquarie, and as they usually represent various parts of each district (sometimes many parts), they may be taken as giving a fair idea of the vegetation in general. A few descriptions are compiled from the notes, generously given to me, of other ecologists, and such matter from earlier writers as appears of moment has been embodied, but in the last two cases the sources are always indicated. As for the material from which the descriptions of the communities have been drawn up, it consists almost entirely of notes including photographs taken by me in the field.

1) Evolution by means of Hybridization, 1916. The Hague.

Chapter II.
History of Botanical Investigation.

1. The Period of Voyages of Discovery in the South Pacific and of Investigations by Botanists from abroad.

The voyages of Captain Cook. Leaving out of the question knowledge of the plants such as the Maoris possessed, New Zealand botany commences with the landing of Sir JOSEPH BANKS and DR SOLANDER at Poverty Bay on Oct. 8th, 1769, from COOK's famous ship, the Endeavour. COOK remained in New Zealand waters until March 31st, 1770. Collecting was confined to a few places on the coast of North Island from Poverty Bay to the Bay of Islands and to Queen Charlotte Sound and Admiralty Bay in South Island. Altogether some 360 species of vascular plants were gathered, most coming from Tolaga Bay (160 spp.) and Queen Charlotte Sound (220 spp.).

COOK set sail on his second voyage on April 9th, 1772, accompanied by J. R. FORSTER and G. FORSTER, the son, as naturalists. At the Cape of Good Hope, A. SPARMANN was engaged as assistant naturalist. On March 26th, 1773, COOK put into Dusky Sound, remaining until May 1st, during which period the first investigation of the actual South Island flora was made, that of Queen Charlotte Sound being almost identical with that of the most southern part of North Island. May 18th to June 7th was spent in Queen Charlotte Sound. The total number of vascular plants collected was 190, a poor result considering the great opportunities for collecting. The botanists ascended none of the mountains at Dusky Sound and their rich high-mountain flora remained unknown, except for two or three specimens probably collected by some of COOK's officers who partially climbed one of the mountains. Twelve months after their return, the FORSTERS conjointly issued their *Characteres Genera Plantarum*, in which, with other genera, 31 belonging to New Zealand are described and figured. In 1786 G. FORSTER published his *Florulae Insularum Australium Prodromus,* 170 of the species mentioned therein being from New Zealand.

From Vancouver's voyage (1791) to the publication of the Flora Antarctica (1847). Captain VANCOUVER on his way to north-west North America, put into Dusky Sound for nearly 3 weeks. A. MENZIES, the surgeon, collected ferns, mosses and liverworts. The greater part of his specimens were described by Sir W. J. HOOKER and beautifully figured in the *Musci Exotici* (1818—20) and the *Icones Filicum.*

It was not until 1824 that further botanical investigations were made, when from the French corvette, "Coquille", Lieut. D'URVILLE and A. LESSON made a collection of plants at the Bay of Islands. The colonization of New South Wales had for some time past made itself felt. Whaling and sealing were pursued with vigour on the New Zealand coast; missions had been established since 1814, and settlement in the north was gradually extending. C. FRASER of the Sydney Botanic Garden, in 1825, collected a few plants at the Bay of Islands, and probably he had previously received one or two species from Macquarie Island.

In 1826 ALLAN CUNNINGHAM, His Majesty's Botanist at Port Jackson, spent four months botanizing from the Bay of Islands to Hokianga and in the neighbourhood of Whangaroa. Aided by natives and missionaries, he did valuable work, collecting 300 species, many of them new, together with ample duplicates. The following year, D'URVILLE again visited New Zealand, this time as commander of the "Astrolabe". He was accompanied, as before, by LESSON. Collections were made at various localities on the north coast of South Island, and at a few places in North Island from Tolaga Bay to the Bay of Islands. The botanical results were published in 1832 by A. RICHARD in his *Essai d'une Flore de la Nouvelle Zélande*[1]) which dealt also with the previous French collections and those of the FORSTERS.

R. CUNNINGHAM, brother of ALLAN CUNNINGHAM, Superintendent of the Sydney Gardens, made extensive collections during 5 months of 1833 near Whangaroa, the Bay of Islands and Hokianga. The material collected by himself and his brother forms the basis of a series of papers by A. CUNNINGHAM published in the *Companion to the Botanical Magazine*, Vol. 2., and continued in the *Annals and Magazine of Natural History* (the popular name of that publication) under the title *Florae Insularum Novae Zelandiae Praecursor*[2]). In 1839 J. C. BIDWILL visited New Zealand, penetrating to the centre of North Island and ascending Ngauruhoe, in the neighbourhood of which he discovered typical alpine plants. Later on, he revisited the Colony, and, in Nelson, added to the knowledge of the high-mountain flora. From 1840—42 E. DIEFFENBACH, naturalist to the New Zealand Company, travelled through much of North Island, ascended Mount Egmont, and, in South Island, spent some time on the coast of Marlborough[3]). The flora of Banks Peninsula was first investigated in 1840—41 by E. RAOUL, surgeon to the "Aube" and "Allier". He discovered many new spermophytes, and in 1844 his finely illustrated *Choix de Plantes de la Nouvelle Zélande* appeared, in which are enumerated all the species then known.

---

1) This excellent publication, even yet of moment, is the first dealing with the flora as a whole. The 380 species include 211 spermophytes, 51 pteridophytes and 118 lower cryptogams.

2) This important work contains descriptions of nearly all the then known species and includes 639, of which 394 are spermophytes, 95 pteridophytes and 150 lower cryptogams, an increase of 259 species since RICHARD's work.

3) Considering his excellent opportunities, his collections were scanty. In his *Travels in New Zealand* (1843) there is some scattered information as to the vegetation, and one chapter deals briefly with the flora. He also visited Chatham Island, but gathered only 12 species.

The years 1839—40 mark a most important phase in the history of New Zealand botany, in the searching examination of the Lord Auckland and Campbell Islands by the botanists of the French and English Antarctic expeditions. The former, under Admiral D'URVILLE, anchored in Port Ross from the 11th to the 12th of March, 1840. Botanical collections were made by the naturalists, HOMBRON and JACQUINOT, and the Admiral himself. The results were published at intervals from 1845—54 as part of a splendid work *Voyage au Pole sud et dans l'Océanie sur les Corvettes "l'Astrolabe" et "la Zélée" pendant les années 1837 à 1840*[1]).

The English expedition, under Sir JAMES ROSS, spent from Nov. 20th to Dec. 13th at the northern end of the Lord Auckland Group, and from Dec. 13th to Dec. 17th, 1840, on Campbell Island. JOSEPH DALTON HOOKER was botanist to the expedition, and associated with him was LYALL. It is impossible to speak too highly of HOOKER's untiring industry and skill as a collector, for, notwithstanding the visits of several botanists to these islands, and of one well-equipped expedition, but few additional species have been recorded. HOOKER devoted the first volume of his magnificent *Flora Antarctica* (1844) to the description of this wonderful flora that he had been so largely instrumental in making known. The 400 species dealt with comprise 105 spermophytes, 18 pteridophytes, and 277 lower cryptogams. The American WILKE's expedition visited Lord Auckland Island at about the same time as the French, but its botanical discoveries are of little moment.

2. The Period of Colonial Collectors and HOOKER's further Investigations.

General. Although a scattered European population had occupied parts of northern North Island for some 20 years, it was not till 1839 that regular settlement began, and that, one after the other, the Provinces were founded. Thenceforth, the history of botanical research is bound up with that of the development of the Colony and for some time its progress depended on that love for Nature which inspired a few enthusiasts to collect the plants near at hand, or even to undertake distant botanical excursions. But there were none in the young Colony who felt equal to describing their discoveries, nor indeed was there means for local publication. Consequently all sent their collections to HOOKER, and rightly so since he was not only the most competent man to deal with them, but the great resources of Kew and the collections of the earlier explorations were at his disposal.

---

1) There is a folio volume of admirable plates with the names of the species by HOMBRON and JACQUINOT, and two volumes of descriptions, the one (*Muscineae* and *Thallophyta*) by MONTAGNE, and the other (vascular plants) by DECAISNE.

From the arrival of COLENSO in New Zealand to the publication of the Flora Novae-Zelandiae (1853—55). The Rev. W. COLENSO came to New Zealand as a missionary in 1834. His earliest collections were made from North Cape to Whangarei. Between 1842 and 1847, he made many arduous journeys on foot through some of the most difficult country in North Island, accompanied only by a few Maoris. The Ruahine Mountains, the Volcanic Plateau and the high lands of the East Cape district were perhaps the most important of numerous localities visited, since they added many species to the little-known high-mountain flora. At about this time A. SINCLAIR collected on the east of North Island, and somewhat later he added, in Nelson, a good many species to the high-mountain flora. From 1847—51, LYALL, surgeon to the Acheron, then surveying the coast, made important collections chiefly in the south and south-west of South Island. Though he ascended to more than 900 m. in the Fiord district, he does not seem to have gained the alpine belt. Early in the fifties DAVID MONRO commenced his explorations of Marlborough and eastern Nelson, and some of his novelties are recorded in the appendix to the *Flora Novae-Zelandiae*. In 1853 the first volume of this work appeared, dealing with the spermophytes, and two years later, the second volume devoted to the cryptogams. The number of species described, 1767, nearly doubled the last enumeration. The work abounds in information valuable for the present-day student of the flora; but more important still is the classical *Introductory Essay*, in which there is a philosophical discussion on the limits of species, variation, the affinities and origin of the flora and so on, which for lucidity, marshalling of facts, carefully balanced conclusions, and praiseworthy moderation has never been excelled.

From 1855 to the publication of the Handbook (1867). When the *Flora Novae-Zelandiae* appeared nothing was known of the South Island high-mountain flora beyond the results of sparse collecting in the extreme northern mountains, nor of that of the lowlands except at distant points along the coast. The years 1855—67 saw this state of affairs righted, thanks especially to the labours of TRAVERS, HAAST, HECTOR and BUCHANAN.

W. T. L. TRAVERS explored the Nelson portion of the Southern Alps from 1854—60, making important collections which he sent to Kew. HAAST, although primarily a geologist, made extensive collections of plants during his explorations of wide areas for the most part absolutely unknown. The central chain of the Southern Alps and its neighbourhood were the chief areas visited. His collections and observations threw a flood of light upon the high mountain florula, putting it into its true position in the general flora of New Zealand. JAMES HECTOR and J. BUCHANAN explored the rugged mountains of south-west Otago. Their collections were of extreme value not merely for their many novelties, but because they showed south-west Otago to be a well-marked botanical district. The botany of eastern

Otago was, in part, made known by LAUDER LINDSAY of Edinburgh, who
spent 4 months at the task in 1861—62, and afterwards published his
"Observations on New Zealand lichens" (1866) and *Contributions to New
Zealand Botany* (1868). In 1863 F. VON HOCHSTETTER's classical work on
New Zealand appeared. Although mainly geological, it contains much
ecological information, including an excellent account of North Island rain-
forest proper. The same year saw the visit of H. H. TRAVERS, at the
instance of his father, to the Chatham Islands. During a four months'
stay, he formed that collection which, entrusted to VON MUELLER, resulted
in the publication in 1864 of *Vegetation of the Chatham Islands* in which
129 species of vascular plants are enumerated.

Through the representations of DR. KNIGHT and others, the New Zealand
Government, in 1861, arranged with HOOKER to prepare a flora of the
region, on the lines recommended by Sir W. J. HOOKER for a uniform series
for the British Colonies. The part dealing with the vascular plants appeared
in 1864 as Part I of the *Handbook of the New Zealand Flora,* and Part II,
dealing with the *Muscineae* &c., together with an appendix recording recent
discoveries, in 1867. That the work was most excellent, HOOKER's name
is sufficient guarantee, but that descriptions so full and clear were drawn
up, in most cases, from dried material must be a source of wonder to all
using the *Handbook.*

3. The Period of Publications by New Zealand Botanists.

**From the founding of the New Zealand Institute (1867) to the
publication of KIRK's Students' Flora (1899).** In 1867 the New Zealand
Institute was founded by Act of Parliament, which assured it an income
of £ 500[1]). To it were affiliated, in due course, the scientific societies of
the Provinces. There was now ample means of publication for local workers,
who alone can undertake those critical studies which demand observations
from living plants. Nor were such workers wanting, as a glance at the
Bibliography of this volume shows. Indeed matter was awaiting publication
and the first volume of the *Transactions* of the new Society included plant-
geographical essays previously written by certain of those collectors who
had done yeoman service for HOOKER, — COLENSO, HECTOR, MONRO, TRAVERS and
BUCHANAN. Especially important are HECTOR's principles of plant-distribution
in South Island[2]), and COLENSO's phyto-geographical areas of North Island[3]).

The progress during this subperiod was considerable. Year by year
new species of spermophytes were published and (though many are certainly

---

1) This sum was increased to £ 1000 in 1919 and to £ 1500 in 1926.

2) He defines western, central and eastern climatic provinces and 3 vertical belts.
The influence of the Divide is appreciated and stress is laid on the continental climate
of the centre.

3) Six areas and seven vertical belts.

invalid) at a low estimate, the number of species allowed in the *Handbook* was increased by one-half. Localities of which the florulas were virtually unknown were investigated and those but hastily examined were explored afresh; indeed knowledge as to plant-distribution increased fourfold. Floristically, more intensive studies were made of certain critical genera and species. Three works of major importance were published, — *The Indigenous Grasses of New Zealand* (BUCHANAN 1880), *The Forest Flora of New Zealand* (T. KIRK 1889) and *The Students' Flora of New Zealand and the Outlying Islands* (T. KIRK, 1899). Three botanical collectors and authors stand forth conspicuously — T. KIRK, T. F. CHEESEMAN and D. PETRIE. COLENSO, BUCHANAN, TRAVERS and KNIGHT continued their labours, those of COLENSO ceasing only at his death in 1898, he having been an active worker in the field for at least 60 years. Others of the subperiod who have imprinted their mark on New Zealand botany are G. M. THOMSON, J. ADAMS, J. B. ARM-STRONG, T. H. POTTS, A. HAMILTON, R. BROWN ter., and R. M. LAING. In order to gain a chronological view of the progress made the subperiod may be divided into decades, the first terminating with 1877. During this decade lists of species of the following localities were published: Mount Egmont and Marlborough (BUCHANAN 1869), Great Barrier Island (KIRK 1869), the Thames district (KIRK 1880), the Port Hills and adjacent plain (J. F. ARM-STRONG 1870), the northern district of Auckland (BUCHANAN and KIRK 1870). Auckland Isthmus (KIRK 1871—72), Titirangi district (CHEESEMAN 1872), Hot Lakes district (KIRK 1873), Wellington Province (BUCHANAN 1874), Chatham Islands (BUCHANAN 1875), 45 additions to the Otago Flora (G. M. THOMSON 1877). Lists of introduced plants were drawn up by KIRK, J. F. ARMSTRONG and G. M. THOMSON, respectively, for Auckland (1870), Canterbury (1872) and Otago (1875). The Swedish botanist, S. BERGGREN, visited the Colony and made large collections in nearly all the botanical districts.

The second decade ends with 1887. The localities for which lists were published are: Bluff Harbour; 109 additions to the Otago flora, and Islands in the Hauraki Gulf (KIRK 1878); Okarito to the Franz Josef Glacier (HAMILTON 1879); Mt. Pirongia (CHEESEMAN 1879); Canterbury Province (J. B. ARMSTRONG 1880), Stewart Island (PETRIE 1881); Nelson Province (CHEESEMAN 1882); Macquarie Island (SCOTT 1883); 90 additions to vascular plants of Thames subdistrict (J. ADAMS 1884); Stewart Island (T. KIRK 1885)[1]; Mount Te Aroha (J. ADAMS 1885); 131 additional species to the flora of Nelson (KIRK 1887). An account of the naturalized plants near Wellington was published by KIRK in 1877. An important illustrated paper by BUCHANAN on the alpine flora appeared in 1883. G. M. THOMSON dealt with the pollination of New Zealand plants and in the same year (1882).

---

1) Details are given regarding the vegetation, together with an account of the first ascent of Mt. Anglem. A few additions are made to the seedplants and there is a full list of ferns. Mr. TRALL materially assisted the author.

he published a notable paper on the origin of the flora. During this decade some much-needed studies on the following genera were made: *Veronica* (J. B. ARMSTRONG 1881), *Carex* and *Coprosma* (CHEESEMAN 1884, 1887). BUCHANAN's fine folio work on the indigenous grasses appeared in 1880, illustrated with life-size figures. C. KNIGHT (1860—84) published various papers on the lichen flora. The New Zealand University, founded in 1870, began but extremely slowly to influence research, and the early papers on *Algae* of R. M. LAING (1886) may be traced to its influence. G. M. THOMSON's useful book on ferns came out in 1882, and POTTS's charming *Out in the Open*.

The third decade must be extended to April 10th, 1899. The botanical exploration of New Zealand continued. CHEESEMAN visited the unbotanized Three Kings Islands and the almost unknown Kermadec Group, publishing lists of their species and other details (1888—91). KIRK voyaged to the New Zealand Subantarctic Islands in the summer of 1890, putting forth in 1891 an account of the botany of the group (Macquarie Island excepted), that of the Snares and Antipodes Islands being previously unknown. F. R. CHAPMAN, of the same expedition, also contributed some important particulars on their vegetation (1891). J. ADAMS ascended Mounts Te Moehau and Hikurangi describing their vegetation (1889—1897). HAMILTON wrote an account of his visit to Macquarie Island (1895). PETRIE made a number of interesting journeys, collecting copiously[1]. COLENSO was still active. MACMAHON ascended Mt. Stokes. H. J. MATTHEWS collected near L. Wakatipu for his famous garden. F. A. D. COX made fairly full collections of Chatham Islands plants. L. COCKAYNE commenced his botanical explorations in 1887 and has continued them year by year up to the present time[2]. In 1896 PETRIE's important *List of the Flowering Plants Indigenous to Otago* appeared[3]. The same year DIELS's *Vegetations-Biologie von Neu-Seeland* was published, a pioneer work which laid securely the foundations of ecological botany in New Zealand. The study of mosses by R. BROWN and T. N. BECKETT formed a feature of this decade, BROWN, whose love of botany was intense, publishing many supposed new species[4] year by year up till

1) Ruahine Mts., Volcanic Plateau, Clinton Valley, Waimakariri Basin and North Westland, Mt. Hikurangi, many parts of Auckland.

2) For the decade these included: Humboldt Mts., Puketeraki Mts., the neighbourhood of the Waimakariri and its main tributaries, the Seaward Kaikoura Mts., North Westland, the neighbourhood of L. Wakatipu, and various localities from North Canterbury to Foveaux Strait.

3) This represents, in small compass, the wealth of information regarding the species and their distribution that this ardent botanist had acquired during 20 years of careful observation. Especially well had he examined the mountains and valleys of Central Otago which had yielded an amazing harvest.

4) Considering BROWN's advanced age when he commenced publishing, and the extreme disadvantages under which he worked, his results are remarkably good, especially his critical treatment of certain groups of minute plants and his establishment of valid genera.

his death. The veteran COLENSO forwarded large consignments of *Hepaticae* to Kew, and these, determined by STEPHANI, yielded many species new to the flora. Important studies of this decade were the determination of New Zealand Tertiary fossil plants by ETTINGSHAUSEN and NYLANDER's exhaustive account of all known species of New Zealand lichens. In 1897 New Zealand science experienced a great loss in the death of T. KIRK. For about 34 years all his time and energy had been devoted to New Zealand botany. Early on, he became leader of botanical thought in the Colony, and that position he held firmly until his death. The incomplete flora which KIRK had prepared at the instance of the Government appeared in 1909. Though lacking the author's guiding hand, it is a fine piece of work, and one must ever regret its non-completion. His most admirable *Forest Flora of New Zealand* still remains the most authoritative work on the forest trees.

From 1899 to the publication of CHEESEMAN's Manual (1906). The 8 years of this short sub-period show a decrease in contributions to New Zealand botany, but this was more apparent than real, for there was much activity that did not appear until the publication of CHEESEMAN's *Manual,* many collectors[1]) having been busy supplying that author with material.

In 1899, L. COCKAYNE commenced his ecological publications with a paper on the burning and regeneration of subalpine scrub[2]). LAING continued his publications on *Algae* (1900, 5. 6), BROWN his papers on mosses, and CARSE wrote on the botany of Mauku. The influence of the University somewhat increased, as shown by an important paper by A. P. W. THOMAS (Professor of biology Auckland University College) on the prothallus of *Phylloglossum,* and several papers by students of Canterbury College. R. M. LAING and Miss E. W. BLACKWELL wrote a popular book entitled *Plants of New Zealand* (1906), profusely illustrated by excellent photographs. In 1906 CHEESEMAN's *Manual* appeared, he having been employed by the Government for its production. It showed how great the progress in floristic botany had been since the publication of the *Handbook* in 1867, the species of spermophytes having been raised from 935 to 1415. With the appearance of this book the gifted author sprang into the front rank of the floristic botanists of the day. For 20 years it was the mainstay of New Zealand botanists of all grades and it has been a potent factor for botanical advance.

---

1) One of the most active was W. TOWNSON, who made a close examination of western Nelson, hitherto unbotanized, and threw a flood of light on its remarkable florula. F. G. GIBBS did excellent service in many parts of Nelson over a wide area and discovered many novelties. H. J. MATTHEWS, T. H. MACMAHON, H. CARSE, and R. H. MATTHEWS and indeed all the New Zealand botanists and collectors also supplied CHEESEMAN.

2) This was followed by communications dealing with seedlings (1899, 1900, 1901), Plant-geography of Waimakariri (1900), Chatham Island (1902), Subantarctic Islands (1904) and several shorter papers.

From the appearance of the Manual in 1906 to the end of 1916.

The appearance of the *Manual* gave a fresh impetus to research. The former collectors, their ranks increased by younger naturalists, enabled CHEESEMAN almost yearly to bring out papers supplementary to his work[1]), the number of students of the flora &c. also gradually increased. The Government employed L. COCKAYNE to make a series of botanical surveys[2]). The Philosophical Institute of Canterbury organized an expedition to the Lord Auckland and Campbell Islands and published a work in 2 volumes, which *inter alia* contains a full account of the flora and vegetation[3]) of the Subantarctic Botanical Province. W. R. B. OLIVER who with some companions, spent a year on the Kermadec Islands in order to study their natural history, wrote an admirable account of their vegetation and flora. A. H. COCKAYNE broke new ground with a paper — the forerunner of many which appeared later — treating of the effect of burning tussock-grassland. E. CHEELS "Bibliography of Australian, New Zealand and South Sea Island Lichens" (1906) must receive mention.

From 1911 to 1913 inclusive various articles were published, some ecological[4]), others floristic[5]). To some extent the influence of the University increased[6]) and this is specially marked by the appearance of J. HOLLOWAY's paper on the *Lycopodiaceae* — the first of his brilliant series of publications. Another paper novel to New Zealand science was L. COCKAYNE's "Observations concerning Evolution, derived from Ecological Studies in New Zealand".

---

1) These are entitled "Contributions to a Fuller Knowledge of the Flora of New Zealand" (1907—1920): also he published several papers describing new species &c., the last appearing in 1923.

2) The results appear in the following richly-illustrated Reports: Kapiti Island (1907), Waipoua Kauri forest and Tongariro National Park (1908), Stewart Island (1909) and two on Sand-dunes (1909—11).

3) CHEESEMAN and L. COCKAYNE deal respectively with the floristic and ecological botany of the group; LAING with the marine *Algae* and the ecological botany of Campbell Island; PETRIE with the taxonomy of the grasses; and the *Fungi, Hepaticae, Lichenes,* and *Musci* are respectively the work of MASSEE, STEPHANI, LINDAU and BROTHERUS. The origin of the fauna and flora is discussed by CHILTON and CHEESEMAN respectively.

4) L. COCKAYNE and LAING (The Mount Arrowsmith "district"), LAING (The Spenser Mountains), POPPLEWELL (certain parts of the Stewart and South Otago districts), PHILLIPS TURNER (the Waimarino forest and its environs), ASTON (the Tararua Mountains, and effect of introduced mammals on vegetation), PETRIE (denuded Central Otago).

5) CARSE (flora of Mangouni County — partly ecological), ASTON (list of species of Wellington Province), PETRIE (flora of Mount Hector), CROSBY-SMITH (flora of Princess Mountains), TOWNSON (flora of part of North-western district), L. COCKAYNE (lists of species near Franz Josef Glacier and on Clinton Saddle).

6) The following may be cited: Young Stages of *Cyathea* and *Dicksonia* (G. B. STEVENSON); Anatomy of *Lycopodium* (J. H. HOLLOWAY); Fungi of Epiphytic Orchids (T. L. LANCASTER); Anatomy of Subantarctic Plants (Miss HERRIOTT); New Zealand Halophytes (Miss CROSS).

Overseas an increasing interest was taken in New Zealand plants. Goebel, who had visited the Dominion in 1898 and saw a good deal of the vegetation of the Eastern and Western Districts, wrote an important paper on certain *Muscineae*, and his *Experimentelle Morphologie* and *Organographie*, ed. 2. contain many observations regarding New Zealand plants, while *inter alia* he has described the life-history of *Loxsoma*. Diels, who with E. Pritzel had spent some time in the Colony in 1902, published his excellent and suggestive *Jugendformen und Blütenreife* in which many New Zealand plants play a prominent part. Various parts of *Das Pflanzenreich* deal with critical New Zealand genera, especially *Luzula* (Buchenau), *Uncinia* and *Carex* (Kükenthal), *Halorrhagis* and *Gunnera* (Schindler) and *Sphagnum* (Warnstorf). H. N. Dixon commenced a critical study of New Zealand mosses, with a valuable paper on *Dicranoloma*. The embryology of the New Zealand gymnosperms has received considerable attention at the hands of botanists of Chicago University, E. C. Jeffrey of Harvard and others. Miss Gibbs wrote on the female strobilus of *Podocarpus*. Kidston and Gwynne-Vaughan described two Jurassic *Osmundaceae*. Massee continued his excellent papers on New Zealand fungi. Bitter, in his elaborate work on *Acaena*, described many new forms for New Zealand. The studies of Beauverd cleared up several doubtful points regarding the Gnaphaloid *Compositae*. Domin showed the New Zealand species of *Koeleria* to be distinct from any in Europe or South America. D. G. Lillie, biologist on the Terra Nova, made large collections of Jurassic plants in New Zealand, which were dealt with by E. A. N. Arber in his "Eearlier Mesozoic Floras of New Zealand".

The outstanding feature of 1914 was the appearance of Cheeseman's *Illustrations of the New Zealand Flora*, a work in 2 volumes[1]). Destined to be eventually of the most extreme importance was the opening of the Canterbury College Biological Station at Cass, which is situated at the junction of the comparatively dry Eastern and the extremely wet Western Botanical districts an ideal situation for research in the midst of vegetation of many kinds and of a rich flora, while there is example after example of extreme epharmony and many polymorphic swarms of hybrids. The botanical exploration of the Region made some progress and various floristic[2]) anatomical[3]) and ecological[4]) papers appeared. During September a meeting

---

1) It contains 250 plates of life-size drawings illustrating most of the genera. An important part is a list of all previous illustrations. The drawings were prepared at Kew by Miss Matilda Smith under the supervision of W. B. Hemsley.

2) New species &c. (Cheeseman, Petrie), the flora of the Ruahines (Aston), the sub-alpine element of the Banks Peninsula flora (Laing), the flora of Southland (Crosby-Smith), new localities for plants (Aston, Cockayne).

3) Leaf-anatomy of certain trees and shrubs (Miss L. A. Suckling).

4) Rate of growth of the kauri (Cheeseman), vegetation and flora of the lower and upper Routeburn (Poppelwell), ecological study of sanddune plants (Miss E. J. Pegg), vegetation of White Island (W. R. B. Oliver).

of the British Association was held in Australia and in consequence the following botanists took the opportunity of visiting New Zealand: L. H. BAILEY, Miss E. M. BERRIDGE, A. B. RENDLE and Miss E. SAUNDERS. Part II, dealing whit the *Dicranaceae*, appeared of H. N. DIXON's valuable studies in New Zealand bryology.

The most important publication of 1915 was LAING's list of the Norfolk Island flora. The other publications can be divided into floristic[1]), anatomical[2]), and ecological[3]).

Perhaps the year 1916 is distinguished by the commencement of L. COCKAYNE's papers on Floristic Botany which apply different methods[4]) for the recognition of species from those generally in use. Also several ecological papers appeared[5]), HOLLOWAY's splendid work continued, PETRIE and CHEESEMAN added 15 new species to the flora and EASTERFIELD and Mc DOWELL dealt with the chemistry of 2 podocarps. During this same year New Zealand came botanically into special prominence through WILLIS using the distribution of its spermophyte flora to test his now well-known and greatly criticised "Age and Area" theory, all his data being taken from CHEESEMAN's Manual of 1906 — he, himself, never having visited New Zealand. Almost needless to say the paper contains many incongruities and much of the data is now known to be inaccurate.

From the commencement of 1917 to the end of 1927. This subperiod of 10 years has been one of great activity. It has wittnessed the production of HOLLOWAY's work on *Tmesipteris* — destined to become a classic, the appearance of GUTHRIE-SMITH's unique account of the plant and animal ecologie of a limited area, based on the observations of 30 years, also of G. M. THOMSON's great work on the naturalized animals and plants of New Zealand and the appearance in 1921 of the first edition of this book. Further, hybridism amongst the wild plants, on a scale undreamt of has been recognised, and bids fair to revolutionize the conception of many

---

1) New species &c. (CHEESEMAN, PETRIE, COCKAYNE), classification of the forms — mostly cultivated — of *Phormium* (Miss B. D. CROSS), the pteridophytes of Mangonui County (CARSE), new localities for plants (ASTON).

2) The protocorm of *Lycopodium laterale* (HOLLOWAY), prothallia of 3 lycopods (Miss K. V. EDGERLEY), a study of *Nothopanax arboreum* (Miss E. M. PIGOTT).

3) Recent changes in vegetation near L. Taupo (FLETCHER), vegetation of an islet off Stewart Island (POPPELWELL).

4) The author's views on taxonomic methods were published in 1917 in a paper dea· ling with the question of species and varieties, and *experiment* not personal judgment is considered fundamental and most "intermediates" are held to be hybrids. In all, 5 parts of the series have appeared, the last two written in conjunction with H. H. ALLAN.

5) A. H. COCKAYNE (economic ecology of tussock-grassland); ASTON (new vegetation subsequent to the Tarawera eruption); L. COCKAYNE and FOWERAKER (the plant-covering near Cass); POPPELWELL (the vegetation of 2 islets lying off Stewart Island); J. W. BIRD (lianes of Riccarton Bush).

species, while its genetic meaning should be far-reaching. In this regard the visit of J. P. LOTSY to this country and his important lectures at the University Colleges (published by Canterbury University College) must be emphasized. Names new to New Zealand botany have come to the fore and of such those of G. H. CUNNINGHAM and H. H. ALLAN already stand high. In the case of CUNNINGHAM his studies on the indigenous rusts and other fungi are models of what such publications should be, and ALLAN is studying wild hybrids in the field and reproducing them artificially in his garden. Finally the work of botanical exploration has proceeded as never before. In this regard the activity of A. WALL in collecting specimens for the herbarium of Canterbury Museum has been amazing. L. COCKAYNE in connection with research for the Department of Agriculture and the State Forest Service, has investigated much of the vegetation of both Islands — some unknown previously. Nor can the energy of SPEDEN, H. H. ALLAN, GIBBS, W. A. THOMSON, ASTON, J. S. THOMSON, HOLLOWAY, W. MACKAY and G. SIMPSON be overlooked.

By no means second to botanical exploration and publications comes the cultivation of the indigenous plants. The increase in this direction during the subperiod has been highly gratifying. In North Island the cultivation of shrubs and trees is most in vogue, though there are exceptions, but, in proceeding south, conditions become gradually more favourable for high-mountain species, so that the best collections are to be seen in the South Otago district. Of special importance, since they are got together and used for scientific purposes, are those in the suburbs of Dunedin of W. A. THOMSON, J. SCOTT THOMSON and GEORGE SIMPSON, but for wealth of species admirably grown the garden of J. SPEDEN at Gore (SO.) comes first. The most important step of all, however, has been the dedication of a large reserve by the Wellington City Council for the purpose of an "Open-air native plant museum", where a full collection of the indigenous plants is to be established and pieces of the primeval vegetation are to be made artificially, so that both the flora and vegetation will be represented.

The pleasure which the foregoing record of botanical progress gives for the subperiod is grievously marred by the passing away near its close of the two great figures — CHEESEMAN and PETRIE — who, year by year, for more than half a century, had added fact after fact to the knowledge of New Zealand plants. CHEESEMAN's name will ever live in his *Manual of the New Zealand Flora,* which continues the labours of his illustrious predecessors — he no less to be held in highest honour — recorded already in this chapter. As for PETRIE no one can ever ascend to those alpine moorlands of Otago or sail up the lovely Stewart Island, land-locked arms of the Pacific, without being again and again reminded of his splendid pioneer work. Both CHEESEMAN and PETRIE presented their great herbaria to the

Nation, the former being lodged in the Auckland Museum and the latter in the Dominium Museum (Wellington).

In 1917 HOLLOWAY dealt with vegetative reproduction in *Lycopodium*, POPPELWELL gave an account of the vegetation of Haast Pass and environs, FOWERAKER treated of the mat-plants of a shingly river-bed from the ecological-anatomical viewpoint, L. COCKAYNE published a map of the Botanical Districts[1]), and W. R. B. OLIVER gave an excellent account of the flora and vegetation of Lord HOWE Island[2]). PETRIE presented his yearly contribution of new species and TRAILL described the effect of a heavy snowstorm in Stewart Island. It was this year also that L. COCKAYNE's paper — already referred too — on the species question appeared.

The outstanding feature of 1918 was HOLLOWAY's great work on *Tmesipteris*[3]), already mentioned. L. COCKAYNE commenced his researches on the great tussock-grassland plant-formation for the Department of Agriculture[4]), various studies of a more or less ecological character were produced[5]), and PETRIE, CARSE, and L. COCKAYNE published floristic papers, and here comes in A. WALL's excellent paper on the distribution and relationship of the two species of *Senecio* which is a striking example of the efficiency of the methods of the "natural" taxonomy. Also A. W. HILL's paper on *Caltha* in the Southern Hemisphere is important for New Zealand[6]).

The year 1919 was marked by CHEESEMAN's report on the vascular flora of Macquarie Island, HOLLOWAY's continuation of his lycopod studies, and LAING's account of the vegetation und flora of Banks Peninsula and a second, greatly enlarged edition of L. COCKAYNE's *New Zealand Plants and their Story*. Miss BETTS continued her Mineral Belt series, and a number of minor papers appeared[7]).

---

1) This was an attempt to divide the Region into *natural* phytogeographical areas which could be substituted for the unnatural political divisions used hitherto, e. g. Auckland, Taranaki, Canterbury, &c. These districts are now made use of by most New Zealand botanists.

2) This paper has a distinct New Zealand bearing and can be neglected by no student of the flora or the vegetation.

3) LAWSON of Sydney University had published the year previous a preliminary account of the prothallus but this paper HOLLOWAY had not seen. On the other hand, the latter's account of the development of the embryo was altogether new.

4) Altogether 12 articles have been published in the New Zealand Journal of Agriculture (1919—22). Portions of all the tussock-grassland area from north to south of South Island were studied and many mountains ascended to the alpine belt. My assistant W. D. REID rendered extremely valuable aid.

5) POPPELWELL and W. A. THOMSON (the Hollyford Valley), POPPELWELL (2 more Stewart Island islets), Miss M. W. BETTS (ecological-anatomical studies of some Mineral Belt plants).

6) Excellent figures are given of the leaves of the species and those of the Tasmanian, Australian and New Zealand species are of particular interest here.

7) Miss HERRIOTT (the indigenous species persisting in Hagley Park, City of Christchurch); CHEESEMAN and PETRIE (each 3 new species); and CARSE (a variety of *Pteris macilenta*).

HOLLOWAY's detailed and elaborate paper on the structure of the pro-thallus of 5 species of *Lycopodium* was easily the outstanding feature of 1920, the most ambitious of the remaining papers[1]) being WALL's account of the ecology and systematic position of *Ranunculus pauciflorus*. Attention may also be called to L. COCKAYNE's experiments conducted in semi-arid Central Otago for the Department of agriculture in regard to the relative palatability of certain pasture-plants for sheep, and his observations concerning regeneration of depleted land.

The year 1921 stands out conspicuously in the history of New Zealand botany through the appearance of the books of GUTHRIE-SMITH and L. COCKAYNE, already referred to, a second important memoir on *Tmesipteris* by HOLLOWAY and the commencement of G. H. CUNNINGHAM's admirable mycologial publications. The State Forest Service was established by Act of Parliament under the Directorship of L. M. ELLIS. As a part of his world tour E. H. WILSON visited New Zealand, travelling extensively in both Islands. Important systematic changes — long awaited — were the separation of *Hebe* from the unwieldy genus *Veronica* by PENNELL of the New York Botanical Garden and the reinstatement of the family *Winteraceae* by HUTCHINSON of Kew, as also the genus *Wintera* for the' New Zealand species — universally referred to as *Drimys*. The other papers were of little moment[2]), excepting perhaps L. COCKAYNE's Cawthron lecture[3]) and a too brief account by H. B. KIRK on the rate of growth of certain forest trees. Towards the close of the year D. H. CAMPBELL of Leland Stanford University visited the Dominion and collected liverworts in the dicotylous-podocarp forests of Westland and later published a semi-popular account of the New Zealand flora and vegetation. Finally, attention must be called to L. COCKAYNE's discovery of hybridism to an intense degree between certain of the species of *Nothofagus*. In this work W. D. REID took a prominent part.

The years 1922—23 are best taken together, since the *Transactions of the New Zealand Institute*, made up of papers sent for publication in December 1921, did not appear until 1923. Early in 1922 G. M. THOMSON's great work on acclimatization, already referred to, appeared. Economic

---

1) CHEESEMAN (pneumatophores in *Eugenia maire*), PETRIE (description of 5 new species), POPPELWELL (lists of species of Ben Lomond — far from complete — and Hoko-nui Hills), W. MARTIN (list of pteridophytes of Banks Peninsula), Miss BETTS (continuation of anatomy of Mineral Belt species, and a similar study on rosette-plants of tussock-grassland).

2) W. R. B. OLIVER (clearing up certain points in nomenclature), PETRIE and CHEESEMAN (each a short paper with new species), W. MARTIN and A. WALL (each a short paper with new localities for species); the latter of special interest as showing WALL's remarkable activity, the localities ranging from the Hurunui to Foveaux Strait and many at over 1500 m. altitude.

3) This deals with the distribution of vascular plants within the Region.

ecology, as has been seen, was slowly coming into its own, and its advance was evident in LEVY's valuable series of papers appearing in book-form as *The Grasslands of New Zealand* and L. COCKAYNE's articles on the results of his Central Otago experiments conducted for the Department of Agriculture[1]). Another of H. N. DIXON's much-wanted Bulletins on bryology appeared, dealing with the *Dicranaceae* (in part), the *Leucobryaceae*, the *Fissidentaceae*, the *Calymperaceae* and the *Pottiaceae*[2]). CUNNINGHAM's splendid paper[3]) on the rust-fungi of New Zealand was issued and marked him at once as a mycologist of the front rank. L. COCKAYNE broke new ground for New Zealand botany with a paper classifying the wild hybrids on the opportunity afforded the parent species for crossing and enumerating 130 supposed hybrids[4]). HOLLOWAY abandoned for a time his *Lycopodium-Tmesipteris* studies and turned to the *Hymenophyllaceae*[5]). A short but quite important paper was that of CHRISTENSEN on the burial of plants by shingle in an aggrading river-bed, some species being killed outright and others rooting from the trunk &c. and surviving. Miss E. M. HERRIOTT treated of the ecology and anatomy of *Durvillea* and there were a number of short papers floristic[6]) and ecological[7]).

The leading publications of 1924 were a number of papers by G. H. CUNNINGHAM on various families of *fungi* and the second part of HOLLOWAY's studies on the *Hymenophyllaceae;* also a short paper by ERDTMAN

---

1) The history of economic ecology in New Zealand by L. COCKAYNE is published in the *Report of Proceedings* of the Imperial Botanical Conference of 1924 and a list is given of 31 papers, series of papers and books.

2) R. BROWN, ter with rare discrimination — the reader should have seen (as I have many times) the amazing difficulties under which my old and honoured friend worked — founded two new genera — *Hennedia* and *Dendia* — from minute plants and both are considered valid groups, though the former name is changed to *Hennediella* — there being an algal genus called *Hennedia* — and the latter is reduced by BROTHERUS to a subgenus.

3) W. D. REID collected with great judgment much material for this work in the high-mountain belts while with me on the tussock-grassland investigation. Several other papers on fungi by CUNNINGHAM also appeared.

4) These in 1925 COCKAYNE raised to 208, and it was explained that the matter was not one of groups of one or two individuals but of great polymorphic swarms.

5) Two papers have been published, the second in 1924, which deal respectively with life-forms and the phytogeographical distribution. Both contain a wealth of information previously unpublished and procured in the field by the author at all seasons and in many localities and habitats.

6) CHRISTENSEN paid great attention to the flora and vegetation of the Hanmer area for some 9 years, with the result that no part of the high mountains of equal extent is better known. By the ecological method he discovered jordanons of critical species and clearly proved that *Helichrysum Purdiei*, who no one doubted its being an excellent species, was merely one hybrid of a great hybrid swarm.

7) PETRIE and CHEESEMAN (each his usual quota of new species), SAINSBURY (an account of a new locality for the otherwise almost extinct *Pittosporum obcordatum* and a description of the seedling), WALL (on *Raoulia mammillaris*).

of Stockholm proved that pollen of podocarps occurred in Chatham Island peat. W. MARTIN gave a useful, popular account of the indigenous plants in the neighbourhood of Dunedin. YEATES dealt with the root nodules of podocarps. H. H. ALLAN commenced his studies of wild hybrids, bringing strong evidence that *Coprosma Cunninghamii* was merely a small part of the great swarm *C. propinqua* $\times$ *robusta*. GARRATT of Yale produced a pamphlet for the State Forest Service dealing with the macroscopic and microscopic features of 28 New Zealand woods. NANNFELDT showed that the New Zealand species of *Centella* was endemic.

Though the output of botanical publications for 1925 was not large they were on the whole of considerable moment. Foremost comes the 2nd edition of CHEESEMAN's *Manual of New Zealand Flora*, which is not — like the first edition — a critical examination of all available material but, for the most part, is the old edition unaltered, together with the species published since 1906 which the distinguished author considered valid, and the newer knowledge (but not in all cases) added concerning distribution. That more could be expected at the advanced age of the author is out of the question. The work closes worthily a most distinguished career. Next come LOTSY's lectures on evolution considered in the light of hybridization published by Canterbury College a noteworthy departure in regard to a New Zealand University College which it is to be hoped is the forerunner of other high-class University publications. Though LOTSY's lectures[2] deal with evolution &c. in general, there is an introduction on plant-hybridism in New Zealand and a full list of the wild hybrids by L. COCKAYNE.

The remaining publications of 1925 are all of importance. They comprise a paper on strictly orthodox lines on the biogeographical relations of the New Zealand Region, the subject being so presented that the different parts are concise and readily assimilated (W. R. B. OLIVER); a continuation of CUNNINGHAM's studies on *fungi*; a fairly complete account of the vegetation of the Poor Knights Islands (W. R. B. OLIVER[3]); an account of the vegetation and flora of the Mount Cook area (WALL[4]); YEATES gave an excellent account of the nucleolus of *Tmesipteris*[5]); L. COCKAYNE dealt with the occurrence of high-mountain species at a low altitude[6].

---

1) H. H. ALLAN (an account of the introduction and behaviour of spartina-grass in New Zealand), L. COCKAYNE (on a spineless *Discaria*).

2) These have subsequently been translated into both Dutch and German. LOTSY also published in *Genetica* an account of certain red-leaved forms of *Nothofagus* which he had discovered.

3) *Xeronema* — previously known by only one New Caledonian species — was discovered, but the species *X. Callistemon* W. R. B. OLIVER is different, though fairly close.

4) The outstanding feature is records of the species at 2100 m. to 2400 m. altitude.

5) This was his thesis for the Ph. D. degree of the University and was published in the *Proceedings* of the Royal Society.

6) A contribution to the Festschrift of KARL VON GOEBEL.

Towards the close of the year E. C. JEFFREY of Harvard visited New Zealand and collected a good deal of material of hybrids and podocarps for cytological and anatomical research.

In 1926 the publication of matter relative to the wild hybrids began in earnest, the medium of publication — thanks to LOTSY who had seen so many New Zealand hybrid swarms in nature — being *Genetica*. The most ambitious of these papers — the authors inexperienced in this class of work — dealt with a wild *Hebe* community (H. H. ALLAN, G. SIMPSON and J. S. THOMSON)[1] and the hybrids of *Nothofagus* (L. COCKAYNE and E. ATKINSON). ALLAN produced the first two papers of his series "Illustrations of wild hybrids in the New Zealand flora" and an account of the synthesis of the so-called species *Coprosma Cunninghamii* by crossing *C. propinqua* with pollen from *C. robusta*[2]. In addition L. COCKAYNE and H. H. ALLAN have published locally preliminary notes concerning various hybrids and, in a letter to *Nature* deal with the naming of hybrid swarms. J. S. THOMSON and G. SIMPSON have collected splendid material of many hybrid swarms in the South Otago and Fiord districts, W. MACKAY of Greymouth made the interesting discovery of a second *Rubus* hybrid closely allied to *R. Barkeri*, and CARSE has described a hybrid *Cordyline*, a hybrid *Hebe* and a hybrid *Metrosideros*, and WALL a doubtful hybrid *Ranunculus*.

Floristic botany made special progress by the appearance of H. N. DIXON's Part IV of his moss Bulletin which extends to the *Dawsoniaceae*. The usual batch of new species appeared at the hands of different authors and CARSE with the aid of C. CHRISTENSEN of Denmark made some changes in fern nomenclature. CUNNINGHAM continued his fungi studies and SPRAGUE and SUMMERHAYES of Kew removed the species of *Gaya* into the section *Apterocarpa* of *Hoheria*.

A number of important ecological papers appeared of which H. H. ALLAN's on Mount Peel vegetation is the most intensive study so far produced in New Zealand; the same author also dealt with a peculiar coastal scrub and with some striking cases of epharmony. The State Forest Service, which had employed L. COCKAYNE to study the *Nothofagus* forests as a whole, published the first part of his results dealing with the taxonomy of *Nothofagus* and the ecology of the formation. G. SIMPSON and J. S. THOMSON produced an ideal short account of a botanical excursion to an unbotanized mountain range. In fossil botany P. MARSHALL described a new species of *Osmundites* illustrated by excellent microphotographs.

---

1) SIMPSON and THOMSON labelled many of the hybrids *in situ* with copper labels and numbered plants ready for genetic research are now growing in their gardens and in those of L. COCKAYNE and H. H. ALLAN. Nearly all the data in the paper was the work of SIMPSON and THOMSON.

2) ALLAN has now raised seedlings of the F 2 generation and he has crossed *Rubus parvus* with *R. schmidelioides* and raised a plant equivalent to one which has been found in the wild state.

The present year (1927) has been one of considerable botanical activity. To New Zealand botanists it will long be remembered through the visit of G. E. DU RIETZ and his accomplished wife, Mrs. GRETA DU RIETZ. During a stay of about 7 months they visited portions of nearly all the botanical districts, collecting everywhere and at all altitudes the various lichens. This family is greatly in need of revision and Du RIETZ's conclusions — though of necessity the work will occupy several years — will be of the utmost taxonomic, ecological and phytogeographical importance.

Coming to the work of New Zealand botanists CUNNINGHAM continued his researches on *fungi* and produced several notable papers[1]). R. M. LAING issued a much-needed Reference List of New Zealand seaweeds. L. COCKAYNE and H. H. ALLAN published jointly several comprehensive and critical pappers[2]), the most ambitious being one dealing with taxonomic conceptions and methods. ALLAN also continued his work on wild hybrids and, *inter alia,* by crossing two species of *Rubus,* showed conclusively the origin of a curious form of *Rubus,* the parentage of which had previously been quite uncertain. He also concluded his memoir on the vegetation of Mount Peel (E.). A number of other authors issued papers on various subjects[3]). Overseas, A. W. HILL published an important paper on the species and distribution of *Lilaeopsis* and two endemic species were established for New Zealand; and SPRAGUE and SUMMERHAYES critically examined the status of *Fusanus* and restored the genus *Mida* for the New Zealand and Juan Fernandez species.

Part V of H. N. DIXON's important Studies of the New Zealand *Musci* appeared which included the genera from *Aulacopilum* to *Rhacopilum* inclusive. A new departure in botanical research for New Zealand (already highly developed by E. C. JEFFREY of Harvard) was W. P. EVANS's research on the microstructure of the Steventon lignite which showed it to be composed chiefly of coniferous wood but spores also occurred. Finally, as showing the increasing interest in botany in the Dominion, third editions were published of LAING and BLACKWELL's *Plants of New Zealand* and of L. COCKAYNE's *New Zealand Plants and their Story.*

Although the account of botanical investigation put forth in this chapter

---

1) The New Zealand *Lycoperdaceae;* Supplements to the *Uredinales* and *Ustilaginales* of New Zealand.

2) Notes on New Zealand Floristic Botany in which 32 species admitted as indigenous by CHEESEMAN in the Manual, ed. 2, are rejected as being exotic; The Taxonomic Status of the New Zealand species of *Hebe* in which the 86 species of the Manual, ed. 2, are reduced to 67 and 3 new species are added.

3) WALL dealt with the difficult question of the distribution of rare or local plants; Miss T. MURRAY treated of certain *Fungi* parasitic on New Zealand species of *Rubus;* Miss R. PIGOTT discussed the development &c. of *Corynocarpus;* G. M. THOMSON, after a lapse of many years, took up once more his much-needed study of pollination; CROSBY-SMITH gave a short account of the fast vanishing Awarua sphagnum bog and recorded the occurrence of *Donatia.*

cannot claim to be complete, yet it shows there has been remarkable activity since the early days of settlement. It has told of the deep debt New Zealand owes to her splendid botanical pioneers, with WILLIAM COLENSO at their head. The influence of the great HOOKER, friend of all New Zealand botanists, stands forth clearly. It also should have made plain how the New Zealanders themselves have by degrees come to the fore, so that at the present time there are a number of accomplished and, better still, ardent students of plant-life in whose capable hands may be left in all confidence the manifold problems which cry aloud for solution.

Chapter III.
Bibliography.

No space is available for anything approaching a full list of the literature pertaining to New Zealand botany. Papers only describing new species, or such as are purely morphological or phylogenetic as a rule are not cited and many of quite minor importance are omitted. It follows then that the list does not nearly reflect in full the botanical activity of certain workers, especially KIRK, COLENSO, PETRIE, CHEESEMANN, BUCHANAN, G. H. CUNNINGHAM and R. BROWN. On the other hand, a few works dealing with floral relationships with other lands are cited, as also others purely geological or geographical on which, it may be, statements in the book are based. Citations in the body of the work with author's name and date refer to this bibliography. T. N. Z. I. means Transactions and Proceedings of the New Zealand Institute.

ADAMS, J. 1884. On the Botany of the Thames Goldfields. T. N. Z. I. XVI: 385.
— 1885. On the Botany of Te Aroha Mountain. Ibid. XVII: 275.
— 1889. On the Botany of Te Moehau Mountain, Cape Colville. Ibid. XXI: 32.
— 1898. On the Botany of Hikurangi Mountain. Ibid. XXX: 414.
ADAMSON, R. S. 1912. On the Comparative Anatomy of the Leaves of certain Species of Veronica. Journ. Linn. Soc. XL: 247.
ALLAN, H. H. 1923. The Forest Remnants in the Neighbourhood of Feilding. Rep. Aus. Assoc. Adv. Sci. XVI: 402.
— 1924 a. Spartina-grass, and its Introduction into New Zealand. N. Z. Journ. Sc. & Technol. VII: 253.
— 1924 b. On the Hybridity of Coprosma Cunninghamii Hook. f. Ibid. VII: 310.
— 1925. Illustrations of Wild Hybrids in the New Zealand Flora. Genetica. VII: 287.
— 1926 a. A Remarkable New Zealand Scrub Association. Ecology. VII: 72.
— 1926 b. The F 1 Progeny Resulting from Crossing Coprosma propinqua ♀ with C. robusta ♂. Genetica. VIII: 155.
— 1926 c. Illustrations of Wild Hybrids in the New Zealand Flora. Ibid. VIII: 369.
— 1926 d. The surface Roots of an Individual Matai. N. Z. Journ. Sc. & Tech. VIII: 233.
— 1926 e. Epharmonic Response in certain New Zealand Species, and its Bearing on Taxonomic Questions. Journ. Ecol. XIV: 72.

ALLAN, H. H. 1926 f. Vegetation of Mount Peel, Canterbury, N. Z. Part I. — The Forests and Shrublands. T. N. Z. I. LVI: 37.

— 1926 g. As above. Part. II. — The Grasslands and Other Herbaceous Communities. Ibid. LVII: 73.

— 1926 h. Illustrations of Wild Hybrids in the New Zealand Flora. III. Genetica. VIII: 525.

— 1927. An Artificial Rubus Hybrid. T. N. Z. I. LVIII: 51.

— and DALRYMPLE, K. W. 1926. Ferns and Flowering-plants of Mayor Island. Ibid. LVI: 34.

—, G. SIMPSON and J. S. THOMSON. 1926. A Wild Community in New Zealand. Genetica. VIII: 375.

ARBER, E. A. N. 1917. Earlier Mesozoic Floras of New Zealand. Pal. Bull. No. 6. N. Z. Geol. Surv.

ALBOFF, N. 1902. Essai de Flore Raisonnée de la Terre de Feu. Ann. Mus. d. l. Plata.

ARMSTRONG, J. B. 1880. A short Sketch of the Flora of Canterbury, with Catalogue of Species. T. N. Z. I. XII: 324.

— 1881. A Synopsis of the New Zealand Species of Veronica, Linn., with Notes on new Species. Ibid. XIII: 344.

ARMSTRONG, J. F. 1870. On the Vegetation of the Neighbourhood of Christchurch including Riccarton, Dry Bush &c. T. N. Z. I. II. 118.

— 1872 a. On the Naturalized Plants of the Province of Canterbury. Ibid. IV: 284.

ASTON, B. C. 1910 a. Botanical Notes made on a Journey across the Tararuas. Ibid. XLII: 13.

— 1911. List of Phanerogamic Plants Indigenous in the Wellington Province. Ibid. XLIII: 225.

— 1912. Some Effects of Imported Animals on the Indigenous Vegetation. Ibid. XLIV: 19 (Proc.).

— 1914. Notes on the Phanerogamic Flora of the Ruahine Mountain-chain, with a List of the Plants observed thereon. Ibid. XLVI: 40.

— 1916. The Vegetation of the Tarawera Mountains, New Zealand. Ibid. XLVIII: 304.

BEAN, W. J. 1909. Effects of the Winter on Trees and Shrubs at Kew. Kew. Bull.: 232.

BENHAM, W. B. 1902. Earthworms and Palaeogeography. Rep. Aust. Assoc. Adv. Sc. IX: 335.

— 1909. Report on Oligochaeta of the Subantarctic Islands of New Zealand. The Subantarctic Islands of New Zealand I: 251.

BIDWILL, E. 1841. Rambles in New Zealand. London.

BIRD, J. W. 1916. Observations on the Lianes of the ancient Forest of the Canterbury Plains of New Zealand. T. N. Z. I.: 315.

BITTER, G. 1911. Die Gattung Acaena. Vorstudien zu einer Monographie. Stuttgart.

BLANCHARD, E. 1882. Proofs of the Subsidence of a Southern Continent during Recent Geological Epochs. N. Z. Journ. Sc. I: 251.

BROWN, R. 1894. Notes on New Zealand Mosses: Genus Pottia. T. N. Z. I. XXVI: 288.

— 1895. Notes on New Zealand Mosses. Genus Orthotrichum. Ibid. XXVII: 422.

— 1903. On the Musci of the Calcareous Districts of New Zealand, with Descriptions of New Spezies. Ibid XXXV: 323.

BROWN, R. N. R. 1906. Antarctic Botany: Its Present State and Future Problems. Scot. Geog. Mag. Sept.

— 1912. The Problems of Antarctic Plant Life. Rep. Scient. Results Scot. Nat. Antarc. Exp. III: 3.

BUCHANAN, J. 1865. Sketch of the Botany of Otago. Essay, Dunedin Exhibition of 1865. Reprinted 1869 in T. N. Z. I. I.

BUCHANAN, J. 1867. Botanical Notes on the Kaikoura Mountains and Mount Egmont. Rep. Geol. Surv. N. Z. Nr. 4.
— 1869 a. Notes on the Botany of Mount Egmont and neighbourhood, New Zealand. Journ. Linn. Soc. Bot. X: 57.
— 1869 b. Notes on the Botany of the Province of Marlborough, made during a visit there in the months of November, December and January 1866—67. Ibid. X: 63.
— 1873. List of Plants found on Miramar Peninsula, Wellington Harbour. T. N. Z. I. V: 349.
— 1874. Notes on the Flora of the Province of Wellington, with a List of the plants collected therein. Ibid. VI: 210.
— 1875. On the Flowering Plants and Ferns of the Chatham Islands. Ibid. VII: 333.
— 1877 a. On the Botany of Kawau Island. Also Critical Notes on certain Species doubtfully indigenous to Kawau by T. KIRK. Ibid. IX: 503.
— 1879—80. The Indigenous Grasses of New Zealand. Wellington.
— 1882 b. On the Alpine Flora of New Zealand. T. N. Z. I. XIV: 342.
— 1884. Campbell Island and its Flora. Ibid. XVI: 398.
— and KIRK, T. 1870. List of Plants found in the Northern District of the Province of Auckland. Ibid. II: 239. (Introductory remarks by BUCHANAN alone.)
CAMPBELL, D. H. 1923. Australasian Botanical Notes III. New Zealand. Amer. Journ. Bot. X.: 515.
CAMPBELL-WALKER, J. 1877. Report of the Conservator of State Forests. Journ. House of Repr. C. 3.
CARSE, H. 1902 a. On the Occurence of Panax arboreum as an Epiphyte on the Stems of Tree Ferns in the Mauku District. T. N. Z. I. XXXIV: 359.
— 1902 b. On the Flora of the Mauku District. Ibid. XXXIV: 362.
— 1911. On the Flora of the Mangonui County. Ibid. XLIII: 194.
— 1915. The Ferns and Fern Allies of Mangonui County, with some Notes on Abnormal Forms. Ibid. XLVII: 76.
— 1926 a. Botanical Notes, Including Descriptions of New Species. Ibid. LVI: 80.
— 1926 b. Botanical Notes, with Descriptions of New Species. Ibid. LVII: 89.
CHAPMAN, F. 1891. The Outlying Islands south of New Zealand. T. N. Z. I. XXIII: 491.
CHEESEMAN, T. F. 1872. On the Botany of the Titirangi District of the Province of Auckland. Ibid. IV: 270.
— 1873. On the Fertilization of the New Zealand Species of Pterostylis. Ibid. V: 352.
— 1875. On the Fertilization of Acianthus an Cyrtostilis. Ibid. VII: 349.
— 1877. On the Fertilization of Selliera. Ibid. IX: 542.
— 1878. Notes on the Fertilization of Glossostigma. Ibid. X: 352.
— 1880 a. On the Botany of Pirongia Mountain. Ibid. XII: 317.
— 1881. On the Fertilization of Thelymitra. Ibid. XIII: 291.
— 1882. Contributions to a Flora of the Nelson Provincial District. Ibid. XIV: 301.
— 1883. The Naturalized Plants of the Auckland Provincial District. Ibid. XV: 268.
— 1888 a. Notes on the Three Kings Islands. Ibid. XX: 141.
— 1888 b. On the Flora of the Kermadec Islands. Ibid. XX: 151.
— 1891. Further Notes on the Three Kings Islands. Ibid. XXIII: 408.
— 1897 a. On the Flora of the North Cape District. Ibid. XXIX: 333.
— 1897 b. Notice of the Establishment of Vallisneria spiralis in Lake Takapuna, together with some Remarks on its Life-history. Ibid. XXIX: 386.
— 1899. On the Occurrence of Ottelia in New Zealand. Ibid. XXXI: 350.
— 1906. Manual of the New Zealand Flora.
— 1909. On the Systematic Botany of the Islands to the South of New Zealand. The Subantarctic Islands of New Zealand II: 389.
— 1914. The Age and Growth of the Kauri (Agathis australis). T. N. Z. I. XLVI: 9.

CHEESEMAN, T. F. 1919. The Vascular Flora of Macquarie Island. Austral. Antarc. Exped. Scien. Reps. Ser. C. VII: 5.

— 1922 a. Rangitoto Island and its Vegetation. N. Z. Nature Notes: 20.

— 1922 b. The Forest of the Waitakerei Range. Ibid.: 22.

— 1922 c. The Kauri Forest. Ibid.: 22.

— 1925. Manual of the New Zealand Flora, ed. 2. Wellington, N. Z.

— with the assistance of HEMSLEY, W. B. 1914. Illustrations of the New Zealand Flora. Wellington, N. Z.

CHILTON, C. 1909. The Biological Relations of the Subantarctic Islands of New Zealand. Subantarctic Islands of New Zealand. II: 793.

— 1915. Notes from the Canterbury College Biological Station: No. 1 — Introduction and General Description of the Station. T. N. Z. I. XLVII: 331.

CHRISTENSEN, C. E. 1923. On the Behaviour of certain New Zealand Arboreal Plants when gradually buried by River-shingle. Ibid. LIV: 546.

CITERNE, P. 1897. Du genre Acaena. Rev. d. Sc. nat. d. l'Ouest.

COCKAYNE, A. H. 1910. The Natural Pastures of New Zealand. 1. The Effect of Burning on Tussock Country. N. Z. Journ. Agric. I: 7.

— 1911. The spiked blue grass of Australia. Ibid. III: 1.

— 1914. The Surface-sown Grass Lands of New Zealand. Ibid. VII: 465.

— 1916 a. Conversion of Fern-land into Grass. Ibid. XII: 421.

— 1916 b. Notes from the Canterbury College mountain Biological Station: No. 3 — Some Economic Considerations concerning Montane Tussock Grassland. T. N. Z. I. XLVIII: 154.

— 1918 a. The Grass-lands of New Zealand. N. Z. Journ. Agric. XVI: 125, 210, 258 and XVII: 35, 140.

— 1918 b. Some Grassland Problems. Ibid. XVII: 321.

— 1919. Cocksfoot. Its Establishment and Maintenance in Pastures. Ibid. XVIII: 257.

COCKAYNE, L. 1898. On the Freezing of New Zealand Alpine Plants. Notes of an Experiment conducted in the Freezing-Chamber, Lyttelton. T. N. Z. I. XXX: 435.

— 1899 a. An Inquiry into the Seedling Forms of New Zealand Phanerogams and their Development. Part I: Introduction. Ibid. XXXI: 354. — 1899 b. Part II: Description of Seedlings and Notes thereon; l. c. XXXI: 361. — 1900 a. Part III; l. c. XXXII: 83. — 1901. Part IV; l. c. XXXIII: 265.

— 1899 c. On the Burning and Reproduction of Subalpine Scrub and its Associated Plants; with special reference to Arthurs Pass District. Ibid. XXI: 398.

— 1900 b. A Sketch of the Plant Geography of the Waimakariri River Basin, considered chiefly from an Oecological Point of View. Ibid. XXXII: 95.

— 1902. A Short Account of the Plant-covering of Chatham Island. T. N. Z. I. XXXIV: 243.

— 1904. A Botanical Excursion during Midwinter to the Southern Islands of New Zealand. Ibid. XXXVI: 225.

— 1905 a. Notes on the Vegetation of the Open Bay Islands. Ibid.: 368.

— 1905 b. On the Significance of Spines in Discaria Toumatou, Raoul. Rhamnaceae. New Phytol. IV; 79.

— 1906 a. On a Specific Case of Leaf-variation in Coprosma Baueri, Endl. (Rubiaceae). T. N. Z. I. XXXVIII: 341.

— 1906 b. On the Supposed Mount Bonpland Habitat of Celmisia Lindsayi, Hook. f. Ibid. XXXVIII: 345.

— 1906 c. Notes on a Brief Botanical Visit to the Poor Knights Islands. Ibid. XXXVIII: 351.

— 1906 d. Notes on the Subalpine Scrub of Mount Fyffe (Seaward Kaikouras). Ibid. XXXVIII: 361.

COCKAYNE, L. 1907 a. Some Observations on the Coastal Vegetation of the South Island of New Zealand. Part I: General Remarks on the Coastal Plant Covering. Ibid. XXXIX: 313.

— 1907 b. Note on the Behaviour in Cultivation of a Chatham Island Form of Coprosma propinqua, A. Cunn. Ibid. XXXIX: 378.

— 1907 c. On the Sudden Appearance of a New Character in an Individual of Lepto-spermum scoparium. New Phytol. VI: 43.

— 1907 d. Report on a Botanical Survey of Kapiti Island. Wellington.

— 1908 a. A Preliminary Note on Heterophylly in Parsonsia. Rep. Aus. Assoc. Adv. Sc. XI: 486.

— 1908 b. Report on a Botanical Survey of the Waipoua Kauri Forest. Wellington.

— 1908 c. Report on a Botanical Survey of the Tongario National Park. Wellington.

— 1909 a. On a Collection of Plants from the Solanders. T. N. Z. I. XLI: 404.

— 1909 b. Report on the Sand Dunes of New Zealand. Wellington.

— 1909 c. Report on a Botanical Survey of Stewart Island. Wellington.

— 1909 d. The Necessity for Forest-Conservation. Forestry in New Zealand, Dept. of Lands: 85. Wellington.

— 1909 e. The Ecological Botany of the Subantarctic Islands of New Zealand. The Sub-antarctic Islands of New Zealand. I: 182.

— 1910 a. List of the Lichenes and Bryophytes collected in Stewart Island during the Botanical Survey of 1908. T. N. Z. I. XLI: 320.

— 1910 b. On a Non-flowering New Zealand Species of Rubus. Ibid. XLI: 325.

— 1910 c. New Zealand Plants and their Story. Wellington.

— 1911 a. Report on the Dune-Areas of New Zealand, their Geology, Botany and Recla-mation. Wellington.

— 1911 b. On the Peopling by Plants of the Subalpine River-bed of the Rakaia (Southern Alps of New Zealand). Trans. and Proc. Bot. Soc. Ed. XXIX: 104.

— 1912 a. Observations concerning Evolution derived from Ecological Studies in New Zealand. T. N. Z. I. XLIV: 1.

— 1912 b. Some Examples of Precocious Blooming in Heteroblastic Species of New Zealand Plants. Aus. Assoc. for Adv. Sc. XIII: 217.

— 1912 c. Some Noteworthy New Zealand Ferns. Plant World. XV: 49.

— 1916. Notes on New Zealand Floristic Botany, including Descriptions of New Spe-cies &c. T. N. Z. I. XLVIII: 193.

— 1917 a. A Consideration of the Terms "Species" and "Variety", with Special Refe-rence to the Flora of New Zealand. Ibid. XLIX: 66.

— 1917 b. Notes on New Zealand Floristic Botany, including Descriptions of New Species &c. (No. 2). Ibid. XLIX: 56.

— 1918 a. The Importance of Plant Ecology with Regard to Agriculture. N. Z. Journ. Sc. & Tech. I: 70.

— 1918 b. Notes on New Zealand Floristic Botany, including Descriptions of New Spe-cies &c. (No. 3). T. N. Z. I. L: 161.

— 1919 a. An Economic Investigation of the Montane Tussock-Grassland of New Zea-land. No. 1. Introduction. N. Z. Journ. Ag. XVIII: 1. — 1919 b. No. 2. Relative Palatability for sheep of the various Pasture Plants. Ibid. XVIII: 321.

— 1919 c. On the Seedling Form of the Coral-shrub (Helichrysum coralloides [Hook. f] Benth & Hook. f.) Journ. Sc. & Tech. II: 274.

— 1919 d. New Zealand Plants and Their story, ed. 2. Wellington.

— 1919 e. An Economic Investigation of the Montane Tussock-Grassland of New Zea-land. No. 3.

— 1919 f. Increase of the Californian Thistle (Cnicus arvensis) on the Dunstan Range, Central Otago. No. 4. Ibid. XIX: 343. — 1920 a. No. 5. Regeneration of Grassland after Depletion. Ibid. XX: 82. — 1920 b. No. 6. Further Details Regarding the Relative Palatability for sheep of various Pasture-Plants. Ibid. XX: 209. — 1920 c. No. 7. On the Effect of Understocking and Stocking to its full Capacity on a certain Area. Ibid. XX: 337. — 1920 d. No. 8. An Experiment in Central Otago concerning the Relative Palatability for Sheep of various Pasture-Plants. Ibid. XXI: 176. — 1920 e. No. 9. Further Details regarding the Earns-cleugh Palatability Experiment. XXI: 324. — 1921 a. No. 10. The Effect of Spelling a Heavily-grazed Pasture in Central Otago. XXII: 148.

COCKAYNE, L. 1921 b. The Southern-Beech (Nothofagus) Forests of New Zealand. Ibid. XXIII: 353.

— 1921 c. Die Vegetation der Erde XIV. The Vegetation of New Zealand. Leipzig and New York.

— 1921 d. The Distribution of the Vegetation and Flora of New Zealand. The Cawthron Lecture (1919). Nelson.

— 1921 e. An Economic Investigation of the Montane Tussock-Grassland of New Zealand. No. 11. The Grassland of the Humboldt Mountains established since the Burning of their Forest Covering. N. Z. Journ. Agric. XXIII: 137.

— 1922 a. As above. No. 12. The Regrassing Experiments in Central Otago. Part I. An Account of the Depleted Area in Relation to the Experiments. N. Journ. Agric. XXIV: 321; Part II. Objects, Principles and Methods. Ibid. XXV: 1; Part III. Results up to May, 1922. Ibid. XXV: 129.

— 1922 b. Note concerning a spineless wild-irishman (Discaria toumatou). N. Z. Journ. Sc. and Tech. V: 206.

— 1922 c. The Wild Indigenous Plants of the City of Wellington. N. Z. Nature Notes: 11.

— 1922 d. The Plant-life of the Hot Lakes District. Ibid.: 30.

— 1922 e. The Plant-life of the Tongariro National Park and its Environs. Ibid.: 35.

— 1922 f. The Vegetation of a Portion of the "Mineral Belt". Ibid.: 39.

— 1922 g. The Plant-life in the Vicinity of Mount Cook. Ibid.: 48.

— 1922 h. The Vegetation near the Franz Josef Glacier. Ibid.: 52.

— 1923. Hybridism in the New Zealand Flora. New Phytol. XXII: 105.

— 1924 a. The Cultivation of New Zealand Plants. Wellington, N. Z.

— 1924 b. New Zealand Plants for the British Isles. The Garden. LXXXVIII: 615, 632, 646, 660.

— 1925 a. New Zealand Economic Plant Ecology. Rep. Proced. Imp. Bot. Congress: 259.

— 1925 b. On the Occurrence of subalpine Vegetation at a low level in the Fiord Botanical District (New Zealand) and other matters pertaining thereto. Festschrift zum siebzigsten Geburtstage von Karl von Goebel in München: Jena.

— 1925 c. Introduction and List of supposed wild New Zealand Hybrids amongst the Vascular Plants. J. P. Lotsy's Evolution Considered in the Light of Hybridization. Christchurch.

— 1926 a. Monograph on the New Zealand Beech Forests. Part I, The Ecology of the Forests and the Taxonomy of the Beeches. Bull. No. 4, N. Z. State Forest Service.

— 1926 b. Notes on Ecological Field Work in New Zealand. Aims and Methods in the Study of Vegetation: 274.

— 1926 c. Ecological-economic investigation of the New Zealand Nothofagus forests. Ibid.: 330.

— 1926 d. Tussock-grassland investigations in New Zealand. Ibid.: 349.

— 1926 e. An ecological-economic investigation of the New Zealand sand-dunes. Ibid.: 362.

— 1927 a. Monograph on the New Zealand Beech Forests. Part II, The Forests from the Practical and Economic Standpoints. Bull. No. 4. N. Z. State Forest Service.

— 1927 b. New Zealand Plants and Their story, ed. 3. Wellington, N. Z.

Cockayne, L., and Allan, H. H. 1926 a. A Proposed New Botanical District for the New Zealand Region. T. N. Z. I. LVI: 19.

— — 1926 b. Notes on New Zealand Floristic Botany (No. 4). Ibid. LVI: 21.

— — 1926 c. The Present Taxonomic status of the New Zealand Species of Hebe. Ibid. LVII: 11.

— — 1926 d. Notes on New Zealand Floristic Botany, including Descriptions of New Species (No. 5). Ibid. LVII: 48.

— — 1926 e. The Naming of Wild Hybrid Swarms. Nature. CXVIII: 623.

— — 1927. The Bearing of Ecological Studies in New Zealand on Botanical Taxonomic Conceptions and Procedure. Journ. Ecol. XV: 234.

— and Atkinson, E. H. 1926. On the New Zealand wild Hybrids of Nothofagus. Genetica. VIII: 1.

— and Foweraker, C. E. 1916. Notes from the Canterbury College Mountain Biological Station: No. 4 — The Principal Plant Associations in the Immediate Vicinity of the Station. T. N. Z. I. XLVIII: 166.

— and MacKenzie, J. G. 1927. The Otari Open-air Native-plant Museum. N. Z. Journ. Agric.

Colenso, W. 1865. On the Botany of the North Island of New Zealand. Essay for New Zealand Exhibition, Dunedin. Also 1869. T. N. Z. I. I.

— 1881. The Ferns of Scinde Island (Napier). Ibid. XIII: 370.

— 1883. On the large number of Species of Ferns noticed in a small Area in the New Zealand Forests, in the Seventy-mile Bush, between Norsewood and Dannevirke, in the Provincial District of Hawkes Bay. Ibid. XV: 311.

— 1884. In Memoriam. An Account of Visits to and crossing over the Ruahine Range, Hawkes Bay, New Zealand, and of the Natural History of that Region. Napier.

— 1886. On Clianthus puniceus, Sol. Ibid. XVIII: 291.

— 1887. A few Observations on the Tree-ferns of New Zealand; with particular Reference to their peculiar Epiphytes, their Habit, and their Manner of Growth. Ibid. XIX: 252.

— 1895. Notes and Reminiscences of Early Crossings of the romantically-situated Lake Waikaremoana, Country of Hawkes-Bay, of its Neighbouring Country, and of its peculiar Botany; performed in the Years 1841 and 1843. Ibid. XXVII: 383.

Cooke, F. W. 1912. Observations on Salicornia australis. T. N. Z. I. XLIV: 349.

Cotton, C. A. 1922. Geomorphology of New Zealand. Wellington, N. Z.

Crosby-smith, J. 1911. Notes on the Botany of the Lake Hauroko District. T. N. Z. I. XLIII: 248.

— 1914. List of Phanerogamic Plants Indigenous in the Southland District. Ibid. XLVI: 220.

— 1927. The Vegetation of the Awarua Plain. Ibid. LVIII: 55.

Cross, B. D. 1910. Observations on some New Zealand Halophytes. T. N. Z. I. XLII: 545.

— 1915. Investigations on Phormium. Ibid. XLVII: 61.

Cunningham, A. 1837/1840. Florae Insularum Novae Zelandiae Praecursor; or a Specimen of the Botany of the Islands of New Zealand. Companion to the Bot. Mag. II: 223, 327 and 358 and Ann. and Mag. Nat. Hist. I: 210, 376, 455. — II: 44, 125, 205, 356. — III: 29, 111, 244, 314 and IV: 22, 106, 256.

Cunningham, G. H. 1921. The Genus Cordyceps in New Zealand. T. N. Z. I. LIII: 372.

— 1922 a. Clathrus cibarius, the "Bird-cage" Fungus. N. Z. Jour. Sci et Technology. V: 247.

— 1922 b. A Singular Cordyceps from Stephen Island. Trans. Brit. Myc. Soc. VIII: 72.

— 1923 a. Sphaerobolus stellatus, a Fungus with a Remarkable Method of Spore Dissemination. N. Z. Jour. Sci. et Tech. VI: 16.

— 1923 b. The Uredinales, or Rust Fungi, of New Zealand, Part I: Pucciniaceae, Tribe Puccinieae. T. N. Z. I. LIV: 619.

CUNNINGHAM, G. H. 1923 c. Aseroe rubra, an Interesting New Zealand Phalloid. N. Z. Jour. Sci. et Tech. VI: 154.
— 1924 a. The Uredinales, or Rust Fungi, of New Zealand: Supplement to Part I: and Part II. T. N. Z. I. LV: 1.
— 1924 b. Second Supplement to the Uredinales of New Zealand. Ibid.: 392.
— 1924 c. A Revision of the New Zealand Nidulariales, or "Birds-nest" Fungi. Ibid.: 59.
— 1924 d. A Critical Revision of the Australia and New Zealand species of the genus Secotium. Proc. Linn. Soc. N. S. W. XLIX: 97.
— 1924 e. The Ustilaginaceae, or "Smuts" of New Zealand. T. N. Z. I. LV: 397.
— 1924 f. The Development of Gallacea scleroderma. Trans Brit. Myc. Soc. IX: 193.
— 1925 a. The Structure and Development of Two New Zealand species of the Genus Secotium. Ibid. X: 216.
— 1925 b. Gasteromycetes of Australasia II: A Revision of the Genus Tulostoma. Proc. Linn. Soc. N. S. W. L: 245.
— 1925 c. Gasteromycetes of Australasia III: Bovista and Bovistella. 1. c. 367.
— 1926 a. A New Genus of the Hysterangiaceae. T. N. Z. I. LVI: 71.
— 1926 b. Development of Lycoperdon depressum (Fungi). N. Z. Journ. Sc. & Tech. VIII: 228.
— 1926 c. Gasteromycetes of Australasia V. The Genus Calvatia. Proc. Linn. Soc. N. S. W. LI: 363.
— 1927. Lycoperdaceae of New Zealand. T. N. Z. I. LVII: 187.

CURTIS, K. M. 1926 A Die-back of Pinus radiata and P. muricata caused by the Fungus Botridiplodia pinea (Desm.) Petr. T. N. Z. I. LVI: 52.

DENDY, A. 1902. The Chatham Islands: a Study in Biology. Mem. and Proc. Man. Lit. and Phil. Soc. XLVI. (Manchester Memoirs, No. 12: 1.)

DIEFFENBACH, E. 1841. Description of the Chatham Islands. New Zealand Journ.: 125 and 158 and Journ. R. Geog. Soc. XI: 195 and XII: 142.
— 1843. Travels in New Zealand. I. und II. London.

DIELS, L. 1896. Vegetations-Biologie von Neu-Seeland. Engler's Bot. Jahrb. XXII: 201.
— 1905. Über die Vegetationsverhältnisse Neu-Seelands. Ibid. XXXI, Beibl. 79.
— 1906. Die Pflanzenwelt von West-Australien südlich des Wendekreises. Einleitung. Die Grundzüge der Pflanzenwelt von Australien: 1—40.

DIXON, H. N. 1912. On some Mosses of New Zealand. Journ. Linn. Soc. Bot. XL: 433.
— 1913. Studies in the Bryology of New Zealand, Part. I. Bull. Nr. 3. N. Z. Inst: 1.
— 1914. As above, Part II. Ibid: 31.
— 1923. As above, Part III. Ibid: 75.
— 1926. As above, Part IV. Ibid: 153.
— 1927. As above, Part V. Ibid: 239.

DORRIEN-SMITH, A. A. 1910. An Attempt to introduce Olearia semidentata into the British Isles. Kew Bull., No. 4.

DRUDE, O. 1897. Manuel de Géographie Botanique. (Traduit par Georges Poirault.) Paris.

DUNEDIN FIELD CLUB. 1896. Catalogues of the Indigenous and Introduced Flowering Plants Ferns and Seaweeds, occurring in the Dunedin District. Dunedin.

DUSEN, P. 1908. Die tertiäre Flora der Seymour-Insel. Wiss. Erg. d. Schwed. Südpol-Exped., 1901—03. Lief. 3. Stockholm.

EDGERLEY, K. V. 1915. The Prothallia of Three New Zealand Lycopods. T. N. Z. I. XLVII: 94.

ENGLER, A. 1882. Versuch einer Entwicklungsgeschichte der Pflanzenwelt. II. Leipzig.

ERDTMANN, O. G. E. 1924. Studies in Micro-Palaeontology, I—IV. Geol. Fören. För-handl. XLVI: 676.

ETTINGHAUSEN, C. VON. 1887. Beiträge zur Kenntnis der Fossilen Flora Neuseelands. Denk-schr. K. Akad. Wiss. Wien. LII. — 1891. Translation of above. T. N. Z. I. XVIII: 237.

EVANS, W. P. 1927. Microstructure of New Zealand Lignites. N. Z. Journ. Sc. & Tech. IX: 137.

FIELD, H. C. 1890. The Ferns of New Zealand and its immediate Dependencies with Directions for their Collection and Cultivation. Wanganui.

— 1905. Notes on Ferns. Ibid. XXXVII: 377.

FILHOL, H. 1885. Mission de l'Ile Campbell: Recueil de Mémoires &c., rel. à l'observation du Passage de Vénus sur la Soleil III, 2me Partie.

FINLAYSON, A. C. 1903. The Stem-structure of some Leafless Plants of New Zealand, with Especial Reference to their Assimilatory Tissue. T. N. Z. I. XXXV: 360.

FLETCHER, H. J. 1915. Notes on Comparatively Recent Changes in the Vegetation of the Taupo District. Ibid. XLVII: 70.

FORBES, H. O. 1893 a. Antarctica. Nat. Sci. 111: 54. Also Fortnightly Review for May.

— 1893 b. The Chatham Islands and an Antarctic Continent. Nature XLVII: 474.

— 1893 c. The Chatham Islands. Supp. Papers, R. Geog. Soc. 111: 607.

— 1894. Antarctica, a Vanished Austral Land. Fortnightly Rev. LV: 297. Reprint. Ann. Rep. Smithson. Inst. 1896: 297.

FOWERAKER, C. E. 1917. Notes from the Canterbury College Mountain Biological Station, Cass: No. 5 — The Mat-plants, Cushion-plants, and allied Forms of the Cass River Bed (Eastern Botanical District, New Zealand). T. N. Z. I. XLIX: 1.

FORSTER, J. G. 1786 a. Dissertatio Inauguralis Botanico-Medica de Plantis Esculentis Insularum Oceani Australis.

— 1786 b. Florulae Insularum Australium Prodromus. Göttingen.

FORSTER, J. R. and J. G. 1776. Characteres Generum Plantarum quas in Itinere ad Insulas Maris Australis collegerunt, descripserunt, delinearunt. Annis MDCCLXXII to MDCCLXXV. London.

GARRATT, G. A. 1924. Some New Zealand Woods. N. Z. State Forest Service, Profess. Pap. No. 1.

GEPP, A. and E. S. 1911. Marine Algae from the Kermadecs. Journ. Bot. XLIX: 17.

GEYLER, H. T. 1880. Botanische Mittheilungen. Einige Bemerkungen über Phyllocladus. Frankfurt a. M.

GIBBS, L. S. 1911. The Hepatics of New Zealand. Journ. Bot. XLIX: 261.

— 1920. Notes on the Phytogeography of the Mountain Summit Plateaux of Tasmania. Journ. Ecol. VIII: 89, but specially, pp. 92 and 100—106.

GIESENHAGEN, C. 1890. Die Hymenophyllaceen. Flora, Heft 5: 411.

GOEBEL, K. VON. 1906. Archegoniatenstudien (X). Beiträge zur Kenntnis australischer und neuseeländischer Bryophyten. Flora XCVI: 1.

GOVETT, R. H. 1884. A Bird-Killing-Tree. T. N. Z. I. XVI: 364.

GREENSILL, N. A. R. 1903. Structure of Leaf of certain Species of Coprosma. Ibid. XXXV: 342.

GRIFFEN, E. M. 1908. The Development of some New Zealand Conifer Leaves with regard to Transfusion Tissue and to Adaptation to Environment. Ibid. XL: 43.

GUPPY, H. B. 1888. Flora of the Antarctic Islands. Nature XXXVIII: 40. (With note by THISELTON-DYER.)

— Observations of a Naturalist in the Pacific between 1896 and 1899. London.

GUTHRIE-SMITH, H. 1908. The Grasses of Tutira. Ibid. XL: 506.

— 1921. Tutira. The Story of a Sheep Station. Edinburgh.

HAAST, J. VON. 1870. Indroductory Remarks on the Distribution of Plants in the Province of Canterbury (See Armstrong, J. F. 1870). T. N. Z. I. II: 118.

— 1879. Geology of the Provinces of Canterbury and Westland.

HALL, J. W. 1902. Remarks on New Zealand Trees planted at Parawai, Thames, at and subsequent to the Year 1873. T. N. Z. I.: 388.

HAMILTON, A. 1879. List of Plants collected in the District of Okarito, Westland. T. N. Z. I. XI: 435.

— 1895. Notes on a Visit to Macquarie Island. Ibid. XXVII: 559.

— 1904 a. On Abnormal Developments in New Zealand Ferns; with a List of Papers by various Authors on the Ferns of New Zealand. T. N. Z. I. XXXVl: 334.

HASZARD, H. D. M. 1902. Notes on the Growth of some Indigenous and other Trees in New Zealand. Ibid. XXXIV: 386.

HAURI, H. 1916. Anatomische Untersuchungen an Polsterpflanzen nebst morphologischen und ökologischen Notizen. Beih. Bot. Centr. XXXIII: 275.

HAUSSKNECHT, C. 1884. Monographie der Gattung Epilobium. Jena.

HECTOR, J. 1863. Geological Expedition to the West Coast of Otago, New Zealand: Report with Appendix of Meteorological Observations taken on the West Coast of Otago. Otago Prov. Gov. Gaz., Nov. 5th: 435.

— 1869. On the Geographical Botany of New Zealand. T. N. Z. I. I (Part III, Essays).

— 1879. On the Fossil Flora of New Zealand. Ibid. XI: 536.

HEDLEY, C. 1893. On the Relation of the Fauna and Flora of Australia to those of New Zealand. Nat. Sci. III: 187.

— 1895. Considerations on Surviving Refugees in Austral Lands of Ancient Antarctic Life. Proc. R. S. N. S. W. XXIX: 278.

— 1899. A Zoo-geographical Scheme for the Mid-Pacific. Proc. Linn. Soc. N. S. W. XXIV: 391.

HEMSLEY, W. B. 1895. Report on Present State of Knowledge of Various Insular Floras. Challenger Reports, Botany.

HERRIOTT, E. M. 1906. On the Leaf-structure of some Plants from the Southern Islands of New Zealand. T. N. Z. I. XXXVIII: 377.

— 1919. A History of Hagley Park, Christchurch, with special Reference to its Botany. Ibid. LI: 427.

— 1923. Morphological Notes on the Zealand Giant Kelp, *Durvillea antarctica* (Chamisso). Ibid. LIV: 549.

HETLEY, C. 1887/88. The Native Flowers of New Zealand. London.

HILL, A. W. 1918. The Genus Caltha in the Southern Hemisphere. Ann. Bot. XXXII: 421.

— 1927. The Genus *Lilaeopsis:* a Study in Geographical Distribution. Journ. Linn. Soc.-Bot. XLVII: 525.

HILL, H. 1909. On Dactylanthus Taylori. T. N. Z. I. XLI: 437.

— 1911. Rotomahana and District revisited Twenty Three years after the Eruption. Ibid. XLIII: 278.

— 1926. Dactylanthus Taylori. Ibid. LVI: 87.

HOCHSTETTER, F. VON. 1867. New Zealand, its Physical Geography, Geology and Natural History. Stuttgart.

HOLLOWAY, J. E. 1910. A Comparative Study of the Anatomy of Six New Zealand Species of Lycopodium. T. N. Z. I. XLII: 356.

— 1915. Preliminary Note on the Protocorm of Lycopodium laterale R. Br. Prodr. Ibid. XLVII: 73.

— 1916. Studies in the New Zealand Species of the Genus Lycopodium: Part I. Ibid. XLVIII: 253.

— 1917. As above, Part II — Methods of Vegetative Reproduction. Ibid. XLIX: 80.

— 1918. The Prothallus and Young Plant of *Tmesipteris.* Ibid. L: 1.

— 1919. Studies in the New Zealand species of the Genus *Lycopodium:* Part III — The Plasticity of the species. Ibid. LI: 161.

HOLLOWAY, J. E. 1920. As above, Part IV — The structure of the Prothallus in Five species. Ibid. LII: 193.

— 1921. Further Studies on the Prothallus, Embryo, and Young sporophyte of *Tmesipteris*. Ibid. LIII: 386.

— 1923. Studies in the New Zealand Hymenophyllaceae: Part I — The Distribution of the species in Westland, and their Growth-forms. Ibid. LIV: 577.

— 1924. As above, Part II. The Distribution of the species throughout the New Zealand Biological Region. Ibid. LV: 67.

HOOKER, J. D. 1847. Flora Antarctica: I. London.

— 1853/55. Flora Novae-Zelandiae. London.

— 1853. Introductory Essay to the Flora of New Zealand. London.

— 1859. Introductory Essay to the Flora of Tasmania. London.

— 1867. Handbook of the New Zealand Flora. London.

HUTCHINSON, F. Jun. 1902. Notes on the Napier — Greenmeadows Road. Ibid. XXXIV: 409.

HUTCHINSON, J. 1921. The Family Winteraceae. Kew Bull. Nr. 5: 177.

HUTTON, F. W. 1873. On the Geographical Relations of the New Zealand Fauna. Ibid. V: 227 and Ann. Mag., Nat. Hist. Ser. 4. XIII: 25.

— 1876. On the Cause of the former great Extension of the Glaciers in New Zealand. T. N. Z. I. VIII: 383.

— 1884/85. On the Origin of the Fauna and Flora of New Zealand. N. Z. Journ. Sc. II: 1 and 249. Also Ann. Mag. Nat. Hist., ser. 5. XIII and XV: 425 and 77.

— 1899. The Geological History of New Zealand. T. N. Z. I. XXXII: 161.

— 1904. Index Faunae Novae Zealandiae. Introduction: 1.

IHERING, H. VON. 1892. On the Ancient Relations between New Zealand and South America. T. N. Z. I. XXIV: 431.

KIRK, H. B. 1921. On Growth-periods of New Zealand Trees, especially *Nothofagus fusca* and the Totara *(Podocarpus totara)*. Ibid. LIII: 429.

KIRK, T. 1869 a. Notes on Plants observed during a visit to the North of Auckland. Ibid. I: 140.

— 1869 b. On the Botany of the Great Barrier Island. Ibid. I: 144.

— 1870. On the Botany of the Thames Gold-fields. Ibid. II: 89.

— 1871 a. Notes on the Botany of Certain Places in the Waikato District, April and May 1870. Ibid. III: 142.

— 1871 b. On the Occurrence of litoral Plants in the Waikato District. Ibid. III: 147.

— 1871 c. The Flora of the Isthmus of Auckland and the Takapuna District. Ibid. III: 148.

— 1871 d. On the Botany of the Northern Part of the Province of Auckland. Ibid. III: 166.

— 1872 a. On the Flora of the Isthmus of Auckland and the Takapuna District, Part II. Ibid. IV: 228.

— 1872 b. A Comparison of the Indigenous Floras of the British Islands and New Zealand. Ibid. IV: 247.

— 1872 c. Notes on the local Distribution of Certain Plants common to the British Islands and New Zealand. Ibid. IV: 256.

— 1872 d. On the Habit of the Rata (Metrosideros robusta). Ibid. IV: 267.

— 1873 a. On the Naturalized Plants of the Chatham Islands. Ibid. V: 320.

— 1873 b. Notes on the Flora of the Lake District of the North Island. Ibid. V: 322.

— Also in N. Z. Gaz. No. 43 (Sept. 4th).

— 1873 c. On the Botany and Conchology of Great Omaha. T. N. Z. I. V: 363.

— 1876. On a Remarkable Instance of Double Parasitism in Loranthaceae. Ibid. VIII: 329.

— 1878 a. On the Naturalized Plants of Port Nicholson and the adjacent District. Ibid. X: 362.

KIRK, T. 1878 b. On the Botany of the Bluff Hill. Ibid. X: 400.

— 1879 a. Notes on the Botany of Waiheke, Rangitoto, and other Islands in the Hauraki Gulf. Ibid. XI: 444.

— 1879 b. On the Relationship between the Floras of New Zealand and Australia. Ibid. XI: 540.

— 1885 a. On the Flowering Plants of Stewart Island. Ibid. XVII: 213.

— 1885 b. On the Ferns and Fern Allies of Stewart Island. Ibid. XVII: 228.

— 1885 c. Notes on the New Zealand Beeches. Ibid. XVII: 298.

— 1886. Native Forests and State of the Timber Trade. Parts I and II. Append. Journ. House of Representatives, C. — 3.

— 1889. The Forest Flora of New Zealand. Wellington.

— 1891 a. On the Botany of the Snares. T. N. Z. I. XXIII: 426. Also Journ. Bot. (Reprint) N. Z. Journ. Sc. n. s. I: 161.

— 1891 b. On the Botany of Antipodes Island. Ibid. XXIII: 436.

— 1891 c. On the Botany of the Antarctic Islands. Rep. Aus. Assoc. Adv. Sc. III: 213.

— 1893. On Heterostyled Trimorphic Flowers in the New Zealand Fuchsias, with Notes on the Distinctive Characters of the Species. T. N. Z. I. XXV: 261.

— 1896 a. On the Products of a Ballast heap. Ibid. XXVIII: 501.

— 1896 b. The Displacement of Species in New Zealand. Ibid. XXVIII: 1.

— 1897 a. Notes on the Botany of the East Cape District. Ibid. XXIX: 509.

— 1897 b. On the History of Botany in Otago. Ibid. XXIX: 533.

— 1899. The Students' Flora of New Zealand and The Outlying Islands. Welligton.

KÜKENTHAL, G. 1909. Cyperaceae-Caricoideae. Das Pflanzenreich, IV, 20. Leipzig.

KURTZ, F. 1875. Flora der Aucklands-Inseln. Verh. d. Bot. Ver. der Prov. Brandenburg, XVIII: 3.

— 1877. Flora der Aucklands-Inseln. Nachtrag. Ibid. XIX: 168.

LAING, R. M. 1886. Observations on the Fucoideae of Banks Peninsula. T. N. Z. I. XVIII: 303.

— 1895. The Algae of New Zealand: their Characteristics and Distribution. Ibid. XXVII: 297. XXIX: 446.

— 1900, 02, 05. A Revised List of New Zealand Seaweeds, Parts I, II and III. Ibid. XXXII, XXXIV and XXXVII: 57, 327 and 380.

— 1909 a. The Chief Plant Formations and Associations of Campbell Island. The Subantarctic Islands of New Zealand. II: 482.

— 1909 b. The Marine Algae of the Subantarctic Islands of New Zealand. Ibid.: 493.

— 1912. Some Notes on the Botany of the Spenser Mountains, with a List of Species collected. Ibid. XLIV: 60.

— 1915. A Revised List of the Norfolk Island Flora, with some Notes on the species. Ibid. XLVII: 1.

— 1922 a. The Vegetation of Banks Peninsula, with a List of the species (Flowering· plants and Ferns). Ibid. LI: 355.

— 1922 b. The Flora of Banks Peninsula. New Zealand Nature Notes: 41.

— and BLACKWELL, E. W. 1906. Plants of New Zealand. Christchurch. 1927. Ibid. ed. 3.

— and WALL, A. 1924. The Vegetation of Banks Peninsula: Supplement I. T. N. Z. I. LV: 438.

LANCASTER, T. L. 1911. Preliminary Note on the Fungi of the New Zealand Epiphytic Orchids. T. N. Z. I. XLIII: 186.

LAZNIEWSKI, W. W. 1896. Beiträge zur Biologie der Alpenpflanzen. Flora. LXXXII: 224.

LEVY, E. B. 1923 a. The Grasslands of New Zealand. Series I. Principles of Pasture-Establishment. N. Z. Dep. Agric. Bull. No. 107. Wellington.

— 1923 b. As above. Series II. The Taranaki Back-Country. N. Z. Journ. Agric. XXVII: 138, 281.

LINDSAY, R. 1888. Heterophylly in New Zealand Veronicas. Trans. Bot. Soc. Ed. XVII: 243.
— 1891. Presidential Address. Ibid. XX: 193.
— 1898. Hybrid Veronica. Ibid. XXVII: 118.
LINDSAY, W. L. 1865. Relations of the Southern to the Northern Flora of New Zealand. Proc. Brit. Assoc.
LOTSY, J. P. 1925 a. On the Origin of Red-leaved Forms in a Cross of Nothofagus fusca × cliffortioides. Genetica VII: 241.
— 1925 b. Evolution considered in the Light of Hybridization, with Introduction and List of Hybrids by L. Cockayne. Christchurch, N. Z.
LOW, E. 1900. On the Vegetative Organs of Haastia pulvinaris. T. N. Z. I. XXXII: 150.
MARSHALL, P. The Geography of New Zealand. Christchurch.
— 1911. New Zealand and Adjacent Islands. Handbuch der Regional-Geologie. VII: 1.
— 1912. Geology of New Zealand. Wellington.
— 1926. A. New Species of Osmundites from Kawhia, New Zealand. T. N. Z. I. LVI: 210.
MARTIN, W. 1920. Pteridophytes of Banks Peninsula. Ibid. LII: 315.
— 1922. The Flora of Dunedin. New Zealand Nature Notes: 54.
— 1924. Native Plants of Dunedin and Surrounding District. Dunedin, N. Z.
MASSEE, G. 1899, 1907. The Fungus Flora of New Zealand. T. N. Z. I. XXXI and XXXIX: 282 and 1.
— 1909. Fungi and Lichenes (in part). The Subantarctic Islands of New Zealand. II: 528.
MONRO, D. 1865. On the Leading Features of the Geographical Botany of the Provinces of Nelson and Marlborough, New Zealand. Essays for N. Z. Exhibition Dunedin: 6. Reprinted 1869. T. N. Z. I. I.
MONTAGNE, C. 1845. Voyage au Pol Sud. Plantes Cellulaires. Botanique. Paris.
MUELLER, F. VON. 1864. The Vegetation of the Chatham Islands. Melbourne.
— 1874. List of the Algae of the Chatham Islands, collected by H. H. TRAVERS Esq., and examined by Professor JOHN AGARDH, of Lund. T. N. Z. I. VI: 208.
MUELLER, K. 1893. Remarks on Dr. H. VON IHERINGS's Paper "On the Ancient Relations between New Zealand and South America". (Translated from Das Ausland. July 20 th, 1891 by H. SUTER. T. N. Z. I. XXV: 428).
MURRAY, J. 1927. Four Fungi on the Endemic Species of Rubus in New Zealand. Ibid. LVII: 218.
MUSGRAVE, T. 1866. Cast away on the Auckland Isles: a Narrative of the Wreck of the "Grafton" and of the Escape of the Crew after Twenty Months' Suffering. London. (Edited by I. I. SHILLINGLAW.)
NANNFELDT, J. A. 1924. Revision des Verwandtschaftskreises von Centella Asiatica (L.) Urb. Svensk. Bot. Tidsk. XVIII: 397.
OLIVER, W. R. B. 1910. The Vegetation of the Kermadec Islands. T. N. Z. I. XLII: 118.
— 1915. The Vegetation of White Island. Journ. Linn. Soc. — Bot. XLIII: 41.
— 1917. The Vegetation and Flora of Lord Howe Island. T. N. Z. I. XLIX: 94.
— 1921. Notes on Specimens of New Zealand Ferns and Flowering-plants in London Herbaria. Ibid. LIII: 362.
— 1922. Littoral Plant and Animal Communities of Cook Strait. N. Z. Nature Notes: 12.
— 1923. Marine Littoral Plant und Animal Communities in New Zealand. T. N. Z. I. LIV: 496.
— 1925 a. Biographical Relations of the New Zealand Region. Journ. Linn. Soc. Bot. XLVII: 99.
— 1925 b. Vegetation of Poor Knights Islands. N. Z. Journ. Sc. & Tech. VII: 376.
— 1926. New Zealand Angiosperms. T. N. Z. I. LVI: 1.
PARK, J. 1910. The Great Ice Age of New Zealand. Ibid. XLII: 589.
PEGG, E. J. 1914. An Ecological Study of some New Zealand Sand-dune Plants. Ibid. XLVI: 150.

PENNELL, F. W. 1921. "Veronica" in North and South America. Rhodora XXIII: I.

PETRIE, D. 1881. A Visit to Stewart Island, with Notes on its Flora. T. N. Z. I. XIII: 323.

— 1883. Some Effects of the Rabbit Pest. Ibid. I: 413.

— 1885. The rapid Increase of Erechtites prenanthoides D. C. — N. Z. Journ. Sc. II: 454.

— 1896. List of the Flowering Plants Indigenous to Otago, with Indications of their Distribution and Range in Altitude. T. N. Z. I. XXVIII: 540.

— 1903. On the Pollination of Rhabdothamnus Solandri, A. Cunn. Ibid. XXXV: 321.

— 1905. On the Pollination of the Puriri Vitex lucens, T. KIRK. Ibid. XXXVII: 409.

— 1908 b. Account of a Visit to Mount Hector, a High Peak of the Tararuas, with List of Flowering-plants. Ibid. XL: 289.

— 1910 c. On the Naturalisation of Calluna vulgaris, Salisb., in the Taupo District. Ibid. XLII: 199.

— 1912. Report on the Grass-denuded Lands of Central Otago. N. Z. Dept. Agric. Bull. No. 23 (n. s.): 1.

PIGOTT, E. M. 1915. Notes on Nothopanax arboreum, with some Reference to the Development of the Gametophyte. T. N. Z. I. XLVII: 599.

POPPELWELL, D. L. 1912. Notes on the Plant Covering of Codfish Island and the Rugged Islands. T. N. Z. I. XLIV: 76.

— 1913 a. Notes on the Botany of the Ruggedy Mountains and the Upper Freshwater Valley, Stewart Island. Ibid. XLV: 278.

— 1913 b. Notes on a Botanical Excursion to the northern Portion of the Eyre Moun tains. Ibid. XLV: 288.

— 1914. Notes on the Botany of the Routeburn Valley and Lake Harris Saddle. Ibid. XLVI: 22.

— 1915 a. Notes on the Plant Covering of the Garvie Mountains, with a List of Species. Ibid. XLVII: 120.

— 1915 b. Notes of a Botanical Visit to Herekopere Island, Stewart Island. Ibid. XLVII: 142.

— 1916 a. Notes on the Plant-covering of the Breaksea Islands, Stewart Island. Ibid. XLVIII: 246.

— 1916 b. Notes on the Plant-covering of Pukeokaoka, Stewart Island. Ibid. XLVIII: 244.

— 1917 a. Botanical Results of an Excursion to the Upper Makarora Valley and the Haast Pass, supported by a List of the Species observed. Ibid. XLIX: 161.

— 1917 b. Notes on a Botanical Excursion to Long Island near Stewart Island, inclu ding a List of the Species. Ibid. XLIX: 167.

— 1920 a. Notes on the Indigenous Vegetation of the North-eastern Portion of the Hokonui Hills, with a List of Species. Ibid. LII: 239.

— 1920 b. Notes on the Indigenous Vegetation of Ben Lomond, with a List of Species. Ibid. LII: 248.

— and THOMSON, W. A. 1918. Notes of a Botanical Visit to Hollyford Valley and Martin's Bay, with a List of Indigenous Plants. Ibid. L: 146.

POTTS, T. H. 1878. Notes on Ferns. T. N. Z. I. X: 358.

— 1882. Out in the Open: A Budget of Scraps of Natural History gathered in New Zealand. Christchurch.

RAOUL, E. 1844. Choix de Plantes de la Nouvelle Zélande. Paris.

REISCHER, A. 1889. Notes on the Islands to the South of New Zealand. T. N. Z. I. XXI: 378.

RICHARD, A. 1832. Essai d'une Flore de la Nouvelle Zélande. Paris.

RUTLAND, J. 1889. The Fall of the Leaf. T. N. Z. I. XXI: 110.

— 1901. On the Regrowth of the Totara. Ibid. XXXIII: 324.

SAINSBURY, G. O. K. 1923. Notes on Pittosporum obcordatum. Ibid. LIV: 572.

Schenck, H. 1905. Vergleichende Darstellung der Pflanzengeographie der subantarctischen Inseln, insbesondere über Flora und Vegetation von Kerguelen. Wiss. Erg. d. Deut. Tiefsee-Exped. II.

Schimper, A. F. W. 1898. Pflanzen-Geographie auf physiologischer Grundlage. Jena.
— 1903. English Translation of above.

Schlechter, R. 1911. Die Gattung Townsonia. Fedde, Rep. IX: 249.

Schröter, C. and Hauri, H. 1914. Versuch einer Übersicht der siphonogamen Polsterpflanzen. Eng. Bot. Jahr. L: 618.

Scott, J. H. 1883. Macquarie Island. T. N. Z. I. XV: 484.

Simpson, G. and Thomson, J. S. 1926. Results of a Brief Botanical Excursion to Rough Peaks Range. N. Z. Journ. Sc. & Tech. VIII: 372.

Skottsberg, C. 1904. On the Zonal Distribution of South Atlantic and Antarctic Vegetation. Geog. Journ.
— 1915. Notes on the Relations between the Floras of Subantarctic America and New Zealand. Plant World. XVIII: 129.

Smith, W. W. 1904. Plants naturalised in the County of Ashburton. T. N. Z. I. XXXVI: 203.

Speight, R. 1911. The Post-glacial Climate of Canterbury. Ibid. XLIII: 408.
— Cockayne, L., and Laing, R. M. 1911. The Mount-Arrowsmith District: a Study in Physiography and Plant Ecology. Ibid. XLIII: 315.

Sprague, T. A. and Summerhayes, V. S. 1926. The Taxonomic Position of Hoheria. Kew Bull. No. 5: 214.
— 1927. Santalum, Eucarya and Mida. Ibid. No. 5: 193.

Stephani, F. 1909. Hepaticae. Subantarctic Islands of New Zealand. II: 532.

Stewart, J. 1906. Notes on the Growth of certain Native Trees in the Auckland Domain. T. N. Z. I. XXXVIII: 374.

Suckling, L. A. 1914. The leaf-anatomy of some Trees and Shrubs growing on the Port Hills, Christchurch. Ibid. XLVI: 178.

Suter, H. 1891. Notes on the Geographical Relations of our land and fresh-water Mollusca. N. Z. Journ. Sc. I: 250.

Thomson, J. A. 1918. Brachiopoda. Rep. Austr. Antarc. Exped. Zool. IV. pt. 3.

Thomson, G. M. 1875. On some of the Naturalized Plants of Otago. T. N. Z. I. VII: 370
— 1879 a. Notes on Cleistogamic Flowers of the Genus Viola. Ibid. XI: 415.
— 1879 b. On the means of Fertilization among some New Zealand Orchida. Ibid. XI: 418.
— 1881 a. On the Fertilization &c. of New Zealand Flowering Plants. Ibid. XIII: 241.
— 1881 b. The Flowering Plants of New Zealand, and their Relation to the Insect Fauna. Trans. bot. soc. Ed. XIV: 91.
— 1882 a. On the Origin of the New Zealand Flora — being a Presidential Address to the Otago Institute. T. N. Z. I. XIV: 485.
— 1882 b. The Ferns and Fern Allies of New Zealand with Instructions for their Collection and Hints on their Cultivation. Melbourne.
— 1885 a. Botanical Evolution. N. Z. Journ. Sc. II: 361, 409 and 457.
— 1885 b. Introduced Plants of Otago. Ibid. II: 573.
— 1890. Spiny Plants of New Zealand. Nature XLII: 222.
— 1891 a. On some Aspects of Acclimatisation in New Zealand. Rep. Aus. Assoc. Adv. Sc. III: 194.
— 1892. Note on the Cleistogamic Flowers of Melicope simplex. T. N. Z. I. XXIV: 416.
— 1899. On some Peculiar Attachment-discs developed in some Species of Loranthus. Ibid. XXXI: 736.
— 1901 a. Plant-acclimatisation in New Zealand. Ibid. XXXIII: 313.
— 1908. Note on Gastrodia. Ibid. XL: 579.

THOMSON, G. M. 1909. A New Zealand Naturalist's Calendar. Dunedin.
— 1910. Botanical Evidence against the recent Glaciation of New Zealand. T. N. Z. I. XLII: 348.
— 1921. The Naturalization of Animals and Plants in New Zealand. Cambridge.
— 1926. The Pollination of New Zealand Flowers by Birds and Insects. T. N. Z. I. LVII: 106.
TOWNSON, W. 1907. On the Vegetation of the Westport District. Ibid. XXXIX: 380.
TRAILL, W. 1917. Effects of the Snowstorm of the 6th September, 1916, on the Vegetation of Stewart Island. Ibid. XLIX: 518.
TRAVERS, H. H. 1867. Notes on the Chatham Islands (lat. 44° 30' S., long 175° W.). Journ. Linn. Soc. Bot. IX: 135. Also 1869. T. N. Z. I. I: 173.
TRAVERS, W. T. 1865. Remarks on a Comparison of the General Features of the Flora of the Provinces of Nelson and Marlborough with that of Canterbury. Essays for N. Z. Exhibition, Dunedin: 17. Also 1869. T. N. Z. I. I.
— 1869. On Hybridization, with Reference to Variation in Plants. Ibid. I: 89.
— 1870. On the Changes effected in the Natural Features of a New Country by the Introduction of Civilized Races. Ibid. II: 299. — 1871. Part III, l. c. III: 326.
— 1874. On the Spread of Cassinia leptophylla. Ibid. VI: 248.
— 1884. Some Remarks upon the Distribution of the Organic Productions of New Zealand. Ibid. XVI: 461.
TURNER, E. P. 1909. Report on a Botanical Examination of the Higher Waimarino Forest. Wellington.
URQUHART, A. T. 1882. Notes on Epacris microphylla in New Zealand. T. N. Z. I. XIV: 364.
— 1884. On the Natural Spread of the Eucalyptus in the Karaka District. Ibid. XVI: 383.
WALL, A. 1918. On the Distribution of Senecio saxifragoides Hook. f. and its Relation to Senecio lagopus Raoul. Ibid. L: 198.
— 1920. Ranunculus pauciflorus T. Kirk: its Distribution and Ecology, and the Bearing of these upon certain Geological and Phylogenetic Problems. Ibid. LII: 90.
— [1922]. The Botany of Christchurch. Christchurch, N. Z.
— 1923. Raoulia mammillaris Hook. f. Rec. Cant. Mus. II: 105.
— 1925. The Flora of Mount Cook. Christchurch, N. Z.
— 1926. Some Problems of Distribution of Indigenous Plants in New Zealand. T. N. Z. I. LVII: 94.
WALLACE, A. R. 1892. Island Life, ed. 2. London.
WALSH, P. 1893. The Effect of Deer on the New Zealand Bush: A Plea for the Protection of our Forest Reserves. Ibid. XXV: 435.
— 1897. On the Disappearance of the New Zealand Bush. Ibid. XXIX: 490.
— 1899. On the Future of the New Zealand Bush. Ibid. XXXI: 471.
— 1911. The Effects of the Disappearance of the New Zealand Bush. Ibid. XLIII: 436.
WARMING, E., and VAHL, M. 1909. Oecology of Plants. Oxford.
WARNSTORF, C. 1911. Sphagnales-Sphagnaceae. Das Pflanzenreich, Heft 51. Leipzig.
WILLIS, J. C. 1916. The Distribution of Species in New Zealand. Ann. Bot. XXX: 437.
— 1917. The Distribution of the Plants of the Outlying Islands of New Zealand. Ibid. XXXI: 327.
— 1918. The Sources and Distribution of the New Zealand Flora, with a Reply to Criticism. Ibid. XXXII: 339.
— 1919 a. The Flora of Stewart Island (New Zealand): A Study in Taxonomic Distribution. Ibid. XXXIII: 23.
— 1919 b. The Floras of the Outlying Islands of New Zealand and their Distribution. Ibid. XXXIII: 267.

WILLIS, J. C. 1920. Plant Invasions of New Zealand. Ibid. XXXIV: 471.
— 1922. Age and Area. Cambridge.
WILSON, E. H. 1922. Notes from Australasia. II. The New Zealand Forests. Journ. Arn.
     Arboret. II: 282.
— 1923. Northern Trees in Southern Lands. Ibid. IV: 61.
YEATES, J. S. 1924. The Root-nodules of New Zealand Pines. Journ. Sc. & Techn.
     VII: 121.
— 1925. The Nucleolus of Tmesipteris Tannensis. Proc. Roy. Soc. B. XC: 227.

Chapter IV.
Sketch of the leading Physiographical Features of the Region.

1. General.

The New Zealand Botanical Region comprises those islands lying in the south-west Pacific between the parallels of 30^0 and 55^0 S. lat. and 158^0 56' W. and 176^0 W. long. The archipelago, if may be so termed, consists of the following groups of islands, each far distant from the others — the Kermadecs, New Zealand proper, the Chathams and the New Zealand Subantarctic Islands. The total land-area of the Region is about 270,000 sq. km.

New Zealand proper consists of two large islands, North Island and South Island and the much smaller Stewart Island. The above, together with some small islands and islets, including the Three Kings in the north, lie between the 34^0 6' and 47^0 20' parallels S. lat. and the meridians 166^0 30' W. and 178^0 30' W. long.

North Island has an area of 114,740 sq. km., a length of 829 km. and a maximum breadth of 458 km. (Cape Egmont to East Cape), and 324 km. from Tirua Point to Tolaga Bay, but north of lat. 38^0 and south of lat. 40^0 is quite narrow. The area of South Island is 151,120 sq. km., its length 845 km., and its greatest breadth 338 km. (Cape Saunders to Dusky Sound). Stewart Island has an area of 1721 sq. km. and is about 48 km. in length. Taking the land surface as a whole it is long and narrow, the most distant points from the sea being Tokaanu (North Island) 104 km., and 20 km. to the east of Kingston (South Island) 128 km.

The long isolation of New Zealand far from other land masses is a matter of profound significance with regard to the flora. Tasmania, the nearest land of importance, is about 1540 km. distant. The actual Australian continent is somewhat further away (1640 km.). Norfolk Island is 650 km. from North Island, Lord Howe Island 1320 km. and the New Hebrides 1540 km. South America is distant 6900 km. from the Chathams and the latter 600 km. from New Zealand proper. Finally, Antarctica lies 1350 km. from Macquarie Island and 2250 km. from Stewart Island.

A consideration of the ocean-depths in the neighbourhood of the New

Zealand Archipelago both serves to emphasize the isolation of the region and to show how wide-spread would be the effect of a general considerable elevation of the ocean-bed. The 180 m. line follows rather closely the outline of the present main islands and includes the adjacent small islands together with the Three Kings, Stewart Island and the Snares. The 900 m. line conforms closely to the above line on the east, but westwards it extends a considerable distance from the land, while to the south it goes beyond the Lord Auckland Islands. The 1800 m. line includes the whole archipelago except Macquarie and Kermadec Islands, and, extending far to the north-west, it reaches to within comparatively close proximity to the Queensland coast while Lord Howe Island and New Caledonia rise from this broad submarine ridge. Southwards to Antarctica the ocean bed lies between 1800 m. and 3600 m. below the surface, while between Australia on the west and South America on the east the depth is profound.

2. Physical Features of New Zealand proper.

a. North Island.

Mountains. As a rule the land-surface is much broken, hilly, and in parts mountainous. The main range extends from the east of Wellington Harbour to the East Cape. The highest peaks are in the Ruahine and Tararua Mountains, but none reach 1800 m. and few more than 1520 m. The rocks are chiefly mudstones, sandstones and greywacke.

The centre of the island is a volcanic plateau much of which is at an altitude of more than 600 m., but, northwards, gradually becoming lower, it extends to the Bay of Plenty. This area, within recent geological times, has been exposed to powerful volcanic action. The eruptions have been largely explosive and the present surface-pumice is the result. Even yet there is much thermal activity[1]) in the shape of boiling springs, geysers, mud volcanoes, &c., especially on a line connecting White Island, a volcano in the solfatara stage, and Ruapehu. From the highest portion of the Plateau rise the semiactive volcanoes, Ruapehu (2803 m.), Ngauruhoe (2291 m.) and Tongariro (1968 m.). The crater of Ruapehu is filled with ice in which lies a lake of, sometimes, extremely hot water, while small glaciers extend over the crater-rim and descend to comparatively low levels in the gullies.

Mt. Egmont (2521 m.) in Taranaki is an extinct volcano standing far isolated from other mountains; its summit carries perpetual snow. The

---

1) In 1886, the supposed extinct volcano Tarawera, situated on the Volcanic Plateau burst forth forming a rent 19 km. long with a mean width of 108 m. and ejecting light scoria and volcanic dust over an area of 15,000 sq. km.

remaining mountains are not lofty enough to bear subalpine vegetation[1]), nevertheless some of them[2]) show distinct altitudinal belts.

The extreme north of the island consists of a small, narrow, much dissected tableland some 300 m. high, formed of hard igneous and sedimentary rocks. At one time this was disconnected from the mainland, but now is united by a narrow spit of recent and consolidated dunes.

Plains. An extensive plain of marine origin, the Wanganui, lies between the Ruahine Mountains and Mount Egmont extending to Ruapehu where it is over 600 m. altitude and bounded on the south by the coastline. The rock consists of marl enclosing beds of shells. This plain is deeply cut by numerous streams so that, in many places, there is a network of deep gorges, often extremely narrow. River-formed gravel plains, the upper soil of which is frequently extremely rich, occur east and west of the main range (Manawatu, Wairarapa and Hawkes Bay Plains). The Waikato Plain, formed of river-borne pumice, occupies much of southern Auckland; it extends from the Firth of Thames to the R. Waipa. Its surface, rarely more than 30 m. above sea-level, is extremely wet and swampy. Northwards from the Auckland Isthmus there is much low-lying ground.

Rivers. The land throughout is well watered, every gully containing tsi running stream. In many places, the rivers have cut deeply into the surface, so that gorges are a familiar feature. The rivers rising in the high mountains are of a torrential character, but this feature is much less marked than in South Island, while gently flowing streams are more common. Where the rivers have not cut deep beds the adjacent land is liable to be flooded and extensive swamps are so formed (Manawatu, Waikato, Bay of Plenty, Thames, Northern Wairoa, Awanui). The Rivers Waikato and Wanganui are the most important in point of size and drainage area.

Lakes. The largest lakes are on the Volcanic Plateau, the most important being L. Taupo. L. Waikaremoana in the East Cape district lies 600 m. above sea-level. So far as plant-life is concerned it is the natural ponds and shallow lakes met with in many places that are of most importance.

Sea-coast. The coast is about 2152 km. in length. The various outlying islands also fournish coastal conditions. A most important feature of the coast-line is the extensive dune-area of much of the west coast which extends inland in places for 12.8 km. Also, there are considerable dunes in the far north, and north-east on the coast of the Bay of Plenty, and at various places between Poverty Bay and Cape Turnagain.

---

1) Mount Te Moehau in the north of the Coromandel Peninsula, carries a few subalpine plants, but this in relation to frequent wind.

2) The Maungaraki Mts. (900 m.), extending northwards from C. Palliser; the Puketoi Hills (610 m.) in the east of Wellington; the volcanic Mt. Karioi and Mt. Pirongia near the coast of south-west Auckland; the Cape Colville Range; the high land culminating in Mt. Tutamoe (800 m.), south of Hokianga Harbour, and the high land north of the latter with Mt. Raetea (800 m.).

The coast is frequently rocky. The south and south-east coast-line in its southern part consists of cliffs of slaty shale. Further north, low cliffs of soft marl and mudstone occur fronted by a narrow beach with the stony surface worn quite flat. The north-east coast consists at first of steep slaty cliffs, but from Opotiki onwards the land as a rule is low. The shore-line of the Coromandel Peninsula consists mostly of slate cliffs, but volcanic rock is not uncommon. From Auckland to North Cape the coast is much broken and presents a great diversity of stations for plant-life. Low cliffs fringed by a sandy or stony beach are common. There are bold rocky headlands. Many of the rivers have wide mouths, but these are shallow and mud-flats are exposed at low water. The short north coast is frequently precipitious through truncation of the tableland.

Where dunes are absent on the south-west and west coasts there are cliffs, some volcanic (Maunganui Bluff, &c.), others limestone (Kawhia, west of Taranaki &c.), and others of slaty shale (Reef Point, South Waikato Head &c.).

Many miles of coast are without inlets, estuaries &c. With the exception of Wellington and Porirua Harbours in the south, nearly all of any moment are to be found on the east and west coasts of Auckland.

b. South Island.

Mountains. The surface is extremely mountainous. Commencing in the south there are two chains, the one composed of gneisses and granulite on the west, and the other of schist extending from the shore at Dunedin and joining the former between Lakes Wakatipu and Wanaka. The latter is continued in an unbroken line as the Southern Alps to Cook Strait; and its eastern slopes are formed of slaty shales and greywackes. Below the shales &c. on the west the rock is schist, but at low levels occasionally gneiss. Granite occurs in a few places. The loftiest peaks are situated at about the centre of the chain. They vary from some 3000 m. to 3766 m. (Mt. Cook). Proceeding north and south, the range gradually decreases in height, but for a long distance few peaks are lower than 1800 m. Many lofty ranges and spurs extend eastwards for 48 km. or thereabouts from the main Divide. These eastern mountains are especially characterized by the vast masses of unstable debris covering their slopes, locally known as "Shingle-slips".

The snow-line in the Southern Alps is perhaps, on an average, at about 2200 m. altitude, but it is not uniform and varies according to latitude, while also it is lower on the west than on the east. The central part of the range is heavily glaciated, the size of the glaciers being correlated with the altitude of the peaks. The Great Tasman Glacier is about 29 km. long and its terminal face 918 m. above sea-level. On the west, the Fox and Franz Josef Glaciers descend to less than 210 m. altitude.

Glaciers are wanting to the south of lat. 45° and in the north they cease at a little to the north of lat. 43°. The eastern valley glaciers are generally covered with an enormous amount of moraine. The central Southern Alps form an unbroken wall, but to the north and south there are numerous passes, the lowest being the Haast (570 m.).

Besides the Southern Alps there are other lofty ranges. The Kaikoura Mountains are the two parallel ranges in the north-east with several peaks exceeding 2400 m. altitude, but they carry neither glaciers nor perpetual snow, except in small patches; their debris fields are of enormous extent. To the west of the Southern Alps, in the north-west, lie several rugged ranges, which extend from near Greymouth to the north coast with many peaks of 1500 m. altitude. Banks Peninsula, on the east of Canterbury, is formed of much-denuded volcanic rocks, and reaches a maximum height of some 900 m. In addition to the great ranges, there is on the east much hilly land, some undulating, and within the high-mountain area several extensive intermontane basins.

Plains. Gravel plains formed by glacial or snow rivers are a striking feature of South Island topography. The most important are: — The Canterbury Plain (161 km. long by 48 km. wide at its widest); the long narrow Westland coastal plain (200 km. long by 10 km. wide) and the Southland Plain extending from near Lumsden to Foveaux Strait. Flat as the Canterbury Plain appears to the eye, the surface near the foothills of the Southern Alps is more than 457 m. above sea-level. Borings for artesian water shew by the peat-deposits at different depths that there have been several changes in the land-surface during the formation of the plain:

Rivers. The numerous rivers issuing from the glaciers or fed by melting snows, or frequent downpours, are torrential at first, their beds full of huge rocks over which the waters leap and foam. By degrees, the valleys shaped by former glaciers widen and are filled from side to side, it may be, by a flat stony bed over which, in anastamosing streams, the river wanders. Lower down, as the valley widens still more, or when the plain is gained, the river may flow between high permanent terraces that it has built, and frequently there is a series of such at different levels with portions of the ancient flood-plains at their bases. River beds 1 km. or more wide with terraces on either side are a common feature of the valleys and gravel-plains (Fig. 1).

Where a tributary stream in a mountain-valley joins a river the shingle of its bed spreads out as a fan. Such are present at the mouth of almost every gully, sometimes naked and active, at other times plant-clad and passive.

Glaciation. At the height of the New Zealand glacial period great glaciers extended throughout most of the high lands of South Island even as far east as the uppermost part of the Canterbury Plain and, on the west in no few places to the sea. Remains of ice-action are to be seen throughout the Southern Alps, and in the western ranges of Nelson, except where in the drier parts excessive denudation has taken place. Cirques, hanging-valleys, moraines large and small, transported morainic material, roches moutonnées, ice-shorn hillsides, truncated spurs and U-shaped valleys are abundant, and both testify to the extent of the glaciation and provide special growing-places for plants. Glacial lakes are frequent and range from small tarns on moraines to the lakes, many kilometres in length and frequently of great depth, of Canterbury and Otago. In North Island, glaciation, if it occurred at all, was evidently trifling.

Sea-coast. The coast-line (about 2740 km. in length) in general is little broken. Notable exceptions are the Marlborough Sounds and Otago Fiords, the former drowned river-valleys, the latter of glacial origin. Banks Peninsula contains a number of inlets, some much-eroded, submerged, volcanic craters. There are also a few shallow estuaries, more or less closed by sand or gravel spits, whose floors are in part bare at low-water.

In many places, the coast is low and the shore sandy, so that long stretches of dunes occur. Shingly shores are frequent in many parts and correlated with the great river-beds.

Rocky shores and cliffs are frequent at many points on the coast. In this regard, the most noteworthy are the rocky walls of the Otago fiords formed by lofty mountains rising from the water's edge as precipices for 1000 m. or more. These fiords extend inland for many kilometres and some of them almost reach Lakes Te Anau and Manapouri — themselves to all intents and purposes fiords also. Facing Cook Strait, in the north of the island, are also long arms of the sea penetrating the land, but such are not of glacial origin but are submerged river valleys.

Coastal islands are few. The most important are those to the north of Marlborough and in Foveaux Strait. The Open Bay Islands on the west are composed of limestone, and although mere islets possess a remarkable plant-covering.

c. Stewart Island.

Stewart Island lies about 25 km. from South Island from which it is separated by Foveaux Strait, this nowhere more than 48 m. deep.

In shape the island is irregularly triangular. The surface is hilly, much broken and in parts mountainous, the peaks varying in height from 676 m. to 975 m. There is but little truly flat ground. East of the central range the land is low, but broken. At the head of Paterson Inlet a narrow valley, the Freshwater, extends northwards to the Ruggedy Mts., while, westwards, an opening widening out into an ancient dune-area, connects the valley and the west coast. West of Port Pegasus there is some low boggy moorland. The coast is in general rocky. At Mason Bay on the west is an extensive dune-area.

Perhaps the most striking features of the island are Paterson Inlet and Port Pegasus. The former, a broad expanse of lake-like water, irregular in shape, enclosed by hills and dotted with forest-clad rocky islets, extends westward for 17 km. putting forth three diverging arms. Port Pegasus, situated in the south, runs parallel with the south-east coast for about 12 km., its entrance blocked by three islands.

There are a number of outlying rocky islets especially to the east and south-west, while to the north-west is the fairly large Codfish Island. Thirty-two kilometres to the eastward is the flat island of Ruapuke 7.6 km. long by 3.4 km. wide.

3. Physical Features of the Outlying Islands.

The Kermadec Islands. These are four in number. They extend from 29° 15′ S. lat. and 177° 59° W. long. to 31° 24′ S. lat. and 178° 51′ W. long., and are distant about 960 km. from New Zealand. The group is volcanic, but it stands on a submerged plateau, part of a ridge connecting New Zealand with Tonga. Outside the plateau the ocean is 2700 m. deep.

Sunday Island, the largest of the group, 10.3 km. long and 29.25 sq. km. in area, reaches a height of 524 m. It is composed chiefly of pumiceous and other tuffs; lava streams are few. The surface is hilly with many narrow spurs separating deep gullies. These spurs, truncated at the coast, drop as sheer precipices to the water for 200 to 300 m. The greater part of the island is a crater, its rim 55 m. above sea-level in the north but elsewhere averaging over 300 m. There are three small crater lakes. There is a small sandy beach and one of gravel.

Macauley Island, distant 109 km. from Sunday Island, is 2 km. long, 3 sq. km. in area, and its highest point 237 m. above sea-level. Cliffs everywhere fall to the sea.

Curtis Island, 35 km. from the last-named, are two rocky islets with an area of 0.6 sq. km. and the highest point 100 m. The crater-floor contains hot mud, boiling springs and sulphur. (All the above is taken from OLIVER 1910).

The Chatham Islands. These consist of four islands and several detached islets and rocks, lying between the parallels of 43° 35′ and 44° 25′ S. lat., and the meridians of 176° and 176° 55′ W. long. and distant 600 km. from New Zealand proper. Chatham Island, by far the largest of the group, is 967 sq. km. in area and is somewhat the shape of a horseshoe. Generally the land is low but undulating. Much of the interior is occupied by the Te Whanga lagoon which, roughly triangular extends from the north coast southwards for 25 km., and, at its greatest breadth, is nearly 15 km. wide. On the east, it is separated from the sea by a very narrow strip of land broken through at one point. South of the lagoon the island is a compact four-sided block, which, in comparison with the remaining land, looks quite hilly, but its highest part is only 286 m. and the culminating point of the main ridge about 2 m. lower. From this ridge a tableland extends south-wards terminating in abrupt cliffs 182 to 213 m. in depth irregularly cut by small streams. Here and there conical volcanic hills, 152 to 182 m. high, stand out from the flat, northern and central portions of the island.

The extensive coast-line varies from flat ground bordered by dunes or low rocks to the high cliffs of the south and south-west.

Besides the Whanga, there are many other lagoons and lakes, indeed, it is stated that one-third of the surface is occupied by water. Bogs of great extent and depth are a familiar feature both of high and low ground. Small, sluggish streams of peaty dark-brown water are abundant.

Pitt Island, 13.6 km. long by 6 km. across, lies about 22 km. to the south of the main island. Its coast is rocky. The remaining islands (Mangere, South-East Island) are quite small, but the latter rises to 184 m.

The New Zealand Subantarctic Islands. These consist of several distant groups lying between the parallels of 45° 44′ and 47° 43′ S. lat. and 158° 56′ and 179° W. long. The names, distance and direction of each group from the South Cape of Stewart Island are as follows: Snares, 113 km., S. S. W.; Lord Auckland Islands, 348 km., S. by W.; Campbell Islands, 608 km., S. by E.; Macquarie Islands, 1049 km., S. W. by S.; Antipodes Island, 902 km., E. S. E.; Bounty Islands, 902 km., E.

All the islands, excepting the Bounties and the Snares, are chiefly of volcanic origin and, the Bounties excepted, the surface consists in general of a deep layer of peat.

The Snares consist of North-East Island, 1.6 km. long and 0.8 km. wide, which rises perpendicularly on its south side to 131 m. and of four other rocky islets lying to the south-west. The main island has a rocky precipitous coast-line except in one place on the east side where a small stream enters the sea. The island is formed of a pale moderately-coarse muscovite granite.

The Lord Auckland Islands consist of two fairly large islands, Lord Auckland Island, 40 km. long by 27 km. wide in its widest part and Adams Island 24 km. long and 8 km. wide in its widest part, together with a group of small islands to the north and the small precipitous Disappointment Island on the west. An elevation of 360 m. would connect the group with New Zealand proper.

Adams Island is separated from Lord Auckland Island by Carnley Harbour, the site of an old volcano. It is a fairly even ridge, 600 m. high, with a long slope northwards, but southwards descending to the sea in a sheer precipice.

Lord Auckland Island is also high and rises in more than one place to 600 m. Several arms from Carnley Harbour pierce it in the south. On the east are a number of small fiords the result of ice-action, but on the west there is a perpendicular wall of stupendous cliffs.

The islands in the north are separated from Lord Auckland Island by

Port Ross, a land-locked sheet of water. They are quite low but their coasts are rocky. On Enderby Island there is a sandy beach, 8 km. long backed by low dunes. Disappointment Island, some hundreds of metres high with cliffs on all sides is about 3 km. in length.

Rivers of considerable size for so small a land-area fill the valleys of the two larger islands. The watershed of Lord Auckland Island is close to the summit of the western cliffs. There are one or two small mountain lakes.

There is abundant evidence of glacial action, but according to SPEIGHT it is improbable that the islands have been completely covered by ice.

The Campbell Islands consist of a main island (Campbell Island) 48 km. in circumference, but the other members of the group are mere rocks. The northern end of the island rises as a whole to about 300 m., but in the south there are a number of isolated peaks, the highest being from 400 to 500 m. altitude. Two long inlets pierce the land on the east.

The rocks are in part volcanic and in part limestone containing fossils. According to MARSHALL the surface-features are due to glacial action, but there is no evidence that the island was covered by an ice-sheet.

Macquarie Island, according to SCOTT (1883:486), is exceedingly hilly, the hills rising to perhaps 280 m., while numerous tarns lie amongst their hollows. The coast-line consists principally of cliffs with a few shingle beaches. Possibly the island is 30 km. long. The rocks, so far as known, are volcanic. Apparently the island is separated from the rest of the Subantarctic Group by the ocean's depth of no less than 3600 m.

The Antipodes Islands consist of Antipodes Island (8 km. long by 4.6 km. at its widest) and Bollons Island, quite small, but 150 m. high. The surface of the main island is an undulating plateau, Mt. Galloway, the highest point reaching 530 m. The coast consists of high, perpendicular cliffs. The rocks, so far as known, are basalts.

The Bounty Islands are a small group of rocky islets and rocks formed of a pale biotite granite. The largest island is 1 km. long by 0.8 km. wide and 88 m. high. The surface is without a true soil and is polished smooth as glass by thousands of penguins and other birds, which live for part of the year on the island and numerous fur seals. Quantities of guano collect during the breeding season of the birds, but the greater part is removed by the rains of winter.

4. The Soils of the New Zealand Botanical Region.

In what follows merely general and guarded statements are made, based for the most part on the experience of agriculture and horticulture and on rapid field observations. Also the distribution of the plant communities supplies some information — not always reliable — as to the relative fertility of the soil on which they grow.

The soils which occupy the widest areas are pumice, clays of various kinds, loess (in a wide sense), loams, sand and stony debris. Peat, including raw humus, mica-schist soils, calcareous soils, soils derived from basic volcanic rocks and rocks of various kinds are common enough but of more local distribution. Certain soils are purely local, e. g., scoria, sulphur &c., soils near hotsprings &c., salt soils, magnesian soils and soil heavily manured by sea-birds.

Loess soils occur over wide areas in South Island. They have arisen from silt blown from the glacial river-beds; such accumulation and transport still goes on. Loess is frequently mixed with clay derived from the underlying rock or of glacial origin. Loess is an important ingredient of soils to the east of the Divide in South Island.

Pumice soils play a large part in the centre of North Island and the land adjacent. When pure, unweathered and perhaps mixed with scoria, they provide, even in a wet climate merely steppe or desert conditions. When weathered and mixed with humus, pumice soil is "fertile" enough, as the farms of the Waikato, and, in part, those of Hawkes Bay bear witness.

Clay soils of various kinds are common in North, South and Stewart Islands. The extremely abundant greywacke readily weathers into clay. The low hills and undulating ground of lowland Auckland, known as the "gumlands", are covered with a great depth of specially impervious clay deficient in humus which though variable in quality is generally extremely poor. This is especially so with the white clays locally termed "pipeclay". The Stewart Island clays are formed from granite. Glacier clay occurs on mountain slopes and river-valleys. The "fertility" of these clay soils is governed by the drainage-conditions and the percentage of humus. Frequently clay becomes hard and dry during a period of drought and it then offers a most unfavourable station for plant-life.

Mica-schist soils occupy a wide area in the North and South Otago Botanical Districts. They are particularly fertile and contain, when apparently dry, a considerable amount of available moisture.

Calcareous soils are a feature of the Wanganui "coastal" plain. They also occur locally throughout the Region and frequently extend over considerable areas, but the fact that a soil overlies limestone by no means proves that it is truly calcareous.

Sandy soils are frequent on and near the coast, and arise either as blown sand from the shore or from the disintegration of soft sandstone. They are also frequent on the gravel plains.

Alluvial loams form the bulk of the soil of lowland valleys in both Islands.

Humus soils are of widespread distribution. They occur at all altitudes and vary from a thin surface-layer to peat-deposits, 12 m. deep, as in the

4*

Chatham and Lord Auckland Islands. The rain-forest climate is eminently favourable for the production of humus. The subantarctic and wet high mountain climates favour the formation of raw humus and peat.

Volcanic soils, pumice excepted, though of wide occurrence in many parts of the New Zealand region, are generally local and limited in extent. They are specially fertile and the distinction between the vegetation of volcanic and "gumland" soils in Auckland is striking. The other soils mentioned above need no special comment here, but some come into consideration — as do soils of all kinds — when dealing with the plant-communities.

The terms "fertile" and "fertility" have been used in this chapter, and are to be found here and there in other parts of this book. All the same, it is not a really valid ecological expression, for the idea of fertility is derived entirely from the soil in relation to crops, and to the plant in wild nature no soil is fertile or unfertile. In fact, the whole idea of epharmony is opposed to such a term.

Chapter V.
The Climate of New Zealand Proper.

General. Such meteorological statistics as are available are not of much value for ecological plant-geography, nevertheless they give an indication of the general climate to which the plant-communities are exposed and they are of service for comparative purposes. But in a sparsely populated young country such accuracy or thoroughness cannot be expected as in the Old World.

Almost the whole of the data is derived from observations made in the lowlands and there are no statistics concerning the high mountains for altitudes exceeding about 600 m. But, apart altogether from instrumental observations, the members of a plant-community can tell a good deal concerning the various climatic factors on which, in part, its structure and activities depend. So, too, the behaviour of indigenous species abroad and of both indigenous and exotic species in New Zealand furnishes meteorological information of no small value. Finally, my studies of the vegetation carried out at all seasons for some 40 years have enabled me to gain some knowledge of the climate of many localities for which no other data are available.

In what follows only the climate of North Island, South Island and Stewart Island is dealt with, that of the Outlying Islands being discussed when treating of the vegetation and floras of each group. Also, the special details concerning the Botanical Districts are removed from this chapter to their more suitable position in Part IV, and climate is frequently referred to when dealing with the communities.

New Zealand possesses, for the most part, a maritime climate, situated as it is remote from other lands in the widest ocean of the globe with no part of the area more than 128 km. from the sea. There are however marked differences in climate, owing firstly to the region extending through 25 degrees of latitude and secondly to the lofty mountain chains of the main islands lying athwart the prevailing winds. "Aspect" in a wide sense therefore has a remarkable influence both on rainfall and temperature, not only as to average annual amounts, but also in every atmospheric disturbance that passes over the land. Ecologically it is a primary cause of the wide-spread continuous formations — forest and tussock-grassland. Aspect in a narrow sense regulates climate or modifies its effects and so affects both plant-distribution and the composition and structure of communities.

With regard to rainfall, that of South Island, in the west, is extremely high, while, on the contrary, parts of the eastern districts are, in comparison, very dry. North Island has a maximum rainfall — almost a rainy season indeed — in the winter months, but South Island shows a remarkable evenness in its monthly averages. Periods of drought occur at times in the eastern districts, such being commonest in spring and summer, in the north, and in autumn and winter, in the south. Although the total average rainfall, especially in the east, decreases with increase of latitude, yet the number of rainy days is greater in the south than in the north. This arises through the frequent occurrence of atmospheric disturbances in the latitude of the "forties", but the northern districts are under the influence of occasional cyclonic disturbances of tropical origin which travel from north-west to south-east over North Island. Occasionally, extensive "Lows", decreasing northwards, account for much warm and moist weather but do not usually bring about a heavy precipitation. The cyclone track will often pass to the northward of New Zealand; sometimes it crosses as low as Cook Strait; occasionally it comes from the north-east to the East Cape and then passes down the east coast before taking an eastward route under the guiding and controlling influences of the prevailing westerly winds of these higher latitudes. The upper winds are almost invariably westerly and a divergence to a southwesterly direction from the west usually precedes a marked change in weather conditions, which the forecaster values as a guide to the subsequent swing of atmospheric pressure. High pressure, or the anticyclone, may be regarded as the controlling factor of weather conditions and on the edge and between these high pressure systems are found the "Lows".

The frequency of the above disturbances judging from the average of 9 years are as follows: (1) For the cyclone or monsoon of marked intensity — Spring 2; Summer 1.8; Autumn 3.3; Winter 5.3. (2) For the westerly or antarctic low — Spring 6.3; Summer 4.3; Autumn 5.2; Winter 4.7.

Rainfall. The rainfall of New Zealand proper bears a striking relation to the physical configuration of the land, and records gathered throughout the country during a period of more than 60 years present a certain regularity which clearly shows the dominating influence of the mountain ranges. In South Island, the lofty Southern Alps, together with the ranges of the North-western district, lie broadside to the prevailing westerly winds, and on their windward slopes are condensed the vapours which have been gathered by the breezes sweeping over vast stretches of ocean. On the Westland coastal plain, and on the adjacent rugged and precipitous slopes, the rainfall averages from 250 to 500 cm. per annum, while on the leeside of the great mountain barrier the climate is, in comparison extremely dry and, in places, the rainfall is only one tenth of that on the west. **There are in fact two distinct climates that of the west strongly favourable to forest, and that of the east altogether antagonistic to that type of vegetation and in harmony with tussock-grassland** (Fig. 39), but the latter climate, as will be seen, owes in part its character to the wind-factor.

While South Island isohyets stretch east and west, those of North Island are more irregular in form, but demonstrate that the rainfall itself is more regular over the land as a whole and less extreme in a comparison between the different botanical districts. But here again the control of the mountains and plains over precipitation is apparent, the contours of the rainfall areas coinciding more or less with the configuration of the country, the heavier downpours occurring in proximity to Mount Egmont, the central volcanoes, the Dividing Range and the higher summits in general.

The mean annual rainfall of New Zealand proper, as derived from means of representative stations in various parts of the islands, is about 121 cm., but the seasonal falls are far from uniform throughout. The following averages taken from the climatological tables give some idea of the rainfall and its distribution throughout the year for the two main islands, but the first two tables, though useful for comparison with similar statistics for other countries, are of very little phytogeographical moment.

Rainfall (in centimetres).

	Spring	Summer	Autumn	Winter	Annual
North Island	31	25	34	36	126
South Island	30	27	27	31	115

Rainy days (2 mm. or more).

	Spring	Summer	Autumn	Winter	Annual
North Island	45.5	30.3	39.1	47.1	162
South Island	44.4	34.7	36.6	40.9	156

Annual rainfalls at certain representative stations (in centimetres).

North Island	Auckland 61 years	New Plymouth 37 years	Gisborne 36 years	Wellington 56 years
Average	109.8	150.9	118.9	126.4
Maximum	161.9	210.6	163.4	171.9
Minimum	86.9	111.3	66.3	76.3

South Island	Christchurch 37 years	Hokitika 35 years	Dunedin 55 years	Invercargill 18 years
Average	67.5	304.3	95.8	116.8
Maximum	90.3	392.2	138.5	165.3
Minimum	34.4	229.1	56.3	84.4

From the phytogeographical standpoint the number of rainy days is of far greater moment than the amount of the downpour, the following table then is of special interest.

Mean Number of Days with Rain (2 mm. or more).

Locality	Jan.	Feb.	Mar.	Ap.	May	Jun.	July	Aug.	Sept.	Oct.	Nov.	Dec.
Auckland	10.3	9.4	11.0	13.2	18.1	19.1	20.7	19.3	17.6	16.2	11.4	11.4
New Plymouth	12.5	10.5	12.2	14.0	18.0	17.2	19.5	19.2	17.5	18.5	15.7	14.2
Gisborne	9.0	9.7	12.3	12.4	15.6	16.4	16.8	15.7	13.4	11.6	11.6	9.3
Wellington	10.5	9.2	11.7	12.8	16.4	17.3	18.3	17.3	15.5	13.9	12.7	12.0
Christchurch	9.0	7.4	9.3	9.1	10.8	12.1	13.0	11.0	9.9	8.9	9.8	9.1
Hokitika	14.8	10.3	13.5	14.1	15.5	15.0	16.4	16.0	15.3	19.0	13.0	16.3
Dunedin	14.7	11.5	13.2	13.2	14.1	13.0	13.6	13.0	13.0	14.5	14.1	14.8
Invercargill	15.0	10.0	14.0	16.0	17.0	15.0	15.0	14.0	14.0	17.0	18.0	15.0

Annual mean totals: — Auckland 180.4; New Plymouth 189.7; Gisborne 153.8; Wellington 167.7; Christchurch 119.4; Hokitika 179.2; Dunedin 163.3; Invercargill 180.0.

The question of **snow** naturally comes along with that of rain, but as it is discussed from the ecological standpoint when dealing with the high-mountain vegetation in the first chapter of that section, only a brief statement need be made here. In the subalpine and alpine belts of all the islands the winter snow-fall is very heavy and there is a continuous covering for some months, the length of time it remains depending upon aspect and altitude. In the montane belt of South Island there are occasionally heavy falls reaching up to one metre in depth. The sheep-farmer knows all about the relation of his run to snow and divides it into "winter" (snow-free) and "summer (snow for months) country" — the area of the former determining the number of sheep the run can carry. In the lowlands snow

is almost unknown in North Island, but from the North-eastern district
southwards, every few years it may lie for one or two days at sea-level.
On the west snow at sea level is rare.

Temperature. Latitude, insolation, proximity to the ocean, or the large
inland lakes, and height above sea-level are the determining major factors
with regard to temperature. Especially are the oceanic influences a master-
factor with regard to both summer heat and winter cold, upon both of
which they exercise a moderating effect. Indeed, extremes of heat and
cold, such as occur at similar latitudes in the Northern Hemisphere, are
absent throughout New Zealand at every altitudinal belt. The west coast
of South Island lies open to the prevailing westerly winds and is more
humid and equable than the eastern botanical districts which, generally
speaking, possess a more or less continental climate with a considerable
range of temperature. Near the coast of North Island frosts, even on the
grass, are of rare occurrence, but further south, and inland throughout,
they are often experienced. Special details as to temperature are given in
Part IV when treating of the different botanical districts and in the section
dealing with the vegetation of the high mountains.

The meteorological seasons are later than the astronomical. Thus July
is usually the coldest and wettest month in the year, while January is the
driest and warmest. The seasons may be roughly divided as follows: —
spring, — September, October, November; summer, — December, January,
February; autumn, — March, April, May; winter, — June, July, August.
But such divisions are somewhat misleading from the phytogeographical
standpoint, altitude, latitude and aspect being controlling factors with regard
to seasonal changes.

The following means (Centigrade) taken from the climatological tables
give some idea of the temperature of the main islands: —

	Spring	Summer	Autumn	Winter	Year
North Island	12.8°	16.5°	14.1°	9.5°	13.2°
South Island	11.2°	15.2°	11.6°	6.5°	11.1°

Taking the mean maximum and minimum temperatures for the hottest
(January) and coldest (July) months for a number of localities — all at
about sea-level unless height is given — and proceeding from north to
south they are as follows: — Auckland 23° C., 10° C.; Rotorua, (276 m. alt.)
24°, 2.1°; Napier 24°, 5°; Moumahaki (south of EW.) 23.5°, 5.7°; Wellington 21°, 4"; Nelson 23.8°, 2.8°; Hokitika 19.7°, 1.8°; Lincoln (E.) 22.2°, 1.8°;
Queenstown (SO., alt. 301 m.) 19.1°, — 1.1°; Dunedin 19°, 2.8°; Invercargill
19.2°, 1°.

Central Otago (NO.) has the reputation of experiencing the coldest
winters of New Zealand proper and many exaggerated statements have been

made, which, however, are disproved by the inability of any indigenous plant to be cultivated in the open in the colder parts of Europe or North America. All the same, the area in question in certain localities is far colder in winter than most parts of South Island at the same altitude. Thus, at Eweburn on the Maniototo Plain at 420 m. altitude, the average annual minimum for 15 consecutive years is — 12.1° C. and on one year — 16.6° was reached, other low minima for different years being — 12.8°, — 14.4° and — 15°. In North Island, — 12.2° has been recorded at Waiotapu on the Volcanic Plateau at an altitude of about 300 m. Doubtless, in the high mountains lower temperatures are reached but the evidence derived from the cultivation of New Zealand subalpine and alpine plants in Europe shows that a temperature of — 18° C. is more than the majority can tolerate. In Great Britain, New Zealand plants of all kinds can be grown well in parts of Cornwall and Devon, but in England and Scotland generally many are only half-hardy. Mr. C. T. CRAWFORD of St. Andrews, Fife, has sent me a list of New Zealand plants which he cultivates, 76 being perfectly hardy, the lowest shade temperature being nearly — 14° C. and on the grass nearly — 17° C. At the Royal Botanic Garden, Edinburgh, almost any New Zealand high-mountain plant can be successfully cultivated.

In New Zealand itself an exceptional frost, particularly if maintained for a number of days in succession, damages or kills outright many indigenous species. Thus, in 1923, at Queenstown "though there was almost constant frost for six weeks (L. COCKAYNE, *Festschrift zum siebzigsten Geburtstage von Karl von Goebel*, 1925: 77—78) and many supposedly hardy plants were killed, so far as I could ascertain, the thermometer did not fall below — 11° C. But, even if the cold were greater, it certainly cannot have nearly reached — 17° C. for *Eucalyptus Gunnii*, juvenile E. *globulus* and *Pinus radiata* were undamaged". On the other hand, the following species which ascend to above the forest-line were killed or damaged: — *Phormium Colensoi, Weinmannia racemosa, Myrtus pedunculata, Leptospermum scoparium, Nothopanax Colensoi, Gaultheria perplexa, Olearia arborescens, Shawia paniculata* (grows on rock, unprotected by a snow covering at 1200 m. alt. on the Seaward Kaikoura Mountains), *Senecio cassinioides* (never descends to the lowlands) and *Senecio elaeagnifolius*. Also nearly all the purely lowland species were either killed outright or more or less damaged including such as were in their natural habitats.

Just as Central Otago has the greatest winter cold, so is its average and maximum summer temperature greater than in any other part of New Zealand proper. Thus in 1922 the maximum shade temperatures for January and February were respectively 38.8° C. and 36.6°, while from Dec. 1921 to March 1922, inclusive, the temperature on 23 days was over 32.2°, on 47 from 26.6° to 31.7°, on 40 days from 21.1° to 26.1° and on 11 days from 15.5° to 20.5°. Even in May (last month of autumn) a shade temperature of 20.5° may be reached and in September (first month of spring) one of 23.8". The following rather high (for New Zealand) temperatures are recorded occasionally for various localities: — Rotorua 33.3°, Waiotapu 31.1°, Waihi (T.) 30°, Starborough (NE.) 35°; Hanmer 36.8°; Lake Coleridge (E., 366 m. alt.) 30°; Tapanui (SO., 150 m. alt.) 36.6°; and Invercargill 30.5°.

Sunshine. The following table shows the period during which the sun is above the true horizon on the days of midsummer and midwinter: —

Possible sunshine on the	Auckland Hrs. min.	Wellington Hrs. min.	Dunedin Hrs. min.
Longest day	14.40	15.10	15.46
Shortest day	9.38	9.13	8.39

The next table shows the average amount of sunshine at various places in proceeding from north to south. The comparative paucity in the South Otago district (Dunedin and Invercargill) is in part reflected by the high-mountain species at a low altitude and the facility with which that class of plants (indigenous and exotic) can be cultivated.

Auckland		Rotorua		Napier		Wellington		Nelson	
Hours	Min.	Hours	Min.	Hours	Min.	Hours	Min.	Hours	Min.
1.943	55	2.052	10	2.491	33	2.016	49	2.481	46

Blenheim		Hokitika		Lincoln		Dunedin		Invercargill	
Hours	Min.	Hours	Min.	Hours	Min.	Hours	Min.	Hours	Min.
2.154	51	1.924	54	2.087	54	1.663	9	1.600	35

Wind. Wind, especially in certain botanical districts, is a most important ecological factor. Generally speaking its effect becomes more intense the further south one proceeds or the higher one ascends. It is also of great moment on the coast especially on the west and on small islands.

The westerly winds of South Island are of special moment. Striking the western mountain wall, the wind loses its moisture in passing over the high lands and descends on the east as a hot wind sweeping through the river valleys and over the gravel plains, raising transpiration to its maximum. Though this hot **north-west wind** occurs to the east of the Southern Alps as a whole, in places its effect is greatly lessened through interception by ranges to the east of the Divide, so that its maximum strength is limited to the valleys and plains of the North-eastern and Eastern districts, where it even reaches the coast-line, and to the North Otago district. Perhaps its greatest intensity is experienced on the Canterbury Plain and the montane valleys extending westwards from its upper part to the Divide. Such a "Canterbury nor-wester", as it is called, sweeping through the river-gorges bursts with all its fury upon the plain — a hot, dry wind, its progress marked by clouds of sand and silt rising out of the wide beds of the glacial rivers. In the west clouds hang over the distant mountains indicating the rain-storm that is raging there, but over the plain is a clear blue sky while a burning sun strikes down. On plants the leaves hang flaccid, in orchards the trees are stripped of their fruit, everywhere the surface of the ground if unprotected by vegetation becomes dry as dust and the soil of ploughed fields may be blown away. On the dunes sand in clouds is carried back to the sea, sandhills are bodily removed and the rope-like entangled stems of *Desmoschoenus,* metres in length, laid bare. On mountain-passes and exposed ridges the fury of the storm reaches its height, it is impossible to stand upright, small stones are hurled through the air.

The **south-west wind** — of far wider range — is of equal ecologic importance and frequently brings with it squalls of a subantarctic character, leading to snow on the mountains, or even to a heavy downpour, but at other times rain is wanting while a furious gale extending to the southern parts of North Island rages for one or two days at a time. The change from north-west to south-west is quite sudden the temperature dropping many degrees and conditions approximating to those of midwinter may occur in the middle of summer. Obviously such sudden changes are of great physiological importance.

The average velocity of the wind in the following tables is from records of the Robinson anemometer in kilometers per day.

Jan.	Feb.	Mar.	Ap.	May	June	July	Aug.	Sept.	Oct.	Nov.	Dec.
					Auckland						
440	290	285	232	259	254	277	278	290	315	334	301
					Wellington						
490	437	456	460	414	387	371	387	452	559	530	515
					Hokitika						
230	216	211	214	198	187	174	192	229	272	243	228
					Lincoln						
315	292	380	256	213	192	186	214	269	309	315	301

	Auckland (11 years)	Wellington (16 years	Hokitika (16 years)	Lincoln (13 years)
Average per day	288	451	216	262
Maximum velocity for one day .	1558	1920	1108	1547

Part II.

The Vegetation of Primitive and Semi-primitive New Zealand.

Section I.
The Vegetation of the Sea-coast.

Chapter I.
General Observations on the Coastal Vegetation.

Brief account of the coast-line. The coast-line offers most diversified stations for plant-life. Not only does it face the actual ocean for more than 4800 km. but it extends far inland in many places either as shallow estuaries, tidal rivers, drowned valleys or fiords of profound depth. The actual coast may be quite low, more or less hilly, or high mountains may rise precipitously from the water's edge. There are vast stretches of dunes; long lines of cliffs; sandy, muddy, gravelly or shingly shores; low-lying flats exposed to inundation by brackish water; tidal waters where portions of the muddy floor lie bare at low-tide and rocks extending far out into the sea. Lying off certain parts of the coast at various distances are islands differing greatly in size. These offer less complex ecological conditions than does the mainland and, in certain cases, their plant-covering is yet quite virgin. Finally, there are ancient coast-lines where certain maritime species still exist.

Floristic statistics. The coastal species fall into the two categories of those confined, or virtually so, to the shore-line, or its immediate neighbourhood, and those which belong equally to inland formations, some of this latter class being sufficiently abundant to affect in parts the coastal landscape.

The true maritime vascular plants number 185 species, or well-marked varieties, together with 16 groups of hybrids, which belong to 55 families and 107 genera. One hundred and thirty nine of the species are confined to the actual coast, or thereabouts, while 46 occur inland to a limited extent but, with few exceptions, these latter are negligible so far as the general inland vegetation is concerned. Resolved into their phytogeographical elements 134 (72°/₀) are endemic, 24 (of which one or two extend to New Caledonia &c.) Australian, 8 subantarctic South American (2 Australian also), 2 Norfolk Island, and 17 cosmopolitan or sub-cosmopolitan. The larger families and genera and the number of species in each are as follows: (families) *Compositae* 23, *Gramineae* 22, *Scrophulariaceae* 13, *Cyperaceae* 10, *Umbelliferae* 9, *Chenopodiaceae* 8 and *Filices*, *Cruciferae* and *Rubiaceae* 6 each; (genera) *Hebe* 9, *Senecio* and *Poa* 7 each, *Coprosma* 6, *Carex*, *Lepidium*, *Pittosporum* and *Olearia* 5 each.

The following 9 families and 35 genera containing 41 species are purely coastal or almost so in New Zealand, the remaining 144 (nearly 78°/₀) being related to inland species: — (families) *Nyctaginaceae*, *Aizoaceae*, *Corynocarpaceae*, *Tiliaceae*, *Primulaceae*, *Sapotaceae*, *Myoporaceae*, *Cucurbitaceae* and *Goodeniaceae**; (genera) *Ruppia**, *Zannichellia**, *Zostera* (Potamoget.), *Spinifex*, *Atropis*, *Bromus** (Gramin), *Desmoschoenus* (Cyper.— end.), *Hydatella* (Centrolep.), *Leptocarpus** (Restionac), *Macropiper** (Piperac.), *Salicornia*, *Rhagodia**, *Suaeda*[1]), (Chenopod.), *Pisonia* (Nyctaginac.), *Mesembryanthemum*, *Tetragonia* (Aizoac.), *Spergularia* (Caryophyll.), *Clianthus** (Legum.), *Euphorbia* (Euphorbiac.), *Corynocarpus** (Corynocarp.), *Dodonaea** (Sapindac.), *Entelea* (Tiliac.), *Hibiscus* (Malvac.), *Stilbocarpa*, *Meryta* (Araliac.). *Samolus* (Primulac.), *Sideroxylon* (Sapotac.), *Eryngium*, *Apium** (Umbell.), *Ipomaea* (Convol.), *Avicennia* (Verbenac.), *Mimulus* (Scroph.), *Myoporum** (Myopor.), *Sicyos* (Cucurbit.), *Selliera** (Goodeniac.) and *Sonchus*[2]) (Compos.),

The 16 groups[3]) of hybrids (there are certainly more) belong to the following genera: *Asplenium*, *Paratrophis*, *Pittosporum*, *Acaena*, *Plagianthus*, *Hymenanthera*, *Pimelea*, *Pseudopanax*, *Apium*, *Hebe*, *Coprosma*, *Olearia* and *Cassinia*.

---

1) It is well to explain that the figures given here and elsewhere in this book, are based on personal judgement. No two observers would be likely to agree as to whether certain species should be considered coastal or not. Also species of wide distribution are not included in the Australian or South American estimates, notwithstanding their occurrence in those regions, and, again, "wide distribution", "cosmopolitan" &c. are quite loose terms.

2) An asterisk denotes that the species though nearly always coastal does occasionally extend inland. Should a species occur at 100 m. or more altitude on a small island, or a hill adjacent to the coast, it is here considered coastal.

3) Generally the groups are polymorphic, in which case they are called "swarms". Any name, or formula, applies to a whole group or swarm, the individuals, each of which is a hybrid, do not bear the name of the swarm but are each *one* of the swarm. For instance, × *Paratrophis miopaca (P. microphylla × opaca)* is the collective name of the polymorphie group of individual hybrids composing the swarm, each of which is a hybrid belonging to the group × *P. miopaca*. Generally, in taxonomic writings, an individual hybrid bears the name of the group which (the group) also is called a hybrid, but such which of course it cannot be. In many cases, the hybrid individuals are present in great numbers and this applies for the whole region.

With regard to the coastal-inland species it is hardly possible to supply detailed statistics. According a forest, tussock-grassland, or shrubland approaches high-water mark so will a considerable percentage of its species be present. All depends upon the climatic and edaphic conditions of the locality and, in many instances, except close to the water's edge, the special coastal ecological factors may be absent. But, on the other hand, many inland plants tolerate fairly intense maritime conditions, so that more than 100 species thrive on one part or other of the true coast-line. The following, for example are in places sufficiently abundant to be of prime physiognomic importance: *Freycinetia Banksii, Arundo conspicua, Mariscus ustulatus, Cladium Sinclairii, C. junceum, Phormium tenax, P. Colensoi, Urtica ferox, Muehlenbeckia australis, M. complexa* var. *microphylla, Leptospermum scoparium, L. ericoides, Metrosideros lucida, M. perforata, Griselinia lucida, Dracophyllum longifolium, Hebe salicifolia, Pachystegia insignis, Cassinia fulvida, C. Vauvilliersii* and *Gnaphalium trinerve,* but this list could be greatly extended.

General conditions regulating the coastal vegetation. The special conditions to which coastal plants are subject consist of a greater amount of salt in the soil than ordinary land-plants can tolerate, exposure to salt-laden winds which are frequently both violent and of long duration and, in some stations, strong insolation. The coastal climate is generally uniform; frost is absent or trifling in North Island while in South Island too it is of little moment, except on the coast of the Canterbury Plain where it may reach — 9^0 C. Excess of salt in the soil and salt-laden winds are by far the most important of the above factors, and on such depend the characteristic coastal formations and the "adaptations" or capabilities of the species. At the same time, a salt soil is frequently absent, as on dune-areas where the power to tolerate salt winds or indeed violent wind in general is a matter of prime moment. From the above it follows that ground subject to flooding with brackish water or to frequent wetting by sea-spray is the chief home of halophytes, and that other formations will be governed first of all by position with regard to the prevailing wind and its frequency. The composition of coastal shrub associations is distinctly in harmony with the wind-factor. At the base of Bluff Hill, Southland at the spot where the frequent south-west wind strikes with full force, the mixed shrub association of a calmer atmosphere is either absent or replaced by a pure scrub of the xerophytic *Olearia angustifolia.* So too at the water's edge of the inlets of Stewart Island *Senecio rotundifolius* replaces fern-forest, but is itself replaced on the more exposed headlands by *Leptospermum scoparium* (Fig. 2).

The rainfall strongly influences the general vegetation of the shore-line since the number of rainy days determines the presence or absence of forest. Where the maximum of wet days occurs as on the west coast of South Island, rain-forest comes almost to the water's edge. Dune and salt-meadow, special edaphic formations, are but little affected by rainfall and bear their characteristic plants equally in the wettest or driest districts. The semi-subantarctic climate of the South Otago, Fiord and Stewart districts favours plants with subalpine "adaptations" so that there are not only such amongst the true coastal species, but actual subalpine species, rare or absent inland except in the high mountains, may occur close to the sea.

The winter cold of the Canterbury Plain offers an impassable barrier to certain northern plants (e. g. *Mariscus ustulatus, Macropiper excelsum, Dodonaea viscosa, Corynocarpus laevigata*) which, in consequence, have their eastern southern limit on Banks Peninsula, while, on the west, the rainforest climate offers an obstacle of another description, but it permits (e. g. *Freycinetia Banksii, Ascarina lucida, Hedycarya arborea*) certain forest plants to extend further southwards.

The plants themselves play no small part in their own distribution so far as they supply shelter and make soil; trees, shrubs and tussock plants supplying the former and certain coastal ferns and herbs the latter in the form of raw humus or peat. Other factors of local importance receive mention when dealing with the communities.

Chapter II.
The leading Physiognomic Plants and their Life-forms.

Dune plants. *Desmoschoenus spiralis* (A. Rich.) Hook. f. *(Cyperac.)* pingao, is a stout, far-spreading sand-binding sedge. The rhizome, many metres long, is about 1.7 cm. diam., somewhat woody, much-branching and covered with old leaf-sheaths. At first, it creeps close to the surface, but is soon buried and forms eventually a tangle of rope-like stems in the sand. The leaves are in bunches tightly bound together at their bases by the sheaths so that a bulbous mass about 2.5 cm. diam. is formed, but they gradually open out, curving somewhat inwards. The sheath is about 10 cm. long by 5 cm. broad at the base, moderately thick in the centre, translucent and membranous at the margin, and sticky everywhere with a resinous exudation which helps to bind the sheaths together. The blade, 60 cm. long by 7 mm. broad, tapers gradually to a long trigonous point; it is thick, coriaceous, stiff but flexible, concave on the upper and convex on the under surface to that the leaves fit one into the other. The colour is a rather dark glossy green near the base and on the under surface, but the upper surface, especially above, is frequently orange-coloured or reddish, though the leaves, as a whole, viewed from a distance appear yellow. The branches are given off so closely that the separate leaf-bunches touch making semi-tussocks or continuous lines.

Spinifex hirsutus Labill. *(Gramin.)* is a powerful sand-binding grass with an extremely long, much-branched, smooth, hard woody flexible creeping-stem, which puts forth numerous, wiry roots which descend deeply into the sand. The leaf-blade is usually about 47 cm. long by 10 mm. broad and tapers to a fine point; it is thick, coriaceous and flexible and both surfaces are covered thickly with pale adpressed hairs; its sheath is 11.5 cm. long, pale, thick and fleshy. The flowers are dioecious; the male spikes numerous, about 8 cm. long and forming a terminal umbel with sometimes 2 to 3 spikes making a cluster below. The female inflorescence is a large globose head, sometimes 30 cm. diam., composed of 1—2-flowered spikelets each at the base of a long, sharp-pointed, radially spreading spine some 12.5 cm. long.

Carex pumila Thunb. *(Cyperac.)* is a small grass-like sedge having a long, slender rhizome some 3 mm. diam. which gives off, at intervals, bunches of 4 to 6 fully developed leaves. The leaf-blade is thick, coriaceous, flexible, 30 cm. long more or less, glaucous-green and tapers gradually to a fine point; its upper surface is deeply concave through the curving margins. In position, the leaf is erect towards the base, but above it curves so that the apex almost touches the ground. The roots are long and slender. The culms are short, stout and about 15 cm. high. The utricle is large, thick, turgid and about 7 mm. long.

Coprosma acerosa A. Cunn. *(Rubiac.)* is a wiry shrub of the divaricating-form, but more depressed than usual for that class, which makes flattened, orange or yellow, open cushions, or thick mats, of flexible interlacing twigs and has a prostrate, rope-like, main stem about 1.8 cm. diam. which is either buried in the sand or hidden by the shoots. The leaves are small, coriaceous, pale- or yellowish-green, linear, about 7 mm. long, and situated in opposite pairs upon much-reduced branchlets. The roots are extremely long, but short adventitious ones are frequent on the peripheral shoots. The drupe is globose, fleshy, 7 mm. long more or less, translucent and white stained with pale blue.

Cassinia leptophylla (Forst. f.) R. Br., tauhinu; *C. retorta* A. Cunn.; and *C. fulvida* Hook. f., *(Compos.)* are erect, bushy, ericoid shrubs, ranging from 1 m. to 2 m. high, and differing only in the colour of their tomentum and some slight distinctions in leaves and flowers. The main stems are few, naked, not much branched below, but above branch abundantly into slender leafy twigs which finally give off, at a narrow angle, flexible, straight branchlets covered with tomentum, either white *(C. retorta)*, greyish *(C. leptophylla)* or yellow *(C. fulvida)*. The ultimate shoots form close masses. The leaves are linear-obovate *(C. retorta)*, or linear to linear-spathulate, about 3 to 4 mm. long, moderately thick and coriaceous, abundantly tomentose beneath but shining-green above. *C. fulvida* has glutinous branches and its leaf- and stem-tomentum give an almost golden colour to the shrub.

The flower-heads are numerous and in terminal corymbs. The inner involucral bracts have white, radiating tips and so render the inflorescence
conspicuous.

All the species, including the non-coastal, when they grow in close
proximity gave rise to polymorphic hybrid swarms.

Salt-swamp plants. *Leptocarpus simplex* (Murr.) A. Rich. *(Restionac.)*,
oioi, jointed-rush, forms dense tussocks of erect, slender, stiff, wiry but
flexible rush-like stems which vary in harmony with the intensity of the
illumination from dull-green to bright red-orange. There is a rather stout
rhizome, which, at times, growing erect and branching may form a trunk
20 cm. high. The leaves are represented by short, blackish, sheathing scales
which clasping the terete stem at distances of 2.5 to 10 cm. give it a characteristic appearance. The roots are wiry and of medium length. The flowers
are dioecious; the male inflorescence is paniculate with numerous reddish-
brown spikelets, while the female is compacted into rounded glomerules
alternating along the stem.

Juncus maritimus Lam. var. *australiensis* Buchen. *(Juncac.)* is of the
ordinary rush-form and makes dense tussocks about 90 cm. high and 50 cm.
diam. at the base. The dark-green, glossy terete stems and leaves taper
gradually to a pungent but frequently dead brown point.

Plagianthus divaricatus J. R. Forst, *(Malvac.)* is a dark-coloured almost
deciduous shrub of the divaricating life-form. There is a stout main stem
some 5.7 cm. diam. which gives off several branches which passing upwards,
and outwards, finally branch abundantly into short, wiry twigs given off at
a wide angle and closely interlacing, the whole forming a dense, compact,
elastic, rounded mass, a flat mat or an open cushion. According to the
degree of exposure to wind, the periphery may be wiry and close, or twiggy
and open. The naked interior stems are 3 cm. thick or more, twisted, curved
and liane-like in appearance. The leaves on much-reduced branchlets, are
very small, linear to linear-obovate, 7 mm. long and slightly coriaceous.
The flowers are very small, pale yellow or whitish edged with purple,
honey-scented and produced from September to October.

If, when growing on the banks of a tidal river, *Plagianthus divaricatus* meets *P. betulinus*, a swarm of hybrids between thems may occur.
T. Kirk (1899: 70—71) based his species *P. cymosus* on one such hybrid
collected in a Dunedin garden. The swarm is at its fullest development
in the lower Pelorus Valley (SN.).

Avicennia officinalis L. *(Verbenac.)*, manawa, mangrove, is a shrub or
small tree varying in height from 60 cm. to 9 m., or even more. As a tree,
it has a stout, but usually short, main trunk from which a few short primary
branches pass off, spreading outwards and branching some 4 or 5 times so
as to form a round-headed fairly dense crown. The bark is rough, grey
and much furrowed. The ultimate and subultimate twigs are brittle, slender,

much curved and marked with old leaf-scars. The leaves are ovate to ovate-lanceolate, 5 to 10 cm. long, thick, coriaceous, dark-green, rather glossy and with yellow midribs and veins and hoary pubescent beneath and frequently placed more or less vertically. The roots extend for a great distance laterally and this is emphasized by the hundreds of erect branches, (pneumatophores), which project out of the muddy substratum (Fig. 3). These erect roots are from 20 to 30 cm. long on an average; they are straight and taper from the base upwards to a blunt apex.

The inflorescence consists of small heads of 5 to 8 flowers each about 8 mm. diam. The fruit, which is ready to fall by the beginning of January, consists of a capsule some 4.2 cm. long by 2.8 cm. broad and 1 cm. thick; the pericarp is brown and leathery but thin enough for the penetration of sufficient light for the formation of chlorophyll. The embryo, which has no resting-period, emerges early from the seed-coat, so that by the time the fruit is ripe, it completely fills the cavity. At this stage, the embryo consists of the two thick, fleshy cotyledons folded longitudinally, the outer, which is darkgreen on its upper surface, tightly enclosing the inner, and the hypocotyl, which is about 10 mm. long, slightly projecting, while the much shorter epicotyl. which has two pale rudimentary leaves, is pressed close beneath the folded cotyledons. Already the hypocotyl possesses root-rudiments in the shape of small knobs surrounding the swollen apex, while above there is a ring of brownish hairs which project upwards. After falling from the tree the embryo, continuing to increase in size, splits the pericarp along its suture, and first one half and then the other being shed, the green embryo. its cotyledons still tightly folded and the blunt apex of the hypocotyl projecting. lies on its side in the mud, and may be washed hither and thither by the tide. As time goes on, the cotyledons open out by slow degrees; the hypocotyl lengthens and when it is about 2 cm. long and the outer cotyledon is quite raised from the inner the roots are obliquely penetrating the substratum and firmly anchoring the plantlet. Growth now proceeds rapidly, especially that of the roots, but it is a considerable time before the cotyledons become fully flattened out, in fact the plant depends upon their reserve-material and chlorophyll for a considerable period. Thus a young plant with its hypocotyl 4.5 cm. long, its epicotyl 3 cm. long and the first two foliage-leaves 2.5 by 1.5 cm. has the laminae of the cotyledons still considerably folded. Pneumatophores appear quite early, so that a root 50 cm. long arising from a young plant 70 cm. high, may have two of them each about 14 cm. long.

Trees. *Corynocarpus laevigata* J. R. Forst. (*Corynocarpac.*), karaka, (Maori), kopi (Moriori), is an exceedingly handsome small tree 6 m. to 12 m. high with a regular trunk 30 cm. to 60 cm. diam. covered with rather thick bark rough with lenticels, and a dense, rounded, glossy-green head which during March and early April — according to latitude — is covered with showy, fleshy, orange-coloured drupes 2.5 cm. long. The leaves are elliptic-oblong or oblong-obovate, 7.5 cm. to 20 cm. long dark-green, smooth and shining.

Metrosideros tomentosa A. Rich. *(Myrtac.)*, pohutukawa, christmas-tree, occurs in various epharmonic forms, which range from a massive tree, 21 m. high with a trunk 60 cm. to 90 cm. diam., to a small, stiff-stemmed shrub 30 cm. to 60 cm. high, but usually taller. As a tree, it is frequently of an irregular form, especially when its trunk projects more or less horizontally from some coastal cliff. At other times the trunk may be short whith numerous erect trunk-like branches, issuing apparently from the ground, and in some cases growing into one another. The bark is brown, much-furrowed and wrinkled. Many adventitious roots are given off even from quite high up (Fig. 4.), and may form bunches, but those lower down often assume great dimensions and assist materially in anchoring the heavy trunk to a rock-face. The branches are massive and wide-spread and after branching several times finally give off numerous, stout, straight branchlets which bear the decussately arranged leaves and are white with a close covering of tomentum. Boughs, branchlets and leaves form a close head on the tree. The leaves are from lanceolate to broadly oblong, 2.5 to 10 cm. long, darkish-green, very thick, coriaceous and clothed beneath with white tomentum, but this is absent in seedling and juvenile plants. The flowers are arranged in broad, terminal, many-flowered cymes on stout tomentose peduncles and pedicels. The calyx is also tomentose, and functions in protecting the flower-bud. The flowers are dark crimson, so that a tree in full bloom is a magnificent spectacle.

Hybrids occur between *Metrosideros tomentosa* and *M. robusta* and 3 of such are united by Carse under the name × *Metrosideros sub-tomentosa*, but such cannot be types of the whole swarm.

Myoporum laetum Forst. f. *(Myoporac.)*, ngaio, is a small tree, averaging some 6 m. in height, with a trunk 30 cm. or more diam. covered with deeply furrowed bark 5 mm. or so thick. The crown consists of straggling, spreading branches which finally give off numerous, flexible, stout, leafy twigs, viscid at their tips. The individual branch-systems frequently do not touch so that a good deal of light can pass to the ground beneath. The leaves are lanceolate to obovate, 2.6 to 10 cm. long, acute or acuminate, glabrous, soft, flaccid, moderately thin, darkish-green but looking paler than they really are owing to the numerous oil-glands dotted over the surface. The flowers are in clusters of 2 to 6 in the leaf-axils; each is about 13 mm. diam., the petals are white dotted with purple. The drupes are purple, oblong, 8 mm. long and succulent.

The coastal Ferns. The following coastal ferns require notice: *Asplenium obtusatum* Forst. f., *Blechnum durum* (Moore) C. Chr. and *B. Banksii* (Hook. f.) Mett. All grow under identical conditions and are ecologically similar. They thrive best in an equable moist climate with low summer temperature and cloudy skies, and rapidly form raw humus out of their dead parts.

Asplenium obtusatum varies epharmonically in size but is generally a rather large fern. There is a thick rhizome sometimes 30 cm. long and 3.5 cm. diam. The leaves are erect, pinnate with 6 to 20 pairs of close-set sometimes overlapping pinnae, dark-green, linear-oblong, 6 to 30 cm. long, and the stalk is from 5 to 15 cm. greenish to almost black, very stout and channelled above. Large examples have leaves 67 cm. long by 19 cm. broad.

Blechnum durum has a stout rhizome and, in large examples, a distinct trunk 12 cm. or more high and 5 cm. diam. clothed with the old leaf-bases. The leaves which have a short stalk 2.5 to 5 cm. long are arranged in erect, semi-erect or almost flat but recurved rosettes at the summit of the trunk or rhizome. They are lanceolate, dark-green, shining and frequently about 43 cm. long by 4 cm. wide, though much smaller and larger dimensions are common. The pinnae are numerous and close-set, the upper frequently overlapping; at first lanceolate they gradually decrease in length until the lower-most are reduced to rounded auricles. The sporophylls are shorter and narrower than the foliage-leaves.

B. Banksii, an altogether smaller fern than the two preceding, is closely related to *B. durum,* but the leaves are flattened to the substratum. It appears to be confined to rocks or their immediate neighbourhood, whereas *A. obtusatum* and *B. durum* are common plants of forest near the sea and coastal moor.

Chapter III.
The Autecology of the Coastal Plants.

1. Life=Forms.

Trees. The coastal trees number 24; all are evergreen. Excepting *Corynocarpus laevigata, Metrosideros tomentosa* and *Sideroxylon novo-zelandicum,* none exceed 9 m. in height, 6 m. being the average. The tree-form is generally unsuited to coastal conditions, consequently, in exposed stations or on poor soil, no fewer than 21 of the trees do not develop a distinct trunk but, as shrubs only, blossom and ripen abundant fruit.

The following are the life-forms and the number of specis to each: — canopy-tree 11; bushy-tree 7; rhododendron (tree-composite) form 3; araliad form 2; bamboo-like 1.

The tree-trunks, as a rule, are slender and erect, but in the latter respect some are extremely plastic, e. g. the southern tree-composites, which in response to the frequent gales, develop more or less horizontal trunks whose spread may far exceed the vertical height of the tree. *Metrosideros tomentosa,* too, growing out of a cliff-face, extends horizontally, but, when its station is ordinary level ground, the trunk is erect (Fig. 19.) though frequently very short indeed, in which case numerous erect branches function as trunks and the form is that of a gigantic shrub. *Macropiper excelsum* has frequently a much-reduced trunk, but its stems are in a category by themselves, being straight, slender, blackish-purple when young, brownish-purple when mature, marked at distant intervals with leaf-scars and of a bamboo-like appearance. *Dysoxylum spectabile,* though hardly to be included as a true coastal tree, varies from 5 m. to 15 m. in height and from its trunk and thickest branches are put forth, in winter, pendulous panicles, 30 cm. long, bearing the waxy, white flowers.

The coastal trees owe their position chiefly to their inability to tolerate frost. Thus, in passing from north to south, they rapidly decrease in number, until, by the time Foveaux Strait is reached, there remains of the northern trees only *Myoporum laetum.*

The roots of the coastal trees, like those of New Zealand trees in general, often extend semi-horizontally rather than vertically downwards. Frequently they are of great length. The root-systems of *Avicennia officinalis* and *Metrosideros tomentosa* have been described in the last chapter.

Coming now to the leaves of the trees, they may be chararacterized as follows: — Evergreen 24, compound 2, simple 22, broad 24, very large[1]) 6, large 3, medium 15, thin 6, coriaceous 16 (specially thick 6), fleshy 2, glabrous 17 (but 3 hairy only when young), hairy 7 (tomentose beneath 5), glossy 6 (in the case of *Coprosma retusa* this reaches an extreme degree).

Shrubs. The coastal shrubs number 35, of which 20 are mesophytes, 15 xerophytes or semi-xerophytes, 23 erect (tall[2]) 13, of medium height 7, of low stature 3), 12 more or less prostrate (truly so 8) evergreen 32, semi-deciduous 1, flat-stemmed more or less leafless 2.

Their life-forms and the number of species to each are as follows: — (a.) erect shrubs 23, consisting: bushy-shrubs 11, divaricating 3, ericoid 5, flat-stemmed leafless 1, rhododendron-form 2, dracophyllum-form (more or less fastigiate and leaves needle-like) 1; (b.) prostrate shrubs 12, consisting of: stout and straggling 4 (all semi-erect), twiggy open cushion 1, divaricating 2 (both open cushions), mat-plants 4 (stems rooting 2), flat-stemmed leafless 1.

---

1) Throughout the autecological chapters the size of leaves is estimated in the following manner: very large = 20 cm. long; large = 10—20 cm. long; medium = 5—10 cm. long; small = 2.5—5 cm. long; very small = less than 2.5 cm. For all the sizes breadth proportional to length is understood, so that the length is decreased or increased according to increase or decrease of the "normal" breadth. What is aimed at is to get a relative idea of the area of the transpiring surface. Evidently these estimates possess no real exactitude, but, as they are estimated in a uniform manner and by one person, they may serve for purposes of comparison, and no more is claimed than this.

2) Here and elsewhere in this book the relative heights of shrubs are denoted as follows: tall, 3 m. and upwards; of medium stature, 1 m. to rather less than 3 m.; of low stature, less than 1 m.

The stems of most of the shrubs are slender, those of the shrub-composites and *Hymenanthera crassifolia* being the stoutest. This rock-xerophyte has gnarled, stout stems which, apparently rigid, are really quite flexible and hug the under-lying rock as tightly as possible forming stiff mats or low cushions some 30 cm. deep and 1 m. or so diam. The more or less prostrate shrubs have generally flexible stems, those of *Pimelea arenaria* (dune-plant) and *Coprosma Kirkii*[1]) (rock-plant) being especially so. The wiry, reddish stems of *Coprosma acerosa* have been already noted. *Muehlenbeckia Astoni* has stiff wiry, divaricating branches which by interlacing, build on stony shores irregular, rounded cushions 1.5 m. high. The flat, green stems of *Carmichaelia Williamsii* are 1.2 cm. broad and the shrub may attain a height of 5 m. and be virtually a tree. The prostrate *C. Fieldii*[2]), on the contrary, has leafless stems only 2 mm. diam.

Coming next to the leaves of the shrubs, they may be characterized as follows, and to each class is appended the number of species in which it occurs: — simple 33, compound 2 (juvenile *Carmichaelia*), broad 22, narrow 13 (ericoid 11), large 2, medium 14, small 5, very small 14, thin 5, coriaceous 30 (thick, more or less fleshy 5), glabrous 23, hairy 12 (tomentose 9), glossy 4.

Herbs and semiwoody plants. The total number of species of this class is 113 (annuals or biennials 5, perennials 108 — semiwoody 22, herbaceous 86) of which 66 are mesophytes, 47 xerophytes or subxerophytes, evergreen 102, summergreen 6 (excluding annuals &c.), spot-bound 66, wandering 57. As for height[3]), 5 species are very tall, 10 tall, 24 of medium height, 35 of low stature and 39 of very low stature, of which 24 hug the ground or are not more than 4 cm. high.

The life-forms, together with the number of species to each, are as follows: — (1) Annuals or biennials 5, consisting of: erect-branching herb 1 (medium height), erect grass-form 2 (low), prostrate herb 1 (fleshy), moss-like herb 1. (2) Semi-woody plants 22, consisting of: (a.) spot-bound 15, made up of: erect-branching 6, erect tufted 1 (succulent), tufted ferns 2, rush-tussocks 2, prostrate straggling 4 (low mat 1); (b.) wandering 7, made up of: erect-branching 1, prostrate 1, large mat 3, sandbinders 2 (grass 1, herb-like 1). (3) Herbaceous perennials 86, consisting of: (a.) spotbound 46, made up of: erect-branching 3, tufted-ferns 2, tufted grasses 8, tussocks 10 (all grass-form), *Iris*-form 2, rosette-form 14 (3 erect, 10 low, 1 straggling), straggling 7 (4 prostrate, 3 semi-prostrate); b. wandering 40, consisting of: erect 8, tufted 2, mat-form 15, turf-making 8, sand-binders 2 (1 minor

---

1) Probably not a species but merely one or two forms of a hybrid group.

2) Although so far only recorded from the coast-line this may probably extend inland, as does *Dracophyllum pubescens* of the same locality in the Northwestern district.

3) The dimensions used in this book for height of herbs and semiwoody plants are: very tall (over 90 cm.), tall (60—90 cm.), of medium height (30—60 cm.), of low stature (15—30 cm.), of very low stature (less than 15 cm. to 3 cm. and hugging the ground).

only), *Iris*-form 1, rosette-form 3 (2 mat-plants also, but included in such), cushion 1.

Coming now to the leaves (1 species leafless), they may be characterized as follows together with the number in each class: — simple 97, compound 15, very large 2, large 9, medium 23, small 33, very small 45, thin 44, coriaceous 39, fleshy or succulent 29, glabrous 89, hairy 23 (tomentose 2).

A few species demand brief mention. *Plantago Hamiltonii* possesses small, rather broad, thickish, coriaceous, shining green leaves which overlap forming evenly shaped flat rosettes which grow so close together as to form a hard turf. *Gunnera arenaria*, another rosette-plant, makes large circular flat mats. *G. Hamiltonii* is more striking still with its rosettes over 7 cm. diam. of pale, dull, brownish-green leaves with finely toothed margins, stout midrib and veins and long petiole and its far-creeping, fleshy underground stems 5 mm. diam. It grows very rapidly in cultivation, and seems so perfectly fitted for its moist, sandy habitat (first made known by C. M. SMITH) that it is truly remarkable in being one of the rarest species of the New Zealand region. *Eryngium vesiculosum* of salt-meadow has small rosettes of stiff, prickly, grey lanceolate leaves 10 cm. long and by means of stolons it rapidly forms extensive colonies. *Stilbocarpa Lyallii,* of coastal scrub in Stewart Island, travels by means of stout hollow stolons 60 to 75 cm. long which arch above the ground. At a certain stage in the growth of a stolon, a young plant is developed at its extremity which will possess 2 to 3 leaves before the young rhizome bent to the ground by the arching of the stolon will have rooted. Extensive colonies many square metres in area of this striking plant with its great bright-green, long-stalked, orbicular-reniform leaves, are formed in this manner, the stolons passing, in some instances, beneath rocks (Fig. 5), so that plants widely apart may be actually in connection.

There is little of special moment to say concerning the stems of the coastal herbs. In general they are slender. Three of the ferns at times build short trunks with a maximum of some 15 cm. in height, but this is in the moist climate of the south. The rhizomes of the sand-binding *Desmoschoenus* and *Spinifex* attain an amazing lenght. *Urtica australis* growing amongst shingle has a thick, woody prostrate stem. *Euphorbia glauca*, a dune-herb, has unbranched stems more than 1 m. high, stout, terete and marked on the lower two-thirds with old leaf-scars.

The roots, in many cases bear a distinct relation to the habitat. Those of rock, dune and shingle are frequently of great length, but those of salt-swamp or salt-meadow are usually of medium length. The roots of dune species are often copiously provided with hairs to which the sand clings forming an investing layer. *Leptocarpus simplex,* as a plant of certain dunes of North Auckland, forms a stout trunk out of its roots and rhizome after the manner of *Carex secta*, but not nearly so high. In the south this habit seems wanting.

Lianes. The coastal lianes, 6 in number (scramblers 4, winder 1, tendril-climber 1), are of little moment, only *Ipomaea palmata* and *Sicyos australis* are high-climbing plants, the remainder straggle for a metre or so amongst shrubs, grasses or sedges while *Fuchsia procumbens* is often prostrate. The stems of the last-named and of *Tetragonia trigyna* are woody, those of the remainder, though perennial, are herbaceous and that of *Sicyos* is fleshy, juicy and 6 mm. diam.

Excepting *Fuchsia*, which is more or less deciduous, the leaves are evergreen, broad and flat. All are thin except those of *Tetragonia* (fleshy) but those of *Angelica geniculata* and *A. rosaefolia* are waxy beneath. The leaves of *Sicyos* and the last-named are large; those of the remainder are rather small.

Water plants. There are 7 aquatic spermophytes (6 submerged, 1 subject to tidal change) 4 of which have much branched, thin filiform stems and filiform leaves, 2 have grass-like leaves and far-creeping rhizomes and 1 forms dense matted patches of very slender, creeping and rooting stems, but it is uncovered at low-tide.

2. Pollination.

Apart from the pioneer work of G. M. THOMSON (1881a and b) and a few observations by CHEESEMAN, PETRIE and others, there is little to be learnt from botanical literature regarding the pollination of New Zealand plants, nor have I paid attention to the subject; a virgin field thus awaits cultivation so far as present-day methods of investigation are concerned. Here and elsewhere all that can be supplied are a few statistics and general remarks. THOMSON showed that the prevailing belief, as voiced by WALLACE[1] as to insects being strikingly deficient in New Zealand was not correct and that, though butterflies are few, there are many species of moths represented by numerous individuals while, even at that time, the number of known species of beetles was more than 1300. Above all, THOMSON clearly showed what an important part is played by Diptera. Since 1881 thanks to the untiring pioneer researches of G. V. HUDSON and his followers, very many more species of insects capable of pollinating have been discovered, so that there is no longer any question as to there being ample for the duty. That some of the species of butterflies have abundance of individuals may be seen by a visit to the montane tussock-grassland on any sunny summer's day. As for pollination by flies, not only are these insects attracted by special unpleasant odours but the cloying scents of certain species *(Cordyline australis,* spp. of *Clematis* &c.) bring them in great numbers.

Coming now to the actual coastal species and leaving out of consideration the wind-pollinated *Gramineae* and *Cyperaceae*, together with the sub-

1) Geograph. distrib. of Animals 1: 457—464 and Darwinism, Ed. 2: 321. 1889.

merged aquatics and a few other monocotyledons, 78 (57%) of the remaining 135 species are monoclinous and 57 (43%) are diclinous (dioecious 17 species, monoecious &c. 40 species). Theoretically, about 60 per cent possess flowers conspicuous enough to attract insects and some species with small dull-coloured flowers bear them in such abundance as to be conspicuous or they are visited by insects for their honey and a fair number are sweet-scented. From the horticultural standpoint the above estimate of attractive flowers is far too high, there being at most only 30 species acceptable for gardens and but 3 of these — *Xeronema Callistemon, Metrosideros tomentosa* and *Hebe speciosa* — are of the highest class. As for the colours of the attractive flowers, they are as follows: — white including cream 45, yellow 17, lilac to dark purple 16, reddish to crimson 6, pale blue 1. The estimates given in the first edition of this book for honey flowers (42%) may stand for the present. The flowers of *Astelia Banksii, Plagianthus divaricatus* and *Hebe elliptica* are particularly fragrant. No statistics can be given regarding dichogamy, homogamy or methods of pollination but a few special details are presented. Bird-pollination possibly accours in *Metrosideros tomentosa*. The species of *Pimelea* are, as a rule, polygamo-dioecious, but the plants behave as if truly dioecious. *Fuchsia procumbens* has hetero-tristylic flowers as in the other New Zealand species; the stamens of the 3 forms do not vary in length. In *Selliera radicans* (Cheeseman 1877; 542—545) autogamy is impossible and there is perfect correlation between shedding of pollen, trapping of this in the indusium peculiar to the *Goodeniaceae* and eventual protrusion of the stigma and its pollination by a fly.

3. Dissemination.

Too much stress should not be laid upon statistics derived from the so-called "adaptations" or "contrivances" for travel possessed by fruits or seeds, since it stands out clearly from a fairly close study of New Zealand synecology that *it is the community as a whole which moves and not its individuals except in the community itself;* in fact with hardly an exception, long-distance journeys for species, except by extremely short stages, appear impossible. Nevertheless, if "contrivances" for wide dispersal are of any moment, coastal plants above all others should afford evidence of such, since so many of the formations are open and some habitats approximate to waste ground — the habitat favourable above all others for plant-colonization. There also comes in the role of the sea as a motive power.

Another matter, often overlooked is, that in a region like New Zealand subject to violent gales, even extremely large seeds can be carried in the open for long distances. When small stones are whirled high into the air, as is frequently the case in exposed localities, there is hardly any seed which cannot be wind-borne for quite a long way; in other words, no plant need be confined to close proximity to its growing-place for lack of loco-motive power, but the distance reached is to be measured rather by metres than by kilometres.

As for theoretical capability of long-distance travel, those species with fleshy fruits attractive for birds come first. *These number 27 only and but 3 extend throughout both North and South Islands.* With regard to the remainder, it is clear that their edible fruits have had little or nothing to do with their distribution, since 1 is confined to one extremely small area, 3 occur only on the Three Kings Islands, 6 exceed only one degree of latitude, or considerably less, and as for the remaining 14 their distri-bution according to degrees of latitude is: (2^0) 1, (3^0) 1, (4^0) 1, (5^0) 1, (6^0) 3, (7^0) 2, (9^0) 3 and (11^0) 2. Further, field-observations show that it is wrong to assume that, because a fruit fleshy and even conspicuous, it will be eaten by birds, for many such fruits are never touched — a fact brought home to me through collecting seeds in all the vertical belts for many years and at all seasons.

Those species with fruits or seeds having some special apparatus suitable for wind-carriage number only 23, and these are no more benefited thereby than the species with fleshy fruits, but this matter is further discussed when dealing with the autecology of the other belts and the outlying islands.

4. Seasonal changes.

Speaking of New Zealand generally the distinction between the seasons is far less marked than in the North temperate zone. Deciduous trees &c. are not only very few in number but they play, as a rule, little part in the vegetation, and summergreen herbs too are of trifling account. Thus the changes that strike the eye depend chiefly upon blossoming and fruit-ripening. But since so large a proportion of the species produce incon-spicuous flowers and fruits, the effect is generally slight and the winter-aspect of forest, heath, tussock-grassland and dune is not strikingly dif-ferent from that of midsummer. Here as in the other chapters devoted to autecology the matter of flowering receives special prominence, but the treatment is of the briefest. Further, it must be remembered, that the dif-ferences in time of blooming are quite considerable in passing from north to south, or in certain parts from east to west so that much latitude must be given to general statements.

From about the end of May to the beginning of August the maximum period of rest is reached by the coastal plants, and in many of the species no growth of any moment is taking place. But even in the middle of June certain plants of a coastal forest may be in bloom in North Island e. g. *Wintera axillaris, Dysoxylum spectabile, Coprosma grandifolia* and *Leptospermum scoparium.* The vegetation of salt-swamps and salt-meadows, where frosts are frequent, will be more or less browned, while the aerial parts of *Atropis stricta*, *Scirpus americanus* and *S. robustus* will be dead and those of *Salicornia australis* partially so. The rounded bushes of *Plagianthus divaricatus* will have assumed a blacker hue, for then they are either bare of leaves or a few linger in sheltered parts of the shrub. By the end of July, even so far south as Banks Peninsula, *Mesembryanthemum australe* and *Linum monogynum* will be coming into bloom on sheltered cliffs. A little later in the north *Hymenanthera crassifolia*, and *Plagianthus divaricatus* commence to open their flowers, so that the latter fills the air of the saltmeadow with its honey-like fragrance. In the north, too, during September *Pittosporum crassifolium*, *P. umbellatum*, *Pimelea virgata*, *Coprosma retusa* and *C. Kirkii* blossom freely. By the beginning of October flowering commences in earnest and the following coastal species bloom that month: *Desmoschoenus spiralis*, *Carex pumila*, *Ranunculus acaulis*, *Pittosporum Huttonianum*, *Entelea arborescens*, *Angelica rosaefolia*, *Calystegia Soldanella*, *Myoporum laetum* and *Senecio lautus*. November sees many additions to the flowers, and in the south species which hitherto have bloomed only in the north now flower abundantly, but it is not until December that the salt-meadow of the Eastern district is lit up by sheets of canary-yellow *Cotula coronopifolia* and white *Samolus repens* var. *procumbens*. *Mimulus repens* growing in shallow water produces in profusion its showy lilac and yellow blossoms, but they are partly hidden by the foliage. *Atropis stricta*, *Deyeuxia littoralis* and *Scirpus maritimus* will also be in bloom. On coastal rocks of Queen Charlotte Sound *Arthropodium cirrhatum* will be a striking feature; in the north it will have been in flower since November. In the South Otago, Fiord and Stewart districts, banks not far from the sea will be draped by *Gnaphalium trinerve* with its conspicuous bracteate flower-heads. The floral feature of the coast however for the end of this month are the masses of dark-crimson produced by *Metrosideros tomentosa* in the North and South Auckland and the north of the Egmont-Wanganui districts. Almost as striking is the flowering of *Olearia angustifolia* and its hybrids with *O. Colensoi* (= in part *O. Traillii*), while in the Fiord district the closely-related *O. operina* is in bloom. Early in January in the Auckland districts the embryos of *Avicennia* are falling from the trees, and a little later are anchored in the mud. During January and February the various coastal species of *Hebe* will be in bloom and the southern coastal moors be gay with the snowy blossoms of *Gentiana saxosa*. Many of the plants already mentioned continue to flower, and from this time onwards fruits ripen and seeds are shed. Several coastal species are late flowerers and it is not until April that *Olearia angulata*, *O. albida* and *Shawia paniculata* are in full bloom, but generally speaking, except for such species as flower nearly all the year round (*Macropiper excelsum*, *Pisonia Brunoniana*) few species blossom after March, though blooms may be produced at unusual times.

5. Epharmonic modifications.

General.　The striking ecological differences offered by the coastal climate and habitats lead to remarkable epharmonic changes. High winds, strong insolation, salt for above the average in the soil, (it may reach more than 2 per cent according to ASTON) and habitats, ranging from extreme mesophytic to extreme xerophytic, demand such modifications in structure and form, if the plant is to live, that its plasticity may be strained to the utmost. Generally there is some life-form of a species which may be called its "usual" form, though this may not represent its most luxuriant state of growth, but in some instances it is not feasible to declare any particular form "usual", more than one form having a claim to that title. In certain cases, the extremes are so distinct, that were it not for the occurrence of intermediates on intermediate habitats, it would be impossible without experiment to decide as to their status. Frequently, too, the matter is complicated by the presence of persistent juvenile forms, hybrids and jordanons.

Properly speaking, any form of a plant, if beneficial, is epharmonic, and the term fits equally the "usual form" and its modifications, indeed, in some localities one or other of the latter may be the sole representative of an epharmonic series. Here the subject under consideration is briefly dealt with under the following heads.

The tree-form and shrub-form.　Most coastal trees under certain circumstances assume a shrub-form, which may be quite as abundant, or even more so than the tree-form. The main factors concerned are frequent winds from the sea and xerophytic habitats.

Metrosideros tomentosa, as a forest tree, may be 21 m. high, but on the bare scoria of Rangitoto Island (S A.) and similar situations it is dwarfed to a small, stiff-stemmed shrub 30 to 60 cm. high which flowers and produces abundant fruit. *Pittosporum crassifolium* in well-sheltered situations is a bushy-tree some 8 m. high, but its shrub-form is far more common; even in cultivation as a hedge it will flower and fruit when merely 1 m. high or even less. *Pseudopanax Lessonii*, too, is quite as often a shrub as a tree. In a certain coastal wind-scrub (H. H. ALLAN, 1926:73) the dimorphic small tree *Pennantia corymbosa* remains at its juvenile stage as a dense, small-leaved, divaricating shrub.

The prostrate form.　More striking than the common change from tree to shrub is the transformation of these to purely prostrate plants. Thus *Myoporum laetum*, a small, round-headed tree possessing a distinct trunk, is altogether prostrate on certain little islands to the east of northern Auckland, with its branches far-spreading, cord like and twiggy, and did not epharmonic shrubby forms occur elsewhere — lacking experiment — it would probably be considered a true-breeding variety. Various forms of *Coprosma retusa* can be seen in close proximity where a hillside slopes to a stony beach. On rock faces it will hug the rock with far-extending, slender stems; with more shelter on flatter ground it will be a shrub, while on the hillside adjacent there will be the tree-form with a comparatively thick trunk, but its more or less weeping twigs recall those of the prostrate form. This is clear enough from the behaviour of the juvenile when cultivated in good garden soil in a sheltered situation, for it remains prostrate for a considerable time, but finally puts forth erect shoots. The strongly wind-tolerating small tree *Dodonaea viscosa,* when growing on certain shingly beaches is altogether prostrate and even the still more wind-enduring divaricating shrub, *Plagianthus divaricatus,* makes a prostrate, close mat on stony shores exposed to strong insolation. The reverse of this may be artifically induced by cultivation in rich, non-saline soil, the divaricating-form being suppressed, a purely twiggy bushy-shrub taking its place.

Succulence. As for denizens of salt soil generally, a considerable percentage of those on the New Zealand coast are succulent, while, as the statistics for leaves have shown, only those of 40 species are thin out of a total of 108 coastal species. Careful measurements of leaves of 14 species taken by G. SIMPSON and T. S. THOMSON on the shore-line of Otago Harbour — a position far less exposed than when facing the ocean — show in every case increased thickness. The following gives the average increase in thickness for the coastal-form of some of the species: (non-coastal species *Blechnum procerum* 0.01 mm., *Melicope simplex* 0.005 mm., *Hebe salicifolia* var. *communis* 0.0075 mm., *Olearia avicenniaefolia* 0.0075 mm., *Suttonia australis* 0.015 mm., *Muehlenbeckia australis* 0.0125, *Melicytus ramiflorus* 0.0075 mm., *Pittosporum tenuifolium* 0.0125 mm., *Polypodium diversifolium* 0.01 mm., *Cyclophorus serpens* 0.07 mm., *Asplenium flaccidum* 0.01 mm., (coastal species) *Myoporum laetum* 0.0275 mm., *Hebe elliptica* 0.105 mm., Facing the ocean the leaves of *Coprosma retusa* become twice as thick as when growing inland.

Leaves. The coastal climate and xerophytic habitats favour reduction in size of leaves or the presence of small-leaved species (70$^0/_0$ of the species). Rolled &c. leaves are common in the *Gramineae.* More interesting is the behaviour of the leaves of *Coprosma retusa,* which in the prostrate and shrubby forms and on trees where exposed to sun and wind, are comparatively small and rolled, but in the shade are much larger and quite flat.

Contrary to what has just been said, the leaves of certain trees or shrubs on small islands, lying to the east of the Auckland districts, (Rangitoto, Poor Knights &c.) possess leaves much larger than they do on the mainland. T. KIRK (1879: 450 — 452) was the first to point out this unexpected state of affairs and further observations on other islands have both extended the knowledge of the phenomenon and included species in which it is still more marked. The species most concerned are probably *Macro-*

piper excelsum, Melicytus ramiflorus, Suttonia divaricata, Geniostoma ligus-trifolium and *Myoporum laetum,* but KIRK goes so far as to say that the flora of Rangitoto as a whole exhibits "extreme luxuriance of foliage, although its larger members are greatly reduced in stature", and for the Poor Knights I have stated "the arborescent plants exhibit a most remarkable luxuriance of foliage, greater considerably than that of the same species on the main-land" (1906:357). KIRK's explanation of the matter was rather forced and physiologically unsound. As a partial explanation, it may be, that certain of these large-leaved forms are jordanons; indeed, the large-leaved *Macro-piper excelsum* is var. *major* Cheesem., and this is the sole form of Norfolk Island, the Kermadecs and the Three Kings. So, too, the large-leaved *Suttonia* may not be *S. divaricata* but a "new" species. At any rate, matters such as this can only be settled by experiment, and opinions based on herbarium material are not only of no value but scientifically dangerous.

Colour changes. The strong insolation of the coast changes greens into red, orange, brown &c. Thus *Tillaea moschata* has green shoots in the shade, but in bright light, they are red; the redness of the stems of *Lepto-carpus simplex* increases or decreases according to the strength of the illumination; the green leaves of *Cotula pulchella* are brown when not shaded. Indeed so potent is exposure to bright light in altering the colour of leaves and stems, that it is a moot point how far the reddish, yellowish or brownish hue of certain dune species — and their colour is of physiognomic mo-ment — may be considered unfixed epharmonic or truly stable under dif-ferent environments. In the so-called "brown" tussock grasses *(Poa caes-pitosa, Festuca novae-zelandiae),* which for hundreds of miles colour the landscape in South Island, cultivation in my garden has shown that they may remain green for the greater part of the year.

6. Sand-binders.

Certain species, which can live quite well both on semi-stable dunes and "solid" ground, when subject to a gentle sand-burial, elongate their creeping stems after the manner of true sand-binders (e. g. *Arundo conspicua, Poa caespitosa).* Possibly *Acaena novae-zelandiae* var. *pallida* comes into this category for on dunes it extends its area for several metres, thanks to its far-creeping stem tolerating burial by sand.

7. Turf-making.

Low growing rosette-plants which hug the ground under xerophytic conditions, especially those furnished by what is designated further on "coastal moor", may form a dense turf. Thus the far-creeping *Rumex neglectus* of stony shore, where its fair-sized leaves rise considerably above the sub-stratum, on coastal moor makes a true turf. Further cases are cited when dealing with the above habitat.

Cushion-making. Exposure to wind, combined it may be with a dry habitat and strong isolation, leads to the formation of open or dense cushions from certain shrubs, especially those of divaricating-form, e. g. *Plagianthus divaricatus,* the liane *Metrosideros perforata, Leptospermum ericoides, Coprosma propinqua* and species of *Cassinia.* Exceedingly common, especially on exposed slopes are dense cushion-like masses of *Muehlenbeckia complexa* var. *microphylla.* Wind-borne sand frequently fills the mats of an unnamed coastal variety of *Raoulia australis* and, growing through the sand, they rise above the substratum as cushions.

Scrambling lianes. Various coastal species, when growing in the dim light caused by other plants overtopping them, lengthen their stems and become more or less lianoid. The following are examples: *Rhagodia nutans, Salicornia australis* (up to 1 m. or more), *Acaena Sanguisorbae* var. *pallida, Apium filiforme.*

The case of Claytonia. *Claytonia australasica* is a plant of great plasticity occurring under various ecological conditions from the coast to the alpine belt. According to H. H. ALLAN, when growing on loess cliffs in the Eastern district exposed to abundant sea-spray nodules full of starch are developed on its roots.

Forms not specially coastal. Notwithstanding the moulding action of the coastal conditions there are many true coastal species whose form and structure afford no evidence of proximity to the sea, but these are rather plants of sheltered localities where coastal conditions are absent or much modified, and their presence is to be attributed rather to the mildness of the maritime climate than to any coastal adaptations.

Chapter IV.
The Plant Communities.

1. Communities of Salt or Brackish Water.

a. The seaweed communities (by W. R. B. OLIVER).

The rather sharp difference occasioned in the conditions of the littoral belt by the supply of water above and below low-tide mark is reflected in the nature of the algal coverings. Below low-tide level is usually a close growth of fairly tall brown algae with smaller species on and among them. Above this level only small species occur in closed or open formations. In each station is more than one association so that the two facies may conviently be termed "formations".

Large-brown-algae formation. This formation extends from below low-tide mark up to the ordinary limit of low neap-tides. In exposed situations, however, it extends somewhat higher indicating that the daily supply of water is the controlling factor in its distribution. Generally there is a growth of large species of brown algae forming a continuous vegetative covering. Beneath this tier is a lower stratum of smaller algae including many brown and red kinds while the rock-surface itself bears a covering of crustaceous and branched forms of coralline algae. According to the dominant species three principal associations, as below, are to be destinguished.

The Durvillea association is characterized by the presence of *D. antarctica* which is confined in its vertical distribution to the surf zone as a narrow belt generally about 1 m. wide (Fig. 6) with its upper level at low neap-tide mark. The upper margin of this belt is indeed marked by the large, placental holdfasts of the kelp. *Durvillea* itself extends from the Subantarctic Islands northwards as far as Cook Strait on the east to beyond Manukau Heads on the west. Below the *Durvillea* is a growth of smaller brown algae, including *Marginaria Boryana, Lessonia variegata, Carpophyllum maschalocarpum, Cystophora retroflexa,* and *C. torulosa* — all species ranging from 50 cm. to 1 m. high and provided with air vesicles. The undergrowth is composed af algae 20 cm. or so high, common members being *Pterocladia lucida, Zonaria Turneriana, Glossophora Harveyi* and *Stypocaulon paniculatum.* Beneath these on the rock-surface are various species of coralline algae, including branched forms, e. g. *Amphiroa* and *Corallina,* and encrusting species, e. g. *Melobesia* and *Lithothamnion.* Epiphytes are not uncommon, the most conspicuous being *Porphyria subtumens* (on *Durvillaea*) and species of *Corallina* (on *Cystophora retroflexa*).

The Carpophyllum association occurs from North Cape to Stewart Island, where *Durvillea* is absent, from neap-tide mark down. *Carpophyllum maschalocarpum* is dominant and associated with it are *Lessonia variegata, Marginaria Boryana, Cystophora torulosa, C. retroflexa* and, especially in the northern part of New Zealand, *Carpophyllum plumosum, Ecklonia Richardsoniana* and *Sargassum Sinclairii; Xiphophora chondrophylla* forms the upper margin. The undergrowth is similar to that of the *Durvillaea* association, the more common species being *Melanthalia abscissa, Zonaria Turneriana, Caulerpa sedoides, Stypocaulon paniculatum* and *Glossophora Harveyi.* The rock surface supports the usual coralline algae.

Xiphophora chondrophylla usually forms a narrow belt along the upper margin of the *Durvillea* and *Carpophyllum* associations, but if there be a nearly level rock surface at this height, this species may be regarded as the dominant member of a distinct association. On Otago Peninsula where it is well developed it has associated with it *Lessonia variegata* and *Pachymenia lusoria.* Beneath there is an abundant flora (including *Codium adhaerens*) and fauna.

Small emerging-algae formation. Limited to the belt lying above low neap-tide mark the associations here grouped together are characterized by the small size of the species, the usual open nature of the community and the presence of various devices[1]) for conserving moisture whilst exposed to the atmosphere. Various species are dominant in different tide levels and localities but the formation may be conveniently grouped into the following 4 subformations.

The coralline algae subformation is distinguished by the dominance of species of *Corallina*. With *C. officinalis* may be associated in small quantity *Colpomenia sinuosa* and other small algae or *Hormosira Banksii,* so that the two species are fairly evenly distributed[2]).

The small brown or red algae subformation has, as its principal association *Hormosira*[3]) dominant or sometimes pure. *Scytothamnus australis* and species of *Apophloea* also form associations on rocks between tide-marks.

The ulvoid-algae subformation includes two common species — *Ulva rigida* and *Porphyra columbina* — each forming associations and both agreeing in their possession of delicate, flat thalli thinly coated with mucilaginous matter. *Porphyra* inhabits exposed situations but *Ulva* those more sheltered and subject to the influence of fresh water.

The small olive-red-algae subformation embraces two associations which occur high up in the intertidal belt and are made up of mossy algae belonging to the *Rhodophyceae,* though olive-brown in colour. They form low, close growths 2 or 3 cm. high and usually affect shady places. The one *Bostrychia arbuscula* is confined to the south of New Zealand but the other *Carelacanthus spinellus* is more widely distributed.

b. Zostera formation.

There are 2 species of *Zostera* in New Zealand, *Z. nana* and *Z. tasmanica*[4]). *Z. nana* makes a pure association "from about half-tide mark to below low-water level" (W. R. B. OLIVER, 1923:542) on the muddy or even more or less sandy floors of estuaries in New Zealand proper. The brittle rhizomes form an entanglement in the substratum and the short

---

1) In the brown algae a dense outer layer; internal reservoirs in the internodes of *Hormosira Banksii;* the continuous central space filled with water jelly in *Splachnidium rugosum;* the slimy covering of *S. rugosum* and *Codium adhaerens;* the thin mucilaginous coat of *Ulva* and *Porphyra.*

2) In such a case *Corallina* affects wetter situations than *Hormosira,* so that frequently the former is found covering the bottoms of shallow pools while round their margins is *Hormosira.*

3) Fresh or turbid water filling the distended internodes of this plant is apparently not harmful for it luxuriates on reefs in muddy harbours.

4) There appears to be some doubt as to the species. HOOKER referred a specimen from Auckland harbour to *Z. marina* L., remarking it was perhaps *Z. angustifolia* Roth. GRAEBNER (Pflanzenreich. *Potamogetonaceae:* 31), cites only *Z. capricorni* Aschers., as occurring in N. Z., and doubts the occurrence of *Z. tasmanica* (p. 32). CHEESEMAN in both editions of the Manual gives the names as above, but is not certain as to the exact position of the deep-water plant.

grassy leaves when uncovered by water lie flat on the mud. Loose stones are not uncommon on the substratum and are occupied by various small algae. *Zostera tasmanica* is supposed to be never uncovered, but its ecology requires further study.

c. Salt Swamp.

1. Mangrove *(Avicennia)* tidal forest or thicket.

This formation possesses but one species, *Avicennia officinalis*, the life-form of which has been described in chapter I of this Part. It forms girdles or patches of thicket or low forest between tide-marks in shallow estuaries, tidal rivers, sheltered bays and the like. It is restricted to the Auckland district and is to be seen in its greatest luxuriance in the extensive estuaries and tidal rivers of the North Auckland district and the Kaipara subdistrict (Fig. 3.).

The presence of the formation depends chiefly upon the following: — A muddy substratum which, though generally deep, may be quite shallow and overlie rock; absence of frost; warmish water during summer; tide-erosion of insufficient power to uproot the young seedlings.

The substratum generally consists of an extremely soft and sticky mud light brown on the surface but at a little depth black, shiny and of evil odour. In places, very shallow pools of water, left by the retreating tide, lie on the surface of the water-saturated ground, which is everywhere honey-combed by the holes of crabs *(Helice crassa)*, the orifices varying from 1.5 cm. to considerably less, and out of which water flows as one steps on the mud.

When growing in the greatest luxuriance, the trees, 9 m. or more high, form a close belt dull brownish-green in colour, bounding the shore of the estuary or even occupying the centre of the river-bed, the outer limit being determined by the average low-water mark. At high tide, only the crowns of the trees are visible but, at low-water, the muddy floor is exposed and the spreading branches and bare trunks are visible. Everywhere the more or less limpet-covered, asparagus-like pneumatophores[1]) rise up in thousands, projecting from the mud for a height of some 20 cm. on an average. Some are solitary, but others are more or less bunched together. Growing scattered amongst them are usually many young plants of *Avicennia* of all sizes, especially near the shoreward margin of the swamp.

2. The *Salicornia* formation.

This formation is distinguished by the presence of more or less *Salicornia australis* which, when the community is fully developed, forms a

---

1) These not only play a biological part, but strongly oppose tidal erosion and in many places, though very slowly, promote the deposition of mud and other river-borne matter, until eventually the swamp may be replaced by salt-swamp or salt-meadow.

dense carpet (Fig. 7.). It occurs on tidal mud-flats, from a little above the high-water mark of neap-tides to the highest tide-mark of ordinary spring tides, its breadth depending upon the steepness of the ground and the tidal range. The community occurs throughout North Island and South Island, but it is particularly striking in the Sounds-Nelson district.

During spring tides the *Salicornia* is covered daily by the tidal water, the maximum time of submersion of the lowest part of the community being about 225 minutes and the minimum 50 minutes. Each month for a period of 14 days the lower part remains uncovered by water, but the higher part is uncovered for at least 24 days. Obviously, then, the *Salicornia* plants are subject to quite different conditions in relation to the amount of salt in different parts of the community and this, again, must be greatly affected by periods of heavy rain and the varying quantity of fresh water derived from streams flowing into the estuary &c. As for the substratum, sticky mud is hostile to *Salicornia*, and in such it may not grow at all. It especially favours sandy mud or well-drained muddy shingle, i. e. it cannot tolerate stagnation.

The community may be a mere fringe or at its maximum a broad carpet 100 m. or considerably more in breadth. Where fully developed, the plants grow into one another and the carpet is 15 cm., or more, in depth. Near the upper limit there is frequently more or less *Suaeda maritima*.

Where the tidal scour is not too strong the *Salicornia* traps the mud, which in time will accumulate sufficiently to raise the surface so far above the tidal waters as to allow invasion by other halophytes and eventually the etablishment of salt-swamp

For many details in the foregoing account I am greatly indebted to Mr. P. G. MOFFAT, Harbour Master, Port Motueka (SN.). who has compiled a *Tide Range Table* for Port Motueka.

3. *Juncus-Leptocarpus* formation.

This well-marked formation is characterized by the strong dominance of the rush-form and the presence in large quantity of *Juncus maritimus* var. *australiensis* and *Leptocarpus simplex,* or the latter may alone be present.

The floristic composition of the formation consists of only 8 species (families 3, genera 6), of which there occur throughout: *Scirpus americanus, S. robustus, Carex litorosa, Leptocarpus simplex* and *Plagianthus divaricatus*.

The formation occurs on the floors of tidal rivers and estuaries where exposed to the highest spring tides and is present in all the Botanical Districts. Many of the swamps are extensive, e. g. near Havelock (SN.), Collingwood (NW.) and Invercargill (now more or less "reclaimed"), to cite only a few.

The substratum of *Juncus-Leptocarpus* swamp differs from that of

Mangrove in that it is much firmer and less muddy. The soil may be clay, loam, sand or mixtures of these; shells of certain molluscs are abundant and the surface is riddled with the holes of crabs. The community is only covered by water at the highest spring tides and then merely for a brief period. The water with which the ground is saturated is brackish and there is a maximum of salt that can be endured but though no minimum is demanded by the species, yet, if the water be too fresh, the halophytes coming into competition with ordinary swamp-plants (e. g. *Typha, Phormium)* cannot hold their own.

The general physiognomy of the association depends upon the dominant rush-form.   Seen from a distance, it presents a dark even surface, but usually the *Leptocarpus* is slightly taller than the *Juncus,* while the *Scirpus* forms greener patches.   Perhaps 70 to 80 cm. might be considered the usual depth of the vegetation.   Growing on the driest ground, and not exposed to salt-water, except at exceptionally high tides, is a girdle of the dark-coloured rounded bushes of the divaricating-shrub, *Plagianthus divaricatus,* 90 cm. high or less.   Landwards of this comes salt-meadow.   *Leptocarpus* and *Juncus* form pure girdles or clumps, but which is the more salt-tolerating, I do not know, since in different localities, either may form the seaward girdle.   Here and there in the main mass of the association are tussocks of *Carex litorosa.   Cladium junceum* is common in the North Auckland district.   Where deep pools occur, there is abundance of *Scirpus robustus* and with it *S. americanus.*   Near the junction of salt-swamp and salt-meadow, various species of the latter come in, especially *Salicornia australis, Apium filiforme, Selliera radicans* and *Cotula dioica.*

Where the community occupies the shallow portion of a tidal river one bank has probably been eroded and sediment deposited near the other on which, if the tidal scour be not too great, various halophytes become established from seeds brought by birds or the water itself, or vegetative portions deposited by the tide. Later on, some of the soil-particles, continually brought by the water, are held between the plants themselves and their shoots, together with the products of decay, so that the surface is gradually raised and the swamp slowly transformed into salt-meadow. Earlier on, the Mangrove, *Zostera* and *Salicornia* formations, catching and holding the mud, may prepare the requisite conditions for *Juncus-Leptocarpus* swamp. The first plant to appear is *Scirpus americanus,* which, even after the taller members are established, holds its own on the wetter ground, thanks partly to its power of increase by means of the far-spreading rhizomes. Eventually, the taller close-growing *Juncus* or *Leptocarpus* excludes the light and the *Scirpus* is doomed.

The ground on which the formation occurs, if drained, is suitable for occupation by pasture, as in the case of a portion of the great salt-swamp near Invercargill.

4. Various minor communities.

Mimulus repens association. Where sluggish streams flow through salt-meadow, but not in every locality, this association occurs. Usually, there is a fairly deep, flowing portion of the stream and a shallow sluggish part, this latter caused by a sligth overflow. In the shallower part, *M. repens*[1]) may be dense enough to hide the water. Through it may grow some *Scirpus americanus, Triglochin striata* var. *filifolia* and perhaps a little *Cotula coronopifolia.* In the deeper part there is *Scirpus robustus* with an undergrowth of more or less *Mimulus.* Greater stagnation of water soon brings in *Cotula coronopifolia* with floating stems, and the other plants are absent.

Scirpus lacustris association. As the water of the shallow tidal river becomes far less salt, a girdle of the tall rush-like *Scirpus lacustris*[2]), 1.5 m. or so, in height, fringes the bank, but in the South Otago, Stewart and perhaps Fiord districts the association is absent. Although not depending in the least upon the *Scirpus, Plagianthus divaricatus* frequently forms a girdle — often indigenous-induced — on the landward side of the former, its dark hue contrasting with the green of the *Scirpus.*

Brackish-water submerged communities. *Ruppia maritima* forms submerged masses of filiform stems and leaves on the floor of lagoons &c. and slowly-flowing streams in somewhat brackish water.

Althenia bilocularis, a smaller plant of similar life-form, also occurs in some localities on the east of the South Island under identical conditions.

Zannichellia palustris, another ecologically-equivalent species, occupies similar stations to the above.

2. The Salt-meadow Group of Communities.

a. Salt-meadow proper.

This may be defined as a close turf made up of low-growing herbs, out of which rise certain plants of the rush-form and the divaricating-shrub *Plagianthus divaricatus.*

The group of associations comprising salt-meadow contains 19 species (families 11, genera 14) of which the following would be in all primitive salt-meadows: — *Triglochin striata* var. *filifolia, Leptocarpus simplex, Juncus maritimus* var. *australiensis, Chenopodium glaucum* var. *ambiguum, Salicornia australis, Spergularia media, Plagianthus divaricatus, Apium pro-*

---

1) A prostrate perennial herb with stout, succulent, creeping, rooting stems; prostrate or suberect branches and small, obtuse, entire succulent leaves, 4 mm. long.

2) The species also occurs on margins of lakes &c., at some considerable distance from the sea, ascending to 450 m., according to CHEESEMAN (1906: 778).

stratum, A. filiforme and hybrids between the two, *Samolus repens* var. *pro-cumbens, Selliera radicans, Cotula coronopifolia*[1]).

Salt meadow is common along the coast from north to south of the main islands but rare in Stewart Island. It usually occurs on ground subject to an occasional slight flooding by brackish-water, or on wind-swept slopes &c. liable to occasional drenching by sea-spray, but it may lie quite out of reach of water containing salt in excess. In winter or after heavy rain, pools lie everywhere on flat meadows. The soil may vary from clay to sand. None of the species seem to be dependant upon excess of salt in the substratum, though such is generally the case.

The life-forms of the species are as follows: — divaricating-shrub 1, grass-from 2 (summergreen), rush-form 3, herbs and semiwoody plants 11 (creeping and rooting 7).

The composition of salt-meadow is fairly uniform throughout New Zealand proper, but there is no regularity as to dominance of any particular species. *Juncus maritimus* var. *australiensis, Leptocarpus simplex* or *Plagianthus divaricatus* may be abundant in places, and probably, prior to the settlers' fires, were more plentiful still, but the physiognomy of the meadow depends not on the rush-form, but on the presence of a close and even turf made up of the far-creeping perennial herbs *Selliera radicans, Samolus repens* and *Cotula dioica*. Equally abundant, indeed at times dominant, is the succulent *Salicornia australis*, and the summer-green grass *Atropis stricta* is also very common.

The history of development of salt-meadow is plain enough and may take place in several ways. Frequently it occurs as a succession following salt-swamp. Thus, as the swamp gradually rises owing to the plants themselves forming soil, to the deposition of silt, or to a gradual elevation of the land — there are many causes — the water-demanding species die out, the land-species already present increase their areas of occupation and others join the association; indeed, every transitional stage from swamp to meadow can be sen. Another procession of events is when (as already explained) the individual plants of the *Salicornia* formation (Fig. 7) trap the mud as moved by the tide and gradually increase the height of the substratum, so that *Juncus maritimus* var. *australiensis* can come in and, as the ground becomes still higher, *Samolus, Selliera* &c. Salt-meadow, however, can be established without any succession at all where strong winds carry salt spray inland, and a true

---

1) One or other of the jordanons of *Cotula dioica* is present, but in the south of the South Otago district and Stewart Island this compound species is represented by *C. pulchella* and *C. Traillii*. So, too, the extremely common *Atropis stricta* is represented in the above localities by *A. Walkeri*. Other species frequently present in salt-meadow are *Suaeda maritima* (wettish places), *Eryngium vesiculosum, Atriplex patula* (exotic). In some places the exotic *Lepturus incurvatus* and *Plantago Coronopus* form pure colonies, looking exactly as if indigenous.

salt-meadow may be seen which has arisen from this cause on flattish ground near the summit of high sea-cliffs (e. g. Centre Island, Foveaux Strait).

Burning the rushes &c., consolidation by stock and most of all drainage rapidly alter the habitat, various exotic grasses come in and the primitive community — partly purposely, partly by accident — is converted into pasture.

b. Coastal-moor and allied communities.

Coastal moor proper consists of turf formed by a combination of various ordinary creeping-rooting halophytes, together with certain coastal ferns and low-growing plants of a subalpine-subantarctic character.

The species number about 44 (families 24, genera 32) of which the most important are: (halophytes) *Salicornia australis, Samolus repens* var. *procumbens, Selliera radicans, Cotula pulchella*; (subalpine-subantarctic) *Blechnum durum, Agrostis muscosa, Scirpus aucklandicus, Rumex neglectus, Montia fontana, Myosotis pygmaea* var. *Traillii, Euphrasia repens, Asperula perpusilla, Plantago Hamiltonii.*

Coastal-moor is confined to the south of the South Otago[1]) and the Stewart districts and perhaps the Fiord district. The presence of the community depends upon a sour peaty soil, as substratum, the result of the subantarctic character of the climate plus the nature of the plant-covering[2]). The habitat is also exposed to showers of sea-spray, so that there will be a greater percentage of salt in the soil than in the case of boggy ground in general. Water frequently lies in pools and some of these are permanent.

The life-forms are well in keeping with the habitat, consisting as they do of: — turf-making herbs 23, rosette plants 13 (including 2 ferns), tussocks 4 and cushion-plants 4.

The plant-covering consisted of an even turf of extreme density owing to the plants having their rosettes or leaves pressed close to the ground. Its composition was not uniform throughout, but fairly distinct subassociations could be recognized owing to the dominance respectively of *Selliera, Cotula pulchella* or *Plantago Hamiltonii*, though the combinations intergraded. The round, green cushions of *Euphrasia repens* were everywhere, each about 12 cm. diam. which, when thickly covered with their small, snowy flowers were most striking. *Gentiana saxosa*, too, in full bloom, was equally conspicuous. The physiognomy of the turf depended upon which species was dominant. The dense shining rosettes of *Plantago Hamiltonii*, all

---

1) The most typical moor is on the west of the Bluff Peninsula to the south of Ocean Beach. Unfortunately this unique piece of primitive New Zealand is now so altered by the presence of stock as to be no longer the spot described in my notes *before* it became a grazing-ground. Smaller examples occur on the coast-line of Foveaux Strait and on some of the small islands therein.

2) That is to say, certain species occupy the ground, thanks to the favouring climate, these form peat and species more peat-tolerating still, enter in, and intensify the oxylophytic conditions.

touching, the leaves glossy-green but blackish-brown at the base, covering the ground to the exclusion of all else, were a remarkable sight. Colonies of *Rumex neglectus* made green patches; dominance of *Cotula pulchella* lent a reddish-brown colour to the turf, the tiny starry flowers of *Asperula perpusilla* betrayed its presence, otherwise hardly noticeable. Increasing wetness of the ground brought in abundance of *Montia fontana*, and the reddish *Tillaea moschata* occured in profusion fringing pools; *Blechnum durum* and *Asplenium obtusatum,* their fronds flattened to the soil, were everywhere common; tussocks of *Scirpus nodosus* were occasionally plentiful, or those of *Carex appressa* made a pure growth and, in some places, the still taller Fuegian *C. trifida* occured.

On the low flat sandy ground to the west of the New River Estuary, the association is allied to coastal-moor, on the one hand, and dune-hollow, on the other; *Euphrasia repens, Plantago Hamiltonii,* and *Claytonia australasica,* represent the moor, and *Gunnera arenaria, Geranium sessiliflorum, Selliera, Acaena novaezelandiae, A. microphylla* var. *pauciglochidiata* and its hybrids with *A. novae-zelandiae* and *Raoulia apice-nigra* the dune-hollow[1]).

3. Sea-shore Communities.
a. Sandy Shore[2]).

Wide wind-swept stretches of sandy beach are frequently destitute of plant-life, the lower shore being washed by the sea, while the upper loose sand above high-water mark is the sport of the wind. Specially high tides too extend far beyond the average limit, and the undiluted sea-water is detrimental to the well-being of land-plants. Notwithstanding this, the dune-plants, *Spinifex hirsutus, Desmoschoenus spiralis* and *Carex pumila* creep over the loose sand, or build hillocks.

Where conditions render sand-movement sufficiently feeble a few more plants will be present, but, in any case, the number of species is small. *Ranunculus acaulis,* its tiny leaf-rosettes half-buried, their blades flattened to the sand, often forms small colonies. *Calystegia Soldanella* makes green patches, or isolated plants may be dotted about, as may the grasses *Festuca littoralis* and *Deyeuxia Billardieri* and the stiff-stemmed tussocks of *Scirpus nodosus.* The exotic *Salsola Kali* is fairly common in some places. The

1) In some places *Nertera Balfouriana* (usually a mountain plant) is fairly plentiful, and there also occurs (restricted to a very small area) *Gunnera Hamiltonii* which had not been seen since its discovery 44 years ago until C. S. Smith recently found it in abundance growing on rather moist sandy ground on the seaward side of the Oreti River (SO.).

2) The plant-covering of sandy shores can hardly come into a category by itself but more properly belongs to unstable dune. Nevertheless it seems best to deal with it separately, partly because it is a distinct feature of the coast-line, and partly because it appears to be the sole station for *Atriplex crystallina.*

pale-green, succulent *Atriplex crystallina* occurs sparingly throughout, and forms mats upon the sand. It is abundant on certain beaches in Stewart Island, where it forms a girdle on the upper shore 2 m. wide, the stems half-buried and the plants some 90 cm. apart. In the north the grass *Zoysia pungens* is fairly common.

b. Beach of loose stones.

The formation (if it may be so called) occupying a beach of loose stones is open and includes many more or less distinct communities all of which are marked by a combination of the usual shore-line halophytes, together with shrubby and herbaceous mat-plants, which may be non-coastal.

The number of species found on these beaches is certainly not less than 88 (families 36, genera 63), and may be considerably more. The following are common, wide-spread species: — *Deyeuxia Billardieri, Festuca littoralis, Carex ternaria, Leptocarpus simplex, Phormium tenax, Muehlenbeckia complexa, Salicornia australis, Mesembryanthemum australe, Tetragonia expansa, Scleranthus biflorus, Ranunculus acaulis, Acaena novae-zelandiae, Oxalis corniculata, Linum monogynum, Pimelea prostrata, Samolus repens* var. *procumbens, Apium prostratum, A. filiforme, Calystegia Soldanella, Lobelia anceps,* one or other of the species of *Cassinia, Cotula dioica* and its allies.

The formation is common along the coast, the substratum being supplied in part by shingle carried by currents and tide and cast upon the shore.

Beaches of this class offer different conditions according to size of stones, state of consolidation and amount of sand present. Between sand and gravel there is merely a question of degree, so that coarse sand and fine gravel beaches have much the same plant-covering. Where the stones are large enough to be called shingle, the conditions approximate to those of lowland river-bed, excepting that the latter has usually a greater water-supply and more available soil between the stones, while the spray-factor is absent. But, as will be seen, certain river-bed species do occur on some shores though generally well back from the sea. Sand frequently fills up the spaces between the stones and so dune-plants enter into the associations. In the South Otago, Fiord and Stewart districts the semi-subantarctic climate leads to the formation of peat, which, collecting between the stones, encourages the settlement of species of coastal moor. This is intensified where the substratum is gravel.

The life-forms of the species for the formation as a whole are as follows: mat-plants 33 (woody 14), creeping herbs 8, tussocks 12, erect herbs or semi-woody plants 10, rosette plants 7, grass-like creeping 5, tufted grass 4, sand-binders 2, erect shrubs 4, cushions 2, Yucca-like 1.

Where the boulders are largest, vegetation is the most scanty, the commonest plants being the halophytes, *Apium prostratum, Calystegia Sol-*

danella and *Senecio lautus*, this latter through its strong plasticity being able to respond rapidly to change of conditions. Further from the influence of the waves, there may be *Phormium tenax, Scirpus nodosus,* and excepting south of about latitude 44° *Mariscus ustulatus* and *Lobelia anceps.*

The terrace bounding the shingle shore, being both more stable and less halophytic, possesses a more varied plant-covering. *P. tenax,* either in clumps, or as a continuous belt is a fairly common feature. On Cuvier Island it is mixed with *Arundo*; in South Canterbury it forms for a considerable distance a more or less continuous girdle; at Big Bay and Paringa (South Westland), this belt is so near the sea that, in rough weather, it arrests the floating driftwood. *Muehlenbeckia complexa,* as a mat-plant, is a common feature of terraces or high foreshores, and such mats favour the settlement of other plants e. g. *Rhagodia nutans, Linum monogynum, Senecio lautus,* and at the present time certain introduced species. The following are some of the more common species of shingle or boulder terrace: — *Pteridium esculentum, Scirpus nodosus, Carex testacea, Acaena novae-zelandiae, Oxalis corniculata, Linum monogynum, Euphorbia glauca, Pimelea prostrata, Leptospermum scoparium, Apium prostratum, Dichondra repens, Cassinia leptophylla* or *retorta* according to latitude.

On the terrace of the Nineteen Mile Beach (E.) where the Canterbury Plain and Pacific Ocean meet, as well as many common species of such a habitat, are mats or low cushions of *Raoulia lutescens* and open circular mats — grey and green respectively — of the leafless, rush-like *Muehlenbeckia ephedroides* and a prostrate epharmone of *Carmichaelia subulata.* The *Muehlenbeckia* also occurs on shingly shore near Cook Strait where when sand comes in there are low cushions of a variety (unnamed) of *Raoulia australis* and stiff mats of the usually erect shrubs *Coprosma propinqua* and *Plagianthus divaricatus.* As for the associations of fine gravel in a semi-subantarctic climate the following list gives some idea: *Scirpus aucklandicus, Carex pumila, Urtica australis* (Dog Island, Foveaux Strait), *Muehlenbeckia complexa, Rumex neglectus, Tetragonia expansa, Chenopodium glaucum, Atriplex crystallina, Lepidium tenuicaule* (root very long), *Tillaea moschata, Ranunculus acaulis, Geranium sessiliflorum, Myosotis pygmaea* var. *Traillii, Cotula pulchella, C. Traillii.*

The most important matter connected with the development of the vegetation dealt with above is the role of mat-plants — more especially shrub-mats — in trapping seeds and functioning as seed-beds. At the present time this power is potent in bringing exotic species into parts of the habitat not exposed to sea-spray. Amongst such may be noted *Holcus lanatus,* species of *Bromus, Hordeum murinum, Silene anglica,* various herbaceous *Leguminosae, Anagallis arvensis, Sherardia arvensis, Sonchus arvensis* and other composites.

4. Dune.

General. The vegetation of an extensive dune-area consists of several distinct plant-formations or parts of such, yet all are treated here under one head, partly for convenience and partly because the various habitats are the result of topographical change, each leading in distinct sequence to the next.

The total number of species for the whole dune-area is about 150 (families 50 genera 100) of which some 50 are common. The following are virtually confined to dune-areas, sandy shores or small sandy spots: (shrubs) *Pimelea arenaria, Coprosma acerosa, Cassinia retorta* (but now of wider range through the influence of the settler); (herbs) *Ranunculus recens* (vr.), *Euphorbia glauca, Gunnera arenaria, G. Hamiltonii, Calystegia Soldanella;* (grass-like) *Spinifex hirsutus, Deyeuxia Billardieri, Desmoschoenus spiralis, Festuca littoralis, Carex pumila;* (small rushlike) *Heleocharis neo-zelandica.* Excluding Stewart Island, about 61 species extend throughout. The following species of considerable range, but not of necessity extending throughout, are abundant: *Spinifex hirsutus, Arundo conspicua, Festuca littoralis, Mariscus ustulatus, Scirpus nodosus, Desmoschoenus spiralis, Carex ternaria, C. pumila, Leptocarpus simplex, Cordyline australis, Phormium tenax, Libertia peregrinans, Scleranthus biflorus, Muehlenbeckia complexa* var. *microphylla, Acaena novae-zelandiae, Oxalis corniculata, Pimelea arenaria, Leptospermum scoparium, Gunnera arenaria, Centella uniflora, Lilaeopsis novae-zelandiae, Euphorbia glauca, Coprosma acerosa, Lobelia anceps, Selliera radicans, Cassinia leptophylla, Cotula dioica.*

Dunes are the most common of New Zealand coastal land-forms and occupy no less than 127000 hectares while they extend inland in some places for a distance of 12 km. The most extensive areas are on the west coast of North Island but considerable dunes also occur on the east and north of South Island and the west of Stewart Island, the area for both islands, (9700 hectares) being still considerable.

The plant-covering of the dunes is fairly uniform throughout New Zealand, Lord Auckland and Kermadec Islands excepted, but certain latitudinal changes are evident due partly to climatic and partly to historical causes. *Desmoschoenus, Deyeuxia Billardieri, Festuca littoralis, Euphorbia glauca* and *Coprosma acerosa* occur on most of the unstable dunes. *Spinifex hirsutus* extends but a short distance past lat. 42°; the ecologically-equivalent species of *Cassinia (retorta, leptophylla, fulvida)* are each in the order given the dominant or sole form in the northern Central and Southern botanical provinces. *Pimela arenaria,* common on dunes in both the main islands, is represented in the South Otago district and Stewart Island by *P. Lyallii.* A variety of *Geranium sessiliflorum* is abundant on semi-stable dunes in the South Otago district and Stewart Island, but absent elsewhere. Certain species of *Raoulia* are confined to particular localities and conditions.

Since dune phenomena are the same the world over it would serve no purpose to supply details regarding the making of dunes, nor need anything be said concerning the ecological conditions of the habitats. There is the usual foredune, the dune-complex and the wandering-dune, but there is every transition from an area possessing every dune feature, and with "Sand mountains" 60 m. high or more, to small hillocks on a sandy shore. In some parts of Taranaki and Auckland, cliffs, themselves often consolidated dunes, face the ocean, while at high-water there is no exposed beach at their base. Such are capped by enormous deposits of sand which are not infrequently advancing inland.

Dune-vegetation exhibits a gradual procession of events in harmony with the increasing stability of the substratum, the foredune marking the unstable commencement and the fixed inland sand-hill the stable climax. Stages also occur where a new class of associations branches off that may be merely transitory or become permanent, their persistence depending upon the stability of the dune-area as a whole.

Sand grass dune. This is distinguished by the presence of one or both of the powerful sand-binders, *Spinifex hirsutus* ((Fig. 8) or *Desmoschoenus spiralis,* nor need there be any other species.

In North Island and the north of South Island where both plants are present they rarely intermix, while their characteristic colours, — silvery *(Spinifex),* yellow *(Desmoschoenus),* plainly indicate which dominates, giving the dune also a special physiognomy. *S. hirsutus* does not extend inland for any distance, but *Desmoschoenus* occurs, wherever there is drifting coastal sand. Near Foveaux Strait *Festuca littoralis* is an early comer, and sometimes a primary dune-builder. Occasionally, the tufts of *Spinifex* or *Desmoschoenus* form a close covering, but usually there are more bare patches than vegetation. At certain places on the west coast of the Ruahine-Cook district a very uniform foredune, looking like an artifical embankment, extends for several kilometres at a time covered with *Spinifex* (Fig. 8).

As sand-grass dune becomes more stable, thanks to its occupation by the major sandbinders, *Calystegia Soldanella, Deyeuxia Billardieri, Festuca littoralis* and *Scirpus nodosus* gain a footing and probably also some of the sand-tolerating shrubs. The rarer *Euphorbia glauca* belongs to the same association. In Stewart Island, *Sonchus littoralis* may be present. *Calystegia Soldanella* often covers the sand completely with a shining green mantle. Low dunes of that kind persist for a considerable time, the dune itself acting as a solid obstacle thus causing a protecting wind-trough to be formed between itself and the advancing sand.

Shrub-dune. This is distinguished by the dominance of certain sand-collecting shrubs, together with various erect ericoid shrubs, the sand-collectors being the essential feature.

The dominant species are *Pimelea arenaria* (*P. Lyallii* in South Otago and Stewart districts), *Coprosma acerosa* and one or more (according to the locality) of the species of *Cassinia*. The following are early members of the association: *Deyeuxia Billardieri, Festuca littoralis, Scirpus nodosus* and *Calystegia Soldanella*.

The physiognomy of the association depends upon the open cushion-form and yellow hue of the *Coprosma* (the *Pimelea* is far rarer and frequently wanting) and the grey or yellow *Cassinia* (according to the species).

At a greater distance from the sea, or when exposed to weaker sand-advance or erosive wind-action, the above shrubs will form a closed covering and may be reinforced by others, especially *Leptospermum scoparium*. The great tussocks of *Arundo conspicua* (Fig. 9.), and the huge tussock-like, green masses of *Phormium tenax* and *P. Colensoi* are frequently present, the latter being common in the North Auckland district. There too, *Leptospermum ericoides* and *Leucopogon fasciculatus* not erect but forming spreading mats, are abundant.

Fixed-dune. This, the climax of constructive dune-building, frequently lies far back from the sea, and is of considerable age and perhaps denotes a rising coast. In places where the wind strikes obliquely, or is not frequent, fixed-dune may lie just behind the foredune.

The sand is covered by a much more cohesive humus-containing layer of blackish sand than is that of unstable dune. It is sometimes as much as 30 cm. in depth, formed from the decay of many generations of plants.

At the present time, the vegetation is much modified but three distinct groups of associations stand out distinctly enough — grassland, shrubland and fern *(Pteridium)* heath.

The grassland, no longer primitive, consists of many exotic grasses and *Leguminosae* together with (North Island) the indigenous *Microlaena stipoides* and forms of *Danthonia semiannularis* and *D. pilosa. Zoysia pungens* is common in northern localities, but generally where there is less loam.

The shrubland consists of one or more of the species of *Cassinia* and *Phormium tenax, Arundo conspicua, Microlaena stipoides, Discaria toumatou,* species of *Carmichaelia* and *Scirpus nodosus* are common members in many localities. On the northern shore of Foveaux Strait certain species of *Acaena* are abundant and there are many silvery patches of *Raoulia australis* and *R. apice-nigra;* tussocks of *Poa caespitosa* also are plentiful. *Pimelea Lyallii* replaces *P. arenaria,* and other plants not found on dunes further north are *Geranium sessiliflorum* (a dune variety), *Wahlenbergia congesta* and *Gnaphalium trinerve.*

Hollows and sand-plains. An advancing dune-ridge leaves, in its wake, flat sandy ground which may continue to be further hollowed out by wind-action, until moisture rising from the water-table damps the surface-sand and stops further sand-movement. Such a hollow is quite stable and ready for occupation by non-psammophytes.

The final destiny of such areas does not depend upon their plant-covering but on the stability of the adjacent dunes. Thus one of two things may happen, — there may be an invasion of sand and reversion to dune-conditions, certain associations arising only to be destroyed, or there may be a long stage of stability and a succession of associations culminating in a climax-association, two of which — swamp and swamp-forest being ordinary lowland communities are not dealt with here.

The primary groups of associations. Frequently the first arrival on the damp sand is *Gunnera arenaria* (creeping and rooting herb forming circular patches, 8 cm. to 2 m. diam., of small rosettes of thick, pale-green or brownish leaves flattened to the ground). *Scirpus cernuus, Ranunculus acaulis, Epilobium Billardieranum, E. nerteroides, Lilaeopsis novae-zelandiae* and especially *Carex pumila* are characteristic. *Lobelia anceps, Limosella tenuifolia* and *Myriophyllum Votschii* in North Island and northern dunes of South Island. Should there be no incoming of sand the following common halophytes next occupy this non-halophytic station: the species and hybrids of *Apium, Samolus repens* var. *procumbens, Selliera radicans, Cotula coronopifolia, C. dioica* and one or more of its near allies. Finally, colonies of the rush-like *Leptocarpus simplex* may take complete possession of the ground[1]). Should the hollow not become moist, *Carex pumila* early takes possession, increasing at a great rate through its far-creeping rhizome. A rapid drift favours *Spinifex* which quickly builds multitudes of low, rounded mounds.

The secondary groups of associations. Generally at some distance inland but occasionally just beyond the foredune, the sand-hollows may be closely covered by shrubland with *Leptospermum scoparium* dominant (Fig. 10.). In the North Auckland district, and southwards for some distance, many species of the gumland shrubland are present, especially *Leptospermum lineatum, Pomaderris phylicaefolia, Leucopogon fasciculatus, L. Fraseri, Pimelea prostrata* and abundance of *Cassinia retorta*; also there will be more or less of both species of *Phormium* — but not usually in the same locality — and *Arundo conspicua*. On the northern shore of Cook Strait, *Leptospermum scoparium* often forms a closed association, especially where the surface is wet, when certain swamp-plants will be present, e. g. *Heleocharis Cunninghamii, Carex ternaria, Potentilla anserina* var. *anserinoides* and *Coprosma propinqua*. In the Eastern district, in addition to abundance of *Leptospermum scoparium*, there is more or less *Discaria toumatou*.

---

1) On the north shore of Foveaux Strait *Raoulia australis* is plentiful, *Pimelea Lyallii* builds small sand-filled cushions, rosettes of the sand variety of *Geranium sessiliflorum* are here and there, *Ranunculus recens* (also EW.) occurs locally and in the North Auckland district *Heleocharis neo-zealandica* forms slightly raised patches.

Ancient dunes. Ancient sandhills occur in the South Otago district near the Bluff Estuary and along the lines of the ancient straits of Stewart Island. The Bluff dunes are separated from the estuary by *Sphagnum* bog (now almost gone). They are covered with *Danthonia Raoulii* var. *rubra* and the following non-dune, frequently subalpine plants are present: — *Blechnum penna marina, Lycopodium fastigiatum, Gaultheria perplexa* and *Helichrysum bellidioides.* The Stewart Island dunes bear a heath-like vegetation, including *Leptospermum scoparium* (dominant), *Gleichenia circinata* (abundant), *Pteridium esculentum, Lycopodium ramulosum*, the creeping subalpine podocarp *Dacrydium laxifolium, Cladium Vauthiera, Phormium Colensoi, Enargea parviflora, Leucopogon Fraseri, Pentachondra pumila* and *Coprosma repens.* The high-mountain character of this combination is remarkable.

Dune forest. Forest is rare on the dune-areas. Swampy sand-plain may occasionally contain *Podocarpus dacrydioides* forest. Various coastal and inland-coastal trees occur in dune-gullies and sheltered hollows in the North Auckland district, Kaipara subdistrict, and on the north shore of Cook Strait, e. g. — *Macropiper, Knightia, Corynocarpus* and *Myoporum.* On the lee-slopes of the high western dunes of Stewart Island, there is a luxuriant low forest in which *Griselinia littoralis* and *Metrosideros lucida* dominate. Very similar forest occurs on the dunes in the south-east of the South Otago district. On the ancient dunes near the Ruggedy Mts., Stewart Island, according to POPPELWELL (1913 : 283) there is *Dacrydium cupressinum-Weinmannia* forest. On the dune-area facing Foveaux Strait westwards of the Oreti River there was formerly much forest consisting of close-growing, stunted *Podocarpus totara* and *P. spicatus.* Remnants remain in sheltered hollows, and, even in the open in places where the forest was cut down, it is naturally reinstating itself with *P. totara* in the form of much-branched spreading bushes.

5. Rock and Cliff Vegetation.

a. General.

Taking a comprehensive view[1]) as to the meaning of coastal rock vegetation, perhaps as many as 141 species (families 40, genera 89) may be a fair estimate for the various associations as a whole, but many are far from common and 20 species at the very most occur where there is noticeable

---

1) It is not easy to draw the line as to true coastal-rock formations. Thus, at Jackson Bay, Westland, there is an isolated rock, only to be reached at low-water, 3 to 4 m. high, on the flat top of which grows a collection of forest mesophytes, while in peat, on its face, is a dense mat of *Enargea,* a denizen of wet subalpine, or sometimes lowland forests.

sea-spray, while hardly one species can be called an "obligate"[1]) rock-plant[2]).

Rock-surfaces are common on the coast and vary from high vertical cliffs, their bases washed by the waves, to flat sea-worn rocks hardly raised above the water. According to the nature and aspect of the rock most diverse stations are available for plant-life. On a vertical wall of hard rock with an even surface exposed to a maximum of salt-spray, there are no plants except a few lichens, but a sheltered rock in a rainy locality, with many cracks, crevices and ledges where soil can lodge, will support a rich vegetation.

Although rock associations are at first open, rooting-places are so limited in area, that there may be keen competition between the early comers. The soil-making ability of the plants may have much influence upon the ultimate composition of the association, certain peat-formers being able to furnish a soil of considerable depth and great water-holding capacity. Chasmophytes cannot gain a footing, until there be sufficient soil in the cracks or crevices This may consist of small rock-fragments and sand weathered *in situ,* or blown soil. Accumulations, 2.5 to 5 cm. in depth, are quite frequent. The seedlings must be able to tolerate periods of drought. Succulency, thick coriaceous leaves, prostrate and rosette-forms are frequent and advantageous. Often the roots form a mat, as with certain rock-ferns, while even long roots may at first extend laterally along a fissure.

b. The Mesembryanthemum group of associations.

This group is distinguished by the dominance of *Mesembryanthemum australe* which may be virtually pure or have associated with it various halophytes, or if in a position less exposed to sea-spray, non-halophytes enter the associations. Probably all the really important species for the whole group of associations are mentioned in what follows.

Mesembryanthemum rock associations occur on coastal cliffs throughout all New Zealand proper. The dominant species forms bright green patches the shoots hanging downwards and not rooting, so that they can be raised like a curtain, but on flat or creviced rocks roots are produced. Where exposed to the maximum of spray *Mesembryanthemum* may be the sole plant (Fig. 11), but throughout North Island and South Island it is frequently accompanied by *Salicornia australis, Apium prostratum* and *Senecio lautus.* In North Island and the north of South Island, *Coprosma retusa* as a prostrate shrub

---

1) Perhaps *Arthropodium cirrhatum, Anisotome flabellata, Hebe obtusata* and *Celmisia Lindsayi* come into this category.

2) Besides true coastal chasmophytes, the rock-vegetation includes more inland species than any of the other shore formations. Even lianes are not wanting, their sub-xerophytic structure, such as it is, being of advantage for the severe conditions of the open.

hugging the rock is common and in many places especially in the Auckland, Sounds-Nelson and North-western districts, there are often isolated plants or extensive colonies of the liliaceous herb, *Arthropodium cirrhatum*, conspicuous with its broad, thick, pale-green leaves some 60 cm. in length. The fern *Asplenium obtusatum*, an occasional plant of North Island, becomes far more abundant in proceeding south.

With less spray, other plants enter in and *Mesembryanthemum* becomes of decreasing importance. Rocks of many shapes, the outcome of much weathering, standing at various distances from the water, obviously offer much more varied conditions than a high wall-like cliff, and so possess a richer flora. The following are frequently present: — *Polypodium diversifolium, Cyclophorus serpens, Arundo conspicua* (especially in the north), *Poa anceps* var. *condensata, Agropyrum multiflorum, A. scabrum, Scirpus cernuus, S. nodosus, Cladium Sinclairii* (north), *Leptocarpus simplex, Phormium tenax, Muehlenbeckia complexa, Rhagodia nutans* (but not south of Banks Peninsula), *Colobanthus Muelleri, Lepidium oleraceum, Tillaea moschata* (south), *Linum monogynum, Coriaria sarmentosa, Pimelea prostrata*[1]), *Leptospermum scoparium, Samolus repens* var. *procumbens, Lobelia anceps* (south to Banks Peninsula and Okarito), *Selliera radicans, Lagenophora pumila, Sonchus littoralis,* and one or other of the species of *Cassinia*[2]).

c. Pohutukawa (Metrosideros tomentosa) cliff.

Specially characteristic of northern New Zealand[3]) and magnificent when loaded at about Christmas with their crimson blossoms are the long lines of that massive tree *Metrosideros tomentosa* jutting from the faces of coastal cliffs, its spreading branches extending outwards over the rocky shore, at times almost dipping into the water. The frequently almost horizontal position of the tree, the great weight of trunk and crown, together with the resistance this latter offers to the wind, demand efficient means of fixing

---

1) There are one or two other species of *Pimelea,* including the badly-understood *P. Urvilleana,* but the taxonomy of this group needs careful revision, based on field study.

2) The following are restricted in distribution: — *Hymenanthera crassifolia* (vicinity of Cook Strait to the South Otago district but apparently absent on west coast of South Island); *H. obovata* and its hybrids with the above (both sides of Cook Strait); the series of hybrids of which the so-called „species" *Coprosma Kirkii* is based on specimens from different localities in the north of the North Auckland district (at different points on the North Island coast, but probably in far more localities than have been recorded.) The group, in its entirety is almost certainly a hybrid swarm with *C. propinqua* as one of the parents and perhaps *C. retusa* as the other, but possibly other species are concerned.

3) *Metrosideros tomentosa* is common along the coast of North Island from Three Kings Islands to lat. 39° on the west but not quite so far on the east. Inland it occurs on the shores of several lakes (VP.) and has been reported from Lake Waikare moana (E.C.).

to the rock. This is met by the tree's innate power to put forth abundance of aerial roots (Fig. 4.), which spread over the cliff-face and send down branches deep into the rock.

In fairly sheltered positions the combination of species is considerable but with increase of wind and spray only the true coastal plants remain. *Astelia Banksii*[1]) in pure colonies may be plentiful on rock-faces. *Peperomia Urvilleana*[2]), both epiphytic and rupestral, is frequently abundant. The ferns *Asplenium lucidum, Polypodium diversifolium* and *Cyclophorus serpens* are often common. *Arthropodium cirrhatum* may be so abundant as to cover the rock with greenery for many square metres. Mats of *Poa anceps* var. *condensata* are characteristic and various shrubs and other life-forms are represented[3]).

d. Associations with Phormium Colensoi dominant.

Phormium Colensoi[4]) is a common feature of high or low cliffs of the East Cape and Ruahine-Cook districts, whence it extends along the coast of the Sounds-Nelson and North-eastern districts and by way of the west coast of South Island — *P. tenax* also frequently coastal — to Stewart Island. Certain fairly well-marked communities may be distinguished as follows:

Hebe macroura association. This well-marked association of a more or less open character is distinguished by the abundance of *Hebe macroura*[5]) and the presence of the rather tall, semi-woody plants *Senecio Banksii* and *S. Colensoi.*

The association is common on the coast of the East Cape district and

---

1) The thick, stiff, linear, tapering leaves, 1.4 m. long, dark-green above and silvery beneath are in great tufted masses. Thick masses of dead leaf-bases surround the living leaf-sheaths and function as a waterholding and food-containing apparatus.

2) A succulent herb with glabrous, very thick, orbicular leaves ± 8 mm. long and far-creeping, rooting stems.

3) *Coriaria arborea, Leptospermum scoparium, Pseudopanax Lessonii, Hebe macrocarpa, H. salicifolia, Olearia furfuracea* and *Brachyglottis repanda* may be more or less common, and so too, *Arundo conspicua, Phormium tenax, Rhagodia nutans, Mesembryanthemum australe* with its accompanying halophytes, as well as other coastal forest or scrub plants, according to the degree of shelter.

4) The coastal plant differs somewhat from the subalpine forms. The leaves crowded and bunched together are about 1.4 m. long by 9 cm. broad, the lower half of each is erect, owing to the sides of the blade being pressed together, but the upper half droops downwards. When *P. tenax* is also present there will most likely be more or less hybrids which betray their origin in forms of capsule ranging between the erect blunt capsule of *P. tenax* and the drooping, twisted, acute, longer one of *P. Colensoi*.

5) A shrub of spreading, drooping habit, the stems leafy near the extremities only, the leaves moderate-sized, rather broad and the flowers white and sweet-scented. Ecologically similar is *Hebe salicifolia* var. *Atkinsonii* of the neighbourhood of Cook Strait and the Marlborough Sounds.

with it might be included that of the Sugarloaf[1]) near New Plymouth and adjacent islets (EW.) were it not that the accompanying *Phormium* is *P. tenax*. The substratum is a soft marl on which seeds germinate freely and into which roots readily penetrate.

The dominant *Hebe* may grow through the numerous mats of *Poa foliosa* var. *condensata* and hang down the face of the cliff. Various coastal herbs and semiwoody plants and shrubs are present[2]). In sheltered gullies *H. macroura* is erect and it is associated with one of the jordanons of *H. salicifolia* (with which it is almost certain to cross), tall *Arundo conspicua*, *Phormium Colensoi* and *Coriaria arborea* or *sarmentosa*.

Hebe salicifolia var. Atkinsonii association. This is distinguished by the presence — often in abundance — of the above *Hebe*, together with various halophytes. There are certain distinct subassociations, two of which are dealt with here.

The quasi-mountain subassociation is distinguished by its mountain-like facies due to the presence of common mountain species or to such as resemble plants of that character[3]). The community occurs on cliffs in the south of the Ruahine-Cook district and, the more shady and soil-containing the habitat, the greater the number of species and the denser the growth. *Festuca multinodis* forms dense mats and, in addition to the species of the footnote and the *Hebe*, there is abundance of *Cyclophorus serpens*, *Asplenium flaccidum*, *Polystichum Richardi*, *Mesembryanthemum australe*, *Linum monogynum* and *Wahlenbergia gracilis*. Increase of light leads to the outgoing of many species, but shrubs and certain forest plants[4]) enter in.

The Arthropodium subassociation is distinguished by abundance of *Arthropodium cirrhatum*, *Griselinia lucida* and *Astelia Solandri* — the last

---

1) For a list of the species see COCKAYNE, L. 1916: 207—208. The *Hebe* is closely related to *H. Cookiana*. Also here may come the vegetation of the cliffs (lava conglomerate) to the south and probably the west of Mount Egmont. On these *Phormium tenax* is common; the other species are: *Adiantum affine* (widespread on coastal rocks in general), *Blechnum durum* (the first record for North Island), *Poa anceps*, *Poa* species related to *P. pusilla*, *Scirpus cernuus*, *Macropiper*, *Mesembyanthemum*, *Melicytus ramiflorus*, *Gunnera strigosa* (characteristic), *Apium*, *Samolus*, *Hebe* as above, *Plantago Raoulii*, *P. Masonae*, *Coprosma retusa*, *C. propinqua* × *retusa?*, *Selliera*, *Lobelia anceps*, *Gnaphalium trinerve*, *Senecio lautus*, *Sonchus littoralis*.

2) *Scirpus nodosus*, *Apium prostratrum*, *Samolus* and probably other halophytes, *Lobelia anceps*, *Lagenophora pumila*, *Cassinia leptophylla*, *Senecio Banksii*, *S. Colensoi* and *Sonchus littoralis*.

3) Especially *Festuca multinodis*, *Aciphylla squarrosa*, *Raoulia australis* (in driest stations), *Plantago Raoulii* (form with broad, thick leaves), *Pimelea subimbricata*, *Myosotis Forsteri* (rare), *Craspedia maritima*, *Senecio lagopus* and, of course, *Phormium Colensoi* itself.

4) For instance, *Brachyglottis repanda*. On isolated rocks near Cape Turakirae — RC. (ASTON, 1912: 212, the usually epiphytic orchids *Dendrobium Cunninghamii* (as a thick mat), *Sarcochilus adversus* and *Bulbophyllum pygmaeum* are present.

two commonly forest epiphytes. The community occurs on rock-faces in the Sounds-Nelson and North-western districts, and a similar community, but with *Arthropodium* wanting, extends along the coast of the North-eastern district to some distance south from Kaikoura. Various shrubs[1]), frequently in considerable numbers are also present. In a modified form — the *Hebe* and *Phormium* rare or absent — there occurs near Takaka (NW.) on hard calcareous rock an allied community rich in shrubs, ferns &c.[2]).

e. Coastal-fern association.

This is distinguished by the presence and frequent dominance of *Blechnum durum* and *Asplenium obtusatum* while *B. Banksii* may be common.

The association is of a subantarctic character. It occurs in the South of South Island and Stewart Island where the climate favours the making and accumulation of raw humus but is hostile to the well-being of most species; in consequence the community is clearly defined.

In addition to the ferns, the following are generally present: — *Myosotis albida* (coastal herb with thick, soft, hairy leaves $\pm$ 10 cm. $\times$ 5 cm., in erect rosettes), *Hebe elliptica* more or less prostrate is common, but it is a chasmophyte so not dependant on peat, tussocks of *Poa Astoni* ($\pm$ 30 cm. high, leaves rolled, rigid, green, almost pungent), *Tillaea moschata* (a spray-tolerator). The last named is one of the pioneers and a soil-maker. As the soil becomes abundant various coastal-moor species gain a footing, especially *Scirpus aucklandicus, Gentiana saxosa, Plantago Hamiltonii,* and *Cotula pulchella,* together with the usual halophytes.

f. Rocky shore vegetation.

The vegetation of fairly flat, rocky shores is generally rather a mixture of distinct communities in miniature than a defined association. Thus, owing to unevenness of surface, besides pure rock there are depressions where soil or water can accumulate and tiny salt-meadows, swamps and pools with aquatics are dotted about.

On the rock proper close to the sea, the first plant to appear and perhaps the sole species is *Salicornia australis.* Further from the sea will be *Leptocarpus, Scirpus nodosus* and the usual coastal halophytes. In pools *Cotula coronopifolia* and probably *Lilaeopsis Novae-Zelandiae* will be present, and on the rock still further from the effect of spray — according to their

---

1) *Coriaria sarmentosa, Leptospermum scorparium, Suttonia australis, Shawia paniculata, Brachyglottis.*

2) *Adiantum affine, Polystichum Richardi, Asplenium lucidum, Poa anceps, Scirpus nodosus, Arthropodium* (abundant), *Astelia Solandri, Freycinetia, Peperomia Urvilleana, Macropiper excelsum, Pittosporum cornifolium* (on the ground), *Corynocarpus, Dodonaea viscosa, Hymenanthera obovata, Metrosideros robusta, M. Colensoi, Griselinia lucida, Apium prostratum, Solanum aviculare, Coprosma robusta.*

latitudinal distribution — *Coprosma retusa* and *Hymenanthera crassifolia,* perhaps with *Rhagodia nutans* growing through it. In the South Otago district *Scirpus aucklandicus* and *Tillaea moschata* are common, the latter extending northwards to Cook Strait.

g. Rock associations of local occurrence.

Sand-eroded rock, north coast of Cook Strait. Between the Rivers Waitotara and Wangaehu much rock, at a short distance from the coast, has been cut into fantastic forms by wind-borne sand, an extremely xerophytic station being thus provided. The association consists partly of dune-plants *(Festuca littoralis, Coprosma acerosa)* and partly of semi-subalpine species *(Raoulia australis, Pimelea prostrata).* In some places south of the R. Wanganui are larger, flatter rock-surfaces on which are the remains of perhaps a rather ancientplant-covering consisting of *Myoporum laetum, Dodonaea viscosa,* the small forest-tree *Suttonia australis* and the divaricating, frequently subalpine *Corokia Cotoneaster.*

Where the rock is cut to a base-level a flat stony desert results. The rock-floor is strewn with sand-worn stones and there is a thin covering of small gravel and coarse sand. *Pimelea prostrata* and *Coprosma acerosa* are dotted about everywhere in equal numbers (Fig. 12), and, in the lee of the latter, there is generally a tongue of sand (Fig. 13). Silvery cushions of *Raoulia australis* occur here and there; *Spinifex* and *Desmoschoenus* form lines where the sand is finest; there are colonies of *Zoysia pungens* and yellow cushions of *Scleranthus biflorus.*

Celmisia semicordata association. This occurs a little to the north of lat. 42° (NW.). At one spot near the coast the prevailing westerly wind, hemmed in by two adjacent headlands, strikes the summit of the cliff and its immediate vicinity with especial force. In consequence, the subtending girdle of forest is first dwarfed and then broken through. The rocks at a little distance from the edge of the cliff possess a plant-covering physiognomically more like that of subalpine rocks than of the coast. The moist climate favours the formation of peat on the rock; possibly a certain amount of spray is carried by the wind but it must be soon washed from the soil. The subalpine *Celmisia semicordata* is abundant. Other plants present, frequently subalpine, are: — *Oreobolus pectinatus, O. strictus, Phormium Colensoi, Pimelea longifolia, Cyathodes empetrifolia, Olearia avicenniaefolia* and *Senecio bellidioides.* Several plants of the adjacent heath are present and stunted forest-trees and lianes; the coastal element is represented by *Asplenium obtusatum, Scirpus nodosus, Plantago Hamiltonii* and *Senecio rotundifolius.*

Celmisia Lindsayi association. This is confined to cliffs in the South Otago district from Nugget Point southwards. The cliffs and rocky slopes, where there is soil or debris, are occupied by *Metrosideros lucida* forest,

Leptospermum thicket or *Hebe elliptica-Phormium tenax* scrub. The *Celmisia* association is quite distinct from those. It is situated on the solid rock-face, (Fig. 14) or the youngest debris, and so is the first of a series of associations terminating in forest. *Celmisia Lindsayi* (Fig. 15), forming circular patches 90 cm. or more diam. is dominant. *Anisotome intermedia,* its leaves 76 cm. or more in length, is a companion plant, but seems to prefer the debris as a station. Tussocks of *Poa Astoni,* on the rock, and *Hierochloë redolens,* on the debris, complete the association. Further to the south, the *Celmisia* may be absent, but the *Anisotome* continues on cliffs to beyond Curio Bay.

Cliff-vegetation, west coast of Stewart Island. The vegetation of these cliffs is worthy of mention because it includes *Celmisia rigida* and *Anisotome flabellata,* two endemic Stewart Island plants. *C. rigida* grows in peaty ground on the summit of the cliffs. The *Anisotome* is a chasmophyte confined to rock. Its roots are of an extraordinary length. The pale-green, pinnate, thick, fleshy leaves, 2.5 cm. long, are in small rosettes, which, pressed closely together, are flattened to the rock. Neuter of the species is purely coastal.

Rock-debris associations. The readily weathered greywacke on both sides of Cook Strait, and of the islands therein, forms in many places screes which in their physical conditions greatly resemble the debris fields ("shingle-slips") of the high mountains, but the climate being vastly more equable, and the surface of the screes more stable, the vegetation is far richer, nor does it include any highly specialized species restricted to the habitat.

On the northern shore of Cook Strait the scree has an open plant-covering as follows: *Poa caespitosa* (where most consolidated), *Muehlenbeckia complexa* var. *microphylla* (as a mat), *Acaena novae-zelandiae,* *Epilobium novae-zelandiae, Aciphylla squarrosa, Calystegia Soldanella* and some exotics.

On Kapiti Island (Cook Strait) the screes are frequently quite bare but in places they support an open association the chief members being the above *Muehlenbeckia, Arundo conspicua* and *Cassinia leptophylla.* (Fig. 16.)

On the Sounds-Nelson coast facing Cook Strait, there may be *Myoporum laetum* (far broader than its height), *Dodonaea viscosa* and the *Muehlenbeckia.*

6. Coastal Scrub.

a. General.

By "scrub" is meant throughout this book a close growth of compact shrubs or stunted trees, one or both, in which dense life-forms, particularly the divaricating, the ball-like, and the rhododendron-form (shrub-composite) play a major part. In "shrubland", on the contrary, the community may be more or less open, the life-forms more yielding, less dense and with branches more slender and more twiggy in character. Shrubland when low may quite well be designated "heath".

The number of species, not counting those of stunted forest, is about 41 (families 18, genera 23). The following are especially common on some part or other of the coastline: — *Phormium Colensoi, Freycinetia Banksii, Urtica ferox, Muehlenbeckia australis, M. complexa, Pittosporum crassifolium, Melicope simplex, Melicytus ramiflorus, Dodonaea viscosa, Leptospermum scoparium, Metrosideros perforata, M. scandens, M. lucida, M. tomentosa, Dracophyllum longifolium,* varieties of *Hebe salicifolia, H. elliptica, Myoporum laetum, Coprosma lucida, C. retusa, C. crassifolia, Olearia operina, O. angustifolia, O. Colensoi, O. Solandri* and *Senecio rotundifolius.*

The life-forms and number of species to each fall under the following heads: — Trees or tall shrubs stunted to dense, dwarf or medium shrubform 15, lianoid cushions 4, liane forming entanglements on ground 1, ericoid shrubs 4, rhododendron-form 6, divaricating-form 3, giant tussockform 2, bushy but dense shrubs 4, open rather straggling shrub 1, dracophyllum-form 1.

b. Tree-composite scrub.

There are two associations, the one dominated by *Senecio rotundifolius*[1]) and the other either by *Olearia angustifolia*[2]) or *O. operina*[2]), but with these may be a good deal of the *Senecio.*

The following, apart from the two species of *Olearia,* are the principal constituents of both associations: *Phormium Colensoi, Leptospermum scoparium, Metrosideros lucida, Dracophyllum longifolium, Hebe salicifolia* var. *communis, Hebe elliptica,* × *H. ellipsala, Olearia Colensoi* and *Senecio rotundifolius.*

The Senecio rotundifolius association. This is distinguished by the dominance of *Senecio rotundifolius.* It forms a girdle varying in width according to the wind-velocity, round the inlets and sheltered parts of the

---

1) A tall shrub or even low tree up to 3 m. high or more but frequently prostrate in part. The numerous, stout, rigid, naked branches covered with smooth brownish bark spread widely and form, through their final branching, a rounded head of stiff branchlets covered with the large round leathery leaves, bright shining green above but beneath clad with buff tomentum.

2) *O. angustifolia* and *O. operina* are extremely closely-related species of exactly the same life-forms, being either low trees with a dense, broad, hemispherical crown (Fig. 17) or tall shrubs with more or less erect or far-spreading horizontal branches, furnished with coriaceous, more or less lanceolate leaves ± 8 cm. long, covered beneath with white tomentum and arranged in semi-rosettes at the extremities of the stiff branches. The large flower-heads have with ray-florets, but the disc-florets are yellow in *O. operina* and purple in *O. angustifolia.* The latter when in company with *O. Colensoi* gives rise to a polymorphic hybrid swarm the forms of which almost exactly intermediate between the two are the false species *O. Traillii.*

Stewart Island coast and in portions of the Fiord district, but with a somewhat different composition. It extends from the margin of the rain-forest almost to high-water mark. The soil is often shallow peat overlying soft granite, but when the scrub faces a sandy or gravelly shore and there is no cliff adjacent, the soil is deeper and wetter. The crowns of the *Senecio* intermingle (Fig. 18) and make a flat roof which is penetrated, here and there, by the yellowish fastigiate heads of the *Dracophyllum*, its long, needle-like leaves arranged vertically. Sometimes, close to the front, are thick green masses of *Hebe elliptica*. Still further to the front, but absent over wide areas, is the *Phormium*, its broad, green leaves almost drooping into the water. Just behind the scrub proper, though frequently extending long arms above it, there are in the Stewart district irregularly-shaped trees of *Metrosideros lucida*. Within the scrub there is an entangle-ment of prostrate trunks and branches. The ferns *Asplenium obtusatum* and *Blechnum durum* grow on the peaty floor.

Where the wind strikes with greater violence, *Leptospermum scoparium* appears and increases in amount with increase of wind-intensity, becoming by degrees of lower stature until, as a wind-shorn shrub, it takes the front position on rocky headlands (Fig. 2). On the other hand, where there is complete shelter the *Senecio* sub-association is absent, the mesophytic small forest-trees, shrubs and ferns coming to the front.

The macrocephalous Olearia association. Dominance of *Olearia angustifolia* or of *O. operina* apparently denotes a more exposed station than does that of the *Senecio*. There are two distinct subassociations with *O. angustifolia* or *O. operina* dominant respectively.

Olearia angustifolia coastal scrub is strongly xerophytic, a condition demanded by its position on headlands (Fig. 17) and islets fairly in the track of the frequent, violent, cold subantarctic gales, which, at times spray-laden, strike the plants and saturate the shallow, porous soil[1]). The community is common on the most exposed parts of the coast of the Stewart district; there is also a small piece at the base of the Bluff Hill. Seen from a distance, the scrub is defined by its sage-green colour. Where most exposed, it consists of *O. angustifolia* and *O. Colensoi*[2]) with perhaps a few plants of the hybrids between them and on certain islets *Senecio Stewartiae*. Generally, the shrubs are prostrate, but the branches eventually curve upwards. Within there is a tangle of rigid stems. The coastal ferns are the sole, but distant floor-plants.

---

1) It is astonishing how, in such a station, where nanism would appear the response requisite, thick, stiff, tomentose leaves, in rosettes, rigidity of branches and a dense glo-bular life-form, permit the presence of small trees and, in the case of *O. Colensoi*, broad leaves, 18 cm. long.

2) This may be identical with *O. Lyallii* of the Subantarctic Province, the leaves being far larger than in *O. Colensoi* of the high mountains.

A transition between *Senecio* and *Olearia* scrubs occurs at Mason Bay and on Codfish Island. The general character is much as already described, but the sage-green colour is relieved by green patches of *Griselinia littoralis*, especially at some distance from the sea. Where sufficient light penetrates, there are great colonies of *Stilbocarpa Lyallii*, which spread for many metres, thanks to its vegetative increase by means of arched runners. The bright-green, shining, thin leaves, each 30 cm. or more diam., stand on long stalks, 90 cm. above the ground.

Olearia operina coastal scrub is confined to the Fiord district. It is distinguished by the presence of *O. operina* and the absence of *O. Colensoi* and, when in its more sheltered positions, a greater abundance of forest mesophytic trees or shrubs[1]), which latter may form a distinct girdle. Its northern limit appears to be Big Bay (F.).

c. *Hebe elliptica* scrub or thicket.

At various localities in the Otago fiords, on the shores of Foveaux Strait, in Stewart Island and in a modified form at Otago Harbour, there is a scrub in which *H. elliptica*[2]) dominates, but it is more open where the spreading *H. salicifolia* var. *communis* occurs, and then there will be form after form of × *H. ellipsala*.

At the base of the Bluff Hill, besides *H. elliptica* 1.8—2.4 m. high, *Olearia arborescens*, *Nothopanax arboreum*, *Fuchsia excorticata* and *Aristotelia serrata* are more or less common, and there are × *H. ellipsala* and *Melicytus lanceolatus*. Beneath the shrubs are *Astelia nervosa* var. *sylvestris* and the coastal ferns.

On the larger of the Open Bay Islands, in one place, at any rate, is a scrub of *H. elliptica* (a large-leaved form) 4 m. high, over which climbs *Muehlenbeckia australis* while beneath is *Astelia nervosa* var. *sylvestris*, *Histiopteris incisa* and *Asplenium obtusatum*. The soil is deep, coarse peat.

---

1) Principally *Pittosporum fasciculatum*, *Coriaria ruscifolia*, *Aristotelia serrata*, *Fuchsia excorticata*, *Nothopanax Colensoi* and *Pseudopanax crassifolium* var. *unifoliolatum*. *Hebe elliptica*, *H. salicifolia* var. *communis* and × *H. ellipsala* are extremely common. Just above the salt-swamp at the head of Doubtful Sound the community is about 3.6 m. high. In addition to the species already mentioned, it contains juvenile *Dacrydium cupressinum* and *Podocarpus totara*, *Plagianthus divaricatus*, *Carmichaelia arborea*, *Myrtus pedunculata* and *Suttonia divaricata*. The slender branches are covered with mosses and *Hymenophyllum sanguinolentum* and *Nothopanax Colensoi* is an epiphyte. Such a community is far removed from scrub and is rather incipient forest.

2) *Hebe elliptica* in its „typical" form extends all along the west coast of South Island with its northern limit in North Island west of Mount Egmont. On the east coast of South Island it does not reach lat. 45°, but it reappears in the Marlborough Sounds (rare) and as var. *crassifolia* reaches Kapiti Island and one place on the north shore of Cook Strait.

d. Liane scrubs.

Freycinetia scrub. This is distinguished by a more or less pure close growth of the usually root-climbing *Freycinetia Banksii*, its rigid stems raised but little above the ground and forming an entangled mass. The community occurs in rocky places all along the west coast of South Island and is especially common close to high water-mark in the Fiord district. Its richest development is on the more northerly of the Open Bay Islands (W.), growing on hard calcareous rock. There it is about 2 m. high and of an extreme density. The stout stems each 12 cm. circumference, bearing an abundance of thick, yellowish-green, sword-shaped leaves, are twisted in all directions, and interwoven, thus forming a rigid entanglement. The density is often increased by an admixture of the lianes *Muehlenbeckia australis* and *Calystegia tuguriorum*.

At the bases of sea-cliff gullies of the North-western district a sub-association occurs distinguished by the presence of *Melicytus ramiflorus*, *Coprosma lucida*, *Olearia avicenniaefolia*, *Senecio rotundifolius* and *Senecio elaeagnifolius*, and if the two last are growing in proximity there will also be the hybrids between them.

Muehlenbeckia scrub. In certain localities where hillsides slope shorewards (RC., SN., Banks Peninsula &c.) there may be many cushion-like rounded or pyramidal bushes of *Muehlenbeckia complexa* var. *microphylla* the dark, wiry stems intertwined. Such liane-shrubs are dotted about or intermingle.

Scrub consisting of *M. australis* occurs on the more southerly of the Open Bay Islands (W.). The plants, sometimes mixed with the fern *Histiopteris incisa*, grow into one another making close thickets.

Liane-scrubs frequently denote the existence of former forests, and in the case of the Open Bay Islands afford strong evidence of an earlier connection with the mainland (Cockayne 1905c: 373, 374).

On the beach near Nugget Point (SO.), there is a thicket of the shrubby nettle, *Urtica ferox* together with the ferns *Pteridium esculentum* and *Hypolepis rugosula* and the shrubs *Melicytus ramiflorus* and *Fuchsia excorticata*, *Muehlenbeckia australis* binding the whole into a close mass. A similar scrub, but with fewer nembers, occurs on Centre Island, Foveaux Strait and probably on some of the other islands of the Stewart district.

e. Forest scrub.

Forest scrub consists of communities made up of trees dwarfed by fairly intense coastal conditions, especially wind. Where the latter is particularly intense the scrubs are of extreme density and may be termed "wind-scrub", but this ecological class is not always coastal. Between tall forest and wind-scrub there are many transitions, while the effect of the wind is plainly visible in the shorn crowns of the trees forming a dense

canopy sloping shorewards. A good many kinds of trees[1]), both coastal
and inland enter into forest-scrub communities. Lianes are usually abundant
either binding the dwarf trees closely together or themselves become shrubs
pure and simple. Beneath the thick canopy herbaceous ferns and even
tree-ferns may be present. Here a few special associations are alone dealt with.

The North Cape. Where exposed to frequent wind there is a forest-
scrub with an epharmone of *Phyllocladus trichomanoides*, 1.2 m. to 1.8 m.
high (CHEESEMAN, 1897: 363) together with *Pittosporum umbellatum*, *Melicope
simplex*, *Pseudopanax Lessonii* and the shrubs *Pittosporum pimeleoides* and
Corokia Cotoneaster.

Metrosideros tomentosa scrub. *M. tomentosa*, dwarfed to $\pm$ 1.8 m.
high, is dominant. The community occurs at many points of the coast of
the Auckland districts, especially at the base of wind-swept cliffs (Fig. 21).
For instance, below the cliffs of the Waitakerei Hills, (SA.) growing on a
stony terrace in a shallow clay soil, is a scrub of the following composition:
— *M. tomentosa* (dom.), *Cordyline australis*, *Macropiper excelsum*, *Pittosporum crassifolium*, *Corynocarpus laevigata*, *Pseudopanax Lessonii*, *Coprosma
grandifolia*, *C. retusa*, *Brachyglottis repanda*.

On Rangitoto Island there is a remarkable association of the class under
consideration growing under what appear to be most adverse conditions[2])
M. tomentosa dominates, but the following are more or less abundant: —
Astelia Banksii, *A. Cunninghamii* var. *Hookeriana*, *A. Solandri*, *Metrosideros
robusta*, *Griselinia lucida* and *Senecio Kirkii*. All the foregoing, being
usually or at times epiphytes in their customary habitats, might well be
expected on the bare rock (Fig. 20), but this cannot be said for most of the
other common species, e. g. *Knightia excelsa*, *Coriaria arborea*, *Nothopanax*

---

1) The following may be cited but the list is far from complete: *Podocarpus totara*,
P. spicatus, *Phyllocladus trichomanoides*, *Nothofagus cliffortioides*, *N. Solandri*, *Carpodetus serratus*, *Edwardsia microphylla*, *Melicope simplex*, *Dysoxylum spectabile*, *Coriaria
arborea*, *Pennantia corymbosa*, *Dodonaea viscosa*, *Melicytus ramiflorus*, *Leptospermum
scoparium*, *Metrosideros lucida*, *M. tomentosa*, *Griselinia lucida*, *G. littoralis*, *Dracophyllum longifolium*, *Suttonia australis*, *Myoporum laetum*, *Coprosma retusa*, *Olearia
augustifolia*, *Brachyglottis repanda*.

2) Rangitoto Island is an ancient volcano situated in the Hauraki Gulf, more or less
circular in form, about 6 km. diam. and from its centre rises the scoria cone some 270 m.
high surrounding which is an extensive lava-field made up of blocks of scoria of all
sizes piled upon one another and full of gullies, hollows and chasms. There is no visible
water on the island, the rain passing at once through the open substratum, though some
must be absorbed by the porous rock. Notwithstanding the apparently inhospitable
nature of the habitat, more than 180 species of indigenous spermophytes and pteridophytes occur on the island, and although bare patches are frequent enough, there are not
merely rock-plants present but both open and closed scrub and even forest. Every stage
of plant-colonization also can be seen, from occupation of rock by lichens, liverworts,
mosses and even *Hymenophyllaceae*, to actual forest. But on the island generally the
trees remain at the shrub stage, bloming and fruiting abundantly.

arboreum, Leucopogon fasciculatus, Suttonia australis, Geniostoma ligustri-folium, Hebe salicifolia (very floriferous jordanon), *Coprosma robusta,* and *C. lu-cida.* It is hardly going too far to assert that the vegetation of Rangitoto Island is the most remarkable in the Region and certainly the brief sketch given here accords it scant justice. On White Island[1]) is another association but with the shrubs restricted to only 2 species while the total number of species of vascular plants on the island is only twelve. The *Metrosideros* makes a high scrub at the western end of the island varying from 3 or 4 m. to 6 or 8 m. in height where most luxuriant. The shrubs or low trees grow closely, their dark-green foliage being confined to the roof. Dead branches and twigs attached to the living plants are a conspicuous feature and it is evident that the association has no easy task to hold its own.

Other forest-scrubs. These are mostly wind-scrubs. Thus, at Titahi Bay (north shore of Cook Strait, RC.) there is an admirable example of such a scrub, (H. H. ALLAN, 1926: 72—76) which occurs on a somewhat concave slope slightly sheltered from the wind, only the taller shrubs receiving its full force. The lianoid *Metrosideros perforata* is dominant but growing as a ball-like shrub $\pm$ 1 m. high. and with it are associated a number of stunted trees and shrubs[2]). Various lianes bind the plants together and on the floor are the ferns *Asplenium adiantoides, A. flabelli-folium, A. lucidum* and *Polypodium diversifolium.*

On the isthmus connecting Nugget Point (SO.) with the mainland the wind-scrub is about 1 m. high and made up of species common in the adjacent forest[3]) and so dense that one can walk upon it, nor can the plants be disentangled without using force.

Near the Whangamoa Inlet (SN.) where the greywacke cliffs have been greatly weathered and debris suitable for occupation by shrubs has been provided there is an unusual community which consists of: great tussocks of *Phormium Colensoi* and low rather willow-like bushes of *Dodonaea viscosa.*

---

1) White Island is a small cone in the solfatara stage, 2.4 km. diam., 328 m. high, situated in the Bay of Plenty and distant from the mainland 48 km. From its crater pass off great clouds of steam highly charged with hydrochloric acid. It is the presence of this gas (W. R. B. OLIVER, 1915) that governs the distribution of plant-life on the island, since where the fumes strike there is complete absence of vegetation. In foggy weather, OLIVER graphically states, "it may be said to rain dilute. Hydrochloric acid".

2) *Podocarpus dacrydioides, P. spicatus, Macropiper excelsum, Pennantia corymbosa* (juvenile form), *Melicytus ramiflorus, Coprosma propinqua, C. rhamnoides* and, where the scrub merges into coastal forest, *Myrtus bullata, M. obcordata,* $\times$ *M. bullobcordata* and *Shawia paniculata.*

3) *Podocarpus totara, Wintera colorata, Carpodetus, Fuchsia excorticata, Coprosma propinqua, C. foetidissima* and the lianes, *Muehlenbeckia australis, Raoulia australis, Fuchsia perscandens* and *Parsonsia heterophylla* which bind the shrubs tightly together.

7. Coastal Forest.

a. General.

Coastal forest is distinguished *ecologically* by its low stature and close, more or less wind-swept roof, and *floristically* by the presence, but not invariably, of various trees or other plants belonging to the coastal element. In what follows, not only coastal forest proper is dealt with but certain associations, here called "semi-coastal", linking up with the lowland forest of the particular locality.

It would serve no purpose to supply statistics regarding coastal forest since, as will be seen, no true line can be drawn between such and lowland forest; in fact, in many places, coastal forest differs mainly from that of the adjacent lowlands in the greater abundance of wind-tolerating members which may form a seaward girdle.

As seen in chapter III of this section the coastal trees number 24, but only the following are at all common in any association and some of these are limited to a special part of the coast: — *Macropiper excelsum, Pittosporum crassifolium, P. umbellatum, Dodonaea viscosa, Metrosideros tomentosa, Pseudopanax Lessonii, Myoporum laetum* and *Coprosma retusa*. In addition, certain inland trees[1]) play an important part.

Coastal forest occurs all round the shore of the two main islands and Stewart Island where wind and spray do not forbid. Also, in especially sheltered positions, semi-coastal forest comes close to the beach and in the Western district, even where fairly exposed to the prevailing wind, true lowland podocarp-broad-leaved tree forest flourishes quite near the flat shore, protected merely by a narrow girdle of more or less spray-tolerating trees. On wind-swept islands all the forest may be of a coastal type.

Associations where true coastal trees are most in evidence belong to the Auckland districts and, in proceeding southwards, such trees gradually decrease in number, until from Banks Peninsula to Foveaux Strait only *Myoporum laetum* remains, while in the Western, Fiord and Stewart districts the true coastal tree element is absent[2]).

---

1) *Hedycarya arborea, Carpodetus serratus, Pittosporum tenuifolium, Weinmannia racemosa, Edwardsia microphylla, Melicope simplex, M. ternata,* ✕ *M. tersimplex, Dysoxylum spectabile, Coriaria arborea, Pennantia corymbosa, Alectryon excelsum, Melicytus ramiflorus, Leptospermum scoparium, Metrosideros lucida, Myrtus bullata, Fuchsia excorticata, Nothopanax arboreum, Pseudopanax crassifolium* var. *unifoliolatum, Griselinia littoralis, Dracophyllum longifolium, Suttonia salicina, S. australis, Geniostoma ligustrifolium, Vitex lucens, Coprosma robusta, C. lucida, Olearia rani, Brachyglottis repanda.*

2) Hardly any of the coastal trees can tolerate more than a few degrees, of frost, and when it is remembered that nearly all the genera are palaeotropic, while 6 species extend to countries warmer than New Zealand, it seems highly probable that the coastal treeflorula is but a remnant of one much larger and that the species frequent the shore-line rather on account of the mild maritime climate than through possessing special "adapt-

b. Groups of associations.

1. The *Metrosideros tomentosa* (Pohutukawa) group.

Pohutukawa forest is distinguished by the dominance, subdominance or presence in fair quantity of *M. tomentosa*. *Corynocarpus laevigata* may be in abundance or even dominate.

Forest of this class extends throughout the Auckland districts to the north-west coast of the East Cape district but, though *M. tomentosa* reaches lat. 39° on the west coast of North Island, how far along that coast south-wards from Manakau Harbour the coastal communities under consideration extend, or extended, I do not know.

It is not feasible at this period in the history of New Zealand to give details regarding the original composition of the associations of this group so, only a general account can be attempted, limiting descriptions to what I have personally seen of such forest from the far north of North Island to about lat. 37°.

The forest is lower the tree-trunks more slender and the undergrowth more open than in rain-forest generally. Where wind-swept, the roof is close, but, as the forest depends on shelter, the wind-effect is not strongly marked. *Metrosideros tomentosa* is generally dominant, but *Corynocarpus* is often a most important member and will give a distinct facies. *Leptospermum scoparium* may dominate a subassocation as in gully-forest surrounded by shrubland. The commonest of the remaining coastal trees are *Macropiper, Pittosporum crassifolium, Dodonaea viscosa* and *Pseudopanax Lessonii*. Common non-coastal trees are *Knightia, Beilschmiedia taraire* (in some parts), *Dysoxylum, Hoheria populnea* (in the far north), *Melicytus ramiflorus, Suttonia australis* and *Vitex lucens*. Tree-ferns *(Cyathea dealbata, C. medullaris)* are abundant. *Rhipogonum scandens, Muehlenbeckia australis* and *Freycinetia* are common lianes. *Astelia Banksii*, as a huge epiphyte, is frequent in the forest, and prostrate *Freycinetia, Paratrophis microphylla, Melicope ternata* (also as a tree), *Leucopogon fasciculatus, Geniostoma, Rhabdothamnus,* various species of *Coprosma* and *Brachyglottis repanda* may form the undergrowth. Many of the smaller ferns will be present, especially, — *Adiantum hispidulum, Pteris comans, P. macilenta, P. tremula, Blechnum filiforme* (liane), *Doodia media, Asplenium bulbiferum, A. lucidum, Polystichum Richardii, Dryopteris pennigera, D. velutina, Polypodium diversifolium, Arthropteris tenella* (on rocks), *Cyclophorus serpens*. Near streams, there may be in the far north the fine, large- but thin-leaved herb *Colensoa physaloides*.

---

ations". On the other hand, the almost frostless climate of the west coast of South Island is not unsuited for some of the coastal trees unless the colder summers and excessive rainfall are antagonistic. Perhaps the absence in the Western district of those species exteuding to the North-western may be ascribed to the repopulating by plants of the Westland coastal plain after the glacial period.

In certain parts of the Auckland districts groves or girdles of *Metrosideros tomentosa* (Fig. 19), at times adorn the shore just above high-water mark. Such trees have only short primary trunks from which spring numerous ascending branches, which, copiously branching, form dense heads covered with wind-resisting, thick, green leaves, white beneath with dense tomentum. Frequently, the association may be pure, but at other times certain of the coastal trees may be present.

2. The *Corynocarpus* (Karaka) group.

General. This group is distinguished by the dominance or fair abundance of *Corynocarpus laevigata*, associated with such coastal trees and shrubs and spray-tolerating non-coastal species as belong to the locality where the community is found.

Corynocarpus[1]) forest extends from the North Auckland district to the north of South Island and thence along the east coast to about lat. 43⁰. It is abundant at various places in the Egmont-Wanganui, Ruahine-Cook and Northeastern districts but, in proceeding south, the number of species greatly decreases.

Various associations. *Near Tongaporutu* (north of west coast, EW.) H. H. ALLAN reports that the following northern trees occur in the association: — *Entelea arborescens, Sideroxylon novo-zelandicum, Vitex lucens, Olearia furfuracea* var. *rubicunda* and *O. albida*. The main floor plant is *Asplenium lucidum*, and *Marattia fraxinea* occurs in gullies. As for the bulk of the species it is much same as in the next association.

On the shores of Cook Strait and on Kapiti Island typical *Corynocarpus* forest is still to be seen. Such an association has a somewhat billowy, close roof, the different greens of which denote certain species as, — dark-green, *Corynocarpus;* pale yellowish-green, *Melicytus ramiflorus;* bright pale-green, *Myoporum;* darkish-grey, *Leptospermum scoparium*. An outer girdle of wind-tolerating, spray-tolerating species[2]) faces the shore and a little further from the sea certain less-tolerant species[3]) are present. The

---

1) *C. laevigata* itself extends on the east coast of South Island to Banks Peninsula (nearly lat. 44⁰) but, on the west coast, where its association might well be expected, not quite to lat. 42⁰; for instance TOWNSON (1906: 407) reports only "isolated patches" and a few trees on the banks of the Buller. Also, as the Maoris were accustomed to plant this tree near their villages, doubt must frequently arise as to its occurrence in certain localities not being due to this cause. Few trees spread more rapidly by means of seed, and so its presence in apparently primeval forest does not disprove its being absent in the primitive community.

2) Such are *Macropiper excelsum, Paratrophis opaca, Urtica ferox, Hedycarya arborea, Corynocarpus, Melicytus ramiflorus, Pseudopanax crassifolium* var. *unifoliolatum, Myoporum, Coprosma retusa,* and of these the last two and the *Urtica* are nearest the shore.

3) The principal are *Rhopalostylis sapida, Knightia, Pittosporum tenuifolium, Melicope ternata, Dysoxylum spectabile, Pennantia, Fuchsia excorticata, Nothopanax arboreum, Coprosma grandifolia, Brachyglottis repanda* and *Olearia rani*.

principal lianes are *Blechnum filiforme* and *Freycinetia* (both important floor-plants also). Frequently, the interior of the forest is open. Ferns are common, e. g. *Adiantum affine, A. fulvum,* species of *Pteris, Dryopteris velutina, Polystichum Richardi* and *Asplenium Hookerianum.*

Along the coast of the North-eastern district, though more or less *Corynocarpus* may be present, the association is generally dominated by *Myoporum laetum*[1]), which at times is virtually pure. The remaining trees, if *Macropiper* and *Dodonaea* be excepted, are those of the adjacent lowland forest[2]).

3. The *Myoporum* (Ngaio) group.

This group is distinguished by the dominance of *Myoporum laetum* pure at times, but usually a number of non-coastal trees are present.

Myoporum forest occurs in all the Botanical Districts of South Island, excepting the Western, North Otago and Fiord. Its composition depends upon the degree of exposure to sea-spray, and to the dropping out of the true coastal species in proceeding southwards. For the North-eastern district the association has been already described along with *Corynocarpus* forest.

In the Sounds-Nelson district, facing Cook Strait where the north-west wind strikes obliquely, the trees are dwarfed but furnished with far-extending lateral branches and the members may be only *Myoporum* (dom.), *Macropiper excelsum, Dodonaea viscosa, Coriaria arborea, Coprosma robusta, Solanum aviculare* and *Shawia paniculata.*

In the South Otago district in addition to *Myoporum* — there the sole coastal tree — *Melicytus ramiflorus* is exceedingly plentiful and not infrequently dominates. As for the other species they are merely those of the ordinary lowland forest of the neighbourhood, and have been cited for the coastal forest of the North-eastern district but excluding *Melicope ternata, Dodonaea, Macropiper* and *Alectryon,* all of which, except the first-named being in coastal forest of Banks Peninsula.

4. *Nothofagus* coastal forest.

This is merely the ordinary *Nothofagus* forest of the particular locality greatly dwarfed by wind blowing almost constantly from the same quarter. Where the wind strikes with greater intensity still forest dwarfed to scrub, as already described, is the final stage in its reduction.

---

1) This will tolerate a maximum of sea-spray. Thus, facing Cook Strait in the Sounds-Nelson district, leafy branches come down to the ground and extend into the drift-wood of the shore. The leaves of such shoots are remarkably large and thick being frequently 7.5 m. long $\times$ 2.5 cm. or more broad and about 1.5 mm. thick.

2) For instance: *Carpodetus, Pittosporum tenuifolium, Edwardsia microphylla, Melicope ternata, Pennantia, Alectryon excelsum, Aristotelia serrata, Melicytus ramiflorus, Fuchsia excorticata, Nothopanax arboreum, Griselinia littoralis, Suttonia australis,* and *Urtica ferox* are common in some localities.

Forest of the class under consideration occupies the flat ground near the summit of the wall-like cliffs to the south of Westport. *Nothofagus cliffortioides* is dominant; the other important constituents are: *Phyllocladus alpinus, Nothofagus fusca, Weinmannia racemosa, Metrosideros lucida* and *Dracophyllum longifolium*. The principal lianes are *Freycinetia* and *Metrosideros perforata*. A somewhat similar association occurs on the slopes of the Paparoa Mountains, southwards from the Barrytown coastal plain, where there is every transition from lowland forest so scrub.

5. Coastal forest of certain small islands.

If an island is sufficiently small the vegetation as a whole is more or less affected by the wind, and the forest is stunted and of a coastal character. Here only associations belonging to none of the preceeding groups are dealt with.

The Three Kings Islands. According to CHEESEMAN — the only botanist who has visited this group — there are two distinct tree associations, the one dominated by *Leptospermum scoparium*[1]) and the other by *Meryta Sinclairii*[2]) — an araliad with enormous entire leaves in great rosettes.

The Poor Knights. According to W. R.B. OLIVER (1925 : 379, 380) there are two distinct forest associations on the southernmost of the two islands — the one more ore less primitive (really modified) and the other indigenous-induced. During my two hours or so on the island twenty years ago, I saw only the last-named association and took it for primitive — knowing nothing of former occupation of the island by the Maori. The so-called "primitive" association is, however, far the taller. "Conspicuous members" (loc. cit.) are *Metrosideros tomentosa, Olea apetala* (if it be that species) and *Vitex lucens*. Other species are: (rare on mainland) *Pisonia Brunoniana, Entelea arborescens, Sideroxylon novo-zelandicum;* (common on mainland), *Hedycarya arborea, Melicope ternata, Corynocarpus laevigata, Alectryon excelsum, Melicytus ramiflorus* and *Dysoxylum spectabile*. The undergrowth had been destroyed by pigs and OLIVER noted only *Asplenium flaccidum* and *Macropiper excelsum* var. *major*. In the induced forest (if such, and not merely modified) there is abundance of massive *Cordyline australis* and the tall shrubs, *Suttonia divaricata* (but most likely a different species) and *Hebe Bollonsii*, together

---

1) This occurs on the Great King, an island only about 2.8 km. long, 1.2 km. wide and 300 m. high. The *Leptospermum* is 3.6 m. to 7.5 m. high and growing in its company are *Davallia Tasmani, Paratrophis Smithii, Pittosporum Fairchildii, Alectryon grandis* — these 4 peculiar to the group — and *Cyathea medullaris, Melicope ternata, Hedycarya arborea* and *Coprosma macrocarpa.*

2) This association is restricted to an islet lying between the Great and West Kings which is only 1.2 km. long by 0.8 km. wide. The community is truly tropical in aspect. Not only is *Meryta* dominant but there is luxuriant *Cordyline australis* and as undergrowth the large-leaved *Macropiper excelsum* var. *major* and supra-luxuriant *Pteris comans. Meryta* there varies in height from 3 m. to 4.5 m. according to the degree of shelter.

with *Hymenanthera novae-zelandiae, Geniostoma, Myoporum* and some of
the species already cited for the modified association.

6. Semi-coastal forest.

General. At an indefinite distance from the coast, depending perhaps
in some measure upon salt-laden winds extending inland, and still more upon
the sea having receded comparatively recently, various coastal and semi-
coastal species are present in the ordinary rain-forest proper of the locality.
The most important of these species are *Macropiper excelsum, Dysoxylum
spectabile, Corynocarpus laevigata, Alectryon excelsum, Dodonaea viscosa* and
Myoporum laetum, while the following are common in their respective areas
of distribution: — *Cyathea medullaris, Blechnum filiforme, Polystichum
Richardi, Asplenium lucidum, Pteris macilenta, P. tremula, Freycinetia Banksii,*
(as a ground plant), *Rhopalostylis sapida, Paratrophis microphylla, Melicope
ternata, Tetrapathaea tetrandra* (liane), *Myrtus bullata, Solanum aviculare,
Geniostoma ligustrifolium, Coprosma grandifolia, Olearia rani, Brachyglottis
repanda*. With these will occur the various common podocarps, *Metrosideros
robusta, Elaeocarpus dentatus* and other ordinary lowland forest trees and shrubs.

Such forest is ecologically similar to lowland forest, but it will only
tolerate slight frosts, consequently it occurs in its full development only
in North Island and the northern part of South Island. However, lacking
a good many of its members, it extends on the east coast to Banks Peninsula and on the west coast to Greymouth.

Further to the south on the east coast there is no forest until the
South Otago district is gained and there the semi-coastal tree-association
is marked by the presence of *Myoporum laetum* and particularly the abundance of *Melicytus ramiflorus*, while *Griselinia littoralis, Hoheria angustifolia*
and *Edwardsia microphylla* are characteristic. Semi-coastal forest of this
character is almost a thing of the past. On the west coast, there is from
about lat. 43° southwards almost continuous forest but there is little of
semi-coastal type, ordinary broad-leaved — podocarp communities — with
more or less *Nothofagus* in the Fiord district — coming almost to the shore,
but frequently sheltered in the Western district by a girdle of trees, as already
described, or a belt of tall *Phormium tenax*.

Special. The *Dysoxylum spectabile* (*kohekohe*) association is distinguished by either the dominance or the presence in large numbers of
Dysoxylum spectabile[1]) and usually there is a good deal *Macropiper excelsum*[2])

---

1) As occurring near the coast this is a small tree about 6 m. high or considerably
less with cauliflorous trunk covered with a rather pale smooth bark, spreading tortuous
stout branches and a close head of large pinnate leaves more than 30 cm. long with broad
shining leaflets of vivid green and, in winter, the naked trunk and thick branches bear
drooping panicles ± 30 cm. long of waxy-white flowers.

2) The other species present are the usual trees, shrubs and ferns already cited.
On Stephen Island, there is a good deal of *Coprosma retusa* as a small tree, and the palm,
Rhopalostylis sapida, ascends into the forest roof.

The association was formerly common in many places — either adjacent to the coast or up to 2 km. or more inland — in the south of the Egmont-Wanganui and Ruahine-Cook districts and the north of the Sounds-Nelson and North-west districts. As the community occupied remarkably fertile soil suitable for artificial pasture only remnants, usually much damaged persist.

On the islands in Cook Strait, especially Kapiti and Stephan Island, and on the wind-swept peninsula called Pepin Island (SN.), the association is of a true coastal character with a close roof (Fig. 22) while within the *Dysoxylum* possesses far-extending, twisted, gnarled branches.

Stunted podocarp forest probably occurs, or occurred, in many places. In the northern part of the Western district there is a narrow strip of strongly wind-shorn forest which, between Greymouth and Hokitika, consists of stunted *Podocarpus dacrydioides* and *Dacrydium cupressinum*, 6 to 9 m. high (dominant) and the branches of the first-named are much more widely-extended than usual, also *Carpodetus, Weinmannia racemosa, Aristotelia serrata, Fuchsia excorticata, Suttonia salicina, Coprosma lucida, C. foetidissima, C. areolata. Griselinia lucida* and *Astelia Cunninghamii* are common epiphytes. *Metrosideros perforata* and *Rubus australis* are the principal climbers.

Metrosideros lucida forest, in which the *Metrosideros* is almost the sole tall tree, forms a belt in Stewart Island coming between the coastal scrub and the *Weinmannia*-rimu forest. As undergrowth there are the usual ferns and shrubs (few in number) of the district. Similar forest occurs in the South Otago district on the shore of Foveaux Strait. It is of extreme interest that this association is almost identical with the Lord Auckland Islands' forest — a matter to which I called attention in 1904, and which impressed Sir JOSEPH HOOKER so greatly that he wrote me (dated Nov. 15th, 1904) as follows: "Of all the facts which you have marshalled that of the Rata forest in the Auckland Islands is the most pregnant and of the Antipodean flora the least expected".

On the seaward slopes of the Paparoa Mountains (south of NW.) the forest as a whole is much stunted but — where the composition is that of rain-forest proper — the dark, rounded heads of *M. lucida* stand out at fairly regular intervals above this stunted assemblage of trees, shrubs, tree-ferns &c.

Section II.

The Vegetation of the Lowlands and Lower Hills.

Chapter I.

Introductory Remarks.

General. It is not feasible to fix any arbitrary line of demarcation between the lowland-lower hills' vegetation and flora and those of the high mountains, since not only is the dividing-line governed by latitude but also by aspect, rainfall and other ecological factors. Thus in the Auckland botanical districts, with the exception of one or two summits of the Thames Mountains, all the hills, even when they considerably exceed 600 m. altitude, bear a vegetation of a lowland stamp and true high-mountain plants are wanting. All the same, certain species ascend far higher than others and there are more or less distinct altitudinal belts. For the rest of North Island, high-mountain species do not generally appear until an altitude of at least 800 m. is reached. In South Island, though from 400 m. to 600 m. may be taken as a fair estimate for the dividing-line (hardly such, but rather a belt where one flora merges into the other), no universal delimiting-line can be fixed. Where, particularly in the west and south, the climate is wet and cloudy days abundant, should the edaphic conditions be favourable, quite a number of species, otherwise typically alpine or subalpine, descend to sea-level even so far north as lat. 42°. In Stewart Island, not species merely, but actual subalpine associations, occur at virtually sea-level. Furthermore, the foot-hills, bases of the high mountains and the valleys of South Island, at an altitude of considerably less than 600 m., frequently possess a vegetation subalpine rather than lowland.

The area under consideration comprises North Island as a whole, excepting the actual coast-line, one or two mountain summits in the Thames subdistrict and those portions of the high mountains above an altitude of 800 m. at the very least. In South Island, the lowland-lower hills' belt is much more restricted, its greater part follows the coast-line, though it passes inland by way of the lower slopes of the mountains and the numerous valleys, so that, in places, there is lowland vegetation almost to the foot of the actual Divide. In Stewart Island, the line of demarcation may be put down at about 450 m., though it is considerably less in the south.

Notwithstanding the extensive plains of both the main islands, much of the land-surface is extremely broken and hilly, while in many places deep gorges are a characteristic feature, and in others wide shingly river-beds, so that there is great diversity of habitats throughout. The summit vegetation of comparatively low hills, if isolated or specially exposed, bears a mountain stamp both in species and physiognomy. A great deal of the lowland vegetation has been not merely modified in the course of settlement but actually wiped from the face of the land. But, except in a few instances, there are sufficient indications to show clearly enough of what the primitive plant-covering consisted, so that it is still possible to give a fairly accurate sketch of primeval lowland New Zealand. Areas, too, remain absolutely in their virgin condition, but such — forest more particularly — are being so rapidly damaged by deer that areas described as still primeval in the first edition of this book are now more or less modified, while associations I saw in their primitive state more than 45 years ago are now ecologically unrecognizable!

Owing to the comparatively mild climate, and the general absence of such intense ecological factors as occur on the coast and the high mountains, the conditions for plant-life are more uniform than in the two latter belts. But the diversity of stations in different parts of the area leads to the presence of many communities and life-forms. Above all, gradual change in latitude and the relation of the vegetation to excessive rain in many parts, or to long periods of drought in others, are fundamental factors in regard to the distribution of species, and the composition, structure, and peculiarities of the communities. It is these extreme conditions which bring down to the lowlands many true high-mountain species, but this is discussed at some length in Section III, Chapter I.

Floristic details. The lowland-lower hills' flora consists of 1059 species and is made up of the following elements to each of which is appended its number of species: — 1. Species confined to the lowland-lower hills' belt, hereafter termed lowland 559; 2. species found both in the lowlands and high mountains 339; 3. species truly high-mountain which occasionally occur in the lowlands 115; 4. coastal species which either extend inland for some distance or occur far from the sea, but only under special circumstances 46. Leaving out of consideration the coastal and purely high-mountain elements the number of species in the lowland flora proper is 898. In addition there are at least 72 groups — many of them large swarms — of wild hybrids.

Considering first the 559 purely lowland species they consist of 103 pteridophytes (*Filices* 91), 10 gymnosperms, 148 monocotyledons and 298 dicotyledons which belong to 88 families and 231 genera, and 380 species (68%) are endemic.

The following 24 families and 103 genera are confined, or almost so, to the lowlands and lower hills, those marked* being endemic: — (families) *Salviniaceae, Psilotaceae, Araucariaceae, Pandanaceae, Sparganiaceae, Palmae, Lemnaceae, Burmanniaceae, Amaryllidaceae, Moraceae* (also coastal), *Piperaceae* (also coastal), *Lauraceae, Monimiaceae, Rutaceae, Meliaceae, Icacinaceae, Sapindaceae* (also coastal), *Elatinaceae, Passifloraceae, Tetrachondraceae, Verbenaceae* (also coastal), *Solanaceae, Gesneriaceae* and *Capri-*

foliaceae (ascends occassionally to the lower subalpine forest); (genera) *Loxsoma*, Cyathea, Hemitelia, Leptolepia*, Lindsaya, Adiantum, Cheilantes, Pellaea, Paesia, Pteris, Doodia, Athyrium, Diplazium, Nephrolepis, Arthropteris, Cyclophorus, Lygodium, Todaea, Marattia* (Filices), *Azolla* (Salviniac.) *Phylloglossum* (Lycopod.), *Tmesipteris, Psilotum* (Psilotac.), *Agathis* (Araucariac.), *Freycinetia* (Pandanac.), *Sparganium* (Sparganiac.), *Paspalum, Isachne, Oplismenus, Simplicia*, Amphibromus, Arundo* (Gramin.), *Mariscus, Fimbristylis, Lepidosperma* (Cyperac.) *Rhopalostylis (Palmae), Lemna* (Lemnac.), *Sporadanthus** (Restionac.), *Rhipogonum, Dianella* (Liliac.), *Bagnisia* (Burmanniac.), *Dendrobium, Bulbophyllum, Earina, Sarcochilus, Spiranthes, Orthoceras, Caleana, Acianthus, Cyrtostylis, Pentalochilus*, Calochilus, Townsonia* (Orchid.), *Peperomia, Macropiper* — also coastal (Piperac.), *Ascarina* (Chloranthac.), *Paratrophis* — also coastal (Morac.), *Elatostema, Parietaria, Australina* (Urticac.), *Persoonia, Knightia* (Proteac.), *Mida* (Santalac.), *Loranthus, Phrygilanthus* (Loranthac.), *Dactylanthus* (Balanophorac.), *Polygonum* (Polygonac.), *Alternanthera* (Amarantac.), *Hedycarya, Laurelia* (Monimiac.), *Beilschmiedia, Litsaea, Cassytha* (Laurac,), *Ackama** (Cunoniac.), *Chordospartium*, Notospartium** (Legum.), *Pelargonium* (Geran.), *Phebalium, Melicope* (Rutac.), *Dysoxylum* (Meliac.), *Pennantia* (Icacinac.), *Hoheria** (Malvac.), *Elatine* (Elatinac.), *Melicytus* (Violac.), *Tetrapathaea** (Passiflor.), *Eugenia* (Myrtac.), *Schefflera* (Araliac.), *Centella, Daucus* (Umbell.), *Olea* (Oleac.), *Geniostoma* (Loganiac.), *Sebaea* (Gentian.), *Calystegia* — also coastal (Convolv.), *Tetrachondra* (Tetrachondrac.), *Vitex, Teucridium** (Verbenac.), *Scutellaria* (Labiatae), *Jovellana* (Scroph.), *Rhabdothamnus** (Gesneriac.), *Alseuosmia** (Caprifoliac.), *Colensoa** (Campan.), *Siegesbeckia, Centipeda, Brachyglottis*, Picris* (Compos.).

The principal families and genera of the lowland flora, together with the number of species to each are as follows: — (families) *Filices* 91, *Orchidaceae* 51, *Cyperaceae* 48, *Compositae* 43, *Scrophulariaceae* 22, *Leguminosae* 20, *Rubiaceae* 17, *Gramineae* 15, *Myrtaceae* 12, *Ranunculaceae* 11, *Umbelliferae, Liliaceae,* and *Onagraceae* each 10, *Pittosporaceae* and *Araliaceae* each 9; (genera) *Coprosma* 13, *Olearia* 12, *Thelymitra* and *Carmichaelia* each 11, *Hymenophyllum* and *Carex* each 10, *Scirpus, Pterostylis, Pittosporum* and *Hebe* each 9, *Blechnum, Cladium* and *Epilobium* each 8, *Trichomanes, Dryopteris, Lycopodium, Juncus* and *Metrosideros* each 7.

Coming next to the distribution of the 559 lowland species in New Zealand proper, 190 occur from about the north of North Island to about the south of South Island and of these 106 extend to Stewart Island; *499 (155 confined thereto)* occur in North Island and of these no less than 105 are found only to the north of lat. 38°, or extend only a short distance south of that line; only 398 (54 confined thereto) occur in South Island, and of these the large number of 120 lie to the north of lat. 42°, or do not extend very far to the south of that line, leaving only *276 species for the South Island lowland flora proper*; finally 120 (5 confined thereto) occur in Stewart Island. Change of altitude brings about considerable change in the flora. Thus 133 species ascend only to between 600 and 900 m. (mostly in North Island, much lower in most of South Island), but of these 110 do not ascend to more than 100 m. and some of these are confined almost to sea-level, while 75 are so rare as to be of no moment with regard to the general vegetation.

The lowland-high mountain element of the flora consists of 339 species (60 families, 158 genera). The following families and genera, absent in the purely lowland element, are represented: — (families) *Marsiliaceae, Balanophoraceae, Portulacaceae, Saxifragaceae, Stackhousiaceae, Plantaginaceae* (coastal also), *Stylidiaceae;* (genera), *Alsophila, Pteridium, Leptopteris* (Filic.) *Pilularia* (Marsiliac.), *Hierochloe, Trisetum* (also coastal), *Deyeuxia, Triodia, Koeleria, Danthonia, Poa, Festuca, Agropyrum* (Gramin.), *Hypolaena* (Restionac.), *Phormium, Enargea, Chrysobactron, Arthropodium* (also coastal), *Microtis, Townsonia* (Orchid.), *Tupeia** (Loranth.), *Montia* (Portulac.), *Scleranthus* (Carophyl.), *Myosurus* (Ranun.), *Sisymbrium* (Crucif.), *Quintinia, Ixerba*, Carpodetus** (Saxifrag.), *Acaena* — also coastal (Rosac.), *Stackhousia* (Stackhousiac.), *Aristotelia* (Elaeocarp.), *Viola (Violac.), Oreomyrrhis, Lilaeopsis, Aciphylla* (Umbel.), *Siphonidium** (Scroph.), *Plantago* — also coastal (Plantag.), *Isotoma* (Campan.), *Oreostylidium*, Forstera* (Stylid.), *Vittadinia, Pachystegia*, Cassinia* — also coastal, *Microseris, Taraxacum* (Compos).

With regard to the distribution of the 339 lowland-high mountain species 138 (40%) occur throughout North and South Islands, 286 in North Island (9 being restricted thereto) but of these 132 are wanting to the north of the Thames subdistrict and by far the greater part appear for the first time (i. e. in proceeding from north to south) in the East Cape, Volcanic Plateau and Egmont-Wanganui districts. South Island possesses 330 species (42 being confined thereto) and Stewart Island 181.

The species of the element under consideration may be roughly divided into the following 3 classes to each of which is appended its number of species: — (1) species most frequently *lowland* 123; (2) those most frequently high-mountain 104; (3) those belonging equally to either class 112. These figures might give the impression that the lowland-high mountain species belong equally to these two floras, but this would be erroneous. To begin with, about 50 per cent. of the species are absent in the Auckland districts, if the Thames subdistrict be excluded; further, 258 species do not extend upwards beyond the lower-subalpine belt, and only 69 and 12 species respectively enter the upper subalpine and alpine belts. The greater part, indeed, of the 339 species attain their extreme altitudinal limit either in that portion of the tussock-grassland area where lowland and high-mountain species mingle or to the montane and subalpine forest, the members of which are exposed to conditions materially different from those of an open subalpine or alpine hillside. Further, many plants of lowland lakes, swamps and bogs find conditions not very different from what they are accustomed in the allied habitats up to about 900 m. altitude. So, too, the conditions offered by rocks and riverbeds from sea-level to 600, or even 900 m. altitude in some localities, are pretty much the same.

Some of the species common to the lowlands and the high mountains, including some of the category which descend to the lowlands only under special circumstances, occur so abundantly both in certain lowland, or coastal, and high mountain formations as to be of great physiognomic importance, especially *Polystichum vestitum*, *Gleichenia circinata*, *Podocarpus nivalis*, *Poa Colensoi*, *P. intermedia*, *P. caespitosa*, *Festuca novae-zelandiae*, *Danthonia Raoulii* var. *rubra*, *Hypolaena lateriflora*, *Phormium Colensoi*, *Nothofagus Menziesii*, *Weinmannia racemosa*, *Discaria toumatou*, *Leptospermum scoparium*, *Metrosideros lucida*, *Dracophyllum longifolium*, *Suttonia divaricata*, *Hebe salicifolia*, *H. Hulkeana*, *Coprosma foetidissima*, *Pachystegia insignis*, *Raoulia australis*, *R. lutescens*, *R. tenuicaulis* and *Cassinia Vauvilliersii*. Although the same species may be present in the subalpine, alpine and lowland belts yet under the local environments the life-form may be altogether different as in the case of *Leptospermum scoparium*, as a low tree in lowland forest and a prostrate, rooting mat on subalpine moor, or *Weinmannia* [racemosa a tall, massive tree in the lowlands and a dense shrub in the subalpine scrub.

Chapter II.
The leading Physiognomic Plants and their Life=forms.

a. Forest Plants.

Tall or medium-sized trees. *Agathis australis* Salisb. (*Araucariac.*), kauri, is 24 to 30 m. high, or more; its trunk massive, straight, columnar, 1 to 4 m. diam. (6.5 m. has been recorded), unbranched for 15 to 21 m., covered with thick, shining, grey bark; its head spreading with the primary branches some 60 cm. diam., and the leaves linear-oblong to ovate-oblong, about 3.7 cm × 1.2 cm., dark-green and thick.

Alectryon excelsum Gaertn. (*Sapindac.*), titoki, is from 12 to 18 m. high with a straight trunk some 60 cm. diam. covered with black bark, leaves unequally pinnate, 12 to 30 cm. long, with 4 to 6 pairs of pinnae which are rather thin, glossy bright-green above and beneath, ovate-lanceolate, acuminate, entire, or obscurely toothed, each 5 to 10 cm. long, rusty pubescent beneath, and showy fruits in much-branched panicles, the open capsule bearing a shining-black, globose seed embedded in a fiery scarlet, granulated, fleshy cup.

Beilschmiedia tairaire (A. Cunn.) Benth. et Hook. f. (*Laurac.*), tairaire, is 15 to 24 m. high with a straight, usually slender trunk covered with reddish or reddish-brown bark; a small, dense crown, and leaves obovate to broadly-oblong, 12.5 cm. long by 7.5 cm. broad, entire, dark-green, slightly glossy, coriaceous, bluishwhite beneath. It is physiognomic mainly in the North Auckland district.

Beilschmiedia tawa (A. Cunn.) Benth. et Hook. f, tawa, is 12 to 24 m. high with a trunk 30 cm. to 1.2 m. diam. covered with smooth, blackish bark, slender branches, willow-like, thin leaves 7 to 8 cm. long, yellowish-green above but glaucous beneath.

Dacrydium cupressinum Sol. *(Podocarpac.)*, rimu, red-pine, is from about 21 to 30 m. high with a straight, massive trunk up to 1.5 m. diam., covered with flaking, thick, dark-brown bark, a rather small, yellowish head of pendulous branchlets and linear, acute, adpressed leaves some 2 mm. long.

Elaeocarpus dentatus (J. R. et G. Forst.) Vahl *(Elaeocarpac.)*, hinau, is from 12 to 18 m. high with a straight trunk 30 to 90 cm. diam. covered with greyish bark, a medium-sized head with the leaves usually confined to the tips of the branchlets, slightly glossy dark-green leaves linear-oblong to obovate-oblong, petiolate, 5 to 10 cm. long, obscurely-toothed, coriaceous, whitish beneath, and fair-sized, drooping, white flowers.

Elaeocarpus Hookerianus Raoul, pokaka, is 6 to 12 m. high with the trunk 30 to 90 cm. diam. covered with pale bark, the leaves linear-oblong, sinuate-crenate, obtuse, 5 to 7.5 cm. long, coriaceous and flowers smaller than in *E. dentatus.*

Knightia excelsa R. Br. *(Proteac.)*, rewarewa, honeysuckle, is 21 to 30 m. high with a straight trunk covered with very dark, smooth bark, and in its fastigiate habit resembling the Lombardy poplar; the leaves are short-petioled, linear-oblong to obovate-oblong, 12 to 20 cm. long, dark-green above, bright-green beneath, distantly, coarsely and bluntly toothed with large ($\pm$ 8.5 mm.) teeth and, in texture almost woody.

Laurelia novae-zelandiae A. Cunn. *(Monimiac.)*, puketea, is sometimes 36 m. hight with a trunk up to 1.8 m. diam. covered with whitish bark, and with wide plank-buttresses at its base, a fair-sized crown, glossy green foliage and more or less oblong leaves 3 to 7.5 cm. long, coriaceous and coarsely toothed.

Metrosideros lucida (Forst. f.) A. Rich. *(Myrtac.)*, southern-rata, is at times 18 m. high, but usually far less, with an irregular trunk up to 1.8 m. diam. covered with papery, pale bark which hangs in long strips, a head of far-extending branches and lanceolate, acuminate leaves 1.8 to 7 cm. long, very coriaceous, glossy bright-green and dotted beneath with oil-glands; there is profusion of bright-crimson flowers.

Metrosideros robusta A. Cunn., rata, northern-rata, is occasionally 30 m. high with a most irregular trunk formed of roots more or less closely united (see epiphytes in the next chapter) covered with bark much as for *M. lucida,* the branches far-spread, the head dense and vivid green, and the leaves of a broad-oblong type about 2.7 cm. long, flat, stiff, rather thick, dark-green but paler beneath; there is profusion of dark-scarlet flowers.

Nothofagus fusca (Hook. f.) Oerst. *(Fagac.)*, red southern-beech, is 18 to 30 m. high with a massive trunk up to 2.5 m. diam. covered with dark, thick, deeply-furrowed bark, a fairly large, bright-green crown and rather thin, bright-green, deeply and sharply toothed leaves, 2.5 cm. long of an ovate type. This species crosses freely with *N. cliffortioides* making the great hybrid swarm, $\times$ *N. cliffusca*.

Nothofagus Solandri (Hook f.) Oerst., black southern-beech, is 12 to 24 m. high with the trunk up to 1.2 m. diam. covered with dark furrowed bark, a rather dense head and linear-oblong, entire, obtuse leaves some 12 mm. long, cuneate at the base, dark-green, coriaceous, whitish-tomentose beneath. It crosses freely with *N. truncata* — a species very close to *N. fusca* — making the hybrid swarm $\times$ N. soltruncata.

Podocarpus dacrydioides A. Rich. *(Podocarpac.)*, kahikatea, white-pine, is up to 36 m. high with a long, straight trunk 0.60 to 1.2 m. diam. covered with greyish-brown bark, a head small out of all proportion to the size of the tree, leaves subulate-lanceolate, acuminate, 6 mm. long, imbricating and adpressed; in autumn the abundant red fruits are physiognomic.

Podocarpus ferrugineus Don, miro; *P. spicatus* R. Br., matai, black-pine; and *P. totara* A. Cunn., totara, somewhat resemble gigantic yew trees; all have short, linear leaves but those of the first two are green and di-stichous, while those of *P. totara* are spirally arranged, yellowish-green and coriaceous. All are tall and massive but *P. totara* may exceed 30 m. in height and 1,8 m. diam. *P. Hallii* is similar to *P. totara* but smaller and with thin bark.

Weinmannia racemosa L. f. *(Cunoniac)*, towai, kamahi, is up to 27 m. high with a massive irregular trunk covered with pale bark, a compact, fairly broad crown, oblong-lanceolate to orbicular-ovate leaves 2.5 to 10 cm. long, coarsely obtusely-serrate and dull or yellowish green and abundant racemes of pinkish-white flowers.

Weinmannia sylvicola Sol. ex A. Cunn., tawhero, resembles *W. racemosa* but it is smaller and the leaves are 3-foliolate or pinnate.

Small trees and tall shrubs. *Ascarina lucida* Hook. f. *(Chloranthac.)*, physiognomic in the Western district, is a low, bushy-tree with almost black bark and extremely glossy green, oblong, serrate leaves 2.5 to 5 cm. long·

Brachyglottis repanda J. R. et G. Forst. (Compos.), rangiora, is up to 6 m. high with numerous, spreading white-tomentose branches, large, thin, wavy or lobed, broadly oblong leaves up to 30 cm. long, dull-green above, milk-white tomentose beneath with conspicuous, raised veins; flower-heads white, sweet-scented in great showy panicles in early spring.

Carpodetus serratus J. R. et G. Forst. *(Saxifrag.)*, putaputawheta, is at most 9 m. high with trunk 15 to 22 m. diam., dense crown of spreading branches, rather thin ovate-oblong, serrate leaves 2 to 4 cm. long, dark-green above the veins and yellowish-green between; flowers white and abundant.

Coprosma foetidissima J. R. et G. Forst. *(Rubiac.),* hupiro, is a graceful twiggy shrub with spreading, slender, pale branches arching towards the ground, rather distant sub-coriaceous, more or less oblong, pale-green leaves 2.5 to 5 cm. long and showy orange drupes. It crosses freely with *C. Colensoi, C. Banksii, C. Astoni* and their hybrids.

Coprosma grandifolia Hook. f., kanono, is a stout bushy shrub up to 4 m. high with rather thin slightly glossy or dull darkish-green, broadly-oblong leaves 10 to 22 cm. long and large reddish-orange drupes. *C. lucida* Forst. f., karamu, is similar but the leaves are obovate, smaller, coriaceous and glossy; also *C. robusta* Raoul is much like the last but with narrower leaves. Apparently all cross amongst *themselves* and *C. robusta* × *propinqua* makes the great swarm × *C. prorobusta.*

Coprosma rotundifolia A. Cunn. is occasionally a small tree 6 m. high with pale, slender spreading branches, distant broadly-oblong, thin, pubescent leaves about 2 cm. long and dull-red drupes in abundance.

Fuchsia excorticata (J. R. et G. Forst.) L. f. *(Onagrac.),* kotukutuku, is more or less deciduous and has a maximum height of 12 m., an irregular, often massive, sometimes semiprostrate trunk covered with papery bark which hangs in strips, thin, more or less lanceolate leaves 5 to 10 cm. long, green above but silvery beneath and black, succulent berries which are freely eaten by the indigenous pigeon.

Griselinia littoralis Raoul *(Cornac.),* papaumu, broadleaf, is round-headed, 15 m. high at most, with a gnarled, irregular, short trunk, rough furrowed bark, and thick, glossy, yellowish-green, ovate or obovate leaves 2.5 cm. to 10 cm. long.

Hedycarya arborea J. et G. Forst. *(Monimiac.),* porokaiwhiri, is at most 12 m. high with a moderately stout trunk, pale-grey furrowed bark, dark-green, shining, coriaceous, oblong leaves 5 to 10 cm. long, the midrib brown, yellowish-green flowers and bright-red drupes.

Hoheria angustifolia Raoul; *H. populnea* A. Cunn.; and *H. sexstylosa* Col. *(Malvac.),* houhere, lacebark, are graceful with slender, erect trunks, tough bark (can be pulled off in long strips) and abundant showy, white flowers. The leaves of *H. angustifolia* are thin, linear-oblong, spinulose-toothed, ± 5 cm. long; of *H. populnea,* ovate, ± 8 cm. long, coriaceous, toothed; and of *H. sexstylosa* lanceolate, up to 12 cm. long, acuminate, deeply-toothed and thicker than those of *H. angustifolia.* Between the last-named and *H. sexstylosa* is a hybrid swarm. The species bloom at different times.

Melicytus ramiflorus J. R. et G. Forst. *(Violac.),* mahoe, is 5 to 9 m.

high with rounded, dense crown, trunk 30 to 60 cm. diam., leaves oblong-lanceolate, 5 to 12 cm. long, bluntly serrate, green or yellowish-green and berries small, abundant, violet in colour and produced on the naked twigs.

Myrtus bullata Sol. ex A. Cunn. *(Myrtac.)*, ramarama, usually a bushy shrub but sometimes a small tree 6 m. high, has leaves strongly blistered above, reddish-brown, broadly-ovate or orbicular-ovate, 5 to 10 cm. long, concave beneath. It crosses freely with *M. obcordata* forming the great polymorphic hybrid swarm, $\times$ *M. bullobcordata.*

Nothopanax arboreum (Forst. f.) Seem. *(Araliac.)*, whauwhaupaku, ivy-tree, is at most 7 m. high with slender trunk, dense rounded head, long-petioled digitately 5- to 7- foliolate leaves with leaflets more or less oblong, 7.5 to 17 cm. long, dark-green, shining, coriaceous, serrate and black fruits in conspicuous compound umbels. It occasionally crosses with *Pseudopanax crassifolium* var. *unifoliolatum.*

Paratrophis microphylla (Raoul) Ckn. *(Morac.)*, turepo, milk-tree, is 4.5 to 12 m. high with trunk seldom more than 30 cm. diam., bark whitish, much lenticelled, and coriaceous dark-green oblong-ovate to elliptic leaves, $\pm$ 2.5 cm. long. The juvenile of the divaricating life-form is far more physiognomic than the adult.

Pennantia corymbosa J. R. et G. Forst. *(Icacinac.)*, kaikomako, is about 10.5 m. high with slender trunk, close crown, leaves oblong to obovate, 2.5 to 12 cm. long, dark-green, irregularly toothed or sinuate, obtuse and abundant white flowers. The juvenile of the divaricating-form is also physiognomic.

Pittosporum eugenioides A. Cunn. *(Pittosporac.)*, tarata, lemonwood, is 6 to 12 m. high with slender trunk, pale bark, rounded crown, glossy-green, rather thin, undulate, elliptic leaves, 5 to 10 cm. long, and showy compound umbels of fragrant yellow flowers.

Pittosporum tenuifolium Banks et Sol. ex Gaertn., kohuhu, is the name applied to various trees of similar form, making, with the numerous hybrids between them, a huge linneon. The slender trunk is 6 to 9 m. high with black bark, and bears a small dense crown; the leaves are oblong to elliptic-ovate, 2.5 to 5 cm. long, palish-green or mottled with brown, rather thin, glabrous, undulate and acute.

Plagianthus betulinus A. Cunn. *(Malvac.)*, lowland ribbonwood, is deciduous, with trunk occasionally 90 cm. diam., dark brown smooth bark, dense twiggy crown, leaves ovate to ovate-lanceolate 2.5 to 7 cm. long, bright-green, thin, slightly hairy, coarsely-toothed, acuminate and bears conspicuous masses of small, yellowish flowers.

Pseudopanax crassifolium (Sol. ex A. Cunn.) C. Koch var. *unifoliolatum* T. Kirk *(Araliac.)*, horoeka, lancewood, is 6 to 12 m. high with a straight, furrowed naked trunk up to 45 cm. diam., small rounded crown, leaves linear to linear-obovate 7.5 to 12 cm. long, dark-green, stiff, coriaceous, entire or ± toothed and usually obtuse.

Rhopalostylis sapida (Sol. ex Forst. f.) Wendl. et Drude, *(Palmae)*, nikau, nikau-palm, is a tuft-tree 1.8 to 7.5 m. high with a greenish, smooth, slender trunk 14 to 22 cm. diam. marked with pale rings of old leaf-scars, 2.5 cm. apart and radiating upwards and outwards near its apex, a crown of large, shining, dark-green pinnate leaves 1.2 to 2.4 m. long, the midrib green and very stout but its secondary branches yellow, the leaflets numerous, linear-ensiform 60 to 90 cm. long or more.

Schefflera digitata J. R. et G. Forst. *(Araliac.)*, pate, is 3 to 7.5 m. high with straight, slender, spreading branches, long-petioled (5 to 20 cm.), digitately 7- to 10-foliolate leaves with the leaflets oblong-lanceolate or obovate-lanceolate, 7.5 to 17 cm. long, thin, soft, glabrous, sharply serrate and acuminate, and large umbels of small, black, fleshy globose fruits.

Suttonia australis A. Rich. *(Myrsinac.)*, mapou, is about 6 m. high, of dense twiggy habit with black bark (red or purple on twigs), leaves oblong to obovate-oblong 2.5 to 5 cm. long, pale dull-green, subcoriaceous, glabrous, undulate, obtuse, and small black fruits.

Suttonia salicina Hook. f., toro, is 5 to 9 m. high with slender trunk, black bark, rounded crown, leaves linear, 7.5 to 21 cm. long, glossy, yellowish-green, smooth, glabrous, subcoriaceous, and small yellowish-white flowers on the naked twigs.

Wintera axillaris Forst. f. *(Winterac.)*, horopito, is 3.6 to 7.5 m. high with slender trunk rarely 30 cm. diam., black bark, twiggy crown, leaves elliptic-oblong, 5 to 12 cm. long, glossy, dark-green, glaucous beneath, subcoriaceous, glabrous. *W. colorata* (Raoul) Ckn. is hardly a tree, its leaves are shorter, thicker and more glaucous beneath than those of *W. axillaris* and blotched purple above. It ascends to the subalpine belt. Apparently the 2 species cross freely.

Other conspicuous species. *Astelia (Liliac.)* includes five large tussock-like herbs with green, coriaceous, long, linear, grass-like leaves usually about 2.5 cm. wide, two of which *(A. Solandri, A. Cunninghamii)*, perched high on the branches of lofty trees, resemble gigantic nests of birds. *A. nervora* Banks et Sol. ex Hook. f. var. *sylvestris* Ckn. et Allan is abundant on the forest floor and the var. *grandis* (Hook. f. ex T. Kirk) Ckn. et Allan on much wetter ground, and *A. trinervia* T. Kirk, kauri-grass, more than 1 m. high, makes dense thickets in Auckland forests and those in the north-west of the North-western district (according to A. W. WASTNEY).

Divaricating-shrubs (Fig. 31) are those having more or less wiry or rigid stems which branch at a wide angle and are densely interlaced. Usually the leaves are very small. Many species of *Coprosma* (Rubiac.), *Melicope simplex* (Rutac.), *Melicytus micranthus* (Violac.), *Nothopanax anomalum* (Araliac.) and *Suttonia divaricata* (Myrsinac.) are familiar examples;

also the persistent juvenile forms of various trees — some already noted — are greatly in evidence.

Ferns (Filices), as a whole, are of particular physiognomic importance. The most conspicuous are the tree-ferns, especially *Cyathea dealbata* Sw., silver tree-fern; *C. medullaris* Sw., black tree-fern, sometimes 15 m. high and the fronds 6 m. long; *Hemitelia Smithii* Hook; *Dicksonia squarrosa* Sw., weki, slender and not particularly tall; and *Dicksonia fibrosa* Col. (Fig. 23). Of the smaller ferns *Blechnum discolor* (Forst. f.) Mett. is specially physiognomic, occurring as it does in far-extending dense colonies. The linear-lanceolate fronds 30 to 60 cm. long, glossy-green above but pale beneath are tufted at the apex of the short trunk and so make a deep goblet.

Gahnia (Cyperac.) contains 7 forest species which form huge green grass-like tussocks culminating in those 3.6 m. high of *G. xanthocarpa*.

Woody lianes strongly accentuate the tropical appearance of lowland forest and are particularly abundant in rain-forest proper. In many places their "ropes" hang swinging from the tree-tops, solitary or intertwined. Those of *Rubus australis* (Rosac.), *Rhipogonum scandens* (Liliac.) and *Muehlenbeckia australis* (Polygonac.) are of prime physiognomic importance, those of the first-named covered with rough, brown bark and, at times, 8 cm. diam., and of the second smooth, black and jointed.

Freycinetia Banksii (Pandanac.), of *Pandanus*-form, clothes great trunks with its rooting stiff stems and yellow-blotched sword-like leaves. The root-climbing species of *Metrosideros* also play a most important part in draping trunks and tree-fern stems. *Clematis indivisa*, when in bloom, forms dazzling sheets of white on low trees and tall shrubs.

b. Plants of shrubland, heath, swamp, bog, grassland and rock.

Blechnum procerum (Spreng.) J. C. Anders. *(Filices)*, with its great bright-green or dark-green pinnate leaves 1.2 m. long and broad in proportion arching downwards, covers steep slopes for many square metres at a time; to this fern, indeed, do certain river-gorges owe much of their value as tourist resorts.

Cordyline australis (Forst. f.) Hook. f., *(Liliac.)*, ti, tikauka, cabbage-tree, is a typical tuft-tree with straight, erect naked trunk some 6 m. high, but occasionally much higher, and 30 to 60 cm. diam., fissured rough bark, green, coriaceous, glabrous, acuminate, ensiform leaves about 90 cm. long by 5 cm. broad closely tufted near the apex of the trunk, the inner erect and semi-erect, the outer drooping, and immense drooping or erect panicles 60 to 120 cm. long by 30 to 60 cm. through, of numerous crowded white flowers of a luscious rather cloying scent.

Coriaria arborea Lindsay *(Coriariac.)*, tree-tutu, is a bushy shrub or small tree $\pm$ 7 m. high with stiff stems, glossy-green, rather thin, acute, ovate leaves $\pm$ 5 cm. long and drooping racemes some 15 cm. long of small, black, succulent fruits. *C. sarmentosa* Forst. f., tutu, is similar but summer-green only. The genus consists in New Zealand of many jordanons and hybrids in profusion.

Gleichenia circinata Sw. *(Filices)* has a slender far-creeping rhizome and erect bright-green fronds 30 to 60 cm. high, or more, of the usual *Gleichenia*-form, the horizontal pinnae one above the other, which stand side by side more or less interlaced for many square metres at a time.

Hebe salicifolia (Forst. f.) Pennell *(Scrophulariac.)* consists of a number of jordanons all of which are erect bushy-shrubs, some reaching nearly 4 m. in height, with straight, slender branches, slightly yellowish-brown smooth bark, bright-green, rather thin, almost glabrous, acuminate, lanceolate leaves frequently about 10 cm. long, but differing according to the variety, and slender racemes, $\pm$ 12 cm. long of numerous, fragrant, lilac-tinged flowers.

Hypolaena lateriflora Benth. *(Restionac.)* forms close masses of extremely slender, much-branched, wiry, flexuous, interlacing, leafless stems $\pm$ 70 cm high.

Leptospermum scoparium J. R. et G. Forst. *(Myrtac.)*, manuka, red tea-tree, tea-tree, is sometimes a small tree, but its physiognomic status is that of an erect shrub of more or less fastigiate habit, finally 3.6 m. or more high, with a stout main-stem, reddish-brown bark which may hang in ribbons, slender more or less vertical branches which give off numerous close-growing twigs bearing abundant, small dark-green, aromatic, coriaceous, pungent, lanceolate to ovate leaves 5 to 12 mm. long, greyish in the mass, and abundant white flowers $\pm$ 1 cm. or more diam. *L. ericoides* A. Rich., kanuka, white tea-tree — "white" and "red" for the two species of *Leptospermum* refer only to the colour of the wood — is similar to *L. scoparium* but considerably taller and its leaves are not pungent and its flowers smaller but in great profusion.

Phormium tenax J. R. et G. Forst. *(Liliac.)*, harakeke, New Zealand flax, in a wide sense is a tall *Iris*-like herbaceous plant forming somewhat bunched-together tussock-like masses of erect or partly drooping leaves 1 to 2.5 m. or more long by 6 to 12 cm. broad, rather dull-green above, but somewhat silvery beneath, coriaceous tough, flexible, but the individuals of the linneon differ much in all these points; the margin is frequently stained brown. The major part of the leaf is flat, but at a greater or shorter distance from the base the two halves of the blade are equitant making a kind of petiole. There is a stout leaf-bearing, creeping rhizome which branches near the apex. The scape, stout, reddish-purple, often more or less glaucous with wax and, raised above the foliage, bears numerous dull-red flowers 3 to 5 cm. in length which are succeeded by dark-coloured erect or semi-erect capsules, 5 to 10 cm. long. When in company with *P. Colensoi* polymorphic hybrids are present.

Pteridium esculentum (Forst. f.) Ckn. *(Filices)*, rau-aruhe, bracken, is probably rather a variety of the cosmopolitan *P. aquilinum*, but as it is confined to the Southern Hemisphere, it is convenient to name it as above. Its life-form is too well-known to need description. The leaves are frequently more than 1.75 m. long.

Tussock-grasses are amongst the most striking features of the vegetation and of these *Festuca novae-zelandiae* J. B. Armstg. and *Poa caespitosa* Forst. f. rank with forest-trees as of physiognomic importance. Their life-form resembles the head of an old-fashioned besom. Culms and leaves are tightly bunched together and, extremely tight at the base, are looser above and spread or droop laterally to some extent. The apical portions of the leaves are frequently dead and there is also much dead material within the tussock. More widely-spread, but never filling the whole landscape, are the great tussocks of *Arundo conspicua* Forst. f., toetoe, which greatly resemble those of pampas-grass (*Gynerium argenteum* Nees) in habit, leaf and inflorescence. The leaves are long and narrow, coriaceous, flat or involute and the nerves strongly developed. The rhizome is short but stout. The panicle, raised above the foliage on a tall, stout stem, consists of many yellowish drooping branches and renders the plant most conspicuous.

Chapter III.
The Autecology of the Lowland Plants.
1. Life-Forms.

Here, unless stated otherwise, only the 559 species of spermophytes and pteridophytes which are confined to the lowlands and lower hills are dealt with, since their autecology should reflect more closely the ecological conditions than would that of the total lowland-lower hills' flora embracing, as it does, many species common on the high mountains.

a. Trees.

There are 87 species of trees of which 4 are very tall[1]), 7 tall, 14 of medium height and 62 of low stature, but at least 39 are epharmonically shrubs and in some 10 cases it is a matter of opinion merely whether they should be classed as such or as trees; nearly all are evergreen, only 7 species being deciduous or semi-deciduous; perhaps 9 may be considered semi-xerophytes.

The following are the life-forms and the number of species to each: — Tuft-trees 10 (tree-ferns 7), canopy-trees 34, bushy-trees 26, fastigiate tree 1, araliad-form 5, leafless juncoid 5, rhododendron-form 6.

---

1) Few accurate measurements are available. For *Agathis*, trees exceeding 30 m. are common and Cheeseman states it may reach 51 m. (1914: 2 for Plate 84). *Dacrydium cupressinum* and *Podocarpus dacrydioides* not infrequently reach more than 30 m., and a tree of the latter measured by E. P. Turner was nearly 60 m. high.

Regarding the tree trunks those of no less than 50 species rarely exceed 30 cm. diam., and in some instances are much less; those of 18 species range from 30 to 60 cm. or rather more; and those of 10 species generally exceed 90 cm., 9 of them having excessively massive trunks[1]).

In some species the base of the trunk is frequently more or less swollen and buttressed[2]). In tall and medium-sized trees the trunk is frequently unbranched for about two-thirds the height of the tree, a condition arising through suppression of the early branches owing to the close growth of the saplings and their competitors. In the greater part of the species the bark is thin, or fairly so, but in about 18 species it is 12 mm. or more in thickness.

The degree of density and of spread of the tree-crowns is largely a matter of environment. A very considerable proportion are dense and rounded while but few have really far-spreading branches. *Podocarpus dacrydioides* has a very sparse, narrow crown, small out of all proportion to the size of the tree. *Knightia excelsa* is almost as fastigiate as the Lombardy poplar and it shows up distinct from the general forest-mass. The head of *Metrosideros robusta*, though rounded and somewhat wide-spreading, is not dense owing to the leafy portion of each branch not intermingling.

The roots of the trees rarely descend deeply but spread out horizontally for a great distance and the uppermost, in some species, are only half-buried or, at times, raised considerably above the ground-surface. Nodules[3]) are abundant on the rootlets of the *Coniferae*, *Libocedrus* excepted. *Rhopa-*

----

1) The diameter of the trunk, as also the height of the tree, is partly a matter of epharmony. Thus the podocarps of the Volcanic Plateau are, as a rule, of considerably lower stature than those of the Western district, but of far greater girth. In Stewart Island the average both for height and diameter of trunk reaches the minimum. *Agathis* at times possesses by far the most massive trunk of any New Zealand forest-tree, a diameter of 7.2 m. having been recorded, but trunks of this character are usually hollow, but a diameter of from 1.5 to 3 m. is not uncommon. A diameter of 1.5 m. is met with in *Podocarpus dacrydioides* (Fig. 24) and *Dacrydium cupressinum* and the trunk of *P. totara* attains still greater dimensions (Fig. 25.) Nor are massive trunks confined to the *Coniferae*, e. g., that of *Laurelia novae-zelandiae* may reach 1.8 m. diam., and 1.2 m. to 1.5 m. is not an unusual diameter for the trunk of *Nothofagus fusca* or *N. truncata*.

2) True *plank-buttresses* occur in *Laurelia novae-zelandiae* (1.8 m. high by 12.5 cm. thick) and in the lowland-high mountain *Nothofagus fusca* and *N. Menziesii*, and the lowland-lower hills *N. truncata*, and *rounded buttresses* in the case of several trees, particularly *Podocarpus dacrydioides* and *Beilschmiedia tawa*.

3) YEATES (1924: 121—124) has shown that the nodules are lateral roots the apical growth of which has been checked and that they are occupied by a fungus, while, "out of several hundreds of nodules examined representing about twenty species . . . only one nodule could definitely be said to contain bacteria similar to those figured by Spratt and McLuckie." The tubercle-bearing plants constantly grow side by side with vigorous trees of species unprovided with tubercles. Nor is the growth of *Agathis* and the podocarps at its maximum under wet forest conditions, but rather in good, well-drained garden soil, so that explanations of the presence of nodules based on their being "adaptations" seem far fetched.

lostylis gives off short reddish stiltroots at 30 cm. or more from the base of the trunk. The tree ferns have a thick mantle of aereal roots forming the periphery of their trunks. The base of the trunk may thus be much swollen, e. g., in *Cyathea dealbata* it may measure 46 cm. diam., while it is not until a height of 56 cm. from the ground is reached that the trunk is of the general diameter of 15 cm. In leaning trunks of certain tree-ferns large wedge-shaped masses of aereal roots frequently jut out from the undersurface. Pneumatophores are strongly developed in *Laurelia novae-zelandiae*[1]) and to a lesser degree in *Eugenia maire.*[2])

Excepting some of the small trees—especially *Cyathea medullaris, Cordyline*, species of *Pittosporum, Ascarina lucida, Aristotelia serrata*, species of *Hoheria, Leptospermum scoparium* and several species of *Coprosma* — the rate of growth of the trees in general is very slow[3]), but it varies epharmonically to an astonishing extent[4]). In fact, by far the greater number of species demand for their maximum growth far more light than the forest-interior in

---

1) Extending from the plank-buttresses for several (perhaps many) metres on the wet forest-floor are the raised, laterally-flattened roots of *Laurelia* from which solid upright broad or conical projections ± 12 cm. high, usually covered with bryophytes, project out of the substratum of shallow water or soft mud.

2) In *Eugenia* there are small cylindrical root-branches — presumably pneumatophores — first described by CHEESEMAN (1920: 9, 10). Apparently from what I have seen of these organs, they show every transition from ordinary roots which enter the ground to those described by the above-named botanist as erect; those horizontal in direction are a common feature.

3) Figures as usually supplied are more or less misleading, since nothing is said as to the environment of the tree of which the measurement is given and on this all depends. The following in this paragraph are a selection from a number of trees the annual rings of which were accurately counted by H. B. KIRK (to each the age and diameter is given): — *Libocedrus plumosa*, 252 years, 30.2 cm. diam; 143 years, 32.2 cm. diam. *Agathis australis*, 238 years, 39.4 cm. diam. *Dacrydium Kirkii*, 373 years, 44.0 cm. diam. *D. Colensoi*, 479 years, 45.8 cm. diam.; 400 years, 43.6 cm. diam.; 109 years, 21.3 cm. diam.; 101 years, 15.5 cm. diam. *D. cupressinum*, 374 years, 44.6 cm. diam.; 258 years, 472 cm. diam.; 315 years, 47.8 cm. diam. *Podocarpus totara*, 425 years, 111.6 cm. diam.; 343 years, 41.4 diam. The various species of *Nothofagus* grow more rapidly than any other tall tree and the following measurements, taken from an open stand of yung trees, give some idea of rate of growth: — 46 years old, 25 m. high, 45 cm. diam. (breast high); 39 years, 20 m., 41 cm.; 46 years, 24 m., 36 cm.; 51 years, 24.9 qm., 35 cm. The following are rates of growth of certain young trees within *Leptospermum scoparium* shrubland: — *Agathis australis*, 12 years old, 35 cm. high, 8 mm. diam.; 16 years, 1.5 m., 15 mm.; 32 years, 3 m., 20 mm.; 8 years, 27.5 cm., 4 mm.; *Phyllocladus trichomanoides*, 26 years, 2.1 m., 14 mm.; 22 years, 1.5 m., 10 mm.; 15 years, 1.3 m., 10 mm.; *Knightia excelsa* 10 years, 2.7 m., 14 mm.

4) *Nothofagus cliffortioides* (in bright light), 9 years, 3.6 m., 37 mm.; 7 years, 5.4 m., 51 mm.; 11 years, 5.5 m., 60 mm., (in shade alongside the above), 17 years, 1.5 m., 17 mm.; 34 years, 6 m., 58 mm.; *Nothofagus Menziesii* (in fairly bright light on outskirts of forest), 16 years old, 87 mm. × 70 mm. diam.; (in forest interior), 43 years old, 48 mm. × 47 mm. diam.; (in excessive shade within the forest according to G. SIMPSON and J. S. THOMSON), 36 years old, 2.4 m. high, 15.6 mm.; 57 years, 3.6 m., 26.5 mm.

general affords and, after attaining a certain size as saplings, they "linger" rather than develop properly, e. g. *Elaeocarpus Hookerianus*, 70 years, old, 6 m. high and 63 mm. diam.

The leaves of the trees may be characterized as follows: — evergreen 70, deciduous or semi-deciduous 7, compound 19, simple 63, broad 67, narrow 15, very large 10, large 12, medium-sized 29, small 17, very small 14, leafless except in juvenile 5, thin 34, coriaceous 48, glabrous 62, hairy 20 (tomentose beneath 8), waxy beneath 6, glossy 20. Most of the coriaceous-leaved species are not xerophytes, the stout texture of leaf being rather in relation with the evergreen habit. *Myrtus bullata* stands almost alone in its curious reddish-brown broadly-ovate leaves, 2.5 to 5 mm. long, concave beneath, but above, the surface between the veins is raised in blisters, but blistered leaves — a character unknown in those of the parents or in the genus — also occur in some forms of that vast hybrid swarm, $\times$ *Nothofagus cliffusca*. With regard to deciduous leaves, those of the species of *Edwardsia* do not depend upon decrease in temperature but the leaves fall just before blooming takes place and new ones rapidly develop as soon as this is over, but this behaviour is not uniform for all the jordanons.

b. Shrubs.

There are 82 species of shrubs of which 75 are spot-bound, 7 wandering, 46 mesophytes, 36 xerophytes or subxerophytes, 67 evergreen, 3 deciduous or semi-deciduous, 12 virtually leafless when adult; [while, as for height, 23 are tall (8 epharmonically trees), 34 of medium height and 25 of low stature. Though none pass through a long-persisting juvenile stage, 12 (as early juveniles) have a leaf-form different from that of the adult.

The following are the life-forms of the shrubs and the number of species to each: — erect bushy-shrub 26 (more or less dense 22, open 3, fastigiate 1) divaricating 17 (partly so 3, switch-shrubs 2), straggling 10 (erect 7, semi-prostrate 3), switch-shrubs 8 — excluding the 2 divaricating (erect 7, spreading 1), rhododendron-form 7, ball-like 5 (1 partly so), tuft-shrub 3 (ferns 2), mat-form 1, erect creeping and branching 4 (all lycopods), bamboo-form 1.

The branchlets (excluding ferns and *Cordyline*) may be divided into: moderately-stout 16, stiff or wiry 35, slender or twiggy 28. The bark is in nearly all cases quite thin. The trunk of *Blechnum Fraseri* varies from about 30 to 70 cm. in height and is not thicker than a moderately stout walking-stick. From its base numerous runners pass off extending horizontally just beneath the surface of the ground and at a distance of some 10 cm. a young plant is produced which develops first a trunk and then runners, so that extensive, dense colonies are formed by vegetative increase alone. The trunk of *Todaea barbara* is short but massive and thus quite unlike the tall, slender trunk of a typical tree-fern.

Some of the shrubs, especially the xerophytes, root deeply.   The species of *Hebe* readily produce adventitious roots from near the bases of the stems, so that a shallow-rooting mass of roots is formed.

The leaves of the shrubs may be characterized as follows: — compound 5, simple 65, leafy (very small) only in juvenile or wanting 12, broad 50, narrow 20, thin 33, coriaceous 37, glabrous 47, hairy 23 (tomentose 15), and with regard to size 1 is very large, 7 are large, 15 of medium size, 13 small, 34 very small.

c. Herbs and semi-woody plants.

Land-plants.   There are 299 species of herbs and semi-woody plants (exclusive of aquatics but including ferns) of which 184 are spot-bound, 115 wandering, 14 hygrophytes, 260 mesophytes, 25 xerophytes, 15 annuals, 284 perennials (semi-woody 25, summer-green 44, evergreen 215); while as for height 34 are very tall, 45 tall, 74 of medium stature, 53 of low stature and 93 either hug the ground or are less than 15 cm. high.   No fewer than 65 are pteridophytes and 122 monocotyledons, so that only 37 per cent. are dicotyledons.

The life-forms and the number of species to each are as follows:—(1.) *Annuals and biennials* 15 consisting of: tufted fern 1. lufted grass 1, mat-form 2 (rooting 1), erect-branching 9, semi-erect 2. (2.) *Perennials* 284 con-sisting of: (a.) semi-woody 25, which include (α.) wandering 8, consisting of erect-branching 5, rooting-mat 1, rooting-straggling 2, (β.) spot-bound 17, consisting of ferns with short trunk 7, tufted grass 1. erect rosette-plant 2. erect-branching 6 and procumbent 1; and (b.) herbaceous perennials 259 which include (α.) wandering 107, consisting of: erect creeping and rooting 44, (ferns 28, grass-form 7, rush-form 7, iris-form 1, saprophyte[1]), erect-branching creeping and rooting 6 (all herbs), semi-erect branching creeping and rooting 2 (both herbs), prostrate-branching creeping and rooting 5 (herbs 2, *Utricularia*-form 3), rooting-mat 50 (ferns 7, rush-form 1, grass-form 1, herbs 41); and (β.) spot-bound 152, consisting of: erect tufted-form 46 (ferns 21, grass-form 10, rush-form 12, iris-form 3), semi-erect tufted-form 2 (all of grass-form), tussock-form 27 (grass-form 19, rush-form 8) mat-form 5 (grass 4, rush 1), summergreen bulb-form 2 (both grass-like), earth-orchid form 42, semi-erect branching 5 (all herbs), low rosette-form 7 (all herbs), erect rosette-form 2 (both herbs), erect branching 12 (all herbs) and straggling-branching 2 (both herbs).

---

1) The saprophyte *(Bagnisia Hillii)* according to CHEESEMAN (1909: 141, 142) is found in decaying leaves and humus at the base of *Podocarpus dacrydioides.* There is a branching, fleshy rhizome 5 to 10 cm. long and 1 to 1.5 mm. thick. Leaves are wanting and the vegetative parts are colourless. One-flowered peduncles, 5 mm. to 15 mm. in length, are given off from the axils of the minute fleshy bracts. The flower is rose-pink and about 1.5 cm. long by 7 mm. broad.

Taking the leaves (19 are leafless) of all the classes of herbs and semi-woody plants together, their characteristics and the number of species to each, are as follows: — compound 72, simple 208, broad 176, narrow 104, coriaceous or fairly thick 76, fleshy or succulent 4, thin 200, glabrous 217, hairy 63 (tomentose 12). As for their relative dimensions 23 are very large, 33 large, 43 of medium size, 71 small and 110 very small.

Water-plants. New Zealand, notwithstanding its wealth of inland waters, possesses remarkably few aquatic vascular plants. Here, though both the purely lowland species and those ascending to the high mountains are dealt with, only 16 come into consideration. Really no line can be drawn between aquatic and swamp plants and the species of *Myriophyllum* included here could quite well be transferred to the latter class, while several already dealt with as land-plants could come here. But, even were the widest view taken of water-plants the list would number less than thirty.

The 16 species fall into the 2 divisions — free-floating (5 species) and rooting (11 species). The former consists of the highly differentiated *Azolla rubra*, 2 species of *Utricularia* (1 with floating shoots $\pm$ 60 cm. long) and 2 species of *Lemna*. The latter are represented by 4 entirely submerged species (rush-form 3, ribbon-form 1), 2 partly submerged with their upper portions extending out of the water (*Myriophyllum*) and 5 with floating leaves (*Potamogeton* 3, *Callitriche* 1, and the rare grass, *Amphibromus fluitans*).

d. Lianes, epiphytes and parasites.

Lianes. Here not merely the purely lowland species are dealt with, but all that occur in the lowland-high mountain belt. The New Zealand lianes are of peculiar interest since they are not only numerous and of great physiognomic importance, but many bear the true tropical stamp while the climbing *Myrtaceae* are autochthonic. The total number of lianes is 45 belonging to 16 families and 22 genera; 11 are scramblers, 12 root-climbers, 13 winders and 9 tendril-climbers while 29 are woody (27 endemic), and 16 herbaceous or semi-woody (ferns 8).

The scramblers consist of 2 ferns, 1 woody grass, 5 species of *Rubus*, 1 switch-shrub (*Carmichaelia gracilis*), *Fuchsia perscandens* and 1 dimorphic shrub-composite (*Helichrysum dimorphum*). The least differentiated members of this class are but little removed from many woody plants of the forest-interior which lenghten their internodes considerably and assume a spindling habit. For example, one seeing the bushy-shrub form of *Fuchsia perscandens* in the open would hardly connect it with the lianoid coils of the same species on the forest floor. *Gleichenia microphylla* is merely a faculative liane, thanks to it fortuitously possessing plagiotropous pinnae suitable for climbing purposes. So, too, the flexuous stems of the coastal-inland *Angelica geniculata*, too slender to stand erect, make that species a liane. More lianoid is the rare cupressoid *Helichrysum dimorphum* with its flexible cord-like, unbranched climbing-stem and its eventual close head of leaf-bearing twigs. But the highest degree of differentation is shown by the different species of *Rubus*, since their leaves, more or less reduced to midribs furnished with hooked prickless, are special climbing organs which cling to anything they touch, and using the shrubs and comparatively low trees of the forest undergrowht as their path upwards, eventually gain the forest canopy where in the bright light they produce flowers and fruit. Their main stem, at first extremely slender, gradually thickens and becomes covered with rough bark while an ultimate diameter of 12 cm. ore more is quite common. The leaves of the scrambling lianes and the number of species for each class are as follows: — compound 8, simple 3, broad 9, narrow 2, coriaceous or thick 7, thin 4, glabrous 8, hairy 3, very large 2, large 4, small 2, very small 3.

The *root-climbers* consist of 5 ferns, 1 of the *Pandanaceae* and 6 of the *Myrtaceae*. They are not confined to trees and shrubs as their hosts but cover the faces of rocks and frequently creep over the ground. In the case of the ferns, it is hardly feasible to separate the climbing from the epiphytic, because many which have no connection with the ground are really climbing epiphytes. *Polypodium diversifolium* (stem thick, green, fleshy) is equally a ground-herb, a liane and an epiphyte. *Blechnum fili-forme, Arthropteris tenella, Polypodium pustulatum* and *P. novae-zelandiae* are almost obligate lianes. *Freycinetia Banksii* and the species of *Metrosideros* are specially imporant root-climbers. The former has a terete, hard, woody, rigid stem, 2.5 cm. or so diam. fastened to the host by stout roots which pass quite round a slender trunk and finally branching copiously cling most closely to the bark. The leaves are sword-like, 60 cm. or more long, thick coriaceous, glabrous and may quite hide the tree-trunk. The species of *Metrosideros* have stout, woody climbing stems which ascend the loftiest trees and are at first fastened tightly to the host by numerous aereal roots. Lateral branches are given off freely, which at first soft and fleshy, grow rapidly but put forth no roots until woody, the apical portion being rootless and having only partially developed leaves. The climbing shoots are pressed more or less tightly against the bark, *M. perforata* being most marked in this regard, the small, thick, roundish leaves pressed tightly against the bark forming typical leaf-mosaics. Finally, in all the species, lateral non-climbing shoots are developed which can bear flowers and fruit, their leaves more or less distinct from the juvenile. The main climbing stem, in the highest-climbing species, eventually lose their roots, are held away from the trunk and increase in thickness attaining, in *M. scandens*, a diam. of 15 cm.

The winding lianes are represented by 1 fern, 1 lycopod, 6 woody spermophytes and 5 herbs, of which 4 ascend to the forest canopy, 5 are confined to shrubs or low trees and 4 merely climb over tall grasses or at best low shrubs.

The climbing stems are usually quite slender and so can make use of small shrubs and young trees in order to raise themselves in the first instance, but these are frequently killed by the increasing pressure of the winding stem, and the number that thus succumb must be indeed great. Sometimes for many square metres at a time the black stems of *Rhipogonum scandens* form close entanglements, no trace remaining of the original host-plants.

Lycopodium volubile is an interesting example of the transition of a creeping ground-plant to a winding-liane by way of a scrambling plant. In the case of *Senecio sciadophilus*, a true winding liane, the stems when on the ground put forth long roots in abundance, branch, extend over a considerable area and are, in fact, creeping plants pure and simple. The occasional winding of the root-climber *Blechnum filiforme*, when the support is specially slender, may be cited also. *Rhipogonum scandens* at first puts forth a stout, succulent fast-growing stem from its root-stock, sparingly furnished with a few scale-like dark-coloured leaves, but the mature climbing stem gives off non-climbing branches which bear large, green, oblong, coriaceous leaves. *Lygodium articulatum* forms masses of wiry, slender, brown, extremely tough climbing-stems which wind round one another as well as the branches of the support. These stems are morphologically leaf-spindles of unlimited growht. A frond thus may attain the great length of 40 m. but frequently the liane is confined to shrubs or low trees. The 2 species of *Muehlenbeckia* are extremely leafy, *M. australis* covering low trees with a dense mass of verdure. The stems of the 2 species of *Parsonsia* are slender and pliant frequently winding not only round the host but round themselves. Both species show most remarkable heterophylly, there being three distinct leaf-forms at different stages of the plant's development.

The adult leaves of the winding lianes are, compound 1, simple 12, thick or coriaceous 4, thin 9, glabrous 10, hairy 4, glossy 2, waxy beneath 1, very small 3, small 7, medium 3 and large 1.

The tendril-climbers include only the 8 species of *Clematis* and monotypic *Tetrapathaea tetrandra* (Passiflorac.). All are medium-sized lianes with comparatively slender (except *Tetrapathaea*) much-branching stems and they usually drape low trees and shrubs. *Tetrapathaea* is in habit and appearance very similar to *Parsonsia heterophylla* forming dense masses of flexible, slender branches bearing abundant dark-green, glossy, somewhat coriaceuos, ovate-lanceolate, acuminate leaves some 5 cm. long. The tendrils, at first straight and soft, and sligthly curved near their apices, finally become stout and wiry. The species of *Clematis* form a series of transitions from a thin-leaved mesophyte by way of subxerophytes with much-cut reduced leaves to the leafless juncoid xerophytic *C. afoliata*. The adult leaves are, compound 7, simple 1, coriaceous 2, thin 7, glabrous 6, hairy 2; those of *Tetrapathaea* are glossy and produced in the greatest abundance.

Epiphytes. Here all the epiphytes of the Botanical Region are dealt with. But, as seen when considering the lianes, it is frequently difficult or impossible to decide whether a species is a liane or an epiphyte, so, too, many ground-plants are at times epiphytic especially on trunks of tree-ferns[1]) so that the question of their ecological status becomes a matter of personal opinion. Here the number is kept at a fairly low limit and no plant, unless purely an epiphyte or usually such, is admitted. Strictly an epiphyte is a plant which should be invariably seated upon another. In New Zealand this is true only for certain ferns (mostly *Hymenophyllaceae*), bryophytes, lichens, algae and a few spermophytes, nearly all the latter being also denizens of rocks, while some are more or less common as ground-plants, e. g. *Senecio Kirkii* — a candelabra-like shrub 3.6 m. high — is generally terrestial in kauri forest, but epiphytic or rupestral else-where. Then there are the species so frequently epiphytes during their youth but which, as the waterbalance fails, send down roots to the ground (Fig. 26) which, enveloping their host, finally grow into a solid mass which functions as a trunk. Well-known examples of this class are *Metrosideros robusta* — at first a large shrub high up on a lofty tree —, *Nothopanax arboreum* and *Dracophyllum arboreum* (Chathams) both on tree-fern trunks and at times *Metrosideros lucida*, *Griselinia littoralis* and even the more truly epiphytic *G. lucida*. Various unexpected terrestial species under the influence of a specially wet climate may occasionally become epiphytes, e. g. *Phormium Colensoi*, *Celmisia major*.

Here 50 species (some also dealt with in other classes) belonging to 12 families and 21 genera are considered epiphytes (semi-obligate 32, facultative 13—8 more commonly epiphytes, 5 more commonly ground-plants &c. — partial 5). Their life-forms and the number of species to each are as follows: — filmy ferns 17, tufted ferns 5, creeping ferns 4, iris-like tussocks 3, epiphytic orchid 7 (pseudobulbous 2), shrubs 8 (4 eventually trees). pendulous semi-woody plants 4 (1 frequently erect), prostrate creeping herbs 2.

The vertical distribution of the epiphytes is well-marked[1]) each species having a fairly well-defined upper limit, those which ascend highest being the following: *Hymenophyllum Armstrongii*, *H. sanguinolentum*, *H. villosum*, *Asplenium adiantoides*, *A. flaccidum* (var. with long drooping leaves), *Polypodium diversifolium*, *P. Billardieri*, *P. grammitidis*, *Cyclophorus serpens*, *Lycopodium Billardieri* (Fig. 27), *Astelia Cunninghamii*, *A. Solandri*, the 7

---

1) This has been carefully investigated by HOLLOWAY (1923: 609—616) for the *Hymenophyllaceae* and he arranges them — taking all the species, some of which are only occasionally epiphytes — as follows together with the number of species in each class: — low epiphytes 10, mid-epiphytes 3, high epiphytes 8. With 2 exceptions the high and mid-epiphytes are also low epiphytes.

epiphytic orchids, *Pittosporum cornifolium*, *P. Kirkii*, *Griselinia lucida*, *Metrosideros robusta*.

The presence of the different life-forms in any forest evidently depends upon the conditions offered, while such are largely a matter of climate. Tree-fern trunks with a thick water-holding mantle of aereal roots offer suitable conditions for the germination of many young trees and shrubs and for the ramification of the slender wiry stems of the *Hymenophyllaceae*, some seedlings being commoner there than elsewhere and others again equally at home on the forest-floor. The occurrence of many of the spermophytic epiphytes and the larger pteridophytes is bound up with the prior occupation of trunks and branches by epiphytic bryophytes which, in their turn, occupy their epiphytic station thanks to the water-absorbing capacity of their leaves, the sponge-like cushions, mats or masses they build so rapidly and various special water-holding contrivances. In a short time sufficient soil is formed from the decay of the bryophytes to support seedling epiphytes and these, each according to its specific capability, make more soil from their decaying leaves &c. until a surprising quantity of vegetable matter in various stages of decay accumulates on horizontal bougs and forks of branches amply sufficient to support shrubs and herbs of no mean size (Fig. 27.). Leaning trunks are specially favourable for a close epiphytic covering and more favourable still are the irregular bases of *Metrosideros robusta* (Fig. 28.) and *Weinmannia racemosa* (Fig. 29.), while the slender branches of small trees are particularly suited for occupation by filmy ferns which in a very wet climate completely surround them, as may more or less xerophytic epiphytes in a drier climate, e. g. *Cyclophorus*, epiphytic orchids (Fig. 30.).

The epiphytic species of *Astelia* are ecologically equivalent to the tropical epiphytic *Bromeliaceae* and like them store up water. *Astelia Solandri* is a densely-tufted evergreen herb with numerous, ensiform, coriaceous leaves 90 cm. to 1.2 m. long with black, sheathing, fleshy bases covered with a great quantity of long silky hairs which, even in quite dry weather, hold large quantities of water[1]). The plant forms immense tufted masses high up on the tree-tops, the basal part and earlier leaves decayed and making a usually sopping-wet mass of loose vegetable matter.

---

1) W. R. B. OLIVER has most kindly informed me that he has made the discovery that the thick cuticle consists "for the most part of a substance giving a reaction for cellulose" and on its inner side is a layer of cutin. This is pierced by "peculiar organs" consisting of a double row of three or four cells with dense protoplasmic contents" which connect the "outer cellulose absorbent layer of the cuticle, through the cutinised layer, with the epidermal and hypodermal cell layers". OLIVER appears to be of the opinion that these organs " have an absorptive function". They occur all over the leaf including the basal portion of the water resevoir. Miß M. W. BETTS (1920 b: 305) has described something analogous to the above in *Astelia Cockaynei*, a ground plant of various high-mountain communities.

The plant frequently grows on slender perpendicular trunks in what appears an impossible position for a plant so massive and heavy (Fig. 27.).

Pittosporum Kirkii and *P. cornifolium* are sparsely-branched straggling shrubs with long, flexible branchlets, those of the latter slender and drooping. The leaves of both are coriaceous but those of *P. Kirkii* are extremely thick. *P. cornifolium* is occasionally a ground-plant.

Griselinia lucida is at times almost a bushy-tree with thick, furrowed bark. The leaves are obliquely-oblong, thick, coriaceous, bright-green, glossy and 7.5 to 17 cm. long. Although an obligate epiphyte, or rock-plant, it grows luxuriantly in ordinary garden conditions.

The leaves of the epiphytes and the number of species to each may be characterized as follows: — simple 28, compound 22, very large 7, large 10, of medium size 11, small 11, very small 11, coriaceous or thick 31 (fleshy or succulent 6), thin 19.

Parasites. Here all the New Zealand higher parasites, numbering 16 (11 shrubs, 5 herbs), are dealt with[1]).

The shrubby species differ in height, 1 reaching more than 3 m., 3 more than 1 m., 2 about 90 cm. and 5 not more than 15 cm. In habit 5 are dense bushy-shrubs, 1 open and rather straggling, and of the small species 2 are generally but little branched but the other forms tufted masses. All are hemi-parasites; none are restricted to one host[2]). Biologically these shrubs fall into two classes, the one where the seedling at once sends down a root into the tissues of the host and the other where this does not happen, but a lateral root is developed which eventually branches, and, at intervals, puts suckers into the tissues of the host, while at such points aereal shoots may also arise. To the first class belong the endemic *Tupeia* and palaeotropic *Korthalsella* and to the second class *Loranthus* and *Elytranthe*. In the case of *Loranthus micranthus* the lateral root may follow the course of a branch or it may also descend along the main stem and eventually reach, but not penetrate, the surface of the ground, the descending root resembling in appearance a liane. Occasionally the root-tip may leave the branch altogether, in which case it does not descend to the ground but

---

1) *Phrygilanthus tenuiflorus* and *P. Raoulii* have been seen by no living New Zealand botanist either *in situ* or as herbarium species; so, though included in the above total, they are excluded from what follows.

2) *Loranthus micranthus* occurs on various trees and shrubs, including *Podocarpus totara*, *Rubus australis*, spp. of *Carmichaelia*, *Melicope simplex*, *Melicytus ramiflorus*, *Plagianthus betulinus*, spp. of *Leptospermum*, *Dodonaea viscosa*, and many species of *Coprosma*. It is also to be found on various introduced trees especially the plum, pear, poplar and laburnum and, according to W. MARTIN (1924: 27), the elm, willow, apple, rhododendron and "in particular the hawthorn". *Tupeia antarctica* also has many hosts, e. g. *Loranthus micranthus* and *Elytranthe tetrapetala* (both themselves hemi-parasites), *Pittosporum eugenioides*, *P. tenuifolium*, *Carpodetus serratus*, *Hoheria Lyallii*, *Suttonia australis* and *Shawia paniculata.*

bends back towards the branch. Where the suckers enter the host considerable swellings occur. The species of *Loranthus, Elytranthe* and *Tupeia* are abundantly leafy, those of *Korthalsella* are leafless but the stems are green. The leaves of 5 of the above are of medium size and of 1 small; all are green, glabrous and, except for *Tupeia,* thick and coriaceous. The stems of *Korthalsella Lindsayi* and *K. clavata* are jointed and flattened, the joints 8 cm. or more long by 5 cm. or so broad, coriaceous and dark-green. *K. salicornioides* has also jointed stems but they are succulent, virtually terete and only about 1 mm. diam. Two of the species of *Elytranthe* have extremely showy scarlet flowers.

The herbaceous parasites belong to three distinct classes; all are holo-parasites. The 3 species of *Gastrodia* are earth-orchids with slender, straight, erect, unbranched, pale-coloured stems, differing, according to the species, from 30 to 90 cm. or more in height and with extremely long, brittle, fleshy tuberous roots which are parasitic on the roots of certain forest plants.

Dactylanthus Taylori (Balanophorac.), according to CHEESEMAN (1914: text to Pl. 178), is a root-parasite on *Nothofagus, Hedycarya arborea, Pittosporum eugenioides, Schefflera digitata, Nothopanax arboreum, Suttonia australis, Geniostoma ligustrifolium* and *Coprosma grandifolia.* It consists of a fleshy rounded or shapeless tuberous rhizome, rough with wart-like tubercles, which varies in size according to age from 2.5 cm. up to more than 30 cm. diam. During early autumn numerous, usually dioecious, highly-fragrant flowers, 5 cm. to 15 cm. high, are given off from the rhizome. H. HILL (1926: 89) describes a certain valley-floor near Opepe (VP.) as "a veritable garden-meadow of *Dactylanthus.* Opening out, along the dry floor of the valley for a chain or more, appeared hundreds of flowers in clumps."

Cassytha paniculata is a twining, leafless plant having for its host shrubby species of the Auckland gumlands, e. g. *Leptospermum scoparium, Pomaderris Edgerleyi, Cassinia amoena* &c., to which its slender, very stout, pale stems are attached by means of suckers. These stems pass tightly stretched from plant to plant, so that one may easily trip over them.

e. Persistent juvenile-forms.

There are about 200 species of spermophytes which for a longer or shorter period go through an early stage, or stages, of development markedly different from that of the adult and of these about 165 remain purely juvenile for a considerable period — many years in no few instances — or else this early form is capable of reappearing in some part or other of the plant as a reversion-shoot. Here the phenomenon for New Zealand as a whole is dealt most briefly with.

These 165 species belong to 30 families and 50 genera (10 endemic) and 51 are trees, 82 shrubs, 19 ligneous lianes, 10 herbs and 3 water-plants. Apart from the acquatics, in 106 cases the juvenile is the more mesophytic, in 17 cases the more xerophytic, and in 39 cases there is no appreciable difference.

Perhaps the most noteworthy examples of the phenomenon are those where the persistent juvenile is of a different life-form from the adult and it may be so persistent that in some species a tree may be of one life-form below and of another above. Sometimes both may flower, or the juvenile may bloom before the adult stage is reached[1]). Thus, to all intents and purposes there are two "species" in the one plant. This is more plainly indicated when a flowering juvenile is restricted to a particular habitat which supplies conditions antagonistic to the adult stage, as when the xerophytic divaricating shrubs of the open, *Pittosporum divaricatum* (Fig. 31.) and *Corokia Cotoneaster* grow in a moist forest atmosphere they remain at the bushy-shrub (mesophytic) stage.

There is usually no abrupt transition from juvenile to adult but intermediate forms occur, though generally not till the juvenile has reached its full development. Nor can the juvenile always be treated as *one* distinct entity: on the contrary, there is frequently an early stage — the *seedling* —, a second stage — the *juvenile* — and perhaps a third stage — the *semijuvenile* — which may be long-persisting. The seedling stage is very often amply distinct from the adult but its duration too short for it to be classed as persistent. Such is the case with *Hebe* in general, e. g. *H. Traversii* with mesophytic ciliated petioled seedling leaves, but those of the adult sub-xerophytic glabrous and almost sessile; or the toothed seedling leaves of *H. buxifolia* var. *odora;* or the thin pinnatifid leaves of the whipcord series. These seedling stages are of such limited duration only because the seedlings occupy habitats unsuitable for their continuance, as shown by the permanence of the juveniles of the whipcord series of *Hebe*, of *Discaria toumatou,* and of the species of *Carmichaelia,* when grown in moist air.

This last statement leads up to the point that a distinct relation may frequently be observed between the particular forms (seedling and adult) and their environment. Thus the broad-leaved seedling of *Anisotome filifolia* and the rather grass-like seedlings of the highly xerophytic *Aciphyllae* live under more mesophytic conditions than the adults; the persistent seedling of *Rubus cissoides* (incorrectly accorded the varietal name "*pauperatus*") in the full sunshine remains an entangled mass of smooth green stems which never blossom and prickly leaf-midribs without laminae, but not only in forest does it become a leafy flowering scandent plant, but in the open a

---

1) The following have been observed blooming when juvenile: — *Dacrydium intermedium, D. laxifolium, Podocarpus dacrydioides, Clematis indivisa, Ranunculus Lyallii, Pittosporum divaricatum, Weinmannia racemosa, W. sylvicola, Pennantia corymbosa, Aristotelia fruticosa, Discaria toumatou, Plagianthus betulinus, Hoheria sexstylosa, Edwardsia microphylla, Nothopanax simplex, N. Edgerleyi, Anisotome filifolia, Dracophyllum arboreum* (on reversion-shoots), *Parsonsia heterophylla* and *P. capsularis.*

very little shade brings out laminae and even flowers; it is in moist places close to the ground that mesophytic reversion-shoots appear on shrubby cupressoid *Padocarpaceae,* cupressoid (whipcord) *Hebe* and *Phyllocladus alpinus;* the long, thin, juvenile leaves of *Knightia excelsa* of the forest interior stand out in strong contrast to the "almost woody" (CHEESEMAN, 1906: 606) leaves of the adult, raised high above the general forest roof; the much-cut juvenile leaves of *Schefflera digitata* appear only under hygrophytic surroundings; finally, there are the differences between the juvenile climbing and adult flowering forms of the *Metrosideros* lianes, the latter indeed being also facultative shrubs of the open.

The case of *Edwardsia* comes naturally at this point. In New Zealand there are at least 4 species. The most common, *E. microphylla,* goes through a divaricating shrub-form which persists for about 15 years, and then is succeeded by the ordinary deciduous tree-form. But the xerophytic *E. prostrata,* which comes true from seed, and grows only in a xerophytic habitat, is neither more nor less than a permanent form of juvenile *E. microphylla,* while *E. chathamica* is almost of adult "microphylla" form from its commencement, and *E. tetraptera* never gives a hint of the divaricating-shrub stage.

If an epharmonic explanation can be applied to the foregoing and other cases not cited, there are instances where such cannot be urged. For example, there is the remarkable heterophylly in the well-named *Parsonsia heterophylla,* but even in that case the closely-allied *P. capsularis* is sometimes a persistent semi-juvenile. Nor can the dimorphy or heterophylly of the following be epharmonically explained: *Clematis indivisa, Pseudopanax crassifolium* var. *unifoliolatum* (a contrary example to *Knightia*), *P. ferox* and *P. lineare.* Nor when the life-forms of companion plants are considered is it easy to see why those mesophytic trees with long-persisting xerophytic juvenile forms should possess such developmental differences; so, too, with the small forest-trees and shrubs with juveniles more mesophytic than the adults. In fact, taking a broad view of the question and remembering that certain genera (e. g. *Pennantia, Parsonsia)* exhibit the phenomenon in New Zealand but do not do so elsewhere, it may perhaps be assumed that the dimorphy dates back to a former period when the species experienced ecological conditions different from those they now experience. Also the question of somatic segregation and hybrid origin must not be overlooked.

The matter of reversion-shoots must be briefly considered. Apparently each species has its special limitations as to where they can appear. Thus it is only to the height reached by the persistent juvenile that reversion-shoots appear in a considerable number of species[1]) and above this point

1) These species include *Pseudopanax crassifolium* var. *unifoliolatum, Pennantia, Plagianthus betulinus, Hoheria angustifolia, H. sexstylosa, Pittosporum patulum, P. Turneri, P. obcordatum, Edwardsia microphylla, Helichrysum coralloides* and its allies;

there is a region where intermediates occur. In *Elaeocarpus Hookerianus* reversion-shoots occur at at least 12 m. from the ground; in *Podocarpus dacrydioides* and *Dracophyllum arboreum* they reach the topmost twigs. Again, some plants develop either into adults or remain permanent juveniles which flower and fruit, e. g. *Nothopanax Edgerleyi*, *Dacrydium intermedium*, the species of *Weinmannia* and *Ranunculus Lyallii* (very rarely).

The duration of the juvenile stage differs from that where it lasts for only a few weeks to that persisting for 60 years or more as in the cases of *Podocarpus dacrydioides* and *Elaeocarpus Hookerianus*. But in the last examples and many others the rate of development is in harmony with the degree of illumination.

2. Pollination.

If the species confined to the lowland-lower hills' belt be alone considered omitting the wind-pollinated *Gramineae* &c., the total number of species dealt with here is 372 of which the flowers of 51 per cent. (orchids omitted) may be considered monoclinous and those of 49 per cent. diclinous. But these figures do not sufficiently emphasize the importance of allogamy for, as well as orchids in general, many hermaphrodite species are dichogamous, but few reliable details as to protandry &c. are available.

About 59 per cent. of the 372 species possess — as judged from the highly unsatisfactory human standpoint — fairly attractive flowers, but those of the remainder are both dull in colour and very small, while some included in "attractive" hardly deserve their place. The attractive flowers are of the following colours to each of which is appended the number of species so coloured: — whites 79, yellows 41, blues and purples 48, reds 24, greens 24 and browns 4. In some cases the colours are combined, especially in the *Orchidaceae*, and the more brilliant colour, though less in area than the groundwork, is by far the more striking. The most showy flowers occur in the genera *Elytranthe*, *Clematis*, *Ixerba*, *Edwardsia*, *Pennantia*, *Hoheria*, *Metrosideros*, *Hebe*, *Veronica*, *Colensoa*, *Olearia* and *Senecio*. At least 25 per cent. are more or less sweet-scented but I have insufficient details, as also is the case with regard to nectar-flowers. There is no doubt, however, that insects play a most important part in pollinating the lowland plants. Certain of the species are pollinated in part or exclusively by birds and such is probably the case in the following species and genera some, by the bye, not purely lowland: *Phormium*, *Knightia*, *Clianthus*, *Edwardsia*, *Metrosideros scandens*, *M. lucida*, *M. robusta* and perhaps other species, *Fuchsia excorticata*, *F. perscandens*, *Vitex* and *Rhabdothamnus*. Cleistogamy

---

but it is not safe to be dogmatik, for not enough observations have been made, and in some instances reversion-shoots have been observed considerably above their theoretical limit.

has been recorded by G. M. Thomson in *Viola filicaulis, V. Cunninghamii* and *Melicope simplex*.

3. Dissemination.

In the lowland-lower hills' flora there are only about 13 per cent. of the species which bear berries or drupes likely to be eaten by birds. It is specially noteworthy that only 3 of such species are not forest plants, that is the remainder of the forest species have moved along with the forests and not through the agency of birds, though a few may have been carried by other agencies, while the three exceptions are by no means particularly common plants nor do any of them nearly extend throughout both islands. Further, the trifling effect of bird-carriage is shown still more plainly in that only 24 species suitable for such extend throughout both islands, 8 halt near lat. 38^0 and 18 near lat. 42^0 while the remainder occur in either one locality or in a very limited area.

Far more efficacious for long-distance travel are those disseminules which adhere to the feathers &c. of birds, e. g. the species of *Uncinia* (only 1 throughout) and *Acaena* (several quite local). But other fruits have likewise considerable adhering-power, also the number of all classes of seeds &c. which can be carried in mud by birds is considerable.

With regard to seeds &c. specially suitable for wind-carriage those species with minute disseminules come first (orchids 152 in all and pteridophytes 101) and of such 39 per cent. occur throughout both islands, but for the orchids alone only 21 per cent. Species provided with a more or less efficient flying apparatus *(Clematis, Epilobium, Compositae* &c.) number 52 but only 27 per cent. of these occur throughout both islands and some are purely local. *From the figures of this paragraph it is again clear that special facilities for travel do comparatively little.*

4. Seasonal Changes.

Although there is a distinct response in New Zealand lowland plants to the seasons, yet this is rather the result of long "habit" than of any truly great difference between summer and winter. It is true that in the great majority of species there is a longer or shorter winter resting-period but, at the same time, there are species which find that season eminently suitable for the development of flowers. Thus during winter in such lowland forest as they occur the following are in full bloom: — *Dysoxylum spectabile, Nothopanax arboreum*, various species of *Coprosma* especially *C. grandi-folia, C. robusta* and *C. spathulata.* During August a few northern species come into bloom in the North Auckland forest, especially *Edwardsia cha-thamica, Pittosporum cornifolium, Rhabdothamnus Solandri* and *Alseuosmia macrophylla.* Even on the Canterbury Plain, a race of *Edwardsia microphylla* flowers during August. But whenever either of the above species of *Edwardsia* flower just at the time of blossoming the tree loses its leaves only to have them replaced almost immediately. So, too, the deciduous habit of *Hoheria sexstylosa* and *H. angustifolia* is correlated not with winter but with blossoming. On the other hand *Plagianthus betulinus*, which comes into leaf from August to September according to latitude, loses its leaves in the autumn (Fig. 30), as do also either completely or partially *Muehlenbeckia australis, M. complexa, Carmichaelia grandiflora, C. angustata, C. odorata, C. gracilis, Aristotelia serrata, Discaria toumatou, Hoheria glabrata, H. Lyallii, Fuchsia excorticata, F. perscandens, F. procumbens* (coastal), *Olearia virgata, O. laxiflora, O. lineata, O. Hectori, O. odorata, O. fragrantissima* and *Senecio Hectori*.

If September, October and November be considered spring; December, January and February summer, and March, April and May autumn, then we may speak of early spring, mid-spring, late spring, early summer, late summer and autumn, as distinct divisions of the growing-season, each distinguished by certain species coming into bloom. According to those divisions the most floriferous period is late spring together with early summer or roughly speaking from the middle of October to the middle of January, the period opening and closing earlier in the north than in the south. The following are some of the more characteristic species marking the divisions as above: — 1. (Early spring.) *Clematis indivisa, C. foetida, Wintera axillaris, W. colorata, Pittosporum eugenioides, Rubus australis, Melicope ternata, M. simplex, Aristotelia serrata, Leucopogon fasciculatus, Griselinia littoralis, Geniostoma ligustrifolium, Vitex lucens,* various species of *Coprosma, Olearia rani* and *Brachyglottis repanda*. 2. (Mid-spring.) *Agathis australis,* various Podocarpaceae, *Freycinetia Banksii, Pittosporum tenuifolium, Astelia nervosa, Leptospermum scorparium, Persoonia toru, Olea Cunninghamii, Olearia Hectori* and *Senecio Kirkii*. 3. (Late spring.) Various species of *Nothofagus, Clematis Colensoi, Beilschmiedia tawa, Ixerba brexioides, Carpodetus serratus, Weinmannia racemosa, Pennantia corymbosa, Discaria toumatou, Melicytus ramiflorus, Leptospermum ericoides, Epilobium pallidiflorum, Nothopanax simplex* and *Olearia arborescens*. 4. (Early summer.) *Phormium tenax, Cordyline australis, Dianella intermedia, Elytranthe Colensoi, Knightia excelsa, Weinmannia sylvicola, Hoheria angustifolia, Hoheria glabratra, Metrosideros lucida, M. robusta, Gaultheria oppositifolia, Dracophyllum longifolium, Olea montana, Parsonsia heterophylla, Hebe salicifolia, H. angustifolia*. 5. (Late summer.) *Arundo conspicua, Gahnia xanthocarpa, Astelia Solandri, Rhopalostylis sapida, Hoheria sexstylosa, Nothopanax Edgerleyi, Schefflera digitata, Dracophyllum latifolium, Olearia ilicifolia*. (6. Autumn.) *Astelia Cunninghamii, A. trinervia, Earina autumnale, Bulbophyllum tuberculatum, Hoheria populnea, Metrosideros scandens, Meryta Sinclairii, Pseudopanax crassifolium, Suttonia australis* and *Shawia paniculata*.

The above account of seasonal changes, and so in other parts of the book, is inadequate and unsatisfactory, since the matter cannot be compressed into so small a compass nor can differences due to latitude and habitat be indicated.

5. Epharmonic modifications.

The more uniform conditions of climate ruling in the lowlands than on the coast or in the high mountains lead to less epharmonic change and to greater uniformity in the associations. All the same, the amount of epharmonic modification is striking enough. Here a few examples must suffice.

Life-forms of identical jordanons within and outside the forest. Evidently identical jordanons (simple species, simple varieties of a compound species), which occur both within and outside forest, are subject to very different ecological conditions and differences in their life-forms may well be expected. Thus, trees with long unbranched trunks in forest are usually branched to the ground and densely leafy in the open. Certain tall bushy-shrubs of the open, in dense forest develop long, liane-like stems which branch only near the apex. In certain forests *Coprosma rotundifolia* is of true tree-form, but outside it is a strongly divaricating-shrub or sometimes an open cushion. The root-climbing lianes *Metrosideros scandens* and *M. perforata* in the open are dense bushy-shrubs (Fig. 33) made up of the ultimate climbing shoots. How plastic are these species is shown by the fact that when the shrub-form comes in contact with a vertical rock or tree-trunk ordinary climbing-shoots provided with roots are soon put forth. Perhaps the most striking case is that of *Rubus cissoides*, already referred to, which in forest is an ordinary leafy bramble which flowers and fruits freely, but in the open is a globose or flattened, green, bushy mass of flexible stems which never flower, yet any portion, if fairly shaded, will bear flowers. On the contrary, the epharmonically-similar, but not closely related, *R. subpauperatus* blooms freely when fully exposed to sun and wind. *Clematis afoliata*, too, is leafy in forest, but flowers best in the open. Examples of a similar character are *Pteridium esculentum* as a scrambling liane in low *Leptospermum* forest, *Gleichenia microphylla* with the segments of the sun-leaves so bent as to form small pouches, but the shade-leaves of the same individual quite flat, and *Lycopodium ramulosum* sterile in shade but producing sporophylls in bright light.

Effect of a specially moist atmosphere or wet ground. Various ferns, especially *Blechnum fluviatile*, *B. lanceolatum* and *Dryopteris pennigera* develop short, woody trunks when growing on wet ground in forest. The knees of *Laurelia*, so common in swamp-forest, are absent when the tree grows under dry conditions. In western Fiord district forest not only is *Weinmannia racemosa* raised higher than usual above the ground on massive irregular stilt roots (Fig. 29), but *Nothofagus Menziesii* — "normally" with a solid buttressed base — behaves similarly. *Schefflera digitata*, which usually has finely-toothed juvenile leaves similar to the adult, in an especially moist forest-atmosphere may have them deeply cut after the manner of juvenile *Nothopanax simplex*. *Phormium tenax* growing in dry soil possesses numerous root-hairs, but in very wet soil they are wanting. In the forests of the wettest parts of South Island the leaves of species in general are larger than usual and in some species they are of surprising size, e. g., *Rubus australis, Aristotelia serrata, Melicytus ramiflorus, Olearia arborescens.*

Life-forms of identical jordanons with different life-histories. In the case of one and the same jordanon commencing life either as an epiphyte or a ground-plant the adult form may be quite different. Thus, according as *Metrosideros robusta* commences its career as an epiphyte on a lofty tree or a seedling on the forest-floor, so is the adult, on the one hand, a massive tree with an irregular trunk, but on the other hand, a trunkless bushy-tree. So, too, if *Weinmannia racemosa* commences life as an epiphyte on a tree-fern, or a seedling upon a fallen tree trunk, it ends as a tall tree, but if it commences as a ground-plant it ends as a bushy-tree.

Effect of a specially dry climate. The ultimate shoots of *Hymenanthera alpina, Aristotelia fruticosa* and *Carmichaelia Petriei* become truly spinous; the cushion-form is induced in various semi-woody or shrubby plants of an open habitat; the brown hue of the common tussocks changes to green when they are cultivated in good soil in a forest-climate; fairly tall shrubs are considerably reduced in stature; leaves are greatly reduced in size and attain their maximum thickness; leaves of grasses and sedges flat in moist air are more or less tightly rolled.

The deciduous habit. Very few of the trees or shrubs are truly deciduous under all circumstances, but the more frosty the climate the stronger the habit. For instance, *Fuchsia excorticata, Muehlenbeckia australis, Aristotelia serrata* and probably *Senecio Hectori* fluctuate according to circumstances from truly deciduous to evergreen. Also closely related jordanons may differ in this regard.

Chapter IV.
The Plant Communities.

1. Forest.

a. Introductory.

General. New Zealand as a whole, owing to its high average rainfall of a fairly even distribution all the year round, and its equable climate, was originally covered from north to south with a close mantle of forest, except where the edaphic conditions were antagonistic, or the rainfall insufficient to meet the demand of the various factors influencing transpiration, especially wind. Even yet, settlement notwithstanding, considerable areas are clothed with noble forest, extending in no few places from virtually high water-mark to the subalpine belt. In north-western Auckland, in the East Cape district, on the Volcanic Plateau and the adjacent Wanganui coastal plain, on the slopes of the North Island mountains, on most of the land west of the Southern Alps and on the east up to the average limit reached by the westerly rain, in parts of the South Otago and Stewart districts, forests still exist in no whit different from those visited by the early botanists[1]), or, as yet so little modified by artificially-wild, grazing and browsing animals as to be virtually primeval. Nevertheless, to give an accurate account of the New Zealand forests is no longer possible and, day by day, a reliable presentation of the subject becomes more difficult. Thanks, however, to the assistance of the State Forest Service, and to my connection therewith, during the last few years I have had an opportunity to study the forests more closely than previously, so that the details given here replace for the most part those of the first edition of this work. The main difficulty has been to draw the line between a general account of the forests as a whole and one attempting to deal with the many details which close study has revealed — details which, if too much stressed, would hide the really important facts and give a blurred picture rather than a clear vision of the great tree-community of New Zealand.

The forest, as a whole, comes into the same class as **tropical rain-forest,** but it can be naturally divided into **subtropical rain-forest** — a hygrophytic formation — and **subantarctic rain-forest** — a formation more or less meso-phytic in character, and with *Nothofagus* dominant. Speaking in general terms, the lowland-tree associations — even those of swamps — are mostly sub-tropical rain-forest, the species of *Nothofagus* being absent over wide areas or, if present, only occasionally forming pure associations.

The lowland forest in all its modifications consists of about 385 species (families 65, genera 146); pteridophytes and spermophytes number respectively 95 and 290. The following are the most important families and genera together with the number of species to each: — (families) *Filices* 90, *Rubiaceae* 28, *Orchidaceae* 23, *Cyperaceae* 20, *Podocarpaceae* and *Myrtaceae* each 16, *Compositae* 15, *Liliaceae* and *Pittosporaceae* each 10, (genera) *Coprosma* 25, *Hymenophyllum* 18, *Pittosporum* and *Metrosideros* each 10, *Blechnum* 9, *Uncinia* 8 and *Trichomanes* and *Podocarpus* each 7. Leaving the pteridophytes on one side 89 per cent. of the spermophytes are endemic and most of the remainder Australian. *Taking the woody species alone, all except 1 are endemic.* The following genera, all but 2 being monotypic, are endemic and confined to forest: — *Tupeia, Dactylanthus, Ixerba, Car-podetus, Alectryon, Tetrapathaea, Teucridium, Rhabdothamnus, Alseuosmia*

---

1) Thanks to the Scenery Preservation Act of 1903, many areas in these localities, and elsewhere, have been permanently set aside for the preservation of the indigenous plants and animals. There are also several national parks of great extent and climatic reserves on the mountains that serve a similar purpose.

and *Colensoa*. Other endemic genera, but which occur in other formations, are *Loxsoma*, *Hoheria* and *Brachyglottis*.

The life-forms of lowland-lower hills' forest and the number of species to each are as follows: — trees 108, shrubs 59, herbs 43, grass-like plants 24, semi-woody plants 5, lianes 26 (excluding ferns), epiphytes 17 (excluding ferns), parasites 13, ferns 90 (tree-ferns 8, filmy ferns 25).

So far as physiognomy is concerned, there is a considerable resemblance between the various forest associations throughout their lowland range, but one rather ecological than floristic. This likeness is owing to the following circumstances: — (1.) The presence of podocarps with one or other of their species frequently dominating. (2.) The great abundance of tree-ferns. (3.) The vast number of ground-ferns (Fig. 32) many of which are of wide distribution, and the presence throughout in quantity of *Asplenium bulbiferum*, *Blechnum procerum* and *B. discolor*. (4.) The presence in profusion of the lianes *Muehlenbeckia australis*, *Rhipogonum scandens*, *Metrosideros hypericifolia* and *Rubus australis*. (5.) The density of the undergrowth and the presence therein of certain low trees or tall shrubs, especially *Carpodetus serratus*, *Pittosporum tenuifolium*, *Melicytus ramiflorus*, *Myrtus pedunculata*, *Nothopanax arboreum*, *Schefflera digitata*, *Pseudopanax crassifolium* var. *unifoliolatum*, *Suttonia australis* and *Coprosma robusta*. (6.) The abundance of filmy-ferns, mosses, liverworts and, if not too dense, large foliaceous lichens.

Ecological conditions in general governing lowland-lower hills' forest. Leaving on one side all special details concerning the ecological factors, the following are of prime importance, so far as the lowland-lower hills' communities are concerned: — (1.) A rainfall (except in the case of semi-swamp forest) exceeding 90 cm. distributed more or less equally throughout the year. (2.) A sufficient supply of available water at all seasons in the surface-soil. (3.) A winter temperature not falling below — 10° C. (4.) The peculiar conditions brought about by the plants themselves, which define, in large measure, the wind, light, heat, air-moisture, soil-moisture and humus factors. These last-mentioned conditions come gradually during the development of forest and render possible the various phases of succession. Usually, the rainfall is far greater than as given above, but *with its increase neither the ecological nor the floristic composition of a forest is greatly affected*. The even distribution of rainy days throughout the year is a matter of especial importance, for a drought of several months' duration will actually cause the death of several endemic species of trees. As for the heat factor, each species has its special requirements and limitations, and the degree of cold, cited above, is far greater than many plants can tolerate.

The light relation of each species is fundamental with regard to the structure of forest, and it is the main factor regulating succession, as will be seen further on. Here it need only be pointed out that the species differ greatly in their shade-tolerating capacities, and that they range from those which demand bright light to those which grow only in deep shade, by way of species which tolerate many different degrees of illumination.

New Zealand forest as a whole is most catholic in regard to the nature of the soil it, and perhaps nearly all its members, can occupy, though the latter may have individual "preferences" in certain localities for a particular soil. Forests of much the same floristic and ecological character occupy soils where the underlying rock is greywacke, granite, volcanic rocks of various kinds, calcareous rocks of different categories, and mica-schist; in short, pretty well all the geological formations which form the land-surface of the region. Further, forest occurs on stiff clays, rich alluvial soil, shingly river-bed, ancient morainic deposits, pumice and other volcanic ash, sand-dunes, badly-drained ground and shallow swamps.

The ecological conditions of any forest-area are far from uniform. Much depends upon the topography, — slopes ridges, flats and gullies, each possesses its special plant-covering in harmony with the differences in ground-water, depth of humus, air-moisture, light-intensity and shelter from wind. An interesting case is presented by the forest of the Wanganui coastal plain, where, on the slopes, there is hygrophytic tawa (*Beilschmiedia tawa*) forest and on the ridges a pure semi-xerophytic association of *Nothofagus Solandri*, even if these trees be almost in single file.

Distribution of the species. A census of the forest species shows that North Island contains 357 species (pteridophytes 95, spermophytes 262) but in proceeding from north to south 46 either do not reach lat. 38° or extend but a short distance beyond, South Island 312 (pteridophytes 87, spermophytes 225) of which in proceeding from north to south 69 do not reach or hardly overstep lat. 42° on the west or lat. 44° on the east, and Stewart Island 147. About 150 species extend throughout most of North and South Islands, 110 of which extend to Stewart Island.

With regard to the individual species, each has its latitudinal limit, so that in proceeding from north to south, or *vice versa*, species drop out or come in, but from about lat. 38° there is chiefly a more or less gradual dropping out. A few species (about 8) are confined to the north of lat. 36° or somewhat beyond, e. g., *Microlaena Carsei, Dacrydium Kirkii, Pittosporum pimeleoides, P. reflexum* (and the hybrids between the two last), *Ackama rosaefolia* (greatly resembling juvenile *Weinmannia sylvicola*), that great swarm of hybrids and jordanons — *Alseuosmia* and *Colensoa physaloides*. Then in the Thames subdistrict a considerable change occurs in the forest various common southern plants having joined the community, but they are confined usually to the montane belt, e. g. *Hymenophyllum pulcherrimum, H. peltatum, Trichomanes Lyallii, Lindsaya viridis, Blechnum Patersonii* var. *elongatum, B. vulcanicum, B. penna marina*, the physiognomic *Polystichum vestitum, Phyllocladus alpinus, Uncinia ferruginea, Enargea parviflora, Cor-*

dyline indivisa, Nothofagus fusca (confused by most writers with *N. truncata*), *N. Menziesii, Elytranthe tetrapetala, Weinmannia racemosa, Aristotelia fruticosa, Nothopanax simplex, N. Colensoi, Coprosma foetidissima, C. Colensoi*. Perhaps the most striking feature of this list is the sudden appearance (or halt in their northern march) of trees and shrubs of high physiognomic importance whose structure certainly should not have forbidden their reaching further north.

The neighbourhood of latitude 38^0 south, as emphasized more than once in this book, is a highly critical point in regard to plant distribution, some northern species hardly reaching it, but others overstepping it and extending, for a greater or lesser distance, into the East Cape and Egmont-Wanganui districts. The following is a selection of characteristic forest species belonging to this class: — *Lygodium articulatum, Phyllocladus glaucus, Agathis australis, Persoonia toru, Beilschmiedia taraire* (but as a common species this has gone out long before), *Litsaea calicaris, Ixerba brexioides, Quintinia serrata, Weinmannia sylvicola, Metrosideros albiflora, M. carminea (diffusa), Corokia buddleoides* and *Vitex lucens*. But this discarding of species is balanced, so far as forest is concerned, by the incoming of others, some of them local endemics (e. g. *Polypodium novae-zelandiae. Pittosporum Ralphii, Edwardsia tetraptera, Coprosma tenuifolia* and *Jovellana Sinclairii)*, and the remainder with a more or less wide range to the south, the following being specially important: — *Alsophila Colensoi, Hypolepis Millefolium, Podocarpus alpinus* (mainly subalpine until the North-western district) *Nothofagus Solandri, N. cliffortioides, Elytranthe Colensoi, Hoheria sexstylosa, Coprosma Banksii* (× *C. colbanksii* appears), and *Shawia paniculata*. This forest from lat. 38^0 southwards, so soon as its far northern plants are cast aside, extends not only throughout the remainder of North Island but occupies much of the Sounds-Nelson and North-western districts, the forest communities of these two districts being marked off from those typical of South Island generally by the following North Island species: — (lianes) *Blechnum filiforme, Freycinetia Banksii, Metrosideros scandens, M. perforata, M. Colensoi,* (the palm) *Rhopalostylis sapida* (trees and tall shrubs). *Knightia excelsa, Laurelia novae-zelandiae, Suttonia salicina, Beilschmiedia tawa, Dysoxylum spectabile, Metrosideros robusta, Myrtus bullata,* × *M. bullobcordata, Melicope ternata,* × *M. tersimplex, Olearia rani, Coprosma grandifolia* and *Brachyglottis repanda.* Certainly a few of these species extend to the Western and even the Fiord districts and to Banks Peninsula, but the forests of the last two cannot be grouped along with those of the southern North Island, the Sounds-Nelson and the North-western districts.

General principles governing succession. The most important principle underlying succession in New Zealand forests is the relation of the different species to light; also, the gradual incoming of humus on the forest-floor from decay of leaves, dead wood &c. is of fundamental importance, leading as it does to the formation of seed-beds, and the essential station for the establishment of the all-important hygrophytic fern and bryophyte content of the floor vegetation together with many other species dependent on those conditions which humus supplies. Thus in the forest near Lake Rotoma (VP.), where in 1886 the floor was covered with volcanic ash by the eruption of Mount Tarawera, even yet the customary filmy fern-bryophyte mats are absent. Subsidiary to the foregoing are the individual edaphic and biotic (more especially influence of plant on plant) relations but these make themselves manifest rather in diversity of composition and structure of the undergrowth than in succession as a whole. That is to say, there are minor successions within the actual general successions. The latter, in fact, are governed by the tall trees which for the time are dominant. This statement is made more plain when discussing the life-history of various types of forest.

To observe the actual beginnings of certain classes of primeval forest is hardly possible at the present time, all there is to go on being: (1.) the composition and structure of belts of shrubs or small trees at the margin of forest, (2.) the progress of events in *Leptospermum* shrubland, (3.) young trees or shrubs invading *Pteridium* heath or tussock-grassland, (4.) succession on river-bed or river-terrace culminating in forest, (5.) swamp-succession terminating in forest, and (6) regeneration or reinstatement after forest is destroyed or damaged. Succession after burning, so far as those forests near active volcanoes are concerned, must be very similar to what happened when such volcanoes were far more active than at present, and probably certain pieces of forest on the Volcanic Plateau, or even Mount Egmont, originated after destruction of the original forest by fire, and this must have taken place again and again, even where now there are only extinct volcanoes (the Tarawera eruption is a case in point). *Fire thus may be not an artificial but a natural agent.* In what follows, all of the above cases are taken into consideration, though whether induced successions are identical with natural successions (except perhaps fire) no one can say.

Forest species may be grouped into: (1.) **light-demanding,** (2.) **shade-tolerating,** and (3.) **shade-demanding,** the second class consisting of species which are more or less indifferent in regard to excess of light or shade. To be sure, there are many grades in each class but, for the purposes of a general account of succession, the light-demanding and the shade-tolerating come together, and it is these which form the earliest stages of forest. Thus, the extreme light-demanders which get established will be augmented by species able to endure a little shade, and so on until true forest with its undergrowth of shade-demanding and shade-tolerating species has come into being. This process is by no means rapid, since the majority of the trees, as already explained, are of comparatively slow growth.

The next principle to grasp is that the tall trees, dominating forest and forming its roof, as seedlings and saplings are partly light-demanding and partly shade-tolerating or shade-demanding, the first class being evidently suitable for temporary successions and the other two for climaxes.

The most extreme light-demanding species concerned with the establishment of forest are the shrub (or small tree), *Leptospermum scoparium*, and the fern, *Pteridium esculentum*, but both frequently grow so thickly that, until open spaces come into being, few other species, if any, can gain a footing. Next in importance, though it is difficult to say what part it played in primitive New Zealand, is the small, fast-growing tree, *Aristotelia serrata*, which so commonly forms more or less pure associations after forest is felled but is quite rare in adult forest. The following are other light-demanding or, some of them, shade-tolerating species which become established in bright light: *Histiopteris incisa*, *Paesia scaberula*, *Gleichenia microphylla*, *G. circinata* (Filic.), *Lycopodium volubile*, *L. densum* (Lycop.), *Agathis australis* (Araucariac.), all species of *Podocarpus*, all species of *Dacrydium* (*D. intermedium* and *D. Colensoi* tolerate a good deal of shade), the species of *Phyllocladus* (Podocarpac.), *Gahnia pauciflora*, (Cyperac.), *Cordyline Banksii*, *C. indivisa*, *Dianella intermedia* (Liliac.), all species of *Nothofagus* (Fagac.), *Persoonia toru*, *Knightia excelsa* (Proteac.), the forest species of *Muehlenbeckia* (Polygonac.), *Quintinia acutifolia*, *Carpodetus serratus* (Saxifrag.), species of *Pittosporum* of the "*tenuifolia*" group (Pittosporac.), the species of *Weinmannia*, *Ackama rosaefolia* (Cunoniac.), the climbing species of *Rubus* (Rosac.), the arboreal species of *Edwardsia* (Legum.), *Coriaria arborea* (Coriariac.), *Pennantia corymbosa* (Icaciniac.), *Aristotelia fruticosa* (Elaeocarp.), *Plagianthus betulinus*, the species of *Hoheria* (Malvac.), *Melicytus ramiflorus* (Violac.), *Leptospermum ericoides*, *Myrtus bullata*, *M. obcordata* (Myrtac.), *Fuchsia excorticata*, *F. perscandens* (Onagrac.), *Nothopanax anomalum*, *N. arboreum*, *N. Colensoi*, *Pseudopanax crassifolium* (Araliac.), *Griselinia littoralis*, *Corokia Cotoneaster* (Cornac.), *Cyathodes acerosa*, *Leucopogon fasciculatus* (Epacrid.), *Suttonia australis*, *S. divaricata* (Myrsinac.), *Geniostoma ligustrifolium* (Loganiac.), the species of *Parsonsia* (Apocynac.), *Hebe salicifolia* (Scroph.), *Myoporum laetum* (Myoporac.), *Coprosma lucida*, *C. robusta*, *C. propinqua*, *C. rhamnoides*, *C. arborea*, *C. parviflora* (Rubiac.), *Olearia rani*, *Helichrysum glomeratum* and *Brachyglottis repanda* (Compos.). With regard to some in the above list, and others omitted, their toleration of bright light is usually more or less dependent on the air being fairly moist.

Tree-ferns (Fig. 23), so important in forest undergrowth generally, tolerate a good deal of light, as may be seen where large colonies of *Cyathea medullaris* have been established in the open by means of spores (SA., EW.). But *C. dealbata* is the commonest tree-fern in early stages of forest.

Coming now to actual succession, sufficient is not known and may never be known as to the sequence, composition and duration of the various successions of forest which, by degrees, lead to a more or less stable climax-association. For one thing, so many minor successions in one and the same piece of forest becloud the general progress. Thus it seems to me, though this statement may eventually be disproved, there is rather a general move towards a climax except in the earlier stages of associations of light-demanding and light-tolerating species, than well-marked associations (successions), the one succeeding (replacing) the other.

The first contribution in New Zealand to this fundamental question of succession in forest was my account of the development of kauri forest (1908: 30, 31). Later, in the first edition of this work, I put forward with some hesitation the view that the general podocarp forest of North Island was turning by degrees into a climax with the tawa *(Beilschmiedia tawa)* dominant, and that the podocarps of South Island forest, to the south of lat. 42°, would be eventually replaced by *Weinmannia racemosa*. Independently, E. H. WILSON (1921: 235) has expressed the opinion that the kauri and podocarps were being gradually replaced by broad-leaved dicotylous trees. E. B. LEVY (1923) has added strong proof to my tawa theory and L. M. ELLIS (Director, State Forest Service) has told me of a definite and striking example of replacement by tawa he had observed in the Volcanic Plateau district. But special details regarding the life-history of lowland forest are given below when dealing with the different groups.

Principles upon which the classification of New Zealand forests is based. The primary classification of the forest communities into **subtropical** and **subantarctic** rain-forest, suggested by me (1926 : 7) at first thought seems sound enough. But there are wide areas covered by *mixed* forest where podocarps, the usual broad-leaved dicotylous trees and *Nothofagus* are present in abundance, and where the forest interior is almost, if not quite, as hygrophytic as that of subtropical rain-forest. It also seems easy to separate the forests on altitudinal lines, but here again a difficulty arises, for certain North Island high-mountain forests, extending from 900 m. to 1200 m. altitude, are but little different from some of those of the lowland Fiord district.

It is the secondary divisions which are the main stumbling-block. Naturally two important points are their floristic composition and ecological features. The former certainly depends largely on latitude and local endemism but, ecologically, there is little difference between a North Island and Stewart Island forest; so, too, with their, structure. Two features, however, stand out for subtropical rain-forest: that is, it is composed so far as trees go partly of *Podocarpaceae*, *Cupressaceae* or *Araucariaceae*, and partly of broad-leaved dicotylous trees. Unfortunately, for purposes of classification, there is every intermediate stage between a pure podocarp or kauri community and one made up of dicotylous trees alone. Again, a genetic

classification suggests itself, but here, too, intermediate stages occur inter-
calated in one and the same piece of forest.

In what follows, latitudinal and altitudinal distribution of species, structure, special ecology and life-histories have all been taken into consideration. It would have been easy to have made many subdivisions but I
have carefully avoided doing so. The subject is treated rather in a general
manner, the main object having been to attempt the presentation of an
accurate picture of the remarkable New Zealand rain-forest unblurred by
superfluous details.

b. The forest communities.
1. Subtropical rain-forest of broad-leaved dicotylous trees and conifers.
a. General.

Sub-tropical rain-forest associations possess so many features in common that, in order to avoid repetition, the following details are submitted.

The trees, shrubs and ferns, with a few trifling exceptions, are evergreen.
As viewed from without, the evergreen character of the trees and the
general absence of bright greens gives, when seen from a distance, a sombre
aspect to the forest, while the density of their growth altogether masks
the height of the trees. But a closer view reveals the varied greens and
it is not difficult in some instances to recognize certain species from their
colour alone, especially in low, even forests of dry ground. An outside
view, too, reveals but little of the tropical character of the forest. A few
tree-ferns may raise their crowns of spreading, feathery leaves above the
greenery, or in North Island and the Sounds-Nelson and North-western
districts, nikau palms (Rhopalostylis sapida) peep forth, but that is all. But
push through the bolt of shrubs, or low trees, that may fringe its outskirts,
and the vision within will be novel enough to one acquainted only with
the temperate forests of the northern hemisphere.

Massive trunks, unbranched for many metres, meet the eye, some
covered so thickly with lianes and epiphytes, many of which are ferns and
bryophytes, that their bark is invisible. Open spaces are few or wanting.
Young trees, shrubs of many kinds and tree-ferns 5 to 10 m. tall, growing
in clumps or isolated, closely fill the gaps between the tree-trunks. Rope-
like stems of lianes depend from the forest-roof swinging in the air, or lie
sprawling upon the ground. The bases of the trees are not infrequently
swollen and irregular, while their roots spread far and wide over the surface,
at times half-buried, or, here and there, arching into the air, and covered
with seedling trees an shrubs, ferns of goodly size, mantles of mosses
and liverworts, lichens and sheets of pellucid Hymenophyllaceae. Fallen
trees, in various stages of decay lie everywhere, and these too, hidden
by a garb of water-holding greenery, are the home of seedlings innu-

merable. The actual forest-floor is most uneven; rotting logs, fallen branches, raised roots, ferns frequently with short trunks and mounds of humus covered with bryophytes and filmy-ferns make walking laborious. Progress generally is considerably retarded, too, not merely by the above-mentioned obstacles, or by the close-growing shrubs or spreading branches, but a coarse network of the almost black, stiff stems of the liliaceous liane, *Rhipogonum scandens,* forms entanglements beneath which one is compelled at times to crawl on hands and knees. In other places where there is a good deal of light, the hooked prickles on the midribs of *Rubus australis,* catching a garment, may hold one fast. Furthermore, even on level ground, there are water-courses, here and there, and near these the density of the undergrowth increases, lateral branches from the trees on either side meet and become entangled, the growth of ferns becomes thicker, so that progress is well nigh impossible. In hilly forest, the density of gullies is still more intensified.

The close growth, which I have attempted to describe, is in harmony with the moist, equable climate, but regulated by the density of the forest-roof. Everywhere is the effect of that complex of factors evoked by the forest itself manifest in the plant-forms. Shrubs, which in the open would be rounded and symmetrical, put forth long, slender stems, that, liane-like. lean against other trees and gain support. Young trees have frequently much-reduced lateral branches and long, straight, slender main-stems. In some this habit is hereditary, and thus the curious juvenile form of *Pseudopanax crassifolium* var. *unifolialatum* may be an "adaptation" to the forest-life.

On the trunks of most of the trees that do not shed their bark in great flakes, and right up on the highest branches, are not only an abundance of true lianes and epiphytes (Fig. 27), but seedlings of trees and shrubs, ground ferns of many species and hosts of mosses and liverworts. The trunks of tree-ferns, too, are a favourite station for many plants. Even the slender branches of shrubs may be deeply moss-covered while leaves themselves may be the home of various small bryophytes.

Certain forests are not nearly so dense or hygrophytic as described above, nor do they exhibit so fully the various peculiarities as cited, but such occupy drier ground than usual, and even then show unmistakeably their tropical facies.

In every type of New Zealand rain-forest the vegetation is in several distinct layers (stories), each with a definite light-relation. Where the tallest trees are present the uppermost layer (story) consists of their crowns, in some places those of the *Podocarpaceae* or, in certain associations, *Metrosideros robusta,* one or other of the two species of *Weinmannia, Beilschmiedia tawa, B. taraire* or species of *Nothofagus.* Trees of a medium size form the next layer, while growing in their crowns, as also in those of the upper story, are the flowering parts of the lianes and in forests of North Island type the more massive *epiphytes* (e. g. species of *Astelia, Griselinia lucida Pittosporum cornifolium,* young *Metrosideros robusta).* The upper layer does not, as a rule, make a continuous roof, so the light-relation, but not the wind-relation, of these two highest stories is not very different. The third layer is formed by the smaller trees, tallest shrubs, and tall tree-ferns, while in many North Island forests and the Sounds-Nelson and North-western districts, the nikau-palm is conspicuous. Next comes the fourth layer, consisting of the smaller ferns, prostrate or low shrubs, decumbent lianes, tussocks of *Gahnia* or *Astelia* and young plants of various kinds. Finally, there is the layer of actual floor-plants, which is largely made up of small ferns (mostly *Hymenophyllaceae)* and bryophytes, important genera of which are: (Hepaticae) *Aneura, Symphyogyna, Monoclea, Treubia, Chiloscyphus, Frullania, Mastigobryum, Lepidozia, Schistochila, Trichocolea, Plagiochila, Tylimanthus, Madotheca;* (Musci) *Leucoloma, Dicranoloma, Leucobryum, Leptostomum, Hymenodon, Bryum, Echinodium, Ptychomnion, Weymouthia, Mniodendron, Sciadocladus, Lembophyllum, Distichophyllum, Hypopterygium, Cyathophorum, Rhacopilum, Mniadelphus.*

β. Kauri *(Agathis australis)* - broad leaved dicotylous - tree forest.

(1.) General.

Forest of this class is distinguished by the presence of more or less *Agathis australis* together with certain other tall trees especially *Beilschmiedia tawa, B. taraire* (one or both), *Weinmannia sylvicola, Metrosideros robusta* and some (or all) of the usual podocarps. Certain species, rare or wanting elsewhere, except in the Auckland districts, are present, of which (excluding those belonging particularly to the kauri subassociation the following are common or characteristic: — *Blechnum Fraseri, Lygodium articulatum, Dacrydium Kirkii, Mida salicifolia, M. myrtifolia,* the hybrids between these two, *Litsaea calicaris, Melicytus macrophyllus, Dracophyllum latifolium* and the polymorphic *Alseuosmiae.*

The number of species belonging to the community is 231 (pteridophytes 67, spermophytes 164) which belong to 55 families and 117 genera. Lists of the most important species are given when dealing with the associations &c.

The high value of the timber together with the inflammability of damaged *(not virgin!)* kauri forest, has led to an enormous reduction in area of the community, so that, except where reserved, a dozen or so years from now will see it virtually gone for ever. But, extensive as were these forests of present-day New Zealand, they were but the remains or successors of more ancient communities, their site plainly marked by abundance of kauri-resin or in places tree-trunks beneath the surface of the ground. Why these ancient forests vanished none can say, but CHARLES DARWIN who visited a kauri forest in 1835 writes in his *Voyage of the Beagle* in regard to the fernlands near the Bay of Islands, "Some of the residents think that all this extensive open country originally was covered with forests, and that is has been cleared by fire". Probably this surmise is true enough for certain localities, but it can hardly be accepted as a general principle.

The forest-area, before the interference of man, extended from a line joining Doubtless and Ahipara Bays in the north to the Auckland Isthmus in the south, together with the Barrier Islands and the lower slopes of the Thames Mountains. Kauri forest is essentially an association of the lowlands and lower hills and generally does not ascend to much over 400 m. Nor can it tolerate wet ground, so that its abundant remains in lowland bogs indicates change of level in the land-surface (cf. CHEESEMAN 1897 a: 344).

Partly because of its great monetary value and partly because of the dominating appearance of the lordly kauri, the community, no matter its composition, is generally designated "kauri forest". This, however, is a misnomer, for the kauri itself occurs, either as solitary individuals or as large or quite small groups, situated in the general forest-mass, but clearly defined through the presence of the mighty tree unlike any other and the astonishing uniformity in composition of its associated plants.

(2.) Kauri forest in a wide sense.

General. Although there are changes in regard to latitude, and many differences in the relative abundance of species, the whole kauri-forest community may be considered one association and may be conveniently divided into several subassociations.

As a rule the soil occupied by kauri forest is of poor quality from the agricultural standpoint, a fact emphasised by its occupation after the forest is removed by second-class or usually pasture of a much worse character. But there are various local differences of soil within most of the forest areas and these are reflected by the vegetation, dominance of kauri indicating the least "fertile" soil.

The kauri (Agathis australis) subassociation. This plant-community is distinguished by the dominance of *Agathis australis* and its remarkable assemblage of associated species. The number of kauri trees present range from a close growth of such to isolated trees here and there. The associated species are as follows: — *Cyathea dealbata, Dicksonia lanata* (sometimes extremely common), *Blechnum Fraseri, Gahnia xanthocarpa, Astelia trinervia, Freycinetia Banksii, Mida salicifolia, M. myrtifolia, Weinmannia sylvicola* (juvenile), *Phebalium nudum, Dysoxylum spectabile* (juvenile), *Metrosideros scandens, M. albiflora, Nothopanax arboreum, Dracophyllum latifolium, Leucopogon fasciculatus, Geniostoma ligustrifolium, Coprosma grandifolia, Alseuosmia macrophylla* and *Senecio Kirkii*.

The subassociation owes its very characteristic physiognomy partly

to the dense tussock-thickets of *Gahnia-Astelia* (Fig. 34) and partly to
the form of the kauri itself. Where it extends over a wide area, the undergrowth is not thick. The kauri-trunks, usually shining-grey, but sometimes
reddish, rise up on all sides, as far as the eye can pierce, as massive co-
lums 1 to 3 m. diam., unbranched for 20 m. or more (Fig. 35). Round the
base of each tree is a mound of humus, formed from the shed bark, occupied by small tussocks of *Astelia trinervia*, sprawling *Metrosideris scandens*
and *Senecio Kirkii*. Rising up between the giant trunks may be multitudes
of the straight stems of *Beilschmiedia taraire* thrusting their sparse heads
of foliage up to the lower branches of the kauris (Fig. 36.), or *B. tawa*
with its irregular thicker trunk, bryophyte-covered, may be abundant. Between the trees there will be a rather low, open undergrowth consisting of
(the first three species being dominant): *Cyathea dealbata* (trunks 1 m. or
less), *Dicksonia lanata* (25 cm. high, in colonies, the green fronds arching
outwards, their blackish stems shining with a metallic lustre), juvenile *Weinmannia sylvicola* with yellowish-green pinnate leaves, black-stemmed *Wintera axillaris*, slender *Dysoxylum*, graceful young *Beilschmiedia tawa*, Me-
licytus macrophyllus, *Suttonia salicina*, *Coprosma grandifolia*, stemless *Rhopalostylis*, *Alseuosmia macrophylla*, *A. quercifolia*, *A. Banksii*, *A. linariifolia*,
if they be species, and the hybrid swarm in which all take a part, and
juvenile *Podocarpus ferrugineus*. On the ground will be trailing *Freycinetia*,
straggling *Lygodium* (also winding round the young trees), extensive colonies
of *Blechnum Fraseri*, and in some localities the low straggling shrubs *Pittosporum pimeleoides*, *P. reflexum* and their hybrids, and the fern *Schizaea
dichotoma*. High above all rise up the mighty spreading limbs of the kauris
(Fig. 37) and extending to these lace-like foliage of *Beilschmiedia tawa*, or
the darker, denser heads of *B. taraire*. Lianes are not numerous, an occasional *Freycinetia* or *Lygodium* ascend the *Beilschmiedia* trees, but the kauri
itself, owing to its bark-shedding habit, remains inviolate.

Much more common than such extensive kauri communities are groves,
large or small, or solitary trees, scattered through the forest mass. The
kauri trees generally are 20 m. or more distant and the intervening space is
occupied by an extremely thick *Gahnia-Astelia* thicket containing also *Freycinetia*, *Metrosideros scandens*, *M. albiflora*, *Phebalium nudum*, *Senecio Kirkii*
and all the other species already cited as prominent members of the subassociation. Obviously these shrubs &c. are wanting where the *Gahnia-
Astelia* is densest (Fig. 34 but only *A. trinervia* in the picture).

The taraire (Beilschmiedia taraire) subassociation. In this subassociation the kauri is absent and *B. taraire* dominant. It occurs in its full
development to the north of lat. 36° south. The *taraire* trees are about
15 m. high, 3 m. or so apart and their crowns are small but dense. Generally the roof is fairly open. *Metrosideros robusta* is common and as usual
conspicuous through its most irregular trunk full of hollows filled with humus

differ greatly in their shade-tolerating capacities, and that they range from those which demand bright light to those which grow only in deep shade, by way of species which tolerate many different degrees of illumination.

New Zealand forest as a whole is most catholic in regard to the nature of the soil it, and perhaps nearly all its members, can occupy, though the latter may have individual "preferences" in certain localities for a particular soil. Forests of much the same floristic and ecological character occupy soils where the underlying rock is greywacke, granite, volcanic rocks of various kinds, calcareous rocks of different categories, and mica-schist; in short, pretty well all the geological formations which form the land-surface of the region. Further, forest occurs on stiff clays, rich alluvial soil, shingly river-bed, ancient morainic deposits, pumice and other volcanic ash, sand-dunes, badly-drained ground and shallow swamps.

The ecological conditions of any forest-area are far from uniform. Much depends upon the topography, — slopes ridges, flats and gullies, each possesses its special plant-covering in harmony with the differences in ground-water, depth of humus, air-moisture, light-intensity and shelter from wind. An interesting case is presented by the forest of the Wanganui coastal plain, where, on the slopes, there is hygrophytic tawa (*Beilschmiedia tawa*) forest and on the ridges a pure semi-xerophytic association of *Nothofagus Solandri*, even if these trees be almost in single file.

Distribution of the species. A census of the forest species shows that North Island contains 357 species (pteridophytes 95, spermophytes 262) but in proceeding from north to south 46 either do not reach lat. 38° or extend but a short distance beyond, South Island 312 (pteridophytes 87, spermophytes 225) of which in proceeding from north to south 69 do not reach or hardly overstep lat. 42° on the west or lat. 44° on the east, and Stewart Island 147. About 150 species extend throughout most of North and South Islands, 110 of which extend to Stewart Island.

With regard to the individual species, each has its latitudinal limit, so that in proceeding from north to south, or *vice versa*, species drop out or come in, but from about lat. 38° there is chiefly a more or less gradual dropping out. A few species (about 8) are confined to the north of lat. 36° or somewhat beyond, e. g.; *Microlaena Carsei, Dacrydium Kirkii, Pittosporum pimeleoides, P. reflexum* (and the hybrids between the two last), *Ackama rosaefolia* (greatly resembling juvenile *Weinmannia sylvicola*), that great swarm of hybrids and jordanons — *Alseuosmia* and *Colensoa physaloides*. Then in the Thames subdistrict a considerable change occurs in the forest various common southern plants having joined the community, but they are confined usually to the montane belt, e. g. *Hymenophyllum pulcherrimum, H. peltatum, Trichomanes Lyallii, Lindsaya viridis, Blechnum Patersonii* var. *elongatum, B. vulcanicum, B. penna marina*, the physiognomic *Polystichum vestitum, Phyllocladus alpinus, Uncinia ferruginea, Enargea parviflora, Cor-*

Thanks to an early paper of CHEESEMAN's (1871 : 270), it is possible to give some account of the original vast forest on the Waitakerei Hills, since he examined a portion of it before it had been seriously interfered with.

The dominant tree was *Beilschmiedia tawa* which "probably formed three-fifths of the forest". The other most abundant trees were *Agathis*, *Dacrydium cupressinum*, *Elaeocarpus dentatus*, *Knightia*, *Litsaea calicaris*, *Metrosideros robusta*, *Pittosporum tenuifolium* and *Suttonia australis*. The undergrowth was dense and consisted of *Alseuosmia macrophylla* (abundant), species of *Gahnia*, *Astelia* and *Coprosma*, *Rhipogonum*, *Myrtus bullata* and *Senecio Kirkii; Hymenophyllaceae*, ferns in general, and bryophytes were very plentiful.

The originally extensive forests of the Thames subdistrict, except on the Little Barrier Island, Mount Te Aroha and a few other localities, are altogether gone or much modified. Some light is thrown on the original plant-covering by the writings of T. KIRK (1870 : 89) and J. ADAMS (1884 : 385; 1888 : 32). Without going into details, the forest was much as already described, but *Beilschmiedia taraire* and *Dicksonia lanata* were scarce.

Pukatea (Laurelia novae-zelandiae)-nikau (Rhopalostylis sapida) sub-association. What follows refers only to the Waipoua forest (south of Hokianga Harbour—NA.), but doubtless somewhat similar combinations occurred throughout the plant-association.

The subassociation is confined to moist gulleys. *Beilschmiedia taraire* is rare or absent and *Laurelia novae-zelandiae* the common tree; the palm, generally trunkless, may be so abundant as to dominate. The following are characteristic: — *Dicksonia squarrosa* (tree-fern), *Dryopteris pennigera* (here with a trunk), *Asplenium bulbiferum*, mats of *Hymenophyllum demissum*, *Rhipogonum*, *Elatostema rugosum* (the succulent stems with their bronzy leaves raised 1 m. above the ground and occupying many square metres), creeping juvenile *Rubus schmidelioides* and the araliad shrub or tree *Schefflera digitata*. On shaded banks of streams is *Trichomanes rigidum*, its dark fronds covered with small epiphytic mosses.

Life-history of Agathis-dicotylous forest. To trace the beginnings of "kauri forest" the outskirts of the association must be studied where the latter abuts on the contiguous *Leptospermum* shrubland, or the *Pteridium*, fernland. Information is also to be procured where regeneration or reinstatement is in progress, the kauri forest having been cut down or burnt. In order to find out the successions which the forest undergoes up to its so-called "climax", it is necessary to carefully investigate its interior, and to ascertain the light-demanding and shade-enduring capacity of the principal species. As regards the four leading trees, *Agathis* is strongly ligth-demanding, the two species of *Beilschmiedia* are strongly shade-tolerating and *Weinmannia sylvicola* grows vigorously both in sun and shade.

In many places where the forest and the *Leptospermum* shrubland meet, there is a temporary association (the primary succession), made up of certain shrubland and forest species, including seedling and sapling *Agathis*, the composition of the association being somewhat as follows: — *Cyathea dealbata, Blechnum Fraseri, Loxsoma Cunninghamii* (local), *Lycopodium densum, L. volubile, Gahnia xanthocarpa*, juvenile, *Podocarpus totara, Dacrydium cupressinum, Phyllocladus trichomanoides, Persoonia toru* (specially characteristic), *Knightia excelsa, Weinmannia sylvicola* (juvenile), *Melicytus ramiflorus, Leptospermum scoparium* (dominant), *L. ericoides* (occasionally subdominant), *Nothopanax arboreum, Leucopogon fasciculatus, Geniostoma ligustrifolium, Coprosma robusta, Olearia rani, Brachyglottis repanda* and *Senecio Kirkii*. An interesting feature of this transitional forest is the presence of young kauri and young podocarps in much greater abundance than these occur in the forest-interior.

Where kauri forest has been destroyed on the Waitakerei Hills it is frequently succeeded by *Leptospermum scoparium* which, after it has attained a considerable size, lets in sufficient light for light-requiring seedlings to gain a footing. The succession developing below the Leptospermum is much as already given. *Phyllocladus trichomanoides* is the commonest seedling tree, but there is plenty of *Agathis*. Certain floor-plants are antagonistic to the settlement of trees, e. g. *Schoenus brevifolius, Gleichenia microphylla, Blechnum procerum, B. Fraseri* and *Lycopodium densum*.

As the *Leptospermum* grows taller, much more light is let in and eventually the podocarps, kauri and other trees overtop it and it is doomed. Next the sapling forest-trees form a more or less close canopy and shade-tolerating species enter the community, e. g. the species of *Beilschmiedia, Dracophyllum latifolium* and various shrubs and ferns.

The forest having matured, beneath its roof-canopy shade-tolerating trees have slowly developed, ready, at any moment, to replace any mature tree which falls (Fig. 36). On the other hand, there are no kauri seedlings in the dense forest. The adult kauris after some hundreds of years reach maturity, and by degrees they die and falling are replaced by the then slender trees of *Beilschmiedia taraire* and *B. tawa*, or it may be *Weinmannia sylvicola*, these forming the dominant members of the final succession (climax succession).

γ. Podocarp-broad leaved dicotylous forest of dry ground.

(1.) General.

This class of forest is one in which broad-leaved dicotylous trees play a leading part in many places but in others certain podocarps occur either more or less evenly dotted here and there or make a close subassociation. In other words, the broad-leaved trees form the groundwork of the community within which the podocarps are inserted. In what follows the name of this class is shortened to "podocarp-dicotylous forest."

As for the floristic composition of this group of associations, taking its whole range, it embraces almost all the 385 forest species, but in North Island and the north and north-west of South Island it is far richer in species than further to the south.

The following are the most common of the species which occur *through-out* this class of forest: — *Hymenophyllum rarum, H. sanguinolentum, H. dilatatum, H. demissum, H. scabrum, H. flabellatum, H. tunbridgense, H. multifidum, H. bivalve, Trichomanes venosum, Hemitelia Smithii, Dicksonia spuarrosa, Polystichum hispidum, P. adiantiforme, Dryopteris penni-gera, Asplenium adiantoides, A. bulbiferum, A. lucidum, A. flaccidum, Blechnum discolor, B. lanceolatum, B. procerum, B. fluviatile, Hypolepis rugosula, Adiantum affine, Histiopteris incisa, Polypodium Billardieri, P. grammitidis, P. diversifolium, Cyclophorus serpens, Leptopteris hymenophylloides* (Filices), *Lycopodium Billardieri* (Lycopod.), *Tmesipteris tannensis* (Psilotac.), *Podocarpus Hallii, P. ferrugineus, P. spicatus, P. dacrydioides, Dacrydium cupressinum* (Podocarp.), *Microlaena avenacea* (Gramin.), *Uncinia caespitosa, U. uncinata, U. leptostachya, Carex dissita* (Cyperac.), *Rhipogonum scandens, Astelia ner-vosa* var. *sylvestris* (Liliac.), *Dendrobium Cunninghamii, Earina mucronata, E. autumnalis, Pterostylis Banksii, P. graminea, Corysanthes macrantha* (Orchid.), *Urtica incisa* (Urticac.), *Muehlenbeckia complexa, M. australis* (Polygonac.), *Stellaria parviflora* (Caryoph.), *Clematis indivisa, Ranunculus hirtus* (Ranun.), *Cardamine heterophylla* (Crucif.), *Carpodetus serratus* (Saxifrag.), *Rubus australis, R. schmidelioides* (Rosac.), *Coriaria arborea* (Coriariac.), *Aristotelia serrata, Elaeocarpus Hookerianus* (Elaeocarp.), *Plagianthus betulinus* (Malvac.), *Melicytus ramiflorus* (Violac.), *Leptospermum scoparium, Metrosideros hyperi-cifolia* (Myrtac.), *Epilobium pubens, E. rotundifolium, Fuchsia excorticata* (Onagrac.), *Schefflera digitata, Pseudopanax crassifolium* var. *unifoliolatum* (Araliac.), *Hydrocotyle americana, H. novae-zelandiae* (Umbel.), *Griselinia littoralis* (Cornac.), *Cyathodes acerosa* (Epacrid.), *Suttonia australis, S. divari-cata* (Myrsinac.), *Parsonsia heterophylla* (Apocynac.), *Calystegia tuguriorum* (Convol.), *Coprosma lucida, C. rotundifolia, C. areolata, C. rhamnoides, Nertera dichondrifolia* (Rubiac.), *Erechtites prenanthoides* (Compos.).

The forest under consideration extends, but not continuously, from the extreme north of North Island to the south of Stewart Island, but in the conception of such forest the subassociations of kauri forest, excepting the kauri subassociation, should properly be included. Leaving these out of consideration, podocarp-dicotylous forest of the Auckland districts is alto-gether confined — swamp-forest being excluded — to the montane belt. For the rest of North Island, the community originally occupied most of the soil, its continuity being broken only by areas where the edaphic conditions were unfavourable, particularly swamp and poor soil, but in this regard are many exceptions. In South Island, wide areas in the North-western and Fiord districts are occupied by a mixture of the forest under consideration and that where *Nothofagus* dominates. On the other hand, there are pure podocarp-dicotylous forests in the Sounds-Nelson distrct, wihile in the Western district, from its northern boundary (R. Taramakau) for about 161 km., there are continuous pure forests of this class. On the east of South Island up to the Dividing Range, the rainfall of the North-eastern, Eastern and North Otago districts is not generally sufficient for any class of forest except semi-swamp forest to establish itself naturally, so in these districts forests (here including all classes except the last-named) occur only near the coast where mountainous (Seaward Kaikoura Mountains, Banks Peninsula), or in gullies, or sheltered places, at the base of the foothills (Mount Oxford, Mount Peel, Orari Gorge, Geraldine, Raincliffe, Waimate). In the South Otago district, however, the forest under consideration followed the coast-line (in most places it has been destroyed) and extended inland for a considerable distance in the south. Finally, lowland-montane Stewart Island in large part is occupied by the last-mentioned class of forest.

The life-forms of podocarp-dicotylous forest have been considered when treating of forest in general.

(2.) The podocarp communities.

General. It already has been pointed out that the podocarp content of podocarp-dicotylous forest of dry ground varies from an occasional tree here and there to subassociations or associations of such magnitude that the term "podocarp forest" is no misnomer.

The leading podocarps which form more or less pure stands, and may make even associations, are (1.) the rimu (*Dacrydium cupressinum*), (2.) the totara (*Podocarpus totara* — at low levels, *P. Hallii* — at higher levels usually), (3.) the matai *P. spicatus*. Also the miro (*P. ferrugineus*) and the kahikatea (*P. dacrydioides*) are present but usually in much smaller numbers.

All the podocarps require a good deal of light for their early development and without such their growth is either extremely slow or impossible. *P. totara* is the most light-demanding and probably *P. spicatus* the least. Excess of wind, and bright sunshine, is unfavourable, but *P. totara* can tolerate far more of either than the other species. The astonishing number of nodules on their roots may be a factor in allowing these podocarps to occupy a class of soil not favourable for other tall trees (see E. H. WILSON, 1921:236 and YEATES, 1924:124), but judging from the distribution of the forest-trees in general this is hardly likely, and various species (e. g. *Podocarpus nivalis, Dacrydium Bidwillii*) may occupy xerophytic stations.

Rimu (Dacrydium cupressinum) communities. These are distinguished by the dominance of *D. cupressinum*, but other podocarps, and some broad-leaved trees, are generally present.

The floristic character of the communities changes greatly in proceeding

11*

from north to south, but this is dealt with when treating of the broad-leaved dicotylous-tree communities.

Rimu forest, if it may be so called, extends from the high land south of Hokianga Harbour to Stewart Island, but there *Weinmannia racemosa* is present in equal or greater quantity. In the Auckland districts it does not occur at much below 600 m. altitude and even in the East Cape and Volcanic Plateau districts it is an upland community. In the Western district, at the present time, there is a great deal of rimu forest (Fig. 38) and the rimu is particularly tall, but not of excessive girth.

The physiognomy of the forest needs but little description, since that already given for New Zealand rain-forest in general applies quite well. The one striking and peculiar feature is dependent on the rimu itself with its long, straight trunk crowned by a rather small yellowish-green head of drooping-shoots.

Totara (Podocarpus totara, P. Hallii—at times) communities. "Totara forest" is podocarp forest in which either *P. totara* or *P. Hallii*, or both, are dominant. The association is more xerophytic than that of *Dacrydium cupressinum*, the species being able to occupy dry ground or exposed positions where the latter would perish. At the same time, it must be pointed out that totara to a varying extent is nearly always present in podocarp-dicotylous forest and may equal the rimu in importance, in which case there would be a rimu-totara association.

At the present time it is not possible to state accurately the distribution of those forests where totara dominated or was present in abundance. CAMPBELL-WALKER (1877) in his account of forest distribution in New Zealand defined a "central or totara district", which included all the forest-lands of the East Cape, Volcanic Plateau and Ruahine-Cook districts. Now although his area as a whole certainly contained abundant totara *(Podocarpus totara)* yet there were also rimu, kahikatea and other forest-communities and it is probable that the actual totara associations or subassociations were limited, as now, to the Volcanic Plateau from the north and west of Lake Taupo to the main-trunk line and to portions of the East Cape district. Elsewhere there were most likely rimu-totara and rimu-totara-matai associations. In South Island totara forest appears to have been almost restricted to the Eastern district, although both species are of wide distribution and certainly occurred in considerable quantities in all the districts, except North Otago, while in the Western *P. Hallii* is the dominant species of the lower subalpine forest and in Stewart Island the only one.

At one time totara forest was much more widespread in the Eastern and North Otago districts than is now the case, but how long ago this was who can say? All we know is that totara logs lay on the ground in abundance in Central Otago and parts of Canterbury now treeless. There is no clue as to what caused the destruction of these ancient forests. The story goes that there was a vast forest-fire in pre-European days, but it is almost impossible to see how this could cause such wholesale and absolute destruction or why the fallen logs remained. Still more remarkable is the fact of a forest still, undestroyed, marking the limit of the western rainfall (Fig. 39). The climate, too, where those tree-remains lie, is distinctly too dry for the *natural* occupation by totara forest and I can only conclude with SPEIGHT (1911: 417) that the forest came into existence during a much wetter period than the present. This also would account for the rarity of *Dacrydium cupressinum* on Banks Peninsula, the wet post-glacial period leading to the replacement of a primitive *Nothofagus* forest by one with *D. cupressinum* dominant, but the latter during a subsequent drier period, in its turn, being replaced by *Podocarpus totara*.

Wherever totara forest is situated, its composition is similar to that of other adjacent rain-forest associations. A brief description of the association (now destroyed) at an altitude of some 300 m. near Taumarunui will give some idea of the forest as it occurred in the most extensive area of its distribution.

The soil was pumice mixed near the surface with a good deal of humus. Besides the dominant *P. totara* there was much *P. spicatus*. The podocarps formed the upper tier of foliage; their straight, columnar trunks were a striking feature, those of *P. totara* bearing but a scanty covering of bryophytes, its outer bark hanging in long strips. Some of the other forest-trees were *Knightia excelsa*, *Beilschmiedia tawa* (abundant), *Carpodetus serratus*, *Weinmannia racemosa*, *Pennantia corymbosa*, *Alectryon excelsum*, *Hoheria sexstylosa*, *Melicytus ramiflorus*, *Fuchsia excorticata*, *Nothopanax arboreum*, *Suttonia australis*, *Olea montana* and *Brachyglottis repanda*. Most of the above also occurred as shrubs or young trees of undergrowth in which likewise amongst others, were the following: — *Paratrophis microphylla* (very common), *Aristotelia serrata*, *Myrtus pedunculata*, *Schefflera digitata* (dominant in some places), *Coprosma rotundifolia* and *Rhabdothamnus Solandri*. In some gullies of this particular community were hundreds of young plants of *Schefflera* 30 cm. or so high, associated with various ground-ferns, especially *Dryopteris pennigera*. The chief tree-fern, liane and epiphyte respectively were *Cyathea dealbata*, *Metrosideros hypericifolia* and *Astelia Solandri*. Bryophytes and *Hymenophyllaceae* were plentiful.

The forest of Banks Peninsula, now almost gone, originally consisted largely of a *Podocarpus totara* association, but in places *P. spicatus* was dominant, except on certain slopes facing north and in the subalpine belt. In the deeper gullies, where there were permanent streams, was a wealth of ferns including *Hymenophyllaceae* but according to ARMSTRONG's list (1880: 346), and to recent observations, only 2 or 3 species of the latter were at all plentiful. On many slopes and in the waterless stony gullies, the forest was of a dry character, as evidenced by the abundance of *Pellaea rotundifolia* and *Polystichum Richardi*. Certain negative features help to define the forest as a whole. Thus the following, some of which extend much further to the south in the west of the island were absent: — *Trichomanes reniforme, Lindsaya cuneata, Freycinetia Banksii, Astelia Cunninghamii, Ascarina lucida, Weinmannia racemosa, Metrosideros lucida, Metrosideros scandens, Nothopanax Edgerleyi* and *Suttonia salicina*, while *Rhopalostylis*, though present in a few places, was generally wanting. On the other hand, the elsewhere rare liane *Senecio sciadophilus* is still abundant, *Tetrapathaea tetrandra* is not uncommon, the rare *Teucridium parvifolium* is plentiful, the semi-liane *Microlaena polynoda* is fairly frequent and *Cyathea medullaris, Australina pusilla, Corynocarpus laevigata, Pseudopanax ferox* and *Olearia fragrantissima* were occasionally present.

Matai (Podocarpus spicatus) communities. To what extent *Podocarpus spicatus* formed communities, more or less pure throughout its range, it is now impossible to state. It is rare in many forest-areas and, even where common, hardly forms even a colony.

A portion of the Mount Peel podocarp forest seems according to H. H. ALLAN (1926: 40, 41) to be an exception to the above statement. Thus for the forest of the upper terrace flats he writes "In general *P. spicatus* is dominant with *P. totara* and *P. dacrydioides* in lesser amounts". On the driest ground *P. totara* dominates. Also, even yet, a *P. spicatus* association exists on certain parts of the Southland Plain, the trees close together, rather stunted and their crowns more spreading than usual. According to ROBERTS (Kensington 1909: 52), *P. spicatus* forms small societies on the flats of nearly all the Westland rivers as far south as the Cascade River. He mentions also stunted trees, 2.4 m. diam. with "short bunched trunks dividing into several long, heavy branches".

P. spicatus now extremely rare in Stewart Island, must have been dominant or subdominant for the remains of trees, probably dead for 300 to 400 years, lie abundantly upon the forest-floor, but frequently embraced by the trunk, originally the root, of a mature *Weinmannia racemosa* (Fig. 29).

(3.) The broad-leaved dicotylous tree communities.

General. The forests here dealt with are those occasionally composed almost entirely of broad-leaved dicotylous trees, but generally with more or less podocarps present, which usually play a minor part, so far as tall trees are concerned but, as already shown, they may make communities of different grades within the general forest-mass or even compose the greater part of the latter; or, again, podocarps and dicotylous trees be about equal in number. There are, indeed, no hard and fast lines, so that much of the classification proposed by any one can hardly fail to be of an artificial character, or it may be detailed to excess and useless for practical purposes.

The forest communities under consideration, unlike those of kauri or podocarp, are composed of dominant trees which when young are either shade-tolerating or epiphytic, so that, under certain circumstances they are able to replace the conifers: this is gone into under another head. Thus, the forests they eventually govern are climax successions.

Obviously the distribution of these broad-leaved dicotylous forests are those of lowland-montane forest in general as already described. At one time they occupied an area far greater than now, owing to their having been destroyed, year by year, over wide areas to give place to artificial grassland as described in Part III, whereas forests containing an abundance of podocarps — swamp-forest excepted — are removed much more slowly through saw-milling and their damaged remnants may persist for many years.

Unlike the podocarps, which thrive equally well from the north of North Island to Stewart Island, the dominant broad-leaved trees differ in their frost-tolerating capacity, so that the communities are climatic.

Tawa (Beilschmiedia tawa) communities. The dominant tree is *B. tawa*, already discribed in Chapter II of this section. The undergrowth and species are those of subtropical rain-forest of North Island and the Sounds-Nelson district.

It extends from the north of North Island to the Sounds-Nelson district. In North Island it ascends to 600 m., or more. Cook Strait notwithstanding, the forests on both sides of this apparently natural obstacle to distribution are of similar composition — a minor distinction being the occasional presence in Sounds-Nelson of the herbs *Poranthera microphylla* and *Scutellaria novae-zelandiae*.

Frequently, but in the southern part of its range more especially, tawa forest comes into contact with *Nothofagus* forest, but the latter, in general is confined to the more barren, dry slopes or ridges, while the tawa is in the gullies and the rich alluvial soil of the flat ground.

Seen from a distance, the tawa trees are of a rather unpleasing grey colour; nevertheless, at a close view with the crowns of tree-ferns peeping out of the dense undergrowth with its vivid greens, the scene is pleasing enough.

Unlike the outskirts of kauri or podocarp-dicotylous forest and *Nothofagus* forest, few seedlings of the dominant tree are to be seen there, but under the forest-roof or the canopy of small trees of the next story, in open places there will be seedlings in their hundreds; tawa saplings, too, are a frequent feature of the undergrowth.

Montane tawa forest in the Auckland and East Cape districts contains a good deal of the monotypic *Ixerba brexioides*[1]) and a fair amount of

---

1) A beautiful shade-demanding tree of slow growth, 6 to 12 m. high with dark-green, glossy, serrate, narrow-lanceolate leaves ± 12 cm. long, arranged in whorls near

Quintinia serrata[1]) — both trees of physiognomic importance. On the Mamaku Plateau *Weinmannia racemosa* (not *W. sylvicola* as further north) is a common tree, *Dicksonia fibrosa* (Fig. 23) is an abundant tree-fern and *Alseuosmia macrophylla* is frequent in the undergrowth.

Kamahi (Weinmannia racemosa) communities. These are distinguished by the dominance of *W. racemosa* (already described in Chapter II of this section), or there may be a good deal of *Dacrydium cupressinum* dotted about or in colonies, while there are almost always some of the other podocarps.

Weinmannia forest is of wide distribution and extends from the Mamaku Plateau in the north to Stewart Island, thanks to the frost-tolerating capacity of the tree and its ready establishment. Thus, as already explained it rarely commences life as a terrestial plant, but begins as an epiphyte upon a tree-fern stem, or as a seedling upon some fallen moss-covered trunk. The epiphytic habit is most advantageous in dense forest, for not only does the seedling escape competition with the floor plants, but is in a better position than most with regard to light. As the seedling grows into a tree its roots embrace its host or the fallen tree and eventually growing together function as the base of the trunk (Fig. 29.). Probably *Weinmannia racemosa* is the commonest massive forest-tree in New Zealand, though most who work in the forest would select *Dacrydium cupressinum* for that honour, but rather by reason of its being a timber-tree with which they are specially concerned than because of its relative abundance in forests generally. Frequently, valuable so-called "rimu" milling-forest contains twice as much or considerably more kamahi then rimu.

Weinmannia racemosa forest is both lowland and montane, the species itself ascending into the subalpine belt. As for its composition, that of North Island, and the north-east of South Island (SN.), is similar to that of the podocarp-dicotylous forests already dealt with. But in the North-western and Western districts, certain species rare or wanting elsewhere enter in, particularly *Ascarina lucida*[2]) and *Quintinia acutifolia*[3]), as also much juvenile *Elaeocarpus Hookerianus* of divaricating form[4]). Generally there is more or less *Dacrydium cupressinum* or indeed it may dominate.

---

the apices of the branches and bearing in early summer abundance of white flowers about 2.5 diam. arranged in terminal panicles.

1) A rather smaller tree than *Ixerba* with pale, rather thin, greenish-yellow, narrow-oblong leaves ± 12 cm. long and their margins crinkled.

2) *Ascarina lucida* is a low, bushy-tree with almost black bark and green, extremely glossy, oblong serrate leaves 2.5 to 5 cm. long.

3) *Quintinia acutifolia* is a small, rather fastigiate tree of slender habit and abundant leaves of oblong type 7.5 to 12 cm. long, yellowish with green veins and midrib. The flowers are pale lilac and arranged in many-flowered racemes about 10 cm. long.

4) Other important shrubs etc. of the undergrowth are small *Podocarpus ferrugineus*, *Carpodetus*, juvenile *Pseudopanax crassifolium* var. *unifoliolatum*, *Schefflera digitata* and *Coprosma foetidissima*.

From any lowland-forest association of the eastern and southern parts of South Island and from that of Stewart Island the association is distinguished by the presence in abundance of the lianes *Freycinetia* and *Metrosideros scandens* and the epiphyte *Astelia Cunninghamii* and by the absence of *Pittosporum eugenioides* and *P. tenuifolium*. Bryophytes, though not building cushions, are abundant enough to be of prime physiognomic importance, especially *Weymouthia mollis* and the larger *W. Billardieri* hanging from slender branches or twigs and, on the forest-floor, *Plagiochila gigantea* and other species of the genus, extensive mats of the pale-green *Trichocolea tomentella* frequently glistening with drops of water and species of *Schistochila*. Ferns are extremely abundant on the forest-floor including colonies of *Gleichenia Cunninghamii* (Fig. 40.), abundant *Leptopteris superba* and *Blechnum nigrum* (in the darkest places).

In the South Otago and Stewart districts, the *Dacrydium-Weinmannia* forest is much the same for both districts, so that one description will suffice for the two. Generally some *Metrosideros lucida* is present and, at times, in such abundance, especially near the sea or on hillsides, that it equals the other two trees.

The dominant shrub of the undergrowth is *Coprosma foetidissima* and *C. Colensoi* and *C. Astoni* are common as also the hybrids between the three. *Rhipogonum* is the only important liane. Asteliads, as epiphytes, are absent but *Dendrobium Cunninghamii* is still plentiful and *Griselinia littoralis* replaces its epiphytic relative of the north. *Hemitelia* is the prevalent tree-fern, and, as ground ferns *Blechnum discolor*, *B. procerum* and at times, *Leptopteris superba* make extensive colonies. In the north of the South Otago district *Weinmannia* was virtually absent in the originally extensive forest-area. *Pittosporum tenuifolium*, *P. eugenoides* and *Nothopanax arboreum* are absent in Stewart Island, the place of the first-named being filled by *P. fasciculatum* and of the last by *N. Colensoi*.

Northern rata (Metrosideros robusta) forest. A forest or a minor community of this class is distinguished by *Metrosideros robusta* — that huge tree of most irregular form, owing to its epiphytic origin — its spreading limbs bearing veritable gardens of shrubs, ferns, a pendent lycopod, and one or two great asteliads. Its composition is similar to that of the rimu forest which it has replaced or is replacing, as will be seen below. *Weinmannia racemosa* and *Beilschmiedia tawa* are usually important trees. The base of the northern-rata will be covered with various *Hymenophyllaceae* (e. g. *Trichomanes reniforme* and *Hymenophyllum flabellatum* in dense mats), bryophytes (including cushions of *Leucobryum candidum*) and many seedling trees and shrubs. Where the forest is montane, or upper lowland, there probably will be a good deal of the beautiful *Senecio Kirkii* on the base of the rata, its leaves soft, dark-green and fleshy and the snow-white flower-heads, each 3 cm. diam , in great abundance.

Southern-rata (Metrosideros lucida) forest. This is distinguished by
the dominance of *M. lucida*, but frequently there is an equal amount of
Weinmannia racemosa, and such would be "southern rata-kamahi forest".
The community belongs esentially to South Island, with its chief development
in the North-western, Western, South Otago, Stewart and perhaps Fiord
districts, being in the first two districts an upper lowland and montane
forest and in the others frequently coastal and semi-coastal as well as
montane.

In the Western district at above 450 m. altitude *M. lucida* becomes
dominant and the lowland podocarps gradually decrease in numbers. *Weinmannia racemosa* is so plentiful in places as to dominate. *Quintinia acutifolia* is conspicuous through its somewhat fastigiate habit as a sapling and
the yellowish leaves blotched with purple but pale beneath. Many of the
lowland shrubs and ferns are present. The undergrowth is dense, especially
in gullies. Bryophytes (species of *Gottschea, Schistochila, Aneura, Mniodendron, Plagiochila, Lembophyllum* &c.) and *Hymenophyllaceae* abound. *Leptopteris superba* forms extensive colonies.

At the Franz Josef Glacier, the terminal face of which descends to
213 m., the southern-rata association comes on to the ice-worn rocks at a
few metres from the ice on either side of the glacier. The forest here,
the roof of which has the characteristic billowy appearance, consists principally of the following: — *Metrosideros lucida* and *Weinmannia racemosa*
(the dominant canopy trees), *Carpodetus serratus, Coriaria arborea, Aristotelia serrata, Hoheria glabrata, Melicytus ramiflorus, Pseudopanax crassifolium* var. *unifoliolatum, Schefflera digitata, Griselinia littoralis, Hebe
salicifolia, Coprosma lucida, Olearia arborescens* and *O. avicenniaefolia.*
The pteridophytes include *Hemitelia Smithii* (tree-fern, but here of low
stature), several *Hymenophyllaceae, Hypolepis tenuifolia, Histiopteris incisa,
Blechnum procerum, B. lanceolatum, Asplenium bulbiferum, A. flaccidum,
Polystichum vestitum, Polypodium diversifolium, P. Billardieri* and *Lycopodium volubile.*

In Stewart Island, southern-rata forest, except close to the shores of
the inlets, or on small islands therein, is a montane community. In exposed
places the trunk may be prostrate or semi-prostrate as in Lord Auckland
Island. The undergrowth is that of ordinary Stewart Island forest.

On Mount Peel (E.) according to H. H. ALLAN (1926: 44) *M. lucida* forms
a subassociation on "rocky knolls and slopes with a western aspect", the
tree attaining a height of about 7 m. The undergrowth is sparse and is
partly made up of *Cyathodes acerosa, Coprosma parviflora, C. rhamnoides*
and *Suttonia australis.*

Forest communities of minor importance. *Rewa rewa (Knightia)
association* is by no means a common community. It is recognized even
at a distance by the fastigiate habit of the dominant *K. excelsa* with its

dark-coloured foliage. The association occurs in various parts of North Islands on very steep, barren hill faces, and on the sides of deep, narrow gullies.

Near the Rotoma Saddle, on the steep ridge beyond which lies the R. Tarawera, there is a fine example of a *Knightia* association. Mixed here and there with the *Knightia* are occasional trees of *Beilschmiedia tawa*, but *Leptospermum ericoides* is far more abundant. As for the undergrowth, it is that of ordinary forest of the vicinity but obviously it cannot be rich in bryophytes.

The puriri (Vitex lucens) association is well-marked and distingusihed by the dominance of *V. lucens*. It was evidently once common on volcanic soil in the North Auckland district, but the value of the puriri for fencing posts &c., has led to a great reduction in its area. *Vitex lucens* itself is a most handsome tree, 12 to 18 m. high with a massive, freqently irregular trunk and much-spreading rounded crown with large digitate leaves on petioles some 10 cm. long, and 3 to 5 dark-green, smooth, glossy, oblong leaflets, the largest measuring 7.5 cm. long, or more. Besides *Vitex* the following were common members of the association: — *Podocarpus totara, Beilschmiedia taraire, Dysoxylum spectabile* and *Melicytus ramiflorus*. Asteliads were in abundance on the puriri. This account is inadequate, but I have seen only much-reduced and damaged examples of the association.

On river-banks there are communities of small trees; here associations on deep alluvial soil are alone noted, those of stony river bed being dealt with further on under another heading. The associations often form a fringe along the bank which may or may not be connected with an adjacent forest-mass.

The following are the principal species of the above habitat for all New Zealand proper: — *Blechnum procerum, Dryopteris pennigera, Cordyline australis, Paratrophis microphylla, Muehlenbeckia australis, Beilschmiedia tawa, Carpodetus serratus, Pittosporum tenuifolium, P. Ralphii, Ackama rosaefolia, Rubus schmidelioides* and its var. *coloratus, R. subpauperatus, Edwardsia tetraptera, E. chathamica, E. microphylla, Melicope simplex, Coriaria arborea, Pennantia corymbosa, Aristotelia serrata,* the species of *Hoheria* (subgen. *Euhoheria*), *Plagianthus betulinus, Hymenanthera dentata* var. *angustifolia, Melicytus ramiflorus, M. micranthus, Myrtus bullata, M. ob-cordata,* × *M. bullobcordata, Leptospermum scoparium, L. ericoides, Fuchsia excorticata, F. perscandens* and the hybrids between the last two, *Nothopanax arboreum, N. anomalum, Pseudopanax crassifolium* var. *unifoliolatum, Suttonia australis,* the species of *Parsonsia* and their hybrids, *Teucridium parviflorum, Hebe salicifolia* in a wide sense, *Coprosma robusta, C. propinqua,* × *C. pro-robusta, C. rotundifolia, Olearia Hectori* and *O. fragrantissima.*

The species of *Hoheria* and *Edwardsia* according to their distribution and *Plagianthus betulinus* are specially characteristic. In the East Cape district *E. tetraptera* (little more than a shrub), *Pittosporum Ralphii* and *Hoheria sexstylosa* are companion-plants. *H. dentata* var. *angustifolia* (may be an endemic species) is characteristic of the river-bank association on the Southland Plain and *Olearia Hectori* may occur also. The river-bed forest of the Western district, described further on under another head, is a closely-related association, but one with a quite different life-history.

(4.) The life-history of podocarp-dicotylous broad-leaved forest
of dry ground.

The early beginnings of the forest under consideration have already been explained in the general introduction to forest, so far as my limited knowledge of the subject goes. Also it has been pointed out that it is not possible to give an account of the various successions, for with the exceptions dealt with below, it seems to be rather a gradual process, except in the earlier stages, leading up to a climax than a series of temporary associations terminating in such a climax — but this statement, opposed to accepted ecological teaching, may be due perhaps to my ignorance rather than to the real facts of the case.

As already seen, the two distinct groups of associations or subassociations which stand out are those composed respectively of podocarps and broad-leaved dicotylous trees. Owing to the podocarps being light-demanding they become members of the forest earlier than the bulk of the other trees. Also, theoretically, podocarp forests should have been in existence long before dicotylous trees were evolved. Be this as it may, podocarps, along with certain dicotylous trees and shrubs, are important members of the youngest forest associations (successions). Notwithstanding their slow growth, and because it is little if any slower than that of the tall or medium-sized dicotylous trees, the podocarps reach a height which cannot be overtaken by the later arriving dicotylous competitors, e. g. *Beilschmiedia tawa*. Thus, in young forest, there will always have been a strong podocarp element and, in the early history of those forests of which the present are the direct descendents, the podocarps would be supreme. Nor, even yet, are they readily supplanted by their rivals, for most of the species can exist as "lingerers" for many years, ready to grow with considerable vigour as soon as light is let in to the interior of the forest through the falling of some over-mature tree.

When in the course of its development, the light within the young forest becomes more subdued, the various shade-tolerating trees, shrubs, and ferns put in an appearance. These, so long as the undergrowth does not become too dense, even for them, will grow at a fair pace and, as low trees &c., thrive beneath the forest roof. By degrees, too, with decrease of light, the shade-demanding species will enter the community.

Within the forest there is great competition between the species. Where there is sufficient light, colonies of tree-ferns (these are by no means purely shade-plants) are readily established, and these forbid the presence of seedlings through the dense shade they cast. Favoured by rather dry ground, wide breadths of open forest-floor are rapidly occupied by colonies of *Blechnum discolor* which are hostile to the incoming of *all* seedlings. Certain lianes, particularly *Rhipogonum scandens,* destroy the shrubs and young trees which they embrace and entanglements of naked liane-stems result.

In many North Island forests, hundreds of seedlings of *Beilschmiedia tawa* are to be seen in the more open places, thanks to the high germinating-power of the seeds, the large fruits of which lie where they fall, and the shade-tolerating capacity of the young plants. With but few competitors, the advantage is greatly on the side of the young tawas, some of which grow into saplings; indeed, sapling tawas often form o considerable percentage of the undergrowth, while, beneath them, there may be seedling tawas in profusion. *Forest of this class is potential tawa forest.* In fact, every transition can be observed from pure podocarp forest to that where the podocarps are altogether wanting over a wide area, and where nearly all the tall trees are *Beilschmiedia tawa.*

In certain North Island tree communities the tawa is far less in evidence, but the podocarps — particularly *Dacrydium cupressinum* (rimu) — have an openly-declared enemy in *Metrosideros robusta* (northern-rata), which as an epiphyte — thanks to its minute seeds — so readily gains a footing on the boughs of the rimu. Very soon the humus made by the epiphytic asteliads &c., that is the soil on which the northern-rata's seeds have germinated, becomes insufficient for the rapidly-growing shrub and this puts down roots which, in course of time, reach the ground and eventually crush and kill their host, and growing together form an enormous trunk irregular in shape. This replacement of rimu forest can be seen at every stage of progress in many forest-areas up to the southern limit reached by the northern-rata.

Within the forest a similar phenomenon takes place when tree-ferns are attacked by *Nothopanax arboreum* (also the other small trees of this genus) as an epiphyte and colonies of this small tree are frequent which have originated in this manner.

The establishment of a *Weinmannia racemosa* (kamahi) climax takes place in somewhat the same manner as that of the semiepiphytes cited above. The tree itself begins life, (1.) as a seedling upon the ground, in which case it rarely reaches beyond the shrub-stage, (2.) as an epiphyte on the trunk of a tree-fern, and (3.) the seedling develops upon a fallen tree-trunk. It is the last two cases which concern the incoming of the *Weinmannia* climax-association, for, in both, the light is sufficient for the fairly rapid development of the young tree, both the fallen trunk and the tree-fern indicating an open roof-canopy.

On Banks Peninsula and certain other places where sapling tall trees ready to replace the podocarps are few in numbers, it seems probable that the climax-forest is made up of various shade-demanding or shade-tolerating small trees.

On the slopes of gullies tall trees are absent. There is sufficient illumination for light-demanding species, and such portions of a forest may be considered migratory climaxes. In such, *Fuchsia exorticata* plays an important part; other common members of the community are *Rhipogonum scandens*, *Muehlenbeckia australis*, *Weinmannia racemosa* (as a small bushy tree), *Aristotelia serrata*, *Melicytus ramiflorus* and *Schefflera digitata*.

δ. Podocarp-dicotylous broad-leaved tree forest of wet ground.

General. Though taken here together for convenience' sake two classes of different ecological status are included, the one forest of wide distribution occupying ground of a swampy character and the other that of a boggy nature, more or less sphagnum being usually present, and of local and restricted distribution. To the first category belongs kahikatea (*Podocarpus dacrydioides*) semi-swamp forest, and, to the second, communities called collectively "bog-forest" — a misnomer, to some extent, since one association occurs on fairly dry ground — where the small podocarps, *Dacrydium Colensoi* or *D. intermedium*, dominate.

Kahikatea (Podocarpus dacrydioides) semi-swamp forest. This class of forest is distinguished by the strong dominance of *P. dacrydioides* which may be almost the sole tree.

The flora consists of about 138 species which belong to 41 families and 69 genera but *Filices* 32 species, *Rubiaceae* 13 species and *Cyperaceae* 10 species are the only families of floristic importance and of genera *Coprosma* heads the list with 11 species only.

The species belong to these 3 categories (1.) those which hardly occur in any other community, here termed "semi-obligate"; (2.) those which are very common but thrive equally well in dry ground forest, here termed "facultative"; and (3.) those which are comparatively rare and occur usually in the driest stations such forest can provide. The following is a list of the more common species belonging to classes (1) and (2): — (semi-obligate) *Laurelia novae-zelandiae*, *Eugenia maire*, *Coprosma tenuicaulis*, *C. rigida;* (facultative) *Dicksonia squarrosa*, *Dryopteris pennigera*, *Asplenium bulbiferum*, *Blechnum procerum*, *B. fluviatile*, *Podocarpus dacrydioides*, *P. spicatus*, *Freycinetia Banksii; Microlaena avenacea*, *Gahnia xanthocarpa*, *Rhipogonum scandens*, *Astelia nervosa* var. *grandis*, *Rhopalostylis sapida*, *Paratrophis microphylla*, *Elatostema rugosum*, *Muehlenbeckia complexa*, *Wintera axillaris*, *W. colorata*, *Carpodetus serratus*, *Rubus schmidelioides*, *Melicope simplex*, *Melicytus micranthus*, *Leptospermum scoparium*, *Myrtus pedunculata*, *Nothopanax anomalum*, *Schefflera digitata*, *Pseudopanax crassi-*

folium var. *unifoliolatum, Hydrocotyle americana, Suttonia divaricata, Genio-stoma ligustrifolium, Coprosma robusta, C. spathulata, C. rotundifolia* and *C. areolata.*

Kahikatea forest occurs in the lowland belt from the extreme north of North Island to Foveaux Strait, but is wanting in the North Otago district and Stewart Island (merely a few trees of *Podocarpus dacrydioides*).

The life-forms of the formation, taking all the species together with the number of species to each are as fellows: — trees 31, shrubs 22, lianes 15, epiphytes 27 (ferns 16), herbs &c. 12, grass-like plants 11, herbaceous ferns 12, earth-orchids 6, parasites 2.

The forest under consideration is distinctly a lowland community which attains its greatest development near those large rivers which overflow their banks, though small typical areas occur anywhere on low-lying swampy ground even, indeed, in dune-hollows. Although there are still wide areas of virgin forest, the great demand of late years for the timber of *P. dacry-dioides* for butter-boxes has led to great destruction and in a few years, except in localities difficult of access, the association will be eradicated and the land it occupied turned into pasture, for the soil it occupies, if it can be drained at a reasonable price, is eminently suitable for dairy farms. Like most formations of a wide range, there is considerable floristic change in proceeding from north to south. Unlike rain-forest in general, kahikatea forest is not dependent on a considerable rainfall, nevertheless it shows the general characters of the true rain-forest — but it can hardly be in-cluded as an association of such and is here dealt with as a distinct formation.

The term "semi-swamp" defines the ecological conditions of the forest. Many parts of the floor are too wet for most forest-species, for pools usually abound, out of which project the moss-covered pneumatophores of *Laurelia* should that species be present. But the many fallen tree-trunks offer a position where non-swamp species can be established. Also, the floor, in part, is subject only to periodical flooding, and in places it is always comparatively dry; indeed, fairly typical kahikatea forest is to be found on comparatively well-drained land where forest of dry ground might be expected (Fig. 23). Of course, the conditions for epiphytes are the same as in dry-ground forest, except that some of the trees such specially favour are absent, e. g., *Metrosideros robusta.*

The physiognomy of the formation is unlike that of any other New Zealand forest. The innumerable kahikatea trees stand up, side by side, straight and unbranched for three-fourths of their length, or more, looking like masts of the sailing-ships of other days. Their great height (over 30 m. is common but 60 m. has been recorded) and the closeness of their growth masks their diameter, but many may exceed 1.2 m. The head of branches and foliage is most scanty but in North Island semi-swamp forest it bears a surprising number of huge clumps of *Astelia Solandri* (Fig. 41). Within the forest there is not the usual dense undergrowth of drier subtropical forest, but this is far from being wanting on the drier ground. The most striking feature of the shrubs is the divaricating-form of so many, some of that habit being, indeed, virtually confined to semi-swamp forest. Also, in North Island and the north of South Island, tree-trunks hidden by *Freycinetia*, and the ground closely covered by this liane, are a striking feature.

In order to gain a fair idea of the associations a brief description of certain pieces of forest occurring at different latitudes is next presented.

Near the Northern Wairoa River the trees were more than 24 m. high with their trunks altogether hidden in many cases by *Freycinetia*. The ground in many parts was quite covered with the last-named together with much *Dryopteris pennigera* and *Carex virgata*, the whole forming a close undergrowth out of which rose vast numbers of *Rhopalostylis sapida*, 9 to 12 m. high, their trunks naked and crowns sometimes touching. The general low tree and shrub vegetation was sparse and consisted chiefly of *Paratrophis microphylla*, sparingly-branched *Beilschmiedia taraire*, twiggy *Carpodetus serratus*, slender *Dysoxylum spectabile*, small *Eugenia maire*, low *Laurelia novae-zelandiae*, *Schefflera digitata* and some *Coprosma rigida* and *C. spathulata*. In places there was a moderate amount of *Rhipogonum*. The leading ferns were *Cyathea dealbata* and *C. medullaris* (a little), *Blechnum procerum*, *B. filiforme*, *Polypodium pustulatum* (on raised tree-roots) and *Asplenium bulbiferum*. The soil was extremely wet; pools of water abounded; there was much soft mud. *Podocarpus dacrydioides* was much buttressed and its roots often raised high above the ground.

On the Manawatu coastal plain (RC.) semi-swamp forest was originally greatly in evidence. Happily a few remnants in an almost virgin state are still to be seen. Thus, in a piece of such forest not far from Levin (fortunately a Scenic Reserve), there is a tangle of roots on the wet floor and a dense jungly undergrowth, hardly penetrable, which consists principally of the following: — slender stems of *Freycinetia*, small *Dicksonia squarrosa*, here and there sapling *Laurelia*, *Carpodetus* with naked stems for their lower two-thirds, but above with short flexuous branches. *Suttonia australis* somewhat similar in form, many black stems of *Rhipogonum scandens*, erect, sparsely-branched *Geniostoma*, many small ferns on the raised lateral roots of the trees, tussocks of *Gahnia xanthocarpa*, much large-leaved *Asplenium bulbiferum*, juvenile *Rubus schmidelioides* (on floor and tree roots) and in greater or smaller numbers *Blechnum procerum*, *Asplenium adiantoides*, *Uncinia ferruginea*, *Melicytus ramiflorus*, *Eugenia maire*, *Metrosideros hypericifolia* (on roots), *Suttonia salicina* and *Coprosma grandifolia*. Where the ground is wet rise up huge, much-buttressed trunks of *Laurelia*, its farextending roots putting forth abundant pneumatophores. Finally, there are the straight trunks of *Podocarpus dacrydioides* and their crowns full of *Astelia Solandri* looking like huge birds' nests.

In the North-western district the undergrowth may consist largely of *Wintera colorata, Carpodetus serratus, Nothopanax anomalum, Pseudopanax crassifolium* var. *unifoliolatum, Myrtus pedunculata, Coprosma tenuicaulis* and *C. rotundifolia*. As ground plants, tall *Blechnum procerum, Astelia nervosa* var. *sylvestris* and *Microlaena avenacea* are everywhere and *Nertera dichondraefolia* and *N. depressa* are common. The slender trunks of the smaller trees and shrubs are thickly clad with liverworts. *Dicksonia squarrosa* is the dominant tree-fern and there is some *Hemitelia Smithii*.

In the Western district very large areas of semi-swamp forest occur on the coastal plain. The trunks of *Podocarpus dacrydioides* are draped with *Freycinetia Banksii* and on their boughs are masses of epiphytes[1]). *Blechnum procerum* is abundant in the undergrowth and reaches 1.5 m. in height. All the species cited for North-western semi-swamp forest are present and a number of others, all of which are doubtless more or less common in the forest just mentioned. One of them, *Coprosma tenuicaulis*, extends as far south as Ross, and probably further.

In the Eastern district, at the time of settlement by the European, there were several areas of kahikatea forest at no great distance inland, but all except a small piece near Christchurch[3]) were cut down years ago. At an earlier date, the extensive swamps, now farmland, were occupied by forest, for they contained abundant remains of trees. The above tree-association, the last of its special kind in the world, is still quite vigorous, and although it has been drained, seedling *Podocarpus dacrydioides* are produced by thousands. The species number 69[4]) (trees 22, shrubs 12, lianes 12, parasites 3, herbs 17, ferns 13) and there are 2 hybrid swarms[5]). Besides the dominant *P. dacrydioides* there is abundance of *Elaeocarpus Hookerianus* and some *E. dentatus, P. spicatus* and *P. totara* are also present. *Paratrophis microphylla* is very plentiful. The association is no longer primitive, since it contains certain non-forest species, others of exotic origin, and the relative abundance of the species of the undergrowth is certainly very different from what is was originally.

1) *Astelia Cunninghamii, Dendrobium Cunninghamii, Lycopodium Billardieri* and its var. *gracile* (probably a species), *Asplenium adiantoides.*

2) *Polystichum vestitum, Leptopteris superba, Hemitelia Smithii, Wintera colorata, Rubus schmidelioides, Melicytus lanceolatus, Coprosma parviflora, C. foetidissima.*

3) Part of the forest was generously given to the people a few years ago by its owners, the Deans family, who had religiously preserved it from the earliest days of settlement.

4) An estimate somewhat less than given in *Riccarton Bush* (Christchurch N. Z., 1924), a booklet published by the Riccarton Bush Trustees, since the hybrids are dealt with as species, and certain species not belonging to the forest proper are included.

5) There is a wonderful polymorphic swarm of *Coprosma propinqua* × *robusta*, which has evidently come into being since some of the undergrowth was removed. Also there are many distinct hybrids between *Fuchsia excorticata* (a tree) and *F. perscandens* (a liane).

Life-history of semi-swamp forest. During the continuous occupation of permanent, stationary shallow-water by plants, as described later under the head "Swamp Vegetation", there is a gradual procession of events, the species by their death and decay slowly converting the substratum into loose, dark-coloured soil which, by degrees, becomes occupied by a shrub-association. Later, seedling trees enter the association of which perhaps the first-comer is *Cordyline australis* which can build up that special community, here designated **Cordyline swamp-forest.** Though, undoubtedly, such, in some places, is truly indigenous, it is now generally an indigenous-induced community arising after kahikatea forest has been felled, and, in such a case, there will be a good deal of *Typha*. Indigenous *Cordyline* swamp-forest of North Island is distinguished by the presence of *Astelia nervosa* var. *grandis* in abundance, and the divaricating-shrubs *Melicytus micranthus, Nothopanax anomalum, Coprosma tenuicaulis* and *C. rigida*.

The succession just described, is not always the beginning of kahikatea forest for, as soon as the ground is dry enough for its seeds to germinate and the seedlings to grow, *Podocarpus dacrydioides* puts in an appearance, it being essentially light-demanding. After this, or at the same time, all those members of the final community which are light-tolerant and can become established on the wettest ground, by degrees, will come in and, later, those which live on the trees themselves, or which can grow only on fairly dry ground or require considerable shade.

Semi-swamp forest may also arise from the frequent flooding of dicotylous-podocarp forest, in which case only the water-tolerating species will persist. Between pure kahikatea forest and the adjacent dry forest, if such be present, are various intermediate combinations, and it is rather the tall trees of the latter which are wanting than the plants of the undergrowth, since although pools of water lie everywhere, there is always more or less moderately dry ground. Moreover, fallen logs cumber the ground and these provide a station for the non-swamp tolerating species as also a place where seeds can germinate. In short, the distribution and combination of species within the forest is regulated by the water-content of the soil.

Bog-forest and related communities. Under this head come those communities, generally of small extent, where one or other of the lesser podocarps dominate. For the lowlands there are two groups of associations, dominated respectively by *Dacrydium Colensoi* and *D. intermedium*.

Silver-pine (Dacrydium Colensoi) bog-forest is a small group of associations where *D. Colensoi*[1]) is usually present in considerable quantity. Sphagnum may or may not be a member of the association.

---

1) *Dacrydium Colensoi* is a small cupressoid tree from 6 to 15 m. high with a pyramidal head of slender branchlets covered with scale-like, densely imbricating, very small, thick leaves and a straight trunk ± 60 cm. diam. The juvenile form is furnished with spreading juniper-like, narrow-linear subulate leaves ± 1.2 mm. long, while the

The community occurs, (1.) as a montane association in the Volcanic Plateau district either on the flanks of Mount Ruapehu or in its neighbourhood, and (2.) as a lowland association to the west of the Southern Alps.

The montane bog-forest of the Volcanic Plateau differs in composition in various localities, but its structure and floristic content is usually somewhat as follows: — *Dacrydium Colensoi* is dotted about, its trunk slender and from 30 cm., or less, to about 50 cm. diam., and its crown fastigiate; there is also small *Libocedrus Bidwillii*, stunted *Podocarpus Hallii* and irregularly-branched *Phyllocladus alpinus*, as a small tree, its spreading branches deeply covered with bryophytes. The undergrowth is frequently dense; it consists of *Phyllocladus alpinus*, tussocks of *Gahnia pauciflora*, *Astelia nervosa* var. *sylvestris*, bushy *Weinmannia racemosa*, *Myrtus pedunculata*, *Coprosma foetidissima*, *C. Colensoi* and *Alseuosmia quercifolia* or an unnamed species. The floor is irregular and on it are many bryophytes, together with *Enargea parviflora* and *Libertia pulchella*. Usually, the trees are not more than 12 m. high and the undergrowth 3 m. Generally, the ground is wetter than that of the surrounding forest in which it forms isolated areas, and it may be very wet indeed and with more or less sphagnum.

The silver-pine associations of the Western district can hardly be properly described at the present time, since the greater part has been much modified by the saw-miller[1]). Therefore I am limiting my account to one piece of *virgin* forest I had an opportunity to visit in January 1927.

The association in question varies from *Dacrydium Colensoi*, strongly dominant but mixed with a little *Phyllocladus alpinus* of the same size and habit, to a combination where *Weinmannia racemosa* is far more abundant, the structure more open and the silver-pine larger and with more spreading heads. Where the silver-pine is most abundant, the trunks (bearing small hepatics and small foliaceous lichens) are extremely straight and the heads scanty.

The undergrowth is quite open; it consists of *Myrtus pedunculata* (dominant), *Blechnum procerum*, some *Trichomanes reniforme*, *Phyllocladus alpinus*, an occasional young *Dacrydium cupressinum*, young *Podocarpus dacrydioides*, *P. Hallii*, a fair amount of juvenile *Elaeocarpus Hookerianus*, small *Weinmannia racemosa*, a little *Quintinia acutifolia* and *Coprosma foetidissima*.

semi-juvenile is longpersisting and the leaves are triangular, flat, ± 4 mm. long and more or less distichous. The type was collected from a tree growing in the North Auckland district, which may not be identical with the tree under consideration here, which is *D. westlandicum* T. Kirk.

1) The timber is of extreme durability and greatly in demand for railway-sleepers. This has led to its being taken from the forest long before the latter is being cut by the saw-miller so that in apparently virgin forest all the silver-pine of marketable size has been removed, and where but few trees have been cut down, and the smaller ones left standing, the association can readily be mistaken for virgin forest.

The subsoil is stony, being old water-borne moraine readily given to the formation of iron-pan. The upper soil is peaty and open; pools of water fill the hollows but, though wet enough, water cannot readily be wrung out of the soil by the hand. The surface of the ground is most uneven and covered with many dead twigs and leaves. There is no sphagnum and no seedlings of silver-pine.

As for the origin of silver-pine forest I can say nothing about that of the Volcanic Plateau, nor much that is really pertinent regarding that of the Western district. The only clue (a better knowledge of the virgin association and of the forest which surrounds it should tell far more) is the regeneration of *Dacrydium Colensoi* which is common enough where the forest has been felled.

Where this has taken place and the ground is favourable for the formation of iron-pan, sphagnum-bogs quite primitive in appearance, are rather quickly established. Such are a common feature of the district, and where they occupy a wide area, one is never certain whether they are primitive or induced.

On such induced-bog, sooner or later, various bog-tolerating, light-demanding young trees and shrubs settle down, the first arrival being usually *Leptospermum scoparium* and in its shade various species can be established including, if the light be sufficient, the silver-pine. But the presence of *Leptospermum* is not an essential, and *Dacrydium Colensoi* forms extensive colonies in the full light which eventually, if the far more rapidly-growing *Quintinia acutifolia* did not suppress them would form more or less pure silver-pine forest.

Regenerating bog-forest may also be seen within the general mass of podocarp forest where young silver-pine trees have been left after the large ones have been felled. Such forest, too, perhaps, may sometimes not be regeneration but an early succession. *Leptospermum scoparium* may be dominant and *Dacrydium Colensoi, Podocarpus nivalis, P. acutifolius,* and *Phyllocladus alpinus* plentiful. More or less *sphagnum* will be present.

As to the genetic relation of silver-pine forest to that of dicotylous-podocarp forest, obviously nothing definite can be said. Most likely the silver-pine association is one purely edaphic, perhaps depending, in the first instance, on bog conditions, and with death of the silver-pines and the drying of the ground, it would be replaced by a podocarp or even a *Weinmannia* association. In this case, the isolated silver-pine trees, or clumps, in the general forest-mass would be relics of former bog-forest. That bog-forest was an early succession over wide areas to be followed by podocarp-dicotylous forest seems highly improbable, for the origin and methods of extending the area of the latter are fairly well known, as will be seen for one class when dealing with river-bed forest.

Yellow-pine (Dacrydium intermedium) bog-forest is distinguished by the dominance of *Dacrydium intermedium*, a small tree ecologically equivalent and similar in appearance to *D. Colensoi*, but extending much further to the south. In North Island, there appears to be no lowland yellow-pine association. In South Island, associations occur in the North-western and Fiord districts, and in Stewart Island, where the maximum development of this class of vegetation apparently occurs.

Yellow-pine forest is not confined to boggy ground. Thus, near the shore of Lake Manapouri, an association is met with composed of *Dacrydium intermedium*, small *D. cupressinum, Nothofagus cliffortioides* and *Metrosideros lucida*.

As for the association of the forest in question in the North-western district I have no satisfactory details[1]). Townson speaks of the *species* as being abundant from sea-level to 1200 m. altitude (1906: 421), so it should surely be dominant in places.

The yellow-pine association of Stewart Island extends from the Freshwater Valley to the extreme south of the island, and occurs either on very wet soil or in positions exposed to furious gales.

The association is clearly characterized by the abundance of the slender yellowish-green dimorphic *D. intermedium*, its shoots weeping when juvenile, the presence of great globe-like bryophyte cushions of *Dicranoloma Billardieri* and *Plagiochila gigantea* in extreme abundance measuring 60 cm. $\times$ 60 cm. (Fig. 42), the rich profusion of other mosses and liverworts, and the numerous green tussocks of *Gahnia procera*. The tree-trunks are slender and close together; *Coprosma foetidissima*, so characteristic of the adjacent drier forest, is rare; neither tree-ferns nor ordinary ground-ferns are numerous, though in gullies there is plenty of *Leptopteris superba*. The floor, where the moss-cushions do not become dominant, is quite green or yellowish-green with the thick bryophyte carpet[2]) which also clothes the tree-bases and quite encircles the slender trunks.

---

1) My notes for a certain piece of forest on the lower slopes of Mount Rochfort (NW.), at about 300 m. altitude, give in the sequence as written the following composition &c.: — *Dacrydium intermedium, Gahnia procera, Dracophyllum latifolium, Phyllocladus alpinus, Metrosideros lucida, Weinmannia racemosa, Trichomanes reniforme, Blechnum procerum, Suttonia divaricata, Gleichenia Cunninghamii, Quintinia acutifolia, Dianella intermedia, Nothopanax Colensoi, Elaeocarpus Hookerianus, Astelia Cockaynei.*

2) The following are some of the specially important species: — (Hepaticae) *Aneura eriocaula, A. equitexta, Mastigobryum Mooreanum, Lepidozia Taylori, Plagiochila deltoidea, P. ramosissima, P. strombifolia, Schistochila ciliata, S. nobilis, S. marginata, Trichocolea lanata* and *T. tomentella;* (Musci) *Dicranoloma Menziesii, D. platycaulon, Hypopterygium novae-seelandiae, Lembophyllum cochlearifolium, Mniodendron comatum, M. comosum* and *Sciadocladus Menziesii.* The moss-cushions are filled with a network of roots from the adjacent trees, which thus procure purer water than from the peaty substratum (here root-tubercles as an "adaptation" for "sour" soil would be useless).

The more common woody plants of the undergrowth are: — young *Dacrydium intermedium*, young *Weinmannia racemosa*, *Myrtus pedunculata*, *Nothopanax Edgerleyi*, *N. simplex*, *Griselinia littoralis*, *Cyathodes acerosa*, *Suttonia divaricata* and *Coprosma Colensoi*. Small trees of *Podocarpus Hallii* and *Dacrydium biforme* are often present; *Enargea parviflora* grows in plenty on the bryophyte cushions; the flat-leaved grass *Microlaena avenacea* is plentiful; there are extensive colonies of the plagiotropous fern *Gleichenia Cunninghamii* more than 60 cm. high, while the other common ferns are *Blechnum procerum* and *B. discolor*.

Recently C. M. SMITH made the important discovery of the same association in the southern part of the Fiord District not very far from Foveaux Strait. He wrote me that it occurred "in large tracts west of the Waitutu" and that *Dacrydium biforme* is scattered through the association.

2. Subantarctic Rain-forest.

General. Here all the montane *Nothofagus* associations of South Island are excluded and in North Island only those of the Thames subdistrict receive consideration. These excluded communities are dealt with further on along with the subalpine *Nothofagus* forest. Even from the western portion of the North-western district to the south of the Fiord district, and including the west of the South Otago district, the lowland *Nothofagus* associations are of a subalpine rather than a lowland character, and this becomes intensified as one proceeds south, as will be seen when treating of the forest of the North-western and Fiord districts.

The community under consideration is distinguished by the dominance of one or more species of *Nothofagus,* and the rarity of other tall trees. Nevertheless, especially in the west of South Island, there are forests where the subtropical and subantarctic forest-trees are equal in number, such communities grading gradually into pure forest of either class, as the case may be.

In North Island and the Sounds-Nelson district (South Island), *Nothofagus truncata* and *N. Solandri* dominate, but in the west and south of South Island the dominant trees are *N. Menziesii*, *N. fusca* and *N. cliffortioides,* all of which or one only being present. The hybrid swarms × *N. cliffusca* and × *N. soltruncata* are widely distributed and sometimes dominate small areas.

The floristic composition of lowland *Nothofagus* forest, taking it in its entire range, differs but little from its subalpine congener, as dealt with in Section III. Chapter IV. Special distinctions are noted when discussing the composition of the associations. So, too, the life-forms need no consideration here.

Subantarctic forest is first met with in the Thames subdistrict, whence it follows the dividing-range of North Island and near Cook Strait occurs almost at sea-level. It is found also to some extent in the broken country to the east of the Egmont-Wanganui district. In South Island an association, almost identical with that of the southern Ruahine-Cook district, occurs in the Sounds-Nelson district and extends to the eastern part of the North-western district. To the west of the Tasman Mountains and the Southern Alps, excepting from the R. Taramakau to the R. Paringa, pure or mixed *Nothofagus* forest extends to the south coast, and eastwards passes for a considerable distance into the South Otago district[1]).

The Nothofagus communities may be naturally classified in terms of the dominance of any species of that genus. But a better conception of this class of forest, as a whole, is to be gained as follows from an account of typical portions of the community in proceeding from north to south.

Montane Nothofagus Forest of the Mamaku Plateau. This is distinguished by the dominance of *Nothofagus Menziesii* and *N. fusca*, and the presence of the small tree, *Phyllocladus glaucus;* there is also a little *N. truncata.* *Weinmannia racemosa* dominates in some places and, in others, *Beilschmiedia tawa* is abundant. *Ixerba brexioides,* both juvenile and adult, occur in the undergrowth and *Quintinia serrata* and *Alseuosmia macrophylla* are plentiful. Where the roof-canopy is open there is abundance of young *Nothofagi.*

Besides the species already cited, the following are common: — *Hymenophyllum scabrum, H. flabellatum, H. multifidum, Trichomanes reniforme, Dicksonia squarrosa, Blechnum discolor, Leptopteris superba, Gahnia pauciflora, Astelia nervosa* var. *sylvestris, Wintera colorata, Wintera axillaris, Carpodetus serratus, Elaeocarpus dentatus, Melicytus lanceolatus, Myrtus pedunculata, Nothopanax arboreum, N. Edgerleyi, N. anomalum, Suttonia salicina, Coprosma grandifolia* and *Senecio Kirkii.*

The presence of *Nothofagus* forest on the Mamaku Plateau, together with that of *Phyllocladus glaucus,* according to B. C. ASTON, N. Z. Journ. Agric. XXXII (1926) 369, 374, depends upon the nature of the soil which is a "sandy loam" distinct from the "air-borne sandy silts bearing the typical tawa-rimu forest of the kind most resistent to climatic severity".

Nothofagus Solandri-truncata association. Forest of this class is distinguished by the dominance of *Nothofagus Solandri* and *N. truncata.* Frequently it grows side by side with dicotylous-podocarp forest, the latter occupying gullies and flat ground and the former the ridges or wherever the soil is "poorest".

There are fewer species than in the adjacent dicotylous-podocarp forest. The trees of *N. truncata* generally are of large size — say 24 m. high, or more, and up to 85 cm. or more diam. — but if specially massive they

1) There are some patches of lowland *Nothofagus* forest in the east of the North-eastern district, also at one time there were such on Banks Peninsula, and even yet such forest exists in a number of places in the neighbourhood of Dunedin (recently made known by J. S. THOMSON and SIMPSON); and the vicinity of Catlins River.

usually are more or less decayed; *N. Solandri* is smaller. The undergrowth is often scanty; it consists of species more tolerant of a dry habitat than those in general of the adjacent subtropical rainforest, especially: *Nothopanax arboreum* (of non-epiphytic origin), *Cyathodes acerosa*, juvenile *Weinmannia racemosa* (of non-epiphytic origin), *Leucopogon fasciculatus*, *Geniostoma ligustrifolium*, *Coprosma rhamnoides,* and, where there is abundant light, young *Nothofagi*. Where particularly dry, the usually epiphytic *Astelia Solandri* and *Earina autumnale* are common on the forest-floor and there are frequently carpets of *Trichomanes reniforme*. *Cyathea dealbata* is the common tree-fern. Lianes and large epiphytes are of little moment. Foliaceous lichens are common on tree-trunks.

In some parts of the class of forest under consideration *Dacrydium cupressinum,* or other podocarps, occur in limited quantity, and there are transitions leading to dicotylous-podocarp forest.

The association just described is that of the southern Ruahine-Cook district and the Sounds-Nelson district. An association similar in character was at one time common at the base of the Ruahine Mountains, and that on sandstone ridges in the Egmont-Wanganui district is probably similar.

Lowland Nothofagus forest in the west of the North-western district. The *Nothofagus* forest in this area is by no means uniform in its composition. *N. fusca, N. Menziesii, N. cliffortioides* and some *N. truncata* may all be present, or either of the first two be the sole tree, or *N. cliffortioides* be present in about equal quantity.

The associations occur both in river-valleys and on hillsides, but on the most "fertile" soil there is dicotylous-podocarp forest containing more or less *Nothofagus*.

Owing to the wet climate, the undergrowth is similar to that of the neighbouring dicotylous-podocarp forest, following being characteristic species: — *Alsophila Colensoi, Polystichum vestitum, Leptopteris superba, Quintinia acutifolia* (but not everywhere), *Wintera colorata, Pittosporum divaricatum* (but not everywhere), *Viola filicaulis, Myrtus pedunculata, Nothopanax simplex*, *N. anomalum, Suttonia divaricata, Coprosma foetidissima* and *Nertera dichondraefolia* — all, except perhaps the last, being common lower-subalpine species.

On the floor, mats of *N. dichondraefolia* and bryophytes are common. Large foliaceous lichens (species of *Sticta* &c.) abound on tree-trunks and twigs (e. g. those of *Myrtus pedunculata*). The undergrowth consists of low trees with slender trunks, twiggy shrubs covered with epiphytic bryophytes, occasional tree-ferns (*Hemitelia Smithii, Dicksonia spuarrosa*), and, according to the light-intensity, more or less sapling *Nothofagi*.

Lowland Nothofagus forest of the Fiord district. Generally *Nothofagus Menziesii* is the sole tall tree, but there is usually more or less *N. cliffortioides*. In the western part of the district, there is much dicotylous-podocarp forest, but in the eastern part pure *Nothofagus* forest rules, though, in some localities, podocarps are not altogether absent. From Lake Te Anau southwards, *N. Menziesii* is dominant, though generally mixed with more or less *N. cliffortioides*. Certain species, commonly subalpine, occur in the undergrowth, particularly: — *Hoheria glabrata, Pseudopanax lineare. Archeria Traversii* var. *australis, Phyllocladus alpinus* and *Coprosma ciliata,* but the most important species are *Coprosma foetidissima, Wintera colorata, Nothopanax anomalum, N. simplex*, many hybrids of the swarm *N. simpanomalum, Coprosma Colensoi, C. Astoni* and hybrids between the last two and *C. foetidissima*. On the floor is a deep covering of bryophytes, mats of *Nertera dichondraefolia, Enargea parviflora* and *Libertia pulchella*.

When *Nothofagus fusca* is present, it is rather as colonies than as the dominant of a subassociatien (Fig. 43). But so much of the Fiord district is unknown, that statements based on a few localities are most likely misleading.

Nothofagus Menziesii forest of the South Otago district. This association is similar to the *N. Menziesii* association of the eastern Fiord district of which it is a continuation, but it lacks the true high-mountain element, and it contains fewer mats &c. of liverworts. In places, there is a small amount of *N. cliffortioides*.

The association is mostly confined to the western part of the district, and originally extended from the Longwood Range — where, to some extent, it is generally mixed with podocarps — to its junction with the Fiord forest-covering.

A number of areas of *N. Menziesii* forest occur near Dunedin which have recently been studied intensively by J. SIMPSON and J. S. THOMSON. One on the Silver Peaks is a good many square kilometres in area. On the whole, the species are much the same as those of dicotylous-podocarp forest, but podocarps are wanting. The undergrowth also is far more open. SIMPSON and THOMSON consider the areas as relics of a primitive *Nothofagus* forest and they supply strong evidence supporting this view.

Subantarctic-Subtropical lowland-forest. — Forest in which other dicotylous trees are present as well as one ore more species of *Nothofagus*, together with podocarps, are common troughout the range of *Nothofagus*. They are most abundant in the north and west of South Island, together with the South Otago district. Either the dicotylous-podocarp element may dominate or *Nothofagus*.

Ecologically, such mixed forests are more hygrophytic than pure *Nothofagus* forest and, when the trees of the two classes are in about equal proportion, there is little to distinguish the community from the ordinary subtropical forest of the locality. *Weinmannia racemosa* is generally abundant and may dominate just as in so many dicotylous-podocarp associations.

As to the origin of these mixed forests, speculation comes into play. It is well known that, although the species of *Nothofagus* can grow under more unfavourable conditions (poor soil, moderate rainfall, heavy wind, "sour" ground) than the *tall* podocarps, yet they thrive best with good soil and a moist climate; in fact, there is nothing, so far as their "likes" and "dislikes" go, to hinder them from always growing in subtropical forest. But, where the latter is fully established, there is not light enough in its interior for any species of *Nothofagus* to gain a footing. Once there, however, and if sufficient light is let into the forest, the *Nothofagus* is better able to take advantage of the situation than any other tall tree, thanks to its comparatively rapid growth and its light-demanding nature. In mixed forest in the North-western district, when a podocarp falls, seedlings of one or other of the species of *Nothofagus* generally take possession of the ground. Certainly, if shade-tolerating saplings already occupy the soil, the *Nothofagus* can do little, so that progress towards replacement will be slow enough. Nevertheless, it is not unreasonable to assume that in mixed forest generally *Nothofagus* has a good chance of being the climax. Certain forests show such change in progress, e. g., almost pure *Nothofagus* forest near Lake Te Anau with a few trees of *Dacrydium cupressinum,* or the miserable "suppressed" saplings of *Podocarpus Hallii* in so many *Nothofagus* forests near L. Wakatipu. Indeed such replacement by *Nothofagus* is no uncommon thing. This phenomenon depends upon the rapid growth of a light-demanding tree, just as the *Beilschmiedia tawa* succession depends upon the shade-tolerating habit in the presence of slow-growing, light-demanding podocarps. The same result — eventual dominance — is thus attained by species of opposite ecological properties and requirements!

3. Shrubland and Fernland.

a. General.

By **shrubland** is meant any community in which tall trees are absent and shrubs dominant; and by **fernland** a community consisting chiefly of *Pteridium esculentum.*

Judging from the above definitions two distinct classes of vegetation are to be dealt with here. This is true enough, but the most wide-spread group of shrubland is so intimately connected with fernland — either passing into the other again and again — that they *must* be treated under the same head.

The shrubs which play a part in one or more of the shrubland communities belong to at least 113 species (including 6 hybrid swarms) which fall into 27 families and 40 genera, the largest of which are: (families) *Compositae* 14 species, *Rubiaceae* 12 and *Myrtaceae, Epacridaceae* and *Scrophulariaceae* each 9; (genera) *Coprosma* 12, *Hebe* 9, *Olearia* 8, *Carmichaelia* and *Pimelea* each 5.

At the present time, the greater part of the shrubland is either modified or indigenous-induced, notwithstanding that wide areas look primitive enough. Though a comparative examination of the communities in all parts of the islands places them in the categories of *primitive, modified* and *indigenous-induced,* the first is constantly aped by other two. Thus, there is shrubland and fernland, each apparently a climax-community, but such must be separated into *indigenous-induced* and *primitive,* as in the case of *Leptospermum* shrubland and *Pteridium* fernland, or either may be an early stage of forest, and this, again, induced or primitive.

As for the associations, they may be either climatic or edaphic. The former may be dependant chiefly on latitudinal or altitudinal change, yet it may also be brought about by a certain increase in subantarctic conditions. On the other hand, edaphic shrubland may be extremely restricted in its distribution, as is that of magnesian soil (Fig. 44); or it may be dependant upon the physical character of the substratum, and be of wide range, as in the case of pumice, river-bed, gravel-plain and river-terrace.

Physiognomically (but this of course is really ecological), shrubland falls into the classes of (1.) ericoid-shrubland, (2.) scrub, and (3.) the incipient forest already mentioned, which frequently is ericoid in nature.

With regard to habitat shrubland occurs on almost any class of soil and in all climates, from the wettest to the driest. But, it must be borne in mind that wide areas of both fernland and shrubland are now in existence where such would be absent or extremely rare in primitive New Zealand.

b. *Leptospermum-Pteridium (manuka-bracken) communities.*

α. General.

The communities dealt with under this head are distinguished by the presence of species of *Leptospermum* — usually *L. scoparium* — and *Pteridium esculentum,* either of which may be dominant or pure, or the two may be mixed together. The communities thus range from shrubland to fernland by way of many combinations of the two classes.

The species for the whole group of communities number about 126 (pteridophytes 11, monocotyledons 43, dicotyledons 72), but the estimate could be considerably increased by taking in various species which invade an association at its junction with forest or tussock-grassland. The families and genera number 41 and 73 respectively, the following being the largest: — *Graminae* 15 species, *Orchidaceae* and *Compositae* 13 each, *Cyperaceae* 9; the genus *Thelymitra* 6.

Many of the species play little or no part in the structure of the vegetation but the following are of particular importance in one or more of the associations: — *Pteridium esculentum, Gleichenia microphylla* (Filic.), *Paspalum scrobiculatum, Danthonia pilosa, D. semiannularis* (Gramin.), *Schoenus brevifolius, S. tendo, Gahnia gahniaeformis* (Cyperac.), *Dianella intermedia, Phormium tenax, P. Colensoi* (Liliac.), *Persoonia toru* (Proteac.), *Weinmannia sylvicola* (Cunon.), *Rubus australis* in a wide sense, *Acaena novae-zelandiae, A. sanguisorbae* var. *pusilla,* the hybrids between the last two (Rosac.), *Coriaria sarmentosa* (Coriariac.), *Discaria toumatou, Pomaderris elliptica, P. Edgerleyi, P. phylicaefolia* (Rhamnac.), *Pimelea virgata, P. prostrata* in a wide sense (Thymel.), *Leptospermum scoparium, L. ericoides* (Myrtac.), *Halorrhagis erecta* (Halorrhag.), *Gaultheria antipoda, G. oppositifolia,* the hybrids between the last two (Ericac.), *Cyathodes acerosa, Leucopogon Fraseri, L. fasciculatus, Dracophyllum Urvilleanum, D. subulatum* (Epacrid.), *Hebe macrocarpa, H. salicifolia* in a wide sense, × *H. macrosala, H. diosmaefolia* (Scroph.), *Cassinia leptophylla, C. fulvida, Olearia furfuracea* (Compos.).

Leptospermum - Pteridium communities occur throughout New Zealand proper and are frequently of such extent (many square kilometres at a time) as to dominate the landscape. At the present time their wide range is chiefly due to the action of man, the communities being largely stages of succession (indigenous-induced) following the destruction of forest. Thus, both *Pteridium* and *L. scoparium* are of far greater importance in the general plant-covering than prior to the arrival of the white man. All the same, it is wrong to imagine that similar or allied communities did not occur in primeval New Zealand. On the contrary, during the frequent volcanic outbursts, even now not entirely a thing of the past, much forest must have been set on fire, and bracken and manuka associations come into being then as now; indeed, so recently as 1886, this was the case. Further, there are wide areas of a *Pteridium* association in the neighbourhood of the Otago lakes, near the junction of the comparatively dry and extremely wet districts, the presence of which can best be explained on the supposition of their having gradually replaced forest during a period of decreasing rainfall on the east of the Southern Alps. If the foregoing ideas are correct, then it is clear that man has unwittingly reproduced on a great scale an ancient type of vegetation. But there is this fundamental difference, that in primeval New Zealand the shrubland or fernland would generally revert to forest, whereas, at the present time, this is hindered by the introduced grazing-animal factor and by the purposeful conversion of these waste lands into valuable pasture by means of fire and the grazing of cattle (see Part III).

Coming next to the life-forms of the species, they and the number of species to each are as follows: — trees 3, shrubs 42, herbs 27, semi-woody plants 14, grass-form 23, rush-form 2, lianes 7, parasites 2, ferns 6. Amongst the shrubs are certain trees remaining in the juvenile form; tussocks number 15.

The communities under consideration occur under most diverse conditions. In areas of high precipitation and of the maximum dryness they are equally present. As for soil, it may be alluvial of various kinds, clays, loess, sand, gravel, calcareous or non-calcareous and pumice of different sorts. Though frequently dominant on poor soils this is rather from absence of the competitors which have ousted them from soils more fertile than for any preference for those less fertile. The water-content of the substratum differs greatly according to rainfall and its water-holding capacity, and fluctuates considerably both at different seasons and even brief periods. In certain localities, bog-conditions are present and it is not easy to draw the line between shrubland proper and bog-shrubland.

β. Ericoid shrubland.

1. Leptospermum (manuka) shrubland.

General. The group to be considered under the above name is disting-uished by the dominance of *Leptospermum scoparium* which may be almost pure or have mixed with it various species of forest, fernland or grassland. The species differ considerably according to botanical districts and also decrease in number from the north to the southernmost range of the group.

Leptospermum shrubland — generally indigenous-induced — is physi-ognomic in most parts of New Zealand proper, but is particularly in evi-dence in the Auckland, Volcanic Plateau, south-eastern East Cape and western Sounds-Nelson districts. It varies from low forest, containing many forest-species, which, if let alone, would turn into the type of forest it adjoins, to a dense or open shrubland, the height of which depends partly on the nature of the soil and partly on the time when it was last burned.

At first, the shrubland consists of great numbers of slender young *Leptospermum* of fastigiate habit, growing so close together that the entry of other species is difficult, nor is there sufficient light for the develop-ment of a distinct undergrowth. But, by degrees, as the *Leptospermum* grows, a vast number are suppressed and the survivors — now fairly tall, and eventually small trees — branch, and sufficient light enters for the germination of seeds of trees, shrubs &c., but not for the *Leptospermum* itself, for it is strongly light-demanding. Finally, as the taller-growing species cut off the light, the *Leptospermum* is doomed. Thus, in primitive New Zealand, except in areas adjacent to active volcanoes, more or less per-manent associations of *L. scoparium* must have been rare and confined to swamps, windswept stations, dunes, acid soils, and unfavourable habitats in general. Uninterfered with, its companion plants, if trees, will wipe out the *Leptospermum*, but if burned or cut down, its seedlings arise in their thou-sands and no other species can compete; moreover the seedlings, when only ± 5 cm. high flower and produce viable seed.

In what follows an account is given seriatim of various outstanding groups of associations.

Auckland manuka shrubland. This association is distinguished by the presence of certain species cited below, most of which are absent south of lat. 38°. It frequently intergrades with bog, so that the line between the two may be difficult or impossible to fix. Also, it is greatly modified by man yet, except for the abundance of certain exotic Australian shrubs, the species must be much as they were in the unaltered association and the structure in places not markedly different.

The species number more than 80, the following of which, absent in southern shrubland, are more or less common in this association: — *Lycopodium densum, L. cernuum, Phylloglossum Drummondii, Paspalum scrobiculatum, Schoenus tendo, Lepidosperma laterale, Cordyline pumilio, Thelymitra pulchella, T. pauciflora, T. aemula, T. imberbis, Prasophyllum pumilum, Caladenia exigua, Persoonia toru, Cassytha paniculata, Weinmannia sylvicola, Carmichaelia australis, Pomaderris elliptica, P. Edgerleyi, Leptospermum lineatum, L. scoparium* var. *incanum, Halorrhagis incana, Hebe diosmaefolia, Lagenophora lanata,* and *Olearia furfuracea.*

The association covers much of those deep, clay soils that form so large a part of the surface of the Auckland districts. At one time kauri forest occupied this ground as evidenced by the abundance of fossil resin and even undecayed logs beneath the surface. The clay varies in colour from brick-red, by way of orange and yellow, to almost white, this latter being especially barren. In winter it becomes saturated with water, but, in summer, baked by the sun, it is extremely dry and opens out into many cracks. Hollows and low-lying ground are always filled with peaty water or at least remain permanently wet providing where wettest bog-conditions; here their vegetation is dealt with further on under the heading "bog".

A typical area of this shrubland in its *modified condition* shows many level breadths of *Leptospermum scoparium,* in one or other of its varieties, varying from perhaps 3 m. to a few centimetres in height. Where tall, the plants are crowded with long, straight, bare, blackish stems and small heads of short leaf-bearing branches. Frequently there is a more or less dense undergrowth of the green rushlike stems of *Schoenus tendo* mixed with and overtopped by the yellowish-green pine-like *Lycopodium densum,* or they may grow separately. Probably there will also be some *Pteridium, Pomaderris phylicaefolia* and *Leucopogon fasciculatus.*

The great number of seeds, and their ready germination after fire bring seedlings of *Leptospermum* in their thousands, while at the same time other members of the association reproduce themselves, though in smaller numbers. Where, for example, the *Leptospermum* is 1 m. high, *Dracophyllum Urvilleanum* raises its slender, branched stem 50 to 90 cm. above the dull mass of foliage. So too, *Epacris pauciflora,* but it is not quite so tall. Also the rush-like stems of the *Schoenus* pierce through the close growth, but these not as seedlings but as new shoots from the undamaged rhizome. Sometimes the fire encourages the growth of pure *Pteridium,* for its rhizome being unhurt by fire, new leaves are rapidly put forth which check the establishment of seedlings.

Where the ground is dry and the plants of *Leptospermum* further apart, *Pomaderris phylicaefolia* is abundant, green flat-stemmed tussocks of *Lepidosperma laterale* stand here and there and in some localities *Pomaderris elliptica* and *P. Edgerleyi* (Fig. 45) are characteristic as also *Lycopodium densum* and *L. cernuum*, especially the former. *Gleichenia microphylla* often climbs over the bushes of *Leptospermum*[1]). When the varied forms and colouring of these plants are considered, it can be understood that this type of shrubland is far from being a monotonous spectacle.

When the shrubland is near a forest there are, to the north of the Northern Wairoa River, many young non-flowering trees of *Weinmannia sylvicola* with yellowish-green, long pinnate leaves and frequently young trees of *Knightia;* while occasionally a stunted kauri may be encountered. Near the coast, more especially, *Olearia furfuracea*, most showy with its masses of white flower-heads, gives a special character to the association, and it is often accompanied by the small tree *Persoonia toru*[2]).

Volcanic Plateau (pumice) manuka shrubland. This association is distinguished by the presence, in many parts of *Dracophyllum subulatum* in abundance and more or less *Gaultheria oppositifolia*.

The substratum is pumice with more or less humus on the surface which can retain but little moisture owing to the porosity of the pumice, so notwithstanding a fairly high rainfall and many rainy days, tolerably strong xerophytic conditions are provided. Burning has been rife, but this is only what happened again and again in the days of volcanic activity.

Leptospermum scoparium is usually dominant and with it may be much *L. ericoides* and *Leucopogon fasciculatus;* also *Danthonia semiannularis, Poa caespitosa*[3]) (in places), *Gahnia gahniaeformis, Dianella intermedia, Persoonia toru, Coriaria sarmentosa, Cyathodes acerosa* and *Leucopogon Fraseri*. *Dracophyllum subulatum*[4]) forms an almost pure subassociation which defines the most xerophytic station and poorest soil. Growing near *Gaultheria*

1) The following are also common nembers of the association: — *Lindsaya linearis, Danthonia pilosa, D. semiannularis, Gahnia gahniaeformis, Cordyline pumilio, Dianella intermedia, Drosera auriculata, Leptospermum ericoides, Pimelea prostrata, Leucopogon Fraseri*, and *Cyathodes acerosa.*

2) The following interesting plants are of restricted distribution: — *Gleichenia flabellata* with its semi-horizontal, glossy green pinnae may occupy several sq. metres at a time; *Todea barbara*, its short, massive trunk hidden by the close semi-globular mass of *Osmunda*-like leaves grows in open places amongst *Leptospermum*, it is confined to the far north; *Cassinia amoena* is restricted to the North Cape Hill; *Hebe diosmaefolia* makes pure stands in some localities; *Loxsoma Cunninghamii* grows in shade of *Leptospermum* near margins of forests; *Cassytha paniculata* binds shrub to shrub with its slender, twine-like stems, it is confined to the far north.

3) Probably distinct from the common variety of South Island.

4) An erect dull-coloured shrub, 60 cm. high, with slender, twiggy, fastigiate branches, small, erect, coriaceous, pungent, very narrow leaves 4 to 6 cm. long and deeply-descending roots.

oppositifolia may be *G. antipoda*, in which case hybrids between them will be abundant. At the highest altitude of the plateau (montane), *Phormium Colensoi* is plentiful and there is more or less *Raoulia australis* and *Cassinia Vauvilliersii*[1]).

Solfatara manuka shrubland. Where pumice shrubland comes into close proximity with steam from fumaroles, and the soil contains more or less sulphur &c., the association becomes open and by degrees the species give out, excepting *Leptospermum ericoides* and *Leucopogon fasciculatus*, the former becoming a prostrate rooting shrub, but the latter remaining erect. When surrounding boiling-mud holes, the shrubs are luxuriant and erect. *Lycopodium cernuum* tolerates far more hot steam, heated soil and fumes than any other shrubland species.

Southern North Island manuka shrubland. The shrubland is nearly all indigenous-induced. It is mainly distinguished from that further to the north by the absence of the northern local endemics and so is much poorer in species. When in proximity to forest, it contains various forest-species, in fact, it is an early stage in forest-succession. Some of the constituents of such shrubland are: *Gahnia setifolia*, *G. pauciflora*, *G. gahniaeformis*, *Cladium Vauthiera*, *Dianella intermedia*, various earth-orchids, *Nothofagus Solandri* (if near *Nothofagus* forest), *Clematis Colensoi* or *C. hexasepala*, juvenile *Weinmannia racemosa*, *Rubus australis*, *Coriaria arborea*, *Leptospermum scoparium* (dominant), *Nothopanax arboreum*, juvenile *Pseudopanax crassifolium* var. *unifoliolatum*, *Gaultheria antipoda*, *Dracophyllum Urvilleanum* (occasionally), *Cyathodes acerosa*, *Leucopogon fasciculatus*, *Hebe salicifolia* (one or other of its jordanons), *Coprosma robusta*, *C. lucida*, *C. rhamnoides*.

South Island manuka shrubland. Though, of course, local differences occur, and there is some latitudinal change, most of the South Island communities can be considered together. They occur mostly in the northern,

---

1) The eruption of Tarawera, 42 years ago, covered a vast area of pumice shrubland with more or less thick covering of volcanic ash. Near the centre of eruption the vegetation was buried to a depth of many metres so that quite new ground was available for colonization. Nineteen years after the eruption I paid a special visit to the new ground. Rain, early on, had carved the loose slopes into gully-like channels 9 m. or more deep and these were filled with a close growth of *Arundo conspicua* mixed in places with *Coriaria sarmentosa* this latter forming green patches. The flatter ground contained only plants here and there of which *Danthonia semiannularis* and *Deyeuxia avenoides* var. *brachyantha* were dominant. A species of *Cladonia*, another lichen, a moss or two, *Pteridium*, *Weinmannia racemosa*, *Gaultheria antipoda*, *G. oppositifolia*, and *Raoulia tenuicaulis* were also present. There was a good deal of the introduced *Trifolium repens*, *T. minus*, *Holcus lanatus* and *Hypochaeris radicata*, but they had probably been sown by man. Deducting those species from Aston's list (1916 b.: 309—311) which were re-established from plants damaged but not killed, together with other purely forest species, the number of species established on the really bare ground is only 47 *all* of which came from the immediate neighbourhood.

eastern and central parts of the island, those of the North-western and Western and Stewart districts usually falling into the conception of bog. The Sounds-Nelson manuka shrubland is much the same as that just described for the south of North Island, that near *Nothofagus* forest being practically the same association. Perhaps the chief distinction of the South Island manuka communities is the presence, frequently in abundance, of *Discaria toumatou, Cassinia fulvida*, or it may be *C. Vauvilliersii*, hybrids between the two, and species of *Carmichaelia*.

The comparatively low rainfall, frequent hot winds, and the porous substratum of rounded stones favoured the presence of *primitive* manuka shrubland (now usually modified by burning) on the gravel plains and valleys of the North-eastern and Eastern districts.

The almost pure subsoil of gravel and sand is capped by a varying amount of surface-soil consisting of loess — it may be clayey — mixed with 20 to 50 per cent of stones. Although apparently flat, the surface consists of hollows and ridges. Both *Leptospermum scoparium* and *L. ericoides* are present. Where the shrubs are close, there is a good deal of *Hypnum* on the ground, and to this in part the humus-content of the soil is due. The spinous *Discaria toumatou* and the yellow-leaved *Cassinia fulvida* are common. In open spaces, in the most stony areas, there are abundant silvery mats of *Raoulia lutescens*, some *R. tenuicaulis*, and cushions of *Scleranthus biflorus*.

Magnesian soil shrubland. This association is distinguished by the presence of *Olearia serpentina* — a moderate-sized, small-leaved divaricating shrub closely allied to *O. virgata* — and the dominance usually of *Leptospermum scoparium*. It occupies the lowland-montane portion of the Mineral Belt[1]) where the soil contains more magnesia than most species can tolerate, but where the magnesia is least in evidence there may be stunted *Nothofagus* forest.

Frequently the vegetation is scanty and consists mainly of the species

---

1) The belt is a narrow, frequently stony tract, consisting of peridotite and serpentine rocks, extending for about 96 km. from D'Urville Island to the western part of the Sounds-Nelson district with its widest part (about 5 km.) in the vicinity of the Dun Mountain (Fig. 44.). The vegetation of the belt both in its associations and life forms presents a striking contrast to that of the adjacent luxuriant forests. The transition from forest to Mineral belt vegetation is hardly to be seen, each standing out distinct. Not only does the magnesian soil influence the associations but it changes the life-forms, so that trees beyond its influence are merely shrubs on the belt (the spp. of *Nothofagus, Griselinia littoralis*). Taking the whole altitudinal range of the belt, the following species or well-marked varieties are apparently confined thereto, most, however, being subalpine: — *Poa* sp. related to *P. acicularifolia, Festuca* species hitherto placed in the *F. novae-zelandiae* linneon, *Pimelea Suteri, Myosotis Monroi*, probably an unnamed species of *Hebe, Cassinia albida* var. *serpentina* and *Olearia serpentina* — a certain indicator of the magnesian soil at all altitudes.

already mentioned, together with *Pteridium esculentum, Danthonia pilosa, Corokia Cotoneaster, Cyathodes acerosa, Cassinia leptophylla* and the exotics *Verbascum Thapsus* and *Digitalis purpurea.*

On Red Hill, at an altitude of 150 m. to 360 m. there is a far-richer vegetation and larger florula. The following is a rather full list of the species: — *Blechnum procerum, Pteridium esculentum, Lycopodium varium, Poa Colensoi, Cladium Vauthiera, Schoenus pauciflorus, Gahnia pauciflora, Phormium Colensoi, Cordyline Banksii, Libertia ixioides, Thelymitra* (of the *longifolia* group), *Clematis marata, Weinmannia racemosa, Melicope simplex,* (leaves very small), *Aristotelia fruticosa, Pimelea Gnidia* in a wide sense, *Leptospermum scoparium, L. ericoides, Halorrhagis erecta, Pseudopanax crassifolium* var. *unifoliolatum* (rare), *Nothopanax arboreum, N. anomalum, Corokia Cotoneaster, Griselinia littoralis, Gaultheria antipoda, Cyathodes acerosa, Leucopogon fasciculatus, Dracophyllum longifolium* (rare), *Suttonia chathamica* (rare), *Hebe angustifolia, H. salicifolia* var. *Atkinsonii, Hebe* probably a distinct species, *Hebe* hybrids an enormous swarm now being studied genetically, *Coprosma parviflora, Cassinia albida* var. *serpentina* and *Shawia paniculata.*

The shrubland forms a narrow belt on a ridge and separates two pieces of *Nothofagus* forest. Where at its richest *L. scoparium* is about 4.5 m. high, much-branched or with several naked trunks. Colonies of green-leaved *Gahnia pauciflora*, numerous bushes here and there of *Phormium Colensoi*, and green thickets of the hybrid forms of *Hebe* add brightness to the association, which is increased when the latter and *Pimelea Gnidia* are in bloom. This remarkable community was first discovered by A. W. WASTNEY, who later conducted me to it, but to him the discovery of the wonderful swarm of *Hebe* hybrids is alone due.

2. Cassinia shrubland.

This is dominated by the presence of one or other of the species of *Cassinia*, indeed, it may consist of little else.

Though frequent, both near the coast and inland to the montane or even the lower subalpine belt, the community is nearly always indigenous-induced. In parts of North Island and the Sounds-Nelson district, it occupies wide areas where forest has been felled and burned or burned standing. This extension of its range inland is due to its wind-borne seeds, their strong viability and the rapid growth of the seedlings.

These indigenous-induced communities do not concern this part of the book, but, on certain South Island river-beds, there appears to be modified *Cassinia* shrubland, and some in the montane belt may be of the same character.

In such semi-virgin scrub there is generally little else besides the local species of *Cassinia* and their hybrids, so numerous that they may outnumber the species. There are also the ground-plants of open spaces but such are mostly species of invaded *Raoulia* or tussock-grassland communities.

γ. Pteridium (bracken) fernland.

Fernland is distinguished by the complete dominance of *Pteridium esculentum*. Except where the bracken has been burned repeatedly and heavily grazed by cattle, the members of the association are very few, the most important species being *Coriaria sarmentosa*. This paucity of species arises from the fact that, except in specially exposed situations, *Pteridium* is more or less evergreen in New Zealand, so there is no vernal ground-vegetation as in Europe, indeed the close growth of the leaves inhibits all undergrowth except in open places. Luxuriant fern-heath is frequently more than 1.5 m. high. Such will be quite pure, the floor will be bare, but there will be an abundance of dead fronds.

In the Auckland districts, and other localities, where shrubland and fernland intergrade, the latter may contain more or less shrubs which have come from the former. In the wide girdles of fernland extending round some of the Otago lakes the exotic *Hypericum Androsaemum* and *H. perforatum* are occasionally common, they having entered the association by way of bare ground due to burning.

From the economic standpoint, *Pteridium esculentum* — though indigenous — is far and away the worst weed which the farmer has to combat.

c. Scrub.

Scrub of a subalpine character. This is distinguished by the dominance of shrubs usually found in high-mountain shrubland, particularly species of *Olearia* and divaricating-shrubs.

It occurs in the wettest parts of South Island, principally upon the sides of river-terraces, and is ecologically equivalent to river-terrace scrub of the montane and lower subalpine belts. From the latter, it differs floristically to some considerable extent, and ecologically in its members being upright and not nearly so much tangled together.

Taking old river-bed and terrace scrub of the Western district, the following are the most important species: — *Podocarpus Hallii, P. acutifolius, Muehlenbeckia australis* (liane), *Rubus schmidelioides* var. *coloratus* (liane), *Carmichaelia arborea, Coriaria sarmentosa, Aristotelia fruticosa, Plagianthus betulinus, Nothopanax Colensoi, N. anomalum, Pseudopanax crassifolium* var. *unifoliolatum, Griselinia littoralis, Hebe salicifolia* var. *communis, Coprosma robusta, C. propinqua, C. rugosa, Olearia avicenniaefolia, O. ilicifolia, O. macrodonta* (perhaps *O. arborescens × ilicifolia*) and *O. arborescens* and as undergrowth, *Polystichum vestitum, Blechnum vulcanicum, B. fluviatile, B. penna marina* and occasionally *Rubus parvus*.

In the west part of the South Otago district a scrub, allied to the last, consists chiefly of divaricating-shrubs, including *Olearia virgata*.

In the inland part of the North Otago district there is a good deal of scrub with *Olearia lineata* — a small, twiggy, semi-weeping tree — dominant, accompanied by *Muehlenbeckia complexa* (liane), *Rubus subpauperatus* (liane), *Discaria toumatou*, one or two species of *Carmichaelia*, *Coprosma virescens* and *Olearia odorata;* the exotic *Sambucus niger* and *Rosa Eglanteria* have also joined this association.

4. Water Associations.

Under the above head are included associations of plants specially adapted to the aquatic-life and occupying *permanent* still or flowing fresh-water. Here, only the vascular-plants are dealt with, not because there is any lack of fresh-water algae or bryophytes, but the former have been but little studied as yet, while nothing is known of their synecology. The total number of species is 25 (pteridophytes 4, spermophytes 21), which belong to 12 families and 14 genera, and comprise 5 free-swimming plants, 9 soil-rooted and leaf-floating, 7 submerged and 4 amphibious, but some come into more than one of these classes.

On still waters throughout all New Zealand, Stewart Island excepted, the floating water-fern, *Azolla rubra*, the individual plants only 1.2 to 2.5 cm. long, forms close, red sheets several centimetres thick quite hiding the water-surface, and, which, by the inexperienced, might be mistaken for dry ground. Such an *Azolla* association is generally quite pure, though it may occupy the water-surface of open swamp. The two floating species of *Utricularia* (*U. protrusa*, *U. Mairii*) are extremely rare, the former having been observed only on Lake Tongonge in the far north and Lake Waihi (Waikato) and the latter may be extinct as it was noted only in the Lake Rotomahana which was destroyed by the eruption of Tarawera. *U. protrusa* forms floating colonies 60 cm. or more across.

Where the water is comparatively still and shallow, as near the margins of lakes and slow-flowing rivers, or on the surfaces of ponds and shallow streams, there are frequently wide breadths of the brownish, long-petioled leaves of *Potamogeton Cheesemanii*, their coriaceous, oblong blades some 2.5 cm. long, while, beneath the surface, there are, in abundance, the short-petioled, translucent, ribbon-like leaves generally slowly moving with the current. In similar stations are the amphibious *Myriophyllum elatinoides* and *M. propinquum* forming masses of floating stems with submerged finely-cut leaves, but having also aereal stems with entire leaves. *Callitriche verna* is another species of still water.

The submerged water-plants, some of which have been dealt with as coastal, are as follows: — *Pilularia Novae-Zealandiae, Potamogeton ochreatus,*

P. pectinatus, Zannichellia palustris, Ruppia maritima and with these may perhaps be included *Ranunculus Limosella* and *Glossostigma elatinoides.*

5. Swamp vegetation.

a. General.

Swamp vegetation is composed of tall and medium-sized monocotylous herbs, especially *Typha angustifolia* (in a wide sense) and *Phormium tenax* (in many jordanons and hybrids), together with more or less herbaceous and shrubby dicotyledons which "tolerate", or "prefer", for their growing-place permanent shallow water, overlying loose peaty soil. Such vegetation is the New Zealand representative of the "Niedermoor" of German and the "Fen" and "Carr" of English ecologists.

Except forest, no other class of vegetation has been so greatly altered by man or so completely eradicated, but there are still some examples fairly primitive, whilst comparative studies enable a general idea of New Zealand swamps to be gained, even if it be no longer possible to secure certain details. Perhaps the most interesting feature at the present time is the life-history of commercial *Phormium* "swamps", dealt with in Part III, which are purely indigenous-induced associations but presenting a truly primitive physiognomy.

The swamp-flora, excluding aquatics, consists of 74 species (obligate 37), belonging to 18 families and 37 genera, the largest of which are: (families) *Cyperaceae* 28 species, *Juncaceae* 5 and *Gramineae, Liliaceae, Onagraceae* and *Scrophulariaceae* 4 each; (genera) *Carex* 10, *Scirpus* 6, *Cladium* and *Juncus* 5 each. The following are common species of wide distribution:— *Blechnum procerum* (Filic.), the 2 vars. of *Typha angustifolia* (Typhac.), *Hierochloe redolens, Arundo conspicua* (Gramin.), *Cladium glomeratum. C. teretifolium, Carex virgata, C. secta, C. Oederi* var. *catarractae* (Cyperac.), *Juncus holoschoenus, J. prismatocarpus* (Juncac.), *Cordyline australis, Phormium tenax* (Liliac.) *Leptospermum scoparium* (Myrtac.), *Epilobium pallidiflorum* (Onagrac.), *Gratiola peruviana, Hebe salicifolia* (in a wide sense) (Scroph.), *Coprosma robusta, C. propinqua,* × *C. prorobusta* (Rubiac.), *Olearia virgata* (Compos.).

Swamp communities occur here and there throughout the lowland and montane belts occupying from wide to quite trifling areas. The habitat originates in various ways, gradual occupation of lake by aquatic plants in the first case and swamp-species in the second; or from rivers frequently and regulary overflowing their banks, or from changes in the shore-line cutting off the sea from salt-swamp, whereupon the habitat, as its salt-content lessens, is occupied by fresh-water species, amongst the earliest colonists being *Scirpus lacustris, Typha* and *Phormium*. As time goes on, for swamp generally through the decay of the dead plants, in course of time, considerable accumulation of peat takes place, a sour soil is produced and bog-conditions may follow; but, on the other hand, swamp-tolerating shrubs, and then trees may enter the association, and, as already described, first swamp-forest and then ordinary forest form a climax. Also, of course, there may be the reverse.

Coming next to the life-forms of swamp, they and the number of species to each are as follows: — trees 1, shrubs 10, herbs 21, lianes 1, grass-like plants 17, rush-like plants 21, ferns 3.

There are many changes in the composition of swamp-vegetation in accordance with change in latitude, owing chiefly to the different frost-en-during capacity of the species. As for the associations, many are really successions leading from aquatic vegetation, by way of herb-swamp, shrub-swamp and semi-swamp forest to subtropical rain-forest — the climax. But these stages, as migratory associations are always in existence so they are best treated as distinct entities, which indeed they are.

b. Monocotylous herb swamp.

Raupo (Typha) communities. — This group of associations is distinguished by the dominance of the great, bull-rush (raupo) *Typha angustifolia* (in a wide sense). Frequently there are dotted about in the shallowest water or near the margin stunted plants of *Phormium tenax*. The chief changes in the community as a whole depend upon latitude and altitude.

Raupo associations extend throughout New Zealand — Stewart Island excepted — from sea-level to 750 m. altitude and they occupy soils of many kinds, including sand, so long as they are covered with water all the year round.

The following species abundant in the Auckland districts occur also to some extent in the rest of North Island, but are rare or absent in South Island: — *Sparganium subglobosum* (also rare NW.), *Isachne australis, Heleocharis sphacelata* (occasionally west of South Island and Stewart Island), and *Cladium articulatum*.

A well-developed Raupo swamp in the north consists of close-growing *Typha* 1.8 m. or more tall, with a dense undergrowth of *Isachne australis* its shoots intermingled and *Polygonum serrulatum*[1]). Raised above this stratum, and growing through it, may be an abundance of *Epilobium pallidiflorum*, perhaps 90 cm. high and bearing in summer numerous handsome white or pale rose-coloured flowers 1.2 cm. diam. The floor of the association will be excessively wet, and but for the close undergrowth and decaying vegetable matter, one could hardly penetrate such a swamp. Here and there will be occasional plants of *Eleocharis sphacelata*[2]), the rush-like *Cladium tere-*

---

1) A perennial herb 30—60 cm. high with a creeping, rooting, flexible, hollow stem which finally ascends, and branching but little, bears thin, yellowish-green blotched lanceolate leaves, 8 cm long by 1.5 cm. broad.

2) A perennial rush-like herb with stout, creeping, stoloniferous rhizome and stout, cylindrical, hollow, erect, septate stems 60—90 cm. high.

tifolium, Sparganium subglobosum[1]) and perhaps an occasional example of
Carex secta and *Phormium tenax.* Where the swamp is a trifle drier great
colonies of pure *Cladium articulatum* may appear, perhaps 1.5 m. high, its
stems erect, terete, close together and with pungent leaves of nearly equal
length. Pure masses of *Carex pseudocyperus* var. *fascicularis* 1.2 m. high
may be extremely abundant. Other plants of this association are *Blechnum
procerum, Scirpus inundatus, Schoenus Carsei, Cladium Huttoni, C. glomeratum, Carex virgata, Epilobium chionanthum, E. insulare, Hydrocotyle
pterocarpa*, and, in open water, some of the aquatic plants.

In South Island, raupo swamp contains fewer species, otherwise its
physiognomy and ecology are the same. At an altitude of from 550 to 750 m.
in South Island, where there is a grassland-climate it may contain in addition to *Typha* a girdle of *Carex secta* in the shallower water and also
C. subdola and *C. ternaria.* Change to drier conditions in such swamp, if
there is sufficient rainfall will bring in sphagnum bog.

Phormium communities. — In these, as the name indicates, *P. tenax*
dominates. At the present time quite a wrong estimate might be gained
as to the relative importance of this association as compared with that of
Typha in primeval New Zealand, since many of the extensive areas of
Phormium swamp that now exist have arisen through the draining of raupo-
Phormium-swamp where *Typha* was easily dominant. But areas large and
small persist in many places where the natural transition from pure *Typha*
to dominant *Phormium* can be seen.

Where the *Phormium* is dense, its masses of rhizomes monopolise the
surface and its frequently more or less drooping leaves cut off the light.
But where there are open spaces, the usual swamp-plants of the particular
locality are present. Where the water is fairly deep *Carex secta* or *C. virgata*,
may be abundant and *Blechnum procerum* be important in the undergrowth.
On the drier ground there may be colonies of the great tussocks of *Arundo
conspicua.* Shrubs are generally present more or less, especially *Leptospermum scoparium, Hebe salicifolia, Coprosma propinqua, C. robusta* and
× *C. prorobusta,* these the fore-runners of shrub-swamp. Where there is
open water, there may be colonies of *Azolla* or *Lemna minor,* or floating
Potamogeton, Ranunculus macropus, R. rivularis or *Cotula coronopifolia.*

Phormium tenax and *Arundo conspicua* frequently form narrow girdles
on the margins of rivers. Here too, or elsewhere, where there is muddy
ground subject to frequent wetting, may grow *Tillaea Sinclairii, Myriophyllum
propinquum* (as land-plant), *Elatine gratioloides, Limosella aquatica* and
Gratiolo peruviana.

---

1) A perennial herb 30—60 cm. high with rather slender, creeping rhizome, cord-like
roots with short lateral rootlets and long, bright-green, almost triquetrous leaves
2--3 mm. broad, their bases sheathing.

Nigger-head (Carex secta) communities. Here shock-headed masses of *C. secta* are dominant raised above the water on their "massive" trunks, 60 cm. to 2.4 m. high. Such trunks both raise the living portion of the *Carex* high above the water and afford a station for *Blechnum procerum*, *Hydrocotyle pterocarpa*, *Hierochloe redolens* and seedling shrubs. Nigger-head swamp contains many ordinary swamp-plants and many transitions occur between it and *Phormium* or raupo swamp.

Stewart Island swamps. As far as the swamps of Stewart Island have been investigated, they are closely related to bog. The main point of importance is that the coastal *Leptocarpus simplex* (also occurs in swamp near lake Brunner — NW.) is often dominant making pure colonies. In other places *Cladium glomeratum* and *Carex ternaria* respectively form close masses. Other species of importance are *Blechnum procerum* (stunted), *Carex stellulata*, several common species of *Hydrocotyle*, and where wettest, *Carex secta*.

Montane swamps. Swamps of this class have their greatest development where the snow-rivers have produced deltas at the heads of large lakes (e. g. L. Pukaki, L. Wakatipu), and in some of the intermontane basins.

As in lowland swamps, distribution of the species is governed by the depth of the water, so that *Typha* occupies the deepest part and where more shallow *Carex secta*, *Phormium tenax*, *Cladium teretifolium* or *C. glomeratum* while, in the very shallow water, and wet ground adjacent, there is a varied assemblage[1]), which belongs rather to bog than swamp, certain species of which rarely descend to the lowlands; *Carex Gaudichaudiana* (also lowland) is a specially important constituent.

c. Shrub swamp.

This is to be distinguished by the presence of shrubs in considerable numbers of which *Leptospermum scoparium* is nearly always dominant and frequently forms a close growth of straight, slender stems. *Cordyline australis* is usually an accompanying plant, and as already seen when dealing with forest, it may make a pure low-tree association. Other shrubs of the community are *Carmichaelia divaricata* (NW.), *C. gracilis* very rare — E., NO.), *C. arborea* (W., F.), *Eugenia maire* (North Island), *Pseudopanax crassifolium* var. *unifoliolatum*, *Griselinia littoralis* (west of South Island), *Hebe salicifolia* in a wide sense and its var. *paludosa* (W.), the species of *Coprosma* already cited and *Olearia laxiflora* (NW.).

---

1) *Carex ternaria, Schoenus pauciflorus, Juncus polyanthemos, Luzula campestris* — one or other of its many jordanons, *Rumex flexuosus, Montia fontana, Potentilla anserina* var. *anserinoides, Geranium microphyllum, Viola Cunninghamii, Halorrhagis micrantha, Hydroctyle novae-zelandiae* var. *montana, Oreomyrrhis andicola* in a wide sense, *Mazus radicans, Plantago triandra, Asperula perpusilla, Celmisia longifolia* in a wide sense, *Cotula dioica* var. *montana, Gnaphalium paludosum*.

Shrub-swamp may also be a stage in the retrogressive evolution of swamp from semi-swamp forest, in which case, at first, many forest plants would persist for a time. *Leptospermum* swamp strongly favours the presence of *Sphagnum* and thus is readily transformed into the different ecological category of bog, or the bog-moss, the *Leptospermum* and its accompanying plants, be a relic of a former bog-community.

d. Associations of warm-water, or of hot ground exposed to steam.

According to SETCHELL, various *Schizomycetes* flourish in the numerous hot-water pools of the Volcanic Plateau, but none can endure a higher temperature than 75° C. Where the heat is moderate, there *Cyanophyceae* occur. On shrubs, constantly exposed to steam, the twigs become deeply coated with the orange-red alga *Chroolepus aureus*.

The following ferns and lycopods grow luxuriantly near hot-water streams, their roots in strongly heated soil and their aereal parts exposed to steam: — *Histiopteris incisa, Dryopteris gongylodes, D. dentata, Nephrolepis cordifolia, Gleichenia linearis, G. circinata, Schizaea dichotoma, Lycopodium cernuum* and *Psilotum triquetrum; Histiopteris incisa* and *Lycopodium cernuum* are of extremely common occurrence, the latter constantly hugging the warm ground round steam vents, but the other species are of local distribution.

On the sides of the boiling wells which supply the Boiling River (Otumakokori Stream) *Dryopteris dentata, Nephrolepis cordifolia* and *Gleichenia linearis* grow in great abundance in a hot atmosphere always saturated with moisture[1]).

Hot water swamps contain many ordinary swamps-plants, e. g.: *Typha, Scirpus lacustris, Heleocharis sphacelata, Cladium articulatum, Carex ternaria,* together with *Dryopteris dentata*. The latter, according to KIRK, formerly covered acres of the swamp on either side of the warm river connecting lakes Rotomahana and Tarawera, "its dull green fronds sometimes five feet high and seven inches across but in this state it is usually barren". The same fern grew in abundance also on the "White Terrace", covering "the thin crust overlying the scalding mud and from its erect, rigid habit and the strict sori-laden pinnules presenting a forcible contrast to the luxuriant swamp form".

On the shore of Lake Rotorua at Ohinemutu, growing on heated ground or where there is an excess of certain salts &c. in the soil, is *Fimbristylis dichotoma* and the presence of the following coastal plants may perhaps be correlated with the edaphic conditions: *Bromus arenarius, Scirpus mari-*

---

1) For further particulars see HOCHSTETTER (1867: 402) and KIRK (1873: 336), but the statements as to all these "hot-water" ferns being confined to such a station is now known to be incorrect, only *Nephrolepis* and *Gleichenia linearis* being so restricted. The 2 species of *Dryopteris* are of limited distribution in the North Auckland district and the *Schizaea* is an occasional denizen of the kauri subassociation.

timus, Carex pumila, Leptocarpus simplex, Juncus maritimus var. *austra-liensis* and *Ranunculus acaulis.*

6. Bog Vegetation (Communities of sphagnum and its associated plants).

General. Various kinds of sphagnum are the special feature of bog vegetation and with such are associated species of *Gleichenia, Lycopodium, Cladium, Carex,* also *Hypolaena lateriflora* and species of *Drosera* and *Utricularia.* But the associations differ considerably in various localities and, even sphagnum may occur only to a limited extent, or at times be wanting.

The total number of species in the lowland portion of the formation is about 110 (pteridophytes 13, conifers 2, monocotyledons 42, dicotyledons 43) which belong to 27 families and 67 genera, but 24 species are confined to Stewart Island. The most widely-spread vascular species are the following:— *Gleichenia microphylla* (Filic.), *Cladium teretifolium, C. glomeratum, C. Vauthiera* (Cyperac.), *Hypolaena lateriflora* (Restionac.), *Drosera spathulata, D. binata* (Droserac.), *Leptospermum scoparium* (Myrtac.), *Halorrhagis micrantha (Halorrhag.), Centella uniflora* (Umbel.), and *Pratia angulata* (Campan.).

Wherever there is a soil which remains saturated with water at all seasons, a bog association of some kind will be present. But, although there may be shallow pools here and there, the entire surface must not be permanently covered. According to the average degree of wetness, so will the plant-covering vary, *Sphagnum* bog occupying the wettest and shrub bog the driest ground.

The amount of peat present regulates the species to some degree, but to a lesser extent than would be imagined. Where the peat-content is scanty, pakihi bog occurs and forms a connecting link with *Leptospermum shrubland.* Bog appears to originate in various ways, e. g.: from lake by way of monocotylous-herb swamp; from frequent floods causing swamp in the first place; from water lying in wet hollows; from an excessive rainfall on flat badly-drained ground and from *Sphagnum* settling on the forest-floor in a wet climate.

Obviously an abundant rain-fall, a comparatively low summer temperature and frequent cloudy skies favour bog which is thus a common feature of the west and south of South Island and Stewart Island. In North Island, extensive *Sphagnum* bogs were present on the Waikato Plain, but these were edaphic rather than climatic and arose in swamps caused by overflow of the river and defective drainage. So, too with certain bogs, now reclaimed, on the Canterbury Plain, where the climate is hostile to bog. Hollows in the Auckland gumlands are occupied by bog, and it is a matter of choice whether to call a good deal of the vegetation shrubland or bog.

Well-developed bog shows a distinct succession of vegetation. First, comes *Sphagnum* bog; it is succeeded by various related combinations of species in which rush-like *Cyperaceae* and the xerophytic *Gleichenia circinata* with its plagiotropic pinnae and their pouch-like segments, is dominant. Finally, this is followed by shrubland or low forest, which latter may be replaced by high forest.

The life-forms are varied and include the following: — ericoid-shrub, creeping-shrub with underground stem, rush-form, grass-form, tussock-form, cushion-form, prostrate creeping-herb, *Gleichenia*-form, rosette-herb, summer-green tuberous-form. Generally speaking xerophytic structure is present to a considerable degree. The far-creeping stem of certain species is a most important characteristic, leading, as it does, to enormous vegetative increase and consequent dominance of a particular species over a wide area.

Many bog species, especially in South Island, are also subalpine or alpine, while the bogs of Port Pegasus, Stewart Island are virtually identical with those at 600 m. altitude, and upwards.

The various groups of associations are next to be considered but the treatment, as for most other formations, is far from exhaustive.

Sphagnum-Gleichenia bog. This group is distinguished by the presence of *Sphagnum*, usually in a considerable amount, and the dominance of *Gleichenia circinata* (*G. dicarpa* of CHEESEMAN'S Manual), or *G. alpina*.

The distribution of this class of bog coincides with that of lowland bog in general. It occurs in wet places amongst *Leptospermum* shrubland, in open places in forest (especially west and south of South Island), and as a stage in succession from lake to forest following swamp; even yet, this can be seen in various localities.

The amount of *Sphagnum* varies much in different places. Frequently, there are pale straw-or cream-colored cushions separated from on another by dark, soft peaty ground where coffee-coloured water lies. Or, development has advanced, the cushions grow into one another forming a continuous covering of soft moss saturated with water, except after a period of drought when the surface may be dry. Almost always there are a number of species growing on the moss, some of which will be absent elsewhere on the bog, for *the upper surface of the cushion contains a purer water than bog-soil in general*. Between the *Sphagnum* and its occupants there is a "struggle" for the mastery, for the moss by its upward growth tends to bury any plant whose upward growth is too slow.

The following are common on *Sphagnum*-cushions throughout: — *Blechnum procerum, Gleichenia circinata, Cladium teretifolium, C. Vauthiera, Hypolaena lateriflora, Drosera binata, D. spathulata, Leptospermum scoparium, Halorrhagis micrantha, Centella uniflora*.

Where *Sphagnum* is either absent, or in scattered patches, and the soil consists of soft, wet peat, a low green carpet of *Gleichenia circinata* frequently covers many square metres of ground, and as it can be recognized from a considerable distance, it is the physiognomic indication, par excellence, of the association. In other places, the rush-form is physiognomic, the species being *Cladium teretifolium*, or *C. glomeratum*, plants of wetter ground than the *Gleichenia* from which they stand distinguished by their darker green. In other places again, wide areas are occupied by *Hypolaena lateriflora*, a restiaceous plant with slender, wiry, flexuose, leafless, semi-prostate stems which frequently form thick brownish masses 60 cm. to 1 m. in depth, mixed or unmixed with the *Cladium*.

In the north of the North Auckland district, and the Waikato sub-district, the bamboo-like *Sporodanthus Traversii* (stems erect, up to 3 m. high, crowded, stout, flexible) was originally dominant in certain bogs and made a distinct subassociation.

The bog-vegetation is generally stratified, the fern and sedges, already mentioned, forming the upper layers, or where both occur, the sedges overtop the fern. The ground-layer may consist of low-growing creeping species of *Lycopodium*. This is the case in the Western district where *L. ramulosum*, flattened close to the soil, covers extensive areas. Further north *L. laterale* of similar form, but with more erect stems, is the common bog-species.

On bare places as well as on *Sphagnum* are the tiny red rosettes of *Drosera spathulata* pressed closely to the substratum. Several species of *Utricularia* are common, e. g. *U. delicatula* (NA., SA.), the flowers white with a yellow eye, and *U. novae-zelandia* (North Island, E. in South Island), its flowers pale purple but in South Island generally it is the beautiful *U. monanthos* with bright purple flowers and yellow eye.

The bogs of South Otago, but especially those of Stewart Island, are remarkable for the number of high-mountain plants they contain. For instance, in the latter, there is an association at sea-level equivalent to that of a high-mountain bog, as the following list clearly shows: — *Gleichenia alpina, Microlaena Thomsoni, Oreobolus pectinatus, Carpha alpina* (also SO.), *Chrysobactron Gibbsii, Astelia linearis, Gaimardia ciliata, Caltha novae-zelandiae, Geum leiospermum, Drapetes Dieffenbachii, Actinotus suffocata, Cyathodes empetrifolia* (also SO.), *Dracophyllum rosmarinifolium* (= *D. politum*), *D. Pearsonii, Pentachondra pumila, Liparophyllum Gunnii, Hebe buxifolia, Oreostylidium subulatum* (also SO.), *Donatia novae-zelandiae* (also SO. but rare), *Celmisia argentea* and *C. linearis*. In South Otago bog, *Thelymitra uniflora, Gunnera prorepens* (leaves often almost black) and the wiry *Gaultheria perplexa* are characteristic, but they are of wide range. In both South Otago and Stewart Island, *Danthonia Raoulii* var. *rubra* become established on *Sphagnum-Gleichenia* bog and **tussock-bog** results, which, as the habitat becomes drier, changes into tussock-grassland. Analogous is the occupation of bog in the Western district by the great tussocks of *Gahnia rigida* which may take complete possession but the ralationship here is with forest and not with tussock-grassland.

Shrubland bogs. This group is distinguished by the dominance of *Leptospermum scoparium* accompanied by *Gleichenia*, *Hypolaena* and other plants found in the formation. Even on such bogs, as already described, there is usually more or less *Leptospermum scoparium,* and then it is only a question of time for this to become dominant and shrubland-bog established. Beneath the *Leptospermum* a good many of the bog species continue to thrive, thanks it may be to some epharmonic change. Thus *Hypolaena* and *Gleichenia* become semi-lianes and *Lycopodium ramulosum* grows erect, increases its stature and almost ceases to develop sporophylls. Or shrubland-bog may be distinguished, as in some parts of Stewart Island or South Otago by the presence of the following shrubs either in clumps or dotted about, rather than by the dominance of *Leptospermum:* — *Dracophyllum longifolium, Coprosma parviflora, C. rhamnoides, Hebe buxifolia, Cassinia Vauvilliersii, Olearia virgata.*

Near Lake Manapouri to the east there is an extensive shrubland-bog. There is a great depth of peat. *Dracophyllum Urvilleanum* (one of its varieties) forms so close a mass that nothing else can be seen at even a short distance away. Within the scrub there is *Sphagnum, Hypolaena, Cladium teretifolium, Oreobolus strictus* and other bog species.

A rather remarkable association occurs in those flat, badly-drained areas of the North-western and Western districts known by the Maori name of **"Pakihi".** All that I have studied have been burned repeatedly, and although the vegetation bears a primitive stamp through the absence usually of introduced plants, one can but guess at the composition and arrangement of the primitive vegetation.

The soil-conditions are different from those of true bog. The ground is frequently ancient coastal terrace, and the subsoil consists altogether of stones which through the presence of "iron pan" are impervious to water. The upper soil is clay capped by a layer of peat, usually quite thin, though this may be wanting. Generally, the ground is extremely wet. Where *Leptospermum* does not dominate the covering is a combination of *Cladium teretifolium, C. glomeratum, C. capillaceum* (at times), *C. Vauthiera, Hypolaena lateriflora* and *Gleichenia circinata,* just as ordinary peat-bog, in fact. *Lycopodium ramulosum* is abundant, and there may be *Gahnia* tussocks.

On the pakihis of the North-western district the North Island *Epacris pauciflora* is abundant. *Liparophyllum Gunnii,* and *Gentiana Townsoni* are plentiful. The very local *Siphonidium longiflorum* is common in places. *Sphagnum* is dotted about in certain parts, but is absent over wide areas, nor does it seem to be making peat. *Leptospermum* at all stages of development is everywhere. In the extreme north of the district there is a good deal of *Lycopodium cernuum,* a species otherwise wanting from a little to the south of latitude 38°.

Doubtless much pakihi is indigenous-induced, following the removal of forest, but the native name proves that there was a primitive bog association, yet it was one in which peat was virtually absent.

7. River-bed vegetation.

a. General.

The vegetation of river-bed proper is distinguished by its open character and the presence of mats or low cushions of species of *Raoulia* and *Epilobium* the latter both creeping and erect.

Leaving out of consideration the semi-climax and climax associations, since they belong to other plant-formations, the communities of river-bed proper contain obout 66 species (pteridophytes 3, monocotyledons 9, dicotyledons 54) belonging to 24 families and 38 genera, the only group of any size being the *Compositae* with 15 species. No list is given of the common species as they are all cited in what follows.

The habitat (Fig. 1) here called "river-bed" is characteristic of those shingle plains formed by the various streams which still bear their stony burden from the lofty ranges. These beds extend from the sea-coast far into the mountains. Their greatest development is to the east of the Divide in South Island, though on the west, although shorter, they are still considerable. In North Island river-bed is much less in evidence.

Between the vegetation of river-bed at different altitudinal belts it is not easy to draw the line; ecological differences are slight, and many species occur throughout. Rainfall is the chief factor that governs the combinations; thus in South Island the humid west favours the presence of plants in the lowlands which are purely subalpine in the drier east.

Terraces in most places bound the lowland and montane river-beds, while these latter may be 1.6 km. in width. It is obvious that as the stony plains themselves have been subject to inundation in all parts during their formation, the vegetation-dynamics of river-bed at the present time must be very similar to that of the plains during their construction.

Typical river-bed consists of a more or less flat expanse of stones which vary considerably in size and are mixed with a large but varying proportion of sand and silt. The river-proper wanders from side to side of the bed in anastomosing streams, its path restricted only by the terraces or adjacent mountains slopes. During the frequent floods, the streams may change their course, so turning stable ground into flood-plain, but rendering the abandoned stony bed fit for plant-colonization. Moreover, the stream may cut into its bed, and terraces, such as now exist far from the river's influence, but whose surfaces are ancient flood-plains, be in process of formation. Thus it can be seen that various associations will exist, each marking a certain phase in the development of the land-surface, commencing with the peopling of the bed as the water recedes, and ending with the vegetation of the oldest flood-plain, the climax-association of the east being shrubland or tussock-grassland and of the west forest.

The ecological conditions are comparatively simple. The stony substratum favours rapid drainage, the water-content close to the surface being extremely small. At a depth of about 30 cm. there is always a certain amount of moisture; the stones themselves assist in reducing evaporation; the sand and silt hold water to some extent; and doubtless the water-table is near enough, even on the highest parts of the bed, to allow deep-rooting plants water in abundance. Were this not so, the extensive plantations of the Canterbury Plain, where the rainfall is low, could not exist. At first, there is a total absence of humus and but little of nutritive salts. When the sky is cloudless, the plants are exposed to an extremely bright light and the stones become burning hot. The wind, hemmed in between terraces, or adjacent mountain-slopes, sweeps over the ground with great force, but its effect is mitigated somewhat by the unevenness of the bed and its larger stones, which afford both shelter and shade. On the east, the hot, dry North-west wind not only causes excessive transpiration, but the course of the river is marked by great clouds of silt and finer sand high in the air. The rainfall is a factor of moment though much less than in the case of retentive soils nor does it affect the subterranean water-supply, which is regulated by the downpour on the mountains.

In many localities, the vegetation is much altered, especially by continuous thickets of the introduced *Ulex europaeus* and *Cytisus scoparius*, but, although various foreign species generally occur where these plants do not dominate, much river-bed is still primitive enough to allow accurate conclusions to be drawn as to its original character.

As for the life-forms of the 66 species, 14 are shrubs, 41 herbs or semi-woody plants, 8 of the grass-form, 1 of the rush-form and 2 ferns. Amongst the above are represented mat-forming plants (often circular); prostrate creeping and rooting herbs; low shrubs with leafless green stems; and ericoid shrubs. *Raoulia lutescens* and *Scleranthus biflorus* are cushion plants.

b. Unstable river-bed formation.

Here there is only the one community — the **Epilobium-Raoulia formation** — which is distinguished by the presence of *Epilobium pedunculare*, and in South Island, *E. melanocaulon* and *E. microphyllum* — though these two may be absent at below 300 m. altitude — together with *Raoulia tenuicaulis*.

The formation (it is complete in itself and not an initial succession) occupies that portion of the river-bed subject to frequent flooding and eventually complete overthrowing. However, there must be stability enough to allow the windborne seeds of the *Epilobia* and *Raouliae* to gain a footing, germinate, and produce mature plants. Both *E. pedunculare* and *R. tenuicaulis* have creeping rooting stems which hug the ground. The latter grows rapidly, owing to its mesophytic juvenile form and its reversion-shoots; it forms flat, green or silvery patches 60 cm. or far more diam. (Fig. 46). A few plants of *R. australis* may also be present.

c. The communities of stable river-bed.

General. In the case of stable river-bed the substratum lies beyond the reach of *ordinary* floods. Its vegetation is an initial succession which leads, according to the climate of the locality, to tussock-grassland, shrubland or forest as its climax. The substratum varies considerably according as its surface is stony or silty, the stones, often coated with a dark lichen, may be packed as closely as if paved.

The eastern South Island Raoulia association. The association is distinguished by the presence of mats or low cushions of species of *Raoulia*. They attain their greatest development in the North-eastern and Eastern districts and what follows refers to these alone. Generally, there is far more stony ground than plants. Low, silvery, dense flat cushions of *Raoulia lutescens* are everywhere, their cushion-form arising from a gradual filling of the plant with wind-borne silt. Great dull-green patches of the shrubby *Muehlenbeckia axillaris* are common, and the leafless, grey-coloured rush-like stems of *M. ephedroides*, of similar habit, may be present. *Raoulia australis*, looser in habit than *R. lutescens*, will be abundant both on silty and stony ground. Somewhat older bed may contain large colonies of *Carmichaelia nana*, its flat, leafless, vertical stems close together and a few centimetres tall. The larger, thicker stemmed *C. Monroi*, usually subalpine, may occasionally be present. On the older parts of the bed there will be many large irregular patches of *Raoulia Monroi*, distinguished by its folded, silvery, distichous small leaves. Thickets of *Cassinia fulvida* may abound or of *C. albida* (NE.), and where both species are present there will be a polymorphic hybrid swarm. Patches of *Racromitrium lanuginosum* are frequent. Besides the species cited above the following are more or less abundant: — *Scirpus nodosus, Scleranthus biflorus, Tillaea Sieberiana, Carmichaelia subulata, Oxalis corniculata, Geranium sessiliflorum* var. *glabrum, Epilobium nerterioides, Daucus brachiatus, Leucopogon Fraseri, Gnaphalium japonicum, G. collinum, Cotula perpusilla*.

The later history of the association is dealt with further on under the heading tussock-grassland.

River-bed dune. Low isolated dunes, or short ridges, are frequent here and there on river-bed (E.). They are simply collections of the blown sand and silt. Usually they are fixed by vegetation. The small tree *Edwardsia microphylla* was originally common. *Phormium tenax, Arundo conspicua* and *Cassinia fulvida* are abundant. The species of *Raoulia* are generally absent. The introduced *Lupinus arboreus* has complete possession in some localities.

Groves of trees &c. on eastern river-bed. In some places, even yet, small groves of trees are to be met with, especially at the base of a high terrace. *Edwardsia microphylla* will be dominant, accompanied probably by *Cordyline australis, Pittosporum tenuifolium, Melicytus ramiflorus, Fuchsia* *excorticata, Nothopanax arboreum* and *Griselinia littoralis*, while *Rubus australis* var. *glaber, Muehlenbeckia australis* and *Parsonsia heterophylla* may be abundant as lianes loading the trees with greenery.

Phormium-Arundo-Cordyline swamp and slowly-flowing streams or ponds with aquatic and swamp vegetation are common enough on river-beds, but they are identical with similar associations elsewhere and need no description.

The western South Island Raoulia association. This is distinguished by the absence of many of the eastern xerophytes and the presence of *Mazus radicans, Acaena Sanguisorbae* var. *sericeinitens* and species of *Coriaria* and their numerous, polymorphic hybrids. Frequently the vegetation is closed. Here the association of the Western district is alone considered.

The following are the important species: — *Carex comans, Muehlenbeckia axillaris, Ranunculus foliosus* (sometimes), *Acaena Sanguisorbae* var. *sericeinitens, A. inermis, Coriaria sarmentosa, C. lurida,* ✕ *C. sarlurida, Pimelea prostrata* var. *repens, Epilobium pedunculare, Hydrocotyle novae-zelandiae, Mazus radicans, Veronica Lyallii* (common in montane belt, but rare et sea-level), *Coprosma rugosa, C. brunnea, Nertera depressa, Wahlenbergia albomarginata, Pratia angulata, Helichrysum filicaule, H. bellidioides, Cotula squalida, Raoulia glabra* (sometimes), *R. australis, R. tenuicaulis.*

The station, thanks to the frequent downpour, is mesophytic notwithstanding the coarse, stony substratum. Rocks far larger than on eastern river-bed are present. Hot dry winds are virtually unknown; frosts are never heavy.

On older river-bed shrubs come in and there is a procession of events leading eventually to forest, an account of which is next presented.

Western river-bed forest. This is distinguished by the close growth of low slender trees and the presence of species wanting or rare in the adjacent lowland forest e. g. *Rubus schmidelioides* var. *coloratus, Coriaria arborea, Pennantia corymbosa, Aristotelia serrata, Plagianthus betulinus* and *Coprosma rotundifolia,* as a tree. Here only the association of the Western district receives consideration. The species may be seen from what follows.

The ground is level und traversed by numerous streams. The upper soil consists of humus beneath which is merely river-shingle. The vegetation is in three layers — the floor-plants, the small tree-ferns and shrubs and the low trees. The association is 4.5 to 6 m. high. Slender tree-trunks not exceeding 15 cm. diam. are the rule; they may be erect or more or less leaning and draped with a moos-mantle, while from their branches hangs the pale moos *Weymouthia Billardieri. Coprosma rotundifolia,* elsewhere usually a shrub, is the dominant tree, and it grows in such profusion at times as to make pure stands. Besides the trees already mentioned, the following are common: — *Carpodetus serratus, Melicytus ramiflorus, Fuchsia excorticata, Pseudopanax crassifolium* var. *unifoliolatum* and *Griselinia littoralis. Podocarpus acutifolius* and *Weinmannia racemosa* may occur.

The second tier consists of young forest-trees, the *Coprosma*-form dominating together with small *Dicksonia squarrosa* and *Hemitelia Smithii* and the semi-tree-ferns *Polystichum vestitum* and *Dryopteris pennigera.* On the floor are mosses, liverworts, the liane *Metrosideros hypericifolia* (creeping), *Blechnum procerum* and *B. fluviatile.*

The lianes *Rubus schmidelioides* var. *coloratus, Metrosideros hypericifolia* and *Polypodium diversifolium* are common, the two latter being especially abundant on tree-fern stems. The filmy ferns, *Hymenophyllum scabrum, H. sanguinolentum* and *Trichomanes reniforme* cover the leaning trunks, particularly of *Griselinia littoralis. Polypodium grammitidis, P. Billardieri* and the orchid *Earina mucronata* are fairly common as epiphytes.

At an altitude of some 300 m. on river-bed in the Western district there is an association closely allied to subalpine totara forest, although that of the adjacent slopes is *Weinmannia-Metrosideros* and that of swamps *Podocarpus dacrydioides.*

Podocarpus Hallii, Phyllocladus alpinus (a tree) and *Pseudopanax crassifolium* var. *unifoliolatum* are dominant (Fig. 48) and *Libocedrus Bidwillii* sub-dominant. The forest is low, the trees &c. are erect. The undergrowth consists principally of *Polystichum vestitum, Pittosporum divaricatum, Aristotelia fruticosa, Nothopanax simplex, N. anomalum, Suttonia divaricata, Coprosma rotundifolia, C. propinqua, Olearia ilicifolia* and *O. avicenniaefolia.* There is also some *Wintera colorata, Carpodetus, Pittosporum Colensoi* (Fig. 47), *Pennantia, Myrtus pedunculata* and *Griselinia littoralis. Rubus schmidelioides* var. *coloratus* is the sole liane.

North Island river-bed. Generally the habitat is narrow and the species few. The following are probably the most important: *Epilobium pedunculare, Hebe salicifolia* (in a wide sense), *Veronica catarractae* var. resembling the type, *V. lanceolata, V. diffusa, Raoulia tenuicaulis, Gnaphalium keriense* and *Cassinia leptophylla.*

8. Grassland.

a. General.

The natural grasslands of New Zealand differ essentially from meadows of the Old World. Green, flat-leaved, turf-forming grasses do not rule, in their stead is brown tussock composed of closely-bunched often filiform rolled leaves and slender culms. Frequently, and probably always in the primitive associations, the tussocks grew so closely as to hide such smaller plants as might be present.

Grassland is for more abundant in South Island than in North Island.

It is absent in the north of the island except at the southern boundary of the South Auckland district, its place being taken by shrubland, fernland or bog. Further south, it occurs in the subalpine belt of the high mountains; also, at a lower level, on the flat parts of the Volcanic Plateau, and to the east of the central mountain chain, through, there, most if not all is of indigenous-induced origin. In South Island, it occupies wide areas to the east of the Divide and extends in places from sea-level to perhaps 1500 m. altitude, being indeed the most striking physiognomic feature of the vegetation.

The tussock-grasslands are by no means virgin, yet, to the inexperienced eye, they seem primitive enough. By far the greater part has been modified by sheep-farming since the "fifties" of last century and subjected to the periodical burning which that type of agriculture is considered to demand. Then, from the end of the "seventies" it has been exposed nearly everywhere to the depredations of rabbits but in numbers differing in different areas. Taking all the above into consideration, the marvel is that the tussock-grassland has been so little changed. Certainly, in the lowlands, much has been ploughed and replaced by artificial pastures of European grasses, or by farmlands of short-rotation crops. In some localities, the association is indigenous-induced, it having replaced *Nothofagus cliffortioides* forest after the latter was burned. There is also some evidence already dealt with that considerable areas of apparently primeval tussock-grassland owe their origin to the burning of forest long ago, such evidence being based partly on Maori tradition and partly on the actual presence of burnt trees on the hillsides of Canterbury, Central Otago, and elsewhere.

There are two distinct types of tussock grassland, the one where comparatively small tussock-grasses dominate, here called **"Low tussock-grassland"**, and the other where far larger, more massive tussocks are dominant, here called **"Tall tussock-grassland"**.

In preparing what follows, it has seemed best not to restrict the account of these two formations to their occurrence in the lowlands, but to deal with them up to that level on the mountains where they begin to receive a reinforcement from the true high-mountain species sufficient to alter their floristic and ecological condition.

b. The plant-formations and groups of associations.

1. Low (Festuca-Poa) tussock-grassland.

General. This highly-important plant-formation is distinguished by the presence of the medium-sized tussock-grasses in such quantity as to frequently touch one another — it may be with the tips of the leaves only — and to conceal most of the other species except on a close examination. The dominant tussocks are *Festuca novae-zelandiae*, *Poa caespitosa* and, in some places, *P. intermedia* and, in primitive New Zealand, *Agro-*

pyron scabrum[1]) (in a wide sense). At the present time, as far as I know, there are no examples of the primitive formation, nevertheless by means of comparative studies, carried out over the whole formation, it does not seem impossible to reconstruct the community in one's mind.

The low tussock grassland indigenous flora consists of 216 species (pteridophytes 10, monocotyledons 66 and dicotyledons 140) belonging to 38 families and 104 genera, the largest being, (families) *Gramineae* 36, *Compositae* 35, *Cyperaceae*, *Leguminosae* and *Onagraceae* 11 each, *Umbelliferae* 9, *Rosaceae* 8; (genera) *Poa* and *Epilobium* 11 each, *Carmichaelia* 9, *Deyeuxia* 8, *Carex*, *Acaena* and *Raoulia* 7 each. There are also at least 16 hybrid swarms.

Many of the species included in the above statistics play no part in the structure of the communities, but the following are either of prime-importance in that regard or are wide-spread and abundant: — *Dichelachne crinita, Danthonia pilosa, D. semiannularis, Poa caespitosa, P. Colensoi, P. intermedia, Festuca novae-zelandiae, Agropyron scabrum* (Gramin.), *Carex Colensoi, C. lucida, C. breviculmis* (Cyperac.), *Luzula campestris* (in a very wide sense), *L. ulophylla* (Juncac.), *Chrysobactron Hookeri* (in a wide sense) (Liliac.), *Muehlenbeckia axillaris* (Polygonac.), the 2 vars. of *Scleranthus biflorus* (Caryoph.), *Ranunculus multiscapus* (Ranun.), *Acaena microphylla, A. inermis,* × *A. microinermis, A. Sanguisorbae* var. *pilosa* (Rosac.), *Carmichaelia subulata* (Legum.), the vars. of *Geranium sessiliflorum (Geraniac.), Oxalis corniculata* in a wide sense (Oxalidac.), *Coriaria sarmentosa* (Coriariac.), *Discaria toumatou* (Rhamnac.), *Viola Cunninghamii* (Violac.), *Pimelea prostrata* in a wide sense (Thymel.), *Epilobium Hectori, E. pedunculare* in a wide sense, *E. novae-zelandiae* (Onagrac.), *Hydrocotyle novae-zelandiae* in a wide sense, *Oreomyrrhis Colensoi, O. ramosa, Aciphylla Colensoi, A. squarrosa, Anisotome aromatica, Daucus brachiatus, Angelica montana* — in the primitive formation (Umbel.), *Leucopogon Fraseri* (Epacrid.), *Convolvulus erubescens, Dichondra repens* (Convol.). *Plantago Raoulii, P. spathulata* (Plantag.), the vars. of *Coprosma Petriei, Nertera setulosa* (Rubiac.), *Wahlenbergia albomarginata*, the forms of *W. gracilis* (Campan.), *Vittadinia australis, Celmisia longifolia, Gnaphalium Traversii, G. luteo-album, G. japonicum, G. collinum, Raoulia subsericea, R. glabra, Helichrysum bellidioides, H. filicaule, Craspedia uniflora* in a wide sense, *Cotula squalida, Lagenophora petiolata, L. cuneata, Brachycome Sinclairii, Senecio bellidioides* and *Microseris Forsteri* (Compos.).

---

1) This statement is based on the facts that all the forms of this grass being far more palatable than the *Festuca* or the *Poa* — knowledge recently acquired by experiments carried out by W. D. REID and myself — has been eaten out where not protected, and that when open pasture is fenced securely from grazing animals the *Agropyron* appears once more in true tussock-form. The species itself is a compound one made up of many jordanons, differing greatly in colour and of two types, the one a true tussock and the other creeping and mat-like.

In any consideration of the formation the exotic species cannot be omitted. These number about 74, which belong to 21 families and 54 genera, the most important being the following: *Gramineae* 20 species, *Compositae* 11, *Caryophyllaceae* 8 and *Leguminosae* 6. None of the genera contain more than 4 species[1]).

In the wet North-western district with its forest climate, the formation is confined to stony montane valleys (old flood-plains). In North Island, it occurs to some extent in the montane belt of the Volcanic Plateau.

The life-forms of the formation (216 species) are as follows: trees 2, shrubs 31, tussocks 13, other plants of the grass-form 43, herbs 90, semi-woody plants 30, ferns 7; and 16 of the foregoing are summergreen, 131 mesophytes and 85 sub-xerophytes or xerophytes.

As in the case of forest, the low tussock-grassland formation is present, in its full development, in a wide range of climates and soils, but with this difference, that *forest cannot be established except with a fairly high rainfall and many rainy days, whereas low tussock-grassland has much the same composition and structure in climates ranging from an annual rainfall of 225 cm. to 35 cm.*, nor is this all, for where growing under conditions antagonistic to the natural establishment of forest, violent hot winds may be frequent, as also long periods of drought and a summer-heat constantly above 26° C. in the shade, while 38° is not unknown. Evidently, the mesophytes growing in the protection of the tussocks escape much of these severe conditions, but the tussocks themselves and the shrubs which considerably overtop them (species of *Carmichaelia, Discaria toumatou)* are exposed not only to the semi-arid climate just described but to one of extreme humidity and to every degree of transition from the one to the other. Hence, the old idea — still more or less current — that the tussock-form of these dominant grasses had been evolved in order to combat xerophytic conditions, must be abandoned. And this dictum is strongly supported by the fact, dealt with below, that various turf-making grasses and other life-forms make extensive colonies — no shade or shelter being available — under the dry tussock grassland environment.

This formation is exceedingly uniform in physiognomy, no matter its altitudinal position. When viewed from a distance, the ground appears

---

1) *Festuca rubra* var. *fallax* — has been sown largely, *Anthoxanthum odoratum, Agrostis tenuis, A. alba, Holcus lanatus, Aira caryophyllea, Dactylis glomerata*, frequently sown, *Poa annua, P. pratensis, Rumex Acetosella, Silene anglica, Cerastium glomeratum, C. triviale, Stellaria media, Arenaria serpyllifolia, Sagina apetala, Rosa Eglanteria, Ulex europaeus, Cytisus scoparius, Trifolium arvense, T. repens, T. dubium, Geranium molle, Erodium cicutarium, Linum marginale, Centaurium umbellatum, Myosotis arvensis, Prunella vulgaris, Veronica agrestis, Plantago lanceolata, P. major, Achillea Millefolium. Cnicus lanceolatus, Crepis capillaris, Hypochoeris radicata.* Many of the above merely occur on the bare ground, so they are rather invaders bent on making a new formation than real colonists.

clothed with a yellow carpet so smooth and even as to give the impression that all litter had been swept away with some giant broom. Here and there on the hillsides are dimples on its surface marking gullies or depressions. Such apparent smoothness is quite deceptive. Multitudes of tussocks, each some 40 cm. high, stand everywhere either close and touching or with spaces between. Here and there are solitary specimens, or clumps, of *Cordyline australis* or *Phormium tenax*. In South Island dark coloured bushes of *Discaria toumatou* may be dotted about and species of *Carmichaelia*, their green, erect leafless stems 1 m. high, are not uncommon. Small herbs frequently prostrate or low-growing shrubs occupy the spaces between the tussocks. Some of the more common are: — *Danthonia semiannularis*, *D. pilosa*, *Dichelachne crinita*, *Carex breviculmis*, *Ranunculus multiscapus*, species of *Acaena*, *Geranium sessiliflorum*, *Pelargonium inodorum*, *Oxalis corniculata*, *Epilobium novae-zelandiae*, *Aciphylla squarrosa*, *Hydrocotyle novae-zelandiae*, *Leucopogon Fraseri*, *Dichondra repens*, *Convolvulus erubescens*, *Plantago Raoulii*, *Wahlenbergia gracilis*, *Lagenophora cuneata*, *Vittadinia australis*, *Helichrysum filicaule*, *Celmisia longifolia* and *Gnaphalium collinum*.

Montane grassland is richer in species, certain plants common at a higher altitude being present. The following assist in modifiyng its physiognomy at a near view: — *Poa Colensoi*, *Scleranthus biflorus*, *Acaena Sanguisorba* var. *pilosa*, *Pimelea prostrata* var. *repens*, *Aciphylla Colensoi*, *Plantago spathulata*, *Wahlenbergia albomarginata*, *Brachycome Sinclairii*, *Celmisia spectabilis*, *Raoulia subsericea* and *Senecio bellidioides*.

Life-history of the formation. At the present time shingly riverbeds, fans, lake shores, and gravelly ground in general, present every transition from the migratory river-shingle associations to low tussock-grassland. The agents *par excellence* for having brought about this state of affairs are the mat-plants and low cushion-plants characteristic of the shingly gravelly habitat[1]), foremost in this regard being the species of *Raoulia*[2]) and *Muehlenbeckia axillaris*[3]). These species and others (*e. g. Acaena microphylla, A. inermis)* catch the silt as it blows from the river-bed, so that a new habitat, fairly well provided with moisture, in which seeds readily germinate, is constructed. Likewise, the mats and cushions entrap the

---

1) The first account of this was given by the author (1911 b.) but it dealt with the subalpine belt. Subsequent papers treating of the subject are COCKAYNE and FOWERAKER (1916: 175—76) and FOWERAKER (1917: 4.5.), both relating to montane riverbed; see also COCKAYNE, L. (1926: 356 and fig. 59.).

2) *Raoulia tenuicaulis*, *R. lutescens*, *R. australis*, *R. apice-nigra*, *R. Monroi*, *R. Haastii* and *R. Parkii*.

3) Prostrate shrub with far-creeping, matted, rooting underground-stems, forming large, circular patches; the leaves are very small, oblong to oblong-orbicular, glabrous and more or less coriaceous.

wind-borne or waterborne "seeds" of those species (and indeed of others) which are eventually branded together into low tussock grassland. The presence of large stones, owing to the shelter they afford and the moisture they conserve, also favour the establishment of seedlings, whereas very fine considerably-consolidated debris is antagonistic.

Obviously, the early colonizers of mats &c. must be light-tolerating species, but, except in the case of rapidly-growing large plants, to be shade-tolerating also is an advantage. For, as the seedling tussock-grasses increase in size they cut off the light, so that in time only those species which possess the dual property of tolerating both bright light and more or less shade survive. But the shelter afforded by the incoming tussocks makes the bare patches between them more suitable for seed germination, so that the vegetation by degrees becomes closed. Finally, as there is no uniformity in the habitat — old flood-plain though it be — but many transitions from wet places to those more stony and drier than the average, there comes a distribution of the species in accordance with such differences, the most xerophytic occupying the dry stony ground[1]) and mesophytes on the more silty and moister. In course of time, too, the species themselves modify the habitat through slowly adding humus to the soil.

Besides the species indubitably belonging to the formation there are others of a different class. Thus, an important succession following the river-bed stage is *Discaria toumatou* thicket. Now, though this spinous shrub dotted about the grassland is a characteristic feature, *it must be considered rather a relic of the former shrub-association than a true member of the grassland*, the shrub-association, as it died with age, being replaced by the tussock community. Also, other species may, by invasion, become established here and there, the forerunners of unrelated associations, especially species of *Cassinia*, or if near forest species of *Nothofagus*. So, too, with *Pteridium esculentum* and *Coriaria sarmentosa* — these really members of fern-heath —, *Cordyline australis* and *Phormium tenax*. Aspect plays an all-important part in delimiting these invaders, one which is sunny favouring *Pteridium* and one moister and more shady *Phormium*.

The foregoing sketches only the life-history of those low tussock-grassland associations on ancient river-bed, fans and the like, so that extensive portion of the formation covering hillsides has now to be considered. At the present time, the early, and successive stages of development can no longer be seen. The best clue is to be gained from studying what is taking place on the bare ground — plentiful enough — which has come into being through sheep-grazing, burning the tussocks and the attacks of rabbits.

---

1) A distinct community is confined to the most stony ground — really a piece of the old river-bed association not converted into grassland — consisting chiefly of *Muehlenbeckia axillaris, Carmichaelia nana, Raoulia Monroi* and *Cotula perpusilla*. It is again referred to further on.

These damaged areas — the major part of the formation — are occupied by different species in harmony with the amount of damage done, the nature of the soil and the climate. The effect of the last is so marked, that although low tussock-grassland as a whole is of a uniform physiognomy and composition, the actual procession of successions can hardly have been everywhere the same. This stands out clearly in semi-arid Central Otago (NO.), as will be seen in Part III. However, the general principle seems to be that regeneration — and one may presume the same for primitive establishment — depends upon the ground being occupied, first of all, by mat-plants or low cushion-plants of which the following appear to be the most important so far as montane low tussock-grassland is concerned: — *Muehlenbeckia axillaris, Acaena inermis, Coprosma Petriei, Raoulia subsericea* and *R. glabra.* Badly-damaged tussocks also provide an acceptable station for colonization and so, too, both tall and low-growing open shrubs while, in the latter, many grasses become established and form nuclei for seed-dissemination; *Discaria toumatou* and *Hymenanthera alpina* continually function in this regard.

Primitive low tussock-grassland. At the present time no association, or even portion of such, can claim to be truly primitive, through many areas look as if they were so. Probably the outstanding differences between the primitive and modified grassland are, (1.) that the tussocks were very close together in the primitive formation and that the smaller species were represented by far fewer individuals and belonged rather to stages of succession than to the climax, (2.) that species of higher palatability[1]) for sheep than *Poa caespitosa* or *Festuca novae-zelandiae* were common, especially *Angelica montana*[2]) (now extremely rare), and the varieties of *Agropyron scabrum.* There would also be (3.) far more *Aciphylla squarrosa* in the lowland belt and, in the montane, more *A. Colensoi.* Probably (4.) the invasion of shrubs and of fernland had not commenced.

Modified lowland low tussock-grassland. This association extends on the piedmont alluvial plains and adjacent hillsides from the R. Wairau to the southern limit of the formation. Though essentially belonging to the drier parts of South Island, it occurs to a limited extent where the rainfall is excessive.

Frequently, *Poa caespitosa* and *Festuca novae-zelandiae* — the dominant

---

1) The matter of relative palatability of tussock-grassland plants was studied in considerable detail by W. D. REID and myself, both by experiment and field observations. The results are in perfect accordance with the present composition of low tussock-grassland.

2) C. E. CHRISTENSEN informed me that, where the grassland near Lake Tennyson (NE.) had been fenced from stock and rabbits by a "rabbit-proof" fence, *Angelica montana* was in similar abundance to what it was in the early days, as described by old settlers.

species — occur in about equal numbers but this is at the lower levels. With increase of altitude the *Festuca* rules and the *Poa* is confined to the wetter ground. On the limestone area near Weka Pass (E.) both species occur in equal numbers.

The grassland of the Canterbury Plain (E.) has been greatly modified or swept away by agriculture. As for its flora, it seems clear that the species cited for the formation as common would be abundant. In addition, there are a few species of restricted distribution[1]).

Low tussock-grass of semi-arid habitat. In its most extreme form this association occurs in the valleys and on the lower mountain-slopes of the upper Clutha basin (NO.), and extends to almost the average westerly rainfall. At the present time the species number about 74, but doubtless there were more in the primitive association. An association similar in character, but floristically richer, clothes much of the lower Mackenzie basin-plain and the lower Waitaki valley.

In the virgin association, even in the most arid part, there appears to have been as close a covering of the tussocks of *Festuca novae-zelandiae* and *Poa caespitosa* as in the low tussock-grassland generally. Many forms of *Agropyron scabrum* were abundant[2]). The following are either confined to the association or extremely rare elsewhere: — *Acaena Buchanani, Carmichaelia Petriei, C. curta, C. compacta, Pimelea aridula, P. sericeo-villosa,* a most distinct var. of *Convolvulus erubescens* or an undescribed species and various forms of *Myosotis pygmaea*. At the present time, in many places there is hardly a hint of the original plant-covering, but one of indigenous-induced origin rules in its stead. An account of this remarkable state of affairs is given in Part III.

Montane low tussock-grassland. This falls into two classes of, (1.) where *Festuca novae-zelandiae* is dominant, and (2.) where *Poa intermedia* is extremely abundant and may dominate.

(Class 1.) *Festuca grassland* is common on the greywacke mountains of the North-eastern and Eastern districts. It contains pretty nearly all the

---

1) *Scirpus nodosus, Hypoxis pusilla, Muehlenbeckia ephedroides* (stony ground, a relic from river-bed), *Carmichaelia nana* (as for the last-named but far commoner), *Raoulia Monroi* (associated with *C. nana*, also a river-bed relic), *Cotula Haastii* (far more abundant on the Port Hills and perhaps confined thereto), *C. filiformis* (not seen on the Canterbury Plain since it was collected by HAAST, but occurs in a few places on the Hanmer Plains).

2) This reconstruction of the association has been made possible through research carried out for The New Zealand Department of Agriculture by the author, and especially by an examination of areas fenced from sheep and rabbits, and through a series of experiments carried out by the author on an area of maximum depletion where the regeneration of the pasture could be observed in relation to aspect, altitude, the moisture-content of the soil and competition between species. For a brief account see COCKAYNE, L. (1926: 355—361).

species of lowland tussock-grassland but, in addition, the following, many
of which ascend into the higher belts, are common: — *Hypolepis Millefolium,
Blechnum penna marina, Lycopodium fastigiatum, Deyeuxia avenoides,
Trisetum antarcticum, Triodia Thomsoni, Danthonia setifolia, Uncinia rubra,
Luzula ulophylla* and its hybrids with vars. of *L. campestris, Colobanthus
crassifolius, Stellaria gracilenta, Geum parviflorum, Acaena Sanguisorbae*
var. *pilosa* and its hybrids with related jordanons and other species of the
genus, *Epilobium chloraefolium* var. *verum, E. elegans, Hydrocotyle novae-
zelandiae* var. *montana, Aciphylla Colensoi, Anisotome aromatica, Euphrasia
zelandica, Gentiana corymbifera, Gaultheria depressa, Myosotis australis*
(probably distinct from the Australian species), *Plantago spathulata, Coprosma
Petriei, Wahlenbergia albomarginata, Gnaphalium Traversii, Celmisia spectabilis* and *Helichrysum bellidioides.*

(Class 2.) *Poa intermedia* grassland occurs on the mica-schist mountains (NO., SO.) where, owing to the absence of shingle-slip, grassland
with but few true high-mountain species, ascends on some mountains to
1200 m., or much higher. Its chief characteristics are the presence of *Poa
intermedia* in abundance and a good deal of *Agropyron scabrum* and *Festuca
novae-zelandiae*, and its florula being more that of lowland than of montane
grassland.

2. Tall tussock-grassland.

General. Tall tussock-grassland is characterized by the dominance of
either *Danthonia Raoulii* var. *rubra* or var. *flavescens*[1]) together with certain
low-growing shrubs semi-woody plants and herbs which tolerate wet, sour
soil. At the present time most of such grassland is more or less modified,
while, over wide areas, it has been converted into farms. Ecologically it
falls into two groups of associations dominated respectively by *D. Raoulii*
var. *rubra* and *D. Raoulii* var. *flavescens*.

Its indigenous flora consists of 57 species (pteridophytes 3, monocotyledons 18, dicotyledons 36) belonging to 22 families and 50 genera, the largest
being *Gaultheria* with 3 species only.

Lowland tall tussock-grassland is apparently confined to the South

---

1) Taxonomically, the group of which *D. Raoulii* Steud. is the type is in a state
of great confusion, owing to the two distinct jordanons *D. Raoulii* var. *rubra* Ckn.
(= *D. Raoulii* Steud. in Man. N. Z. Flora, ed. 2, p. 174, in part) and *D. flavescens*
Hook f. and a vast number of hybrids between the two being united together by New
Zealand botanists under the name *D. Raoulii*. There may also be other jordanons in
the mixture. It is not unlikely that *D. Raoulii* Steud. (*D. rigida* Raoul) is identical with
D. flavescens Hook. f. Under the circumstances, in order to avoid confusion, I am here
dividing the group into *D. Raoulii* var. *rubra* Ckn. ined. (= the forms with narrow reddish
leaves — "red-tussock") and var. *flavescens* (Hook. f.) Hack. ex Cheesem. (= the forms
with broad, green leaves — "snow-grass"), since I do not know what is *really* the type
of *D. Raoulii*. Both, I consider valid compound species.

Otago and. Stewart districts. It originally occupied nearly all that part of the Southland Plain where forest was absent but, further north, it evidently clothed much of the lower hill-country. Its upper altitudinal limit is not clearly defined, for it gradually merges into the higher montane and lower subalpine associations of a similar character. In North Island, the tall tussock grassland of the Volcanic Plateau is a closely-related community.

The life-forms may be classified as follows: — shrubs 12, herbs 25, semi-woody plants 5, grass-form 9, rush-form 4, ferns 2.

The physiognomy of the vegetation depends upon the great size of the tussocks, their reddish or green colour, as the case may be, and dense growth; except for an occasional shrub raised above the tussock all else is hidden. Thus, the habitat is far from uniform, and this is reflected in the tier of species subject to violent winds and, at times, strong light, and in the ground layer where the air is still, usually moist and the light comparatively dim.

The Danthonia Raoulii var. rubra (red-tussock) association. This falls into 3 subassociations — (1.) that of the Southland Plain, (2.) the Stewart Island and (3.) the lower montane. All are intimately related to bog and appear to be that formation changed by the incoming of the tall *Danthonia*, its subsequent dominance and the consequent suppression of such species as could not tolerate the diminution of light caused by the close growth of the tussocks.

The following are the principal species of the *Southland Plain subassociation:* — Blechnum penna marina, B. procerum, Danthonia Raoulii var. rubra, Hypolaena lateriflora, Astelia Cockaynei, Herpolirion novae-zelandiae, Thelymitra uniflora, Carmichaelia virgata, Geranium microphyllum, Halorrhagis micrantha, Centella uniflora, Gaultheria depressa, G. perplexa. Cyathodes empetrifolia, Gentiana Grisebachii in a wide sense, Plantago Raoulii in a wide sense, Coprosma parviflora and Lagenophora petiolata. Species of minor importance are Cladium Vauthiera, Oreobolus pectinatus, Gunnera mixta, G. prorepens, Nertera depressa, Helichrysum bellidioides and H. filicaule.

The chief ecological factor is the wet, sour, peaty soil which results from the frequent cold, south-westerly rain. But, as already seen, the tall species are subject to different conditions from those growing more or less close to the ground, in fact the community is a combination of grassland and bog.

The Stewart Island subassociation contains nearly all the above species and, in addition, the following which are either absent or rare in the community last described: — Lindsaya linearis, Lycopodium ramulosum, Microlaena Thomsoni, Gaimardia ciliata, Ranunculus Kirkii, Geum leiospermum, Actinotus suffocata, Liparophyllum Gunnii and Hebe buxifolia in a wide sense. Hypolaena lateriflora and Leptocarpus simplex frequently fill the spaces between the tussocks and exclude the smaller species.

The lower montane subassociation, in addition to many of the bog plants, contains species of drier ground — the habitat ranging from semi-bog conditions to those supplied by fairly well-drained ground — e. g.: *Uncinia rubra, Geranium sessiliflorum* var. *glabrum, Viola Cunninghamii* (also a bog plant), *Aciphylla squarrosa* (also a bog plant), *A. Colensoi, Leucopogon Fraseri, Wahlenbergia albomarginata, Celmisia longiflora, Raoulia subsericea, Craspedia uniflora* in a wide sense, *Senecio southlandicus; Herpolirion* and *Oreostylidium subulatum* are common.

Danthonia Raoulii var. flavescens association. This occurs on the hill-slopes of the northern part of the South Otago district. Fire has almost wiped it out of existence and its former presence is only betrayed by isolated plants of *D. Raoulii* var. *flavescens* or patches here and there which look quite out of place. As to its original composition, I can say nothing, but a well-marked bright-green variety of *Aciphylla squarrosa* — using this name in a very wide sense — is frequently an accompanying plant.

9. Rock vegetation.

General. The group of associations dealt with here is distinguished by the presence of certain species confined, or almost so, to a rock-habitat, together with various epiphytes, root-climbing lianes, xerophytes and sub-xerophytes of the neighbourhood of a rock association, and such mesophytes as epharmonically can colonize rock.

The species number about 193 (pteridophytes 31, monocotyledons 30, dicotyledons 132) which belong to 40 families and 102 genera, the largest being the following: — (families) *Compositae* 32 species, *Filices* 31, *Scrophulariaceae* 15, *Myrtaceae* 9, *Onagraceae* 8; (genera) *Hebe* 11, *Epilobium* and *Metrosideros* 7 each and *Coprosma* 6. The following are common species of wide range: *Adiantum affine, Cheilanthes Sieberi, Asplenium flaccidum, Pteridium esculentum, Polystichum Richardi, Polypodium diversifolium, Cyclophorus serpens* (Filic.), *Dichelachne crinita, Danthonia pilosa, D. semiannularis, Poa caespitosa, Festuca novae-zelandiae, Agropyron scabrum* (Gramin.), *Scirpus nodosus* (Cyperac.), various vars. of *Luzula campestris* (Junc.), *Cordyline australis, Phormium tenax, P. Colensoi* (Liliac.), *Thelymitra longifolia* in a wide sense (Orchid.), *Muehlenbeckia complexa* in a wide sense (Polygonac.), *Scleranthus biflorus* (Caryoph.), *Acaena novae-zelandiae, A. Sanguisorbae* var. *pusilla, Rubus australis* in a wide sense (Rosac.), *Oxalis corniculata* (Oxalidac.), *Leptospermum scoparium, Metrosideros hypericifolia* (Myrtac.), *Epilobium pubens, E. nummularifolium* (Onagrac.), *Griselinia littoralis* (Cornac.), *Dichondra repens* (Convolv.), various vars. of *Hebe salicifolia* (Scroph.), *Coprosma robusta* (Rubiac.), *Wahlenbergia gracilis* in a wide sense (Campan.), *Lagenophora pumila, Shawia paniculata, Helichrysum glomeratum, Brachyglottis repanda* (Compos.).

Rock-vegetation does not play nearly so important a part in the lowland-lower hills belt as in those of the coast and the high mountains, for it is absent in those wide areas, the gravel-plains of both islands, the northern gumlands, the Waikato plain, the Egmont-Wanganui coastal plain, the even pumice land of the Volcanic Plateau, and is but little in evidence on some of the lower hills. By far the greatest development is in river-gorges and narrow valleys, also near some of the large lakes and in the highest part of the lowland belt, but there the associations are usually only a continuation of those at a higher level.

As elsewhere, rock offers most diverse ecological conditions for plant-colonization, the most outstanding differences being due to the degree of wetness of the rock which is, in part correlated with position in regard to illumination. Really, in most associations so-called, there are several distinct communities according to the nature of the substratum, especially with regard to the slope of the rock, the amount and depth of the soil other than pure rock, and the aspect in regard to sun, shade and the prevailing wind. Evidently, then, any attempt to deal here with all the combinations of species is out of the question, all that is attempted is a general account of the principal communities according to latitudinal range and some particulars about special rock-species.

The life-forms of the 193 species are as follows: — shrubs including dwarfed trees and woody lianes 70, herbs and semi-woody plants 77, grass-form 15, rush-form 1 and ferns 30.

Communities of the Auckland districts. The most important rock-plants are as follows: — *Asplenium flaccidum, Polypodium diversifolium, Cyclophorus serpens, Blechnum procerum, Poa anceps, Cladium Sinclairii* (often dominant), *Cordyline Banksii, Phormium tenax, P. Colensoi, Astelia Solandri* (often dominant), *Astelia Cunninghamii* var. *Hookeriana* (abundant on old lava flows), *Bulbophyllum pygmaeum, Earina mucronata, Peperomia Urvilleana, Elatostema rugosum* (wet rocks), *Rubus australis, Coriaria arborea, Leptospermum scoparium, Metrosideros carminea* (sometimes dominant), *M. perforata, Griselinia lucida, Gaultheria antipoda, Cyathodes acerosa, Leucopogon fasciculatus, Hebe salicifolia* in a wide sense, *H. macrocarpa, Nertera Cunninghamii* (wet rocks), *Olearia furfuracea, Celmisia Adamsii* (Thames subdistrict and near Whangarei), *Gnaphalium keriense* (wet rocks), *Brachyglottis repanda.*

Communities for the rest of North Island. On *volcanic conglomerate near L. Rotoma (VP.),* and doubtless elsewhere in the district, there is a well-marked association dominated by the beautiful *Gaultheria oppositifolia, G. antipoda* var. *erecta* and a polymorphic swarm of hybrids between the two. Other species in the association are *Pteridium esculentum, Blechnum procerum, Gahnia gahniaeformis, Danthonia pilosa, Coriaria arborea, Weinmannia racemosa* (stunted), *Dracophyllum Sinclairii* and rather narrow-leaved *Hebe salicifolia.*

In the East Cape district, rocks in the drip of water are clad with a luxuriant growth of *Jovellana Sinclairii*[1]) and probably wide breadths of *Gnaphalium subrigidum* or *G. keriense.* A very common association throughout North Island in river-gorges is that of *Blechnum procerum*[2]), *Cladium Sinclairii* and *Gnaphalium keriense* with shrubs, especially *Hebe salicifolia,* projecting outwards or downwards. On drier rocks, in the full sunlight, the small tuft-tree *Cordyline Banksii* may form close masses accompanied by *Poa anceps* (a most common rock-plant throughout North Island), *Hebe salicifolia, Brachyglottis repanda* and other shrubs.

The vegetation of the soft marl ("papa") cliffs of the many deep gorges of the Wanganui coastal-plain is perhaps the most important rock-association of North Island and represents, on a vaster scale, that briefly described above, where *Blechnum procerum* dominates. Taking the gorge of the Wanganui itself, there are many kilometres of cliff, sometimes sloping and sometimes perpendicular. Above comes low *Beilschmiedia tawa* forest, beneath wich there may be a belt of shrubs and beneath this again the true cliff-covering consisting above of masses of the drooping, smooth, flat, pale-green, grass-like leaves of *Cladium Sinclairii* 1 m. or so long by 2 cm. wide and great breadths of the huge leaves of *Blechnum procerum,* this the physiognomic plant. If the position is specially wet and shaded, pure yellowish-green colonies of *Elatostema rugosum* are characteristic. Nearer the base of the cliff is great abundance of *Gnaphalium keriense* mixed with *Hebe lanceolata* and beneath this again *Adiantum affine* with stunted fronds. *Phormium Colensoi* is abundant in places. *Senecio latifolius* is common where the rock is wet, and, in a few localities, the allied *S. Turneri*[3]) is to be found; also *Ourisia macrophylla* and *Euphrasia cuneata* add to the floral display.

In the Ruahine-Cook 'district the following showy species occur on rock in gorges piercing the Tararua Mountains: — *Carmichaelia odorata, Jovellana repens, Hebe salicifolia* var. *angustissima, Veronica lanceolata,* a species of *Craspedia* with small flower-heads and *Olearia Cheesemanii.*

In the south of the above district, great sheets of *Poa anceps* are specially characteristic of rocks and strongly physiognomic, and near Cook Strait, *Hebe salicifolia* var. *Atkinsonii* is a common member of the association and is also one of the first species to occupy rock. On dry rocks in the open,

---

1) Stem creeping and rooting, finally erect, 15 to 45 cm. tall; leaves thin, pubescent, long-petioled, ovate with blade 2.5 to 7.5 cm. long; flowers rather showy, white spotted purple.

2) The great pinnate leaves 90 cm. to even 3 m. in length project downwards; each keeping so clear of those adjacent that the whole covering makes an ideal mosaic. These close mantles of *B. procerum* are a striking feature of New Zealand everywhere on the banks of moist gulleys and give a most characteristic stamp to the scenery.

3) *S. Turneri* has a long creeping stem and bright-green, cordate leaves 15 cm. long with stalks 15 to 30 cm. It is conspicuous where it occurs, but, so far as known, it is of local distribution though extending from the Wanganui to the Mokau River.

and this applies to much of North Island, the following occur: — *Cyclophorus serpens, Phormium Colensoi, Poa anceps, Leptospermum scoparium, Cyathodes acerosa, Leucopogon fasciculatus* and *Brachyglottis repanda*.

South Island rock communities. *Rock in the Sounds-Nelson district* is frequently distinguished by the presence of *Hebe angustifolia* with which may be associated *Polypodium diversifolium, Asplenium flaccidum*, stunted *Nothofagus Solandri, Muehlenbeckia complexa* var. *microphylla, Coriaria arborea, Melicytus ramiflorus, Leptospermum scoparium, Epilobium pubens, E. junceum, Cyathodes acerosa, Coprosma robusta* and *Hebe salicifolia* var. *Atkinsonii*. On the rocky banks of the Pelorus River, *Hebe rigidula* and *H. divaricata* are common.

In the North-eastern district, generally on limestone but sometimes on greywacke, exposed to the full sunshine is perhaps the most remarkable rock-association of the region with its assemblage of species bearing most striking flowers. It is distinguished by the monotypic *Pachystegia insignis*[1]) which is accompanied by the following though rarely are all present at the same time: — *Hebe Hulkeana*[2]), *Senecio Monroi*[3]), *Linum monogynum*, with its delicate white petals, the golden-flowered *Ranunculus lobulatus*, the silvery-leaved, white-flowered *Celmisia Monroi*, the purple *Notospartium torulosum* or the pink *N. Carmichaeliae*, the pale-lilac *Wahlenbergia Matthewsii* and a number of other species[4]) not restricted to this association though common members.

The volcanic rocks of Banks Peninsula are distinguished by an association made up of *Hebe Lavaudiana* and perhaps *Celmisia Mackaui* (now almost extinct), together with the following: *Polystichum Richardi*, a var. of *Luzula campestris, Phormium tenax, Linum monogynum, Angelica montana*, a species of *Anisotome* related to *A. Enysii, Corokia Cotoneaster, Griselinia littoralis, Cyathodes acerosa, Hebe leiophylla* var. *strictissima, Celmisia longifolia, Shawia paniculata*, and *Senecio saxifragoides* (Port Hills)

---

1) *P. insignis* is a stout low-growing shrub of straggling habit. The sparsely-branching stems are grey or black, exceedingly stiff and bear at the extremities short, open rosettes of 6 to 7 obovate, thick, fleshy leaves each about 9.5 cm. long, shining green above, and clad beneath with dense, felt-like, buff tomentum. The roots pass far down into the rock. The shrub grows on the driest and steepest rock-faces. The flowers are in large hemispherical heads, 5 to 7.5 cm. diam., with yellow disc and white rays; the species apparently consists of several jordanons.

2) An erect shrub about 60 cm. high with slender branches, dark-green shining coarsely-serrate broad leaves 2.5 to 5 cm. long and a much-branched panicle up to 15 cm. long, bearing numerous soft-lilac flowers, but the colour ranges to white.

3) A dense, semi-globose, much-branched shrub up to 90 cm. high, its leaves about 2.5 cm. long, narrow-oblong wrinkled, crenate, thick, green above, white tomentose beneath and terminal corymbs of many golden-rayed flower-heads.

4) *Phormium Colensoi, Angelica montana, Anisotome aromatica* (a distinct jordanon), *Coriaria sarmentosa* and, in places, *Shawia paniculata* and *S. coriacea*, and the hybrids between them.

or *S. lagopus* (Banks Peninsula proper). On rocks in the full sunshine, *Cheilanthes Sieberi* and *Notochlaena distans* are common.

At the lower Waimakariri Gorge (E.) there is a rich rock-vegetation consisting of the usual xerophytic shrubs, but of special interest is the occurrence of the usually coastal *Rhagodia nutans* and *Angelica geniculata* and the strongly xerophytic fern *Gymnogramme rutaefolia*.

Limestone rocks have usually a very open vegetation made up with but few exceptions — e. g. *Asplenium anomodon, Anisotome patula* — of species growing equally well on non-calcareous rock. In the north of the Eastern and south of the North-eastern districts, common members of the association are *Asplenium anomodon, Adiantum affine, Clematis afoliata* — tangled masses of leafless rush-like stems and flowers greenish-yellow, fragrant — and *Olearia avicenniaefolia*. Near Oamaru (NO.) the rocks possess a scanty community consisting of: — *Asplenium anomodon, A. Hookerianum, Epilobium junceum, Anisotome patula*, and in shallow gullies, *Cordyline australis, Discaria, Melicytus ramiflorus, Myoporum laetum* and *Coprosma propinqua*. Further south at Clifton (west of SO.) the rock carries the *Asplenium, Blechnum lanceolatum, Cordyline australis, Hymenanthera alpina, Edwardsia microphylla, Griselinia littoralis, Hebe salicifolia* var. *communis, Pimelea prostrata, Anisotome patula*, and *Coprosma propinqua*. Near Greymouth (south of NW.), in a far wetter climate for limestone than any of the foregoing, there is *Cordyline Banksii, Polypodium diversifolium, Hebe salicifolia, Olearia avicenniaefolia* and *Brachyglottis repanda* (on dry ledges) and *Asplenium lucidum, Epilobium rotundifolium, Lagenophora pumila, Craspedia uniflora* var. and *Gnaphalium Lyallii* (where water drips).

In the semi-arid part of the North Otago district, dry mica-schist rocks are distinguished by the locally-endemic, straggling *Hebe pimeleoides* var. *rupestris* — wiry stems, small glaucous leaves, blue flowers — and the silky-haired *Pimelea aridula,* accompanied by *Cheilanthes Sieberi, Poa intermedia, Agropyrum scabrum, Scleranthus biflorus, Rubus subpauperatus, Discaria toumatou, Corokia Cotoneaster* and *Hymenanthera alpina* (Fig. 82). Increased rain brings in *Leptospermum scoparium*.

Section III.
The Vegetation of the High Mountains.

Chapter I.
General Remarks.

Floristic details. The extremely mountainous character of New Zealand has led to the development of a rich and varied high-mountain flora, no fewer than 511 species never descending — so far as records go — into

the lower lowland-belt, while, in addition, 115 species occur in the lowlands only under special circumstances. There are also the 339 species,
partly dealt with in Section II, Chapter I, which are common to the high-
mountain and lowland belts, so that the high-mountain flora as a whole
numbers 965 species as against 1058 for the lowlands and lower hills, but
if the number of species belonging to the high mountain and coastal elements of the latter be deducted, the high-mountain flora exceeds that of
the lowland-lower hills by 68 species.

If the 511 purely high-mountain species be alone considered, they belong
to only 38 families and 87 genera, but if the 115 occasional lowland but
really high-mountain species be added, the number of families and genera
rise to 47 and 114 respectively. No less than 495 of the 511 (nearly 95°/₀)
species are endemic; but the endemism of many provisionally non-endemics
is highly probable, e. g. *Scirpus crassiusculus, Epilobium tasmanicum* (a considerable mixture), *Myosotis australis* (another mixture), *Plantago Brownii*
(still another mixture) and *Craspedia alpina* — the genus highly polymorphic with its many jordanons and hybrid swarms. In addition to the species,
at least 120 groups of hybrids are present, taking the whole flora into consideration, and many are extremely common, while no less than 42 genera
are concerned.

The following genera are confined to the high-mountain belt (endemic
marked *): — *Marsippospermum, Exocarpus, Hectorella*, Pachycladon*,
Notothlaspi*, Corallospartium*, Swainsona, Pernettya, Mitrasacme, Logania,
Pygmaea*, Phyllachne, Haastia*, Leucogenes*, Ewartia* and *Traversia**. This
list is quite trifling when compared with that of the lowland-lower hills with
its 23 families and 107 genera, but the following two lists bring out more
clearly the striking floristic differences between the two floras, and, at the
same time, with what has gone before, show the special composition of the
high-mountain flora.

1. Genera typical high-mountain which descend to sea-level or thereabouts under special ecological conditions, particularly a semi-subantarctic
climate. *Alsophila, Cystopteris, Triodia, Carpha, Oreobolus, Gaimardia,
Lyperanthus, Adenochilus, Claytonia, Caltha, Drapetes, Actinotus, Pentachondra, Archeria, Liparophyllum, Donatia, Ourisia, Abrotanella* and *Crepis*.

2. Characteristic high-mountain genera with the number of species for
the genus in brackets followed by the number confined respectively to the
high-mountain and lowland belts. (Coastal species are included with lowland, but endemics of the outlying islands are excluded) *Isoetes* (2), 1, 1;
Agrostis (8), 5, 1; *Deyeuxia* (11), 4, 3; *Deschampsia* (5), 2, 0; *Trisetum* (5), 3, 1;
Danthonia (15), 7, 1; *Triodia* (4), 2, 0; *Poa* (25), 8, 6; *Agropyron* (4) 2, 1;
Uncinia (19) 8, 6; *Carex* (52), 19, 16; *Luzula* (13), 9, 1; *Colobanthus* (10), 6, 2;
Ranunculus (45), 30, 14; *Nasturtium* (most likely in part an unnamed en-
demic genus) (7), 5, 2; *Geum* (6), 4, 1; *Acaena* (12), 5, 1; *Pimelea* (16), 8, 4;

Drapetes (4), 3,1; *Epilobium* (38), 14, 9; *Schizeilema* (11), 8, 1; *Aciphylla* (28), 26, 0; *Anisotome* (17), 12, 4; *Cyathodes* (4), 2, 0; *Dracophyllum* (23), 12, 2; *Gentiana* (19), 14, 1; *Myosotis* (32), 24, 4; *Hebe* (67 as defined at present but there are certainly more), 48, 14; *Veronica* (12), 7, 2; *Ourisia* (12), 8, 1; *Euphrasia* (13), 11, 2; *Plantago* (8), 3,1; *Lobelia* (3), 2, 1; *Forstera* (4), 3, 0; *Olearia* (37), 12, 15; *Celmisia* (56), 39, 3; *Raoulia* (23), 17, 1; *Helichrysum* (11), 6, 2; *Cotula* (19), 6, 8; *Abrotanella* (7), 5, 1; *Senecio* (32), 9, 12.

Certain other genera, or families, though they contain few or no high-mountain species play an important part in the vegetation, e. g. *Podocarpaceae, Rubiaceae, Festuca, Phormium, Chrysobactron*, Herpolirion, Nothofagus, Elytranthe, Muehlenbeckia, Stellaria, Drosera, Carmichaelia, Discaria, Geranium, Oxalis, Coriaria, Stackhousia, Aristotelia, Viola, Hymenanthera, Leptospermum, Halorrhagis, Nothopanax, Hydrocotyle, Angelica, Corokia, Griselinia, Gaultheria, Leucopogon, Cyathodes, Epacris, Archeria, Suttonia, Pratia, Wahlenbergia, Lagenophora, Brachycome, Gnaphalium, Cassinia, Craspedia, Microseris* and *Taraxacum*.

The headquarters of the true high-mountain flora of 511 species is in the lofty ranges of South Island, with their 472 species 379, of which (80°/₀) are confined thereto. North Island possesses only 105 species, of which 29 are not in South Island, and Stewart Island 39, of which 9 are confined to that island. The comparative poverty of the North Island high-mountain flora is most likely owing to the small area suitable for occupation as compared with South Island, and the far lower average height of the mountains is another factor concerned. In South Island, each botanical district shows a good deal of local endemism and the area of distribution of many species is small, so that there is a gradual change in the flora in proceeding from north to south. The greatest differences, both ecological and floristical, are in relation to rainfall, there being what may be conveniently styled **wet** and **dry** mountains.

Vertical distribution (the belts of vegetation). Details regarding vertical distribution are not easy to supply. All-important is the average winter snow-line above which the ground for some months is covered continuously with snow. Below this line the covering is not continuous, though at intervals more or less snow lies on the ground throughout the winter. The winter snow-line, and the average periods which snow lies at various altitudes below that line, are correlated with the aspect of the slope, so that alpine species descend much lower on shaded than on sunny faces. Edaphic conditions, hollows, close proximity to shrubs or rocks &c. also contribute their share, e. g. bogs, shingly river-beds and rocks may bring high-mountain species, and even associations, into the montane or even the upper lowland belt (see under next heading). Even on forest-clad mountains many bare patches, containing high-mountain grass or herb vegetation, extend far below the forest-line. Then there is the gradual effect of change in latitude and the differences in distribution on wet or dry mountains, as also the many intermediate stages between such extremes. In fact, each mountain supplies its own special circumstances, and, were the details at my disposal far more accurate, only general statements could be made. As it is, one has to trust in many instances to estimated heights, so that the details given here and further on are essentially approximate.

Commencing at sea-level the belts of vegetation are here styled, *lowland, montane, lower subalpine, upper subalpine* and *alpine.* Obviously, the boundaries of these differ for each botanical district and also in the districts themselves, as also on different parts of the same mountain, so that no actual altitudinal limits can be defined. The chief delimiting factor appears to be the *average length of time the winter snow lies upon the ground,* and in the upper alpine belt of the highest mountains this covering persists far into the summer. This reliance on the different snow-lines as a basis for the delimitation of the belts in supported by the fact that as soon as that part of a mountain is reached certain characteristic species are encountered which are rare or absent at a lower altitude[1]), and also a new vegetation is met with; so, too, though in a somewhat lesser degree, for distribution in regard to the other winter-snow limits.

The average line on any mountain, but different for different aspects, at which the maximum occupation of the ground by winter snow begins marks the commencement of the true **alpine belt,** and this extends either to the perpetual snowfields ore the summit of the mountain. Below this line comes a second one which denotes that there has been a covering of snow for a lesser period than on the alpine belt and this line marks the commencement of the **upper subalpine-belt.** Below the latter, there is usually no continuous covering of snow for more than a week or two at a time, and the average line marking this area is the commencement of the **lower subalpine belt.** Below the latter is an area where snow lies only on an average for a few days at a time the average lowest line of which marks the commencement of the **montane belt,** below which lies the **lowland belt,** in the upper part of which snow lies on an average for a day or two at most, but in the north of North Island, and on its coast snow never falls in this belt.

To supply definite heights for these different altitudinal lines is impossible for, as already seen, there are many factors concerned apart from altitude alone. Consequently, in estimating the floras of the different belts no hard-and-fast limits habe been considered, but rather the distribution of

1) For instance in the North-western district, *Danthonia australis,* which even on mountains to the east may fill hollows where snow lies for a long period; in the Eastern district *Celmisia Haastii* and *C. viscosa;* in the Western and parts of Fiord districts, *Danthonia crassiuscula* and *Celmisia sessiliflora* and in the Fiord district *Ranunculus Buchanani, R. Simpsonii* and *Celmisia Hectori.*

the species and communities. All that can be said here is, that in North Island mountains in general, the *lowland*, or it may be the *lower montane* belt, extends upwards to 700 m. altitude, more or less; that, in the Eastern district the *montane* belt begins at 400 m. or 300 m. in some places; and that in Stewart Island the distribution of high-mountain species and communities is mainly a question of the absence of forest. Even in the Fiord district, a true subalpine vegetation may occur at so low an altitude as 450 m. or lower.

Taking the whole high-mountain flora of 1058 species an estimate for the number of species in each belt is as follows: — *lower-subalpine* 725 (doubtless certain forest species included here belong rather to the montane belt); *upper-subalpine* 496; *alpine* 251, of which perhaps 100 occur in its uppermost part. The lowland-high mountain element plays only a comparatively small part in the upper subalpine and alpine belts with its 130 species (59 are virtually high-mountain) in the former belt and 27 species (15 virtually high-mountain) in the latter. Taking the purely high-mountain species alone, 178 species occur in the montane belt, 361 in the lower subalpine, 366 in the upper subalpine, and 224 in the alpine.

High-mountain plants at sea-level. As already seen, 115 species of high-mountain plants occur at about sea-level, many most characteristic alpine and subalpine species, e. g. to mention a few, *Carpha alpina, Oreobolus pectinatus, Gaimardia ciliata, Astelia linearis, Caltha novae-zelandiae, Carmichaelia Monroi, Drapetes Dieffenbachii, Cyathodes empetrifolia, Coprosma repens, Donatia novae-zelandiae, Celmisia argentea* and *Senecio Lyallii.* There is also a second category, the members of which are nearly or quite numerous enough to class with lowland species, although their distribution may be restricted to a special climate or soil. Such are, *Podocarpus nivalis, Dacrydium Bidwillii, Phyllocladus alpinus, Astelia Cockaynei, Carmichaelia grandiflora, Coriaria lurida, Hoheria glabrata, Pimelia Gnidia, Gunnera dentata, Gentiana Townsoni, Hebe Raoulii, Olearia Colensoi* and river-bed species of *Raoulia.* These lowland-high-moutain plants fall into three principal classes as follows: — (1.) Those which as plants of tussock-grassland find a continuous path by means of that formation to the lowlands. (2.) Those which descend by means of stony river-bed. (3.) Those which are restricted to that part of New Zealand possessing a modified subantarctic climate. To the first two classes belong xerophytes of physically dry stations, and to the third class shrubs of the subalpine-scrub and so-called "bog-xerophytes". Actual high-mountain associations, and not isolated species only, occur at sea-level, and such have already been briefly described in Section II Chapter IV, and veritable high-mountain species, or their equivalents, are present on coastal cliffs (Section I, Chapter IV). As for the causes furthering the presence of high-mountain species at low altitudes something is said under the next head.

The ecological conditions of the high mountains. High mountains the world over are subject to a set of similar and fairly definite conditions which there is no need to discuss here, especially as they are so admirably set forth by SCHRÖTER in his splendid work *Das Planzenleben der Alpen*, 1925—26: 920—1027. On the other hand, those specially affecting New Zealand need brief mention, so far as they are known.

No accurate details are available regarding the high-mountain climate. It is clear however that the species are not attuned to nearly so great intensity of cold as are alpine plants in general. This is clearly brought home by the fact that many species of the alpine belt cannot endure the winter temperature of Kew and hardly any that of Berlin. Probably — 18° C. is more than most can endure. For example, the winter of 1923 did more damage than usual in the neighbourhood of Queenstown (SO., lat. 45°, 310 m. altitude). Though there was almost constant frost for six weeks, and many supposedly hardy plants were killed, so far as can be ascertained the shade temperature did not fall below — 11° C. and this statement is strongly supported by the fact that the exotic *Eucalyptus Gunnii*, juvenile *E. globulus* (adult trees killed or greatly damaged), and *Pinus radiata* were undamaged. On the other hand, various species[1]) which are common above the forest-line were killed or badly damaged, as were the purely lowland and coastal species. Certainly, in the high mountains, a covering of snow stands for a good deal, but many species[2]) tolerate equally both snowy and snowless growing-places, while many of the less hardy shrubs may be only partly buried.

The amount of rain and the yearly number of rainy days is much greater than in the lowlands, while, in addition mist and clouds are frequent. The abundance of vegetation at above 800 m. in Central Otago, below which altitude the induced desert gradually appears, clearly confirms the above statement. But it must not be concluded that in an extremely wet climate xeromorphic plants are an anomaly. On the contrary, too much stress cannot be laid on the effect of short periods of drought. A few fine days in the extremely wet Western district will dry up subalpine streams one might well believe to be permanent, even if they occur on fairly level ground, and make the surface so dry that one has to be most careful regarding

---

1) *Gaultheria perplexa, Leptospermum scoparium, Weinmannia racemosa, Notho-panax Colensoi, Shawia paniculata, Phormium Colensoi, Senecio cassinioides* (high-mountain only) and *Senecio elaeagnifolius*. All the species mentioned in connection with this frost were either cultivated plants in the Queenstown Gardens or growing wild in the immediate neighbourhood.

2) *Ranunculus Grahami*, the rosette species of *Nasturtium* (generic name provisional), *Epilobium rubro-marginatum*, several dwarf *Aciphyllae, Myosotis macrantha, Hebe pinguifolia, H. epacridea, H. Haastii*, cushion species of *Raoulia, Leucogenes grandiceps* and many others.

fire. Where the soil is extremely porous, as on the Volcanic Plateau, during a specially dry period even *Celmisia spectabilis* may be killed by drought. In short, many species obviously depend far more upon precipitation than upon ground-water and, in this regard there is a similarity between rain-forest and herb-field.

The north-west downpour on which depends the distribution of forest on the west of South Island is, until it has crossed the Divide, a warm, snow-melting rain, whereas that from the south-west is cold and generally terminates in snow, succeeded by frost. This latter rain, in South Otago and Stewart Island is a most important factor towards inducing herb-field and bog. Its frequent occurrence, combined with cloudy skies, brings in those subantarctic conditions which even in the lowlands favour high-mountain species, indeed it is not going too far to declare that where the climate is of this character the occurrence of such species is rather a matter of opportunity than of altitude, and given ground to grow upon they will become established and form associations almost anywhere (L. COCKAYNE, 1925: 79, 80). On the slopes of the Southern Alps facing the Canterbury Plain and on the Seaward Kaikoura Mts. much rain comes from the east.

The wind-factor is of great moment, acting as it does both mechanically and physiologically upon the plants, while as an agent of denudation it is of considerable importance. Hemmed in between the high ranges, bounding narrow valleys, the power of the wind attains great intensity. Bare ground is frequently impossible to populate with plants when swept at intervals by furious gales. On exposed ridges and low mountain tops wind forbids a close covering and allows the establishment of dwarf species of plants of the prostrate, cushion and creeping forms, which belong more properly to the alpine belt. The north-west wind of South Island is of peculiar ecological significance. But for its prevalence, there would be forest where tussock-grassland at present rules. When this wind, a true foehn, rages, and the sky is cloudless, under a burning sun transpiration must reach its maximum. So violent is this wind that it is impossible to stand upright on a ridge exposed to its full blast. Even in an extreme forest climate where the wind strikes with its maximum power tussock-grassland may be established yet with forest on either side where the force of the wind is somewhat less (Fig. 49).

Snow is a most important factor in New Zealand high mountains. It has been already shown how the primary division of the vegetation into belts depends upon the average duration of the snow-covering. When the alpine belt is almost bare of snow at the end of January or much later in the Western, Fiord and South Otago districts deep masses of snow, at times resembling glaciers, still lie in gullies and cirques. HAAST (1886: 30) found by actual measurement 12 m. depth of snow on one of the passes of the Southern Alps, and that in a gorge he estimated as being 150 m. deep.

Snow avalanches are extremely frequent, even on the driest mountains, and their effect is great both in destroying vegetation (Fig. 50) and furthering denudation while, as agents of distribution, they bring even living tussocks and shrubs into the valleys where they occasionally become established. In the upper subalpine and alpine belts, the vegetation after the snow has melted looks just as if a steam-roller had passed over the surface, flattened to so the ground are the plants, including tussocks of *Danthonia Raoulii* var. *flavescens* 1.2 m. high. The effect of temporary streams and pools or of snow-water in hollows, is reflected by the presence of special species and combinations of plants. The mechanical effect of a heavy snow-covering upon the subalpine-scrub may be pointed out, but it alone is not responsible for the peculiar life-forms.

The plants themselves once established modify both climate and soil so greatly as to make their own conditions. For example, though in the first instance, soil be the same and climate the same, the ecological circumstances of fell-field, herb-field, forest and scrub are quite dissimilar. The burning of *Nothofagus* forest or subalpine-scrub — a natural circumstance if caused by a volcanic eruption — may lead respectively to replacement by tussock-grassland or by a colony of *Phormium Colensoi* — communities ecologically distinct from the replaced associations. The subantarctic characteristic of a plant's dead parts turning into peat while still attached to the living plant itself is strongly developed in many alpine genera *(Phyllachne, Donatia, Celmisia, Raoulia* &c.) and plays an important part in modifying the habitat.

Aspect is a matter of fundamental importance, leading, as it does, to local climates in close proximity. This is seen on all sunny and shady slopes (the "sunny" and "dark faces" of the shepherds), but it appears at its maximum in narrow mountain valleys where for more than three months yearly the shady side may receive no direct sunshine and the ground remain frozen hard for many days at a time while the opposite slope is quite warm.

Topographical changes are of course of prime importance both with regard to the evolution of vegetation and perhaps of species. At the present time, the great prevalence of mountains composed of greywacke and allied rocks which supply debris in enormous quantity leads to the constant establishment of migratory communities both progressive and retrogressive. Although such are from their nature transient, yet they are constantly being re-established, so that they present permanent habitats where habitat-effects can accumulate and new species arise, if epharmonic response be an agent in evolution. Disintegration of the surface-soil, the result of wind, snow, frost and rain action, is always in progress, especially where tussock-grassland conditions prevail, to that the surface is not even like a meadow but there are low raised mounds of vegetation surrounded by sunken bare patches. Many other details as to habitat-ecology are cited when dealing with the communities.

Repeopling the new ground during the retreat of the glaciers. Even today, glaciated New Zealand, at a very low level indeed, is not a thing of the past. Probably, a fair idea of what many valleys of glaciated West-land were like during the retreat of the ice from the coastal plain and lower mountain slopes is afforded by the Franz Josef and Fox glaciers and their immediate neighbourhood with their terminal faces at 211 m. and 204 m. respectively and distant only a few kilometres from the sea. At the present time, the peopling of the new ground just abandoned by the ice can be observed, together with what has taken place at no distant date, indeed every transition can be plainly seen from bare rock, or moraine a year or so old, to forest. It seems then not unreasonable to conclude that what is happening at the present time is merely a repetition of what occurred throughout the Western district at the conclusion of the New Zealand ice-age.

In the case of the Franz Josef glacier — studied by me in 1910 and 1911 — three habitats are being invaded, namely rock smoothed by the ice, moraine, (both lateral and terminal) and river-bed, the rock being by far the most extensive. At an altitude of about 300 m. close to the abandoned rock (quartzose schist marked by numerous cracks, grooves and notches running parallel to the ice) there is no vegetation, but at a few metres distance from the ice, there are everywhere patches large and small some 2 to 3 m. deep of the moss *Rhacomitrium symphiodon*. This plant clings to the rock with great tenacity; its leaves when wet are spreading and hold much water in their axils, but when dry, they are erect and pressed closely to the stem. When the moss, through its rapid decay has prepared a seed-bed, the chinks in the rock are invaded by vascular plants (Fig. 51), the "seeds" brought by wind from the neighbouring scrub and forest or carried on the rock-surface by water. More than 30 species of pteridophytes and sper-mophytes take part in the invasion[1]).

The transformation of this open succession to one that is closed is a slow process, so that there are many extensive bare patches of consider-able age. Where moraine, even if quite thin is deposited upon the rock, a closed association is quickly produced. Thus Harper Rock, a roche moutonnée still partly embedded in the terminal face of the glacier, was

---

1) The following are the most important: — *Hymenophyllum multifidum* (grows on solid rock and Forms soil), *Lycopodium varium, Deyeuxia pilosa, Poa novae-zelandiae, P. Cockayniana* (forms large mats), *Schoenus pauciflorus* (especially where water lies), *Earina autumnale, Carmichaelia grandiflora, Coriaria arborea, Metrosideros lucida, Gunnera albocarpa* (broad, rooting mats), *Gaultheria rupestris, Dracophyllum longi-folium, Hebe subalpina, Veronica Lyallii, Coprosma rugosa, Celmisia bellidioides, Olearia avicenniaefolia* (almost the first to arrive), *O. ilicifolia, O. arborescens* and *O. Colensoi*. All these species occur in the immediate neighbourhood of the glacier and, except the *Earina*, belong to the subalpine florula. No species, not belonging to the locality has been observed.

quite bare except on the summit where, on a patch of moraine, there was an embryonic scrub of *Arundo conspicua, Carmichaelia grandiflora* and some other plants.

On older moraine-covered rock, at some distance back from the rock now being invaded, is a broad belt of tall scrub consisting of the shrubs already mentioned (subalpine-scrub species), together with rain-forest species, especially *Asplenium bulbiferum, Blechnum lanceolatum, Polystichum vestitum, Histiopteris incisa, Carpodetus serratus, Weinmannia racemosa, Melicytus ramiflorus, Fuchsia excorticata, Coprosma lucida* and *Coprosma foetidissima*. Within, is more or less *Metrosideros lucida*, i. e. the association is potential southern-rata forest. Such an association forms the next belt which extends upwards perhaps to the scrub-line and marks a comparatively recent advance of the ice.

According to BELL (1910: 5) "probably not more than 150 years ago", the glacier extended 820 m. northwards down the valley, depositing on its retreat extensive terminal moraines. On these, and the old river-bed, can be seen vegetation at different stages of formation, the climax, so far, being a scrub about 3.6 m. high (Fig. 52). On older moraine still, there is *Metrosideros lucida* forest but the climax-association of the valley is dicotylous-podocarp forest with *Dacrydium cupressinum, Podocarpus ferrugineus, P. Hallii* and the ordinary trees, shrubs, tree ferns *(Hemitelia Smithii)*[1] and ferns of Western district rain-forest proper.

At the greatest extension of the glaciers in the Pleistocene period, both on the extreme east and west of the Southern Alps, there would be peaks and slopes, not necessarily of great altitude, free from ice and still harbouring the Pleistocene high-mountain flora many of the species of which, judging from Pleistocene fossils from another locality, we may conclude were identical with, or closely resembled, those of today. But as such havens of refuge would be of limited extent, the struggle for existence would greatly reduce the number of species. Some certainly, if the glacier-extension were due to elevation of the land and not to increased cold[2]), would migrate to the Canterbury Plain &c., where, under semi-desert conditions, they might, as DIELS was the first to point out, epharmonically assume more xerophytic structure and the xeromorphic life-forms (the divaricating, flat-stemmed leafless, cupressoid &c.) have arisen, while on the eastern mountain-slopes the drought-conditions would be still more severe and sufficient to have evolved the great *Raoulia* cushions. With the retreat of the glaciers on the west forest would quickly become established on the coastal

---

1) *Hemitelia Smithii* also occurs in the forest above the glacier.

2) The presence of a tree-fern, *Hemitelia Smithii*, in one valley of the Lord Auckland Islands, is a pretty sure sign, as SPEIGHT has pointed out to me, that the ice-period of that group was not due to increased cold.

plain[1]) and following at the heels of the retreating ice seize on the river-valleys and mountain slopes as described above.

On the east of the Divide, occupation by plants of the morainic matter of valley floors can be readily observed at considerably less than 900 m. altitude near the terminal faces of certain large glaciers. In such a habitat-complex, many species can be established as colonists including both those of the high mountains and those of the montane belt. Also, as already described, the process of occupation of river-bed, fans, gravel-plains, and even clay hillsides, can be daily seen in operation, and the important role noted which is played by mat-plants and cushion-plants in providing seed-beds.

Chapter II.
The leading Physiognomic Plants and their Life-forms.

Forest plants. *Nothofagus* (Fagac.), southern-beech, consists in New Zealand of 5 species (all endemic) and at least 2 great hybrid swarms, × *N. cliffusca* and × *N. soltruncata*. *N. cliffortioides* (Hook. f.) Oerst., mountain southern-beech, ranges epharmonically from a tree 6 to 15 m. high to a stunted, spreading, gnarled shrub. The tree (the general form) has a straight trunk, 20 to 60 cm. diam. covered with moderately smooth bark about 4 mm. thick. Where not crowded, the tree is symmetrical with numerous wide-spreading branches to its base which branch abundantly in a more or less distichous manner. The final twigs bear, on their flanks, the numerous hard, stiff, coriaceous, small, glossy, dark-green, ovate, ovate-oblong, or rounded-ovate leaves 4 to 18 mm. long which are clothed beneath with white adpressed hairs. The distichous arrangement of branchlets and leaves gives the appearance of close horizontal layers of foliage one above the other. The flowers are monoecious, the staminate being extremely abundant, and, when a tree is in full bloom, quite showy from their red colour. *Nothofagus Solandri* (Hook. f.) Oerst., black southern-beech, is a lowland tree closely related to *N. cliffortioides* but taller and more massive. *Nothofagus Menziesii* (Hook. f.) Oerst., silver southern-beech, when lowland or montane, is a tall and massive tree, but in the subalpine belt it approximates to *N. cliffortioides* in size, and like it may occur merely as a shrub. The trunk of large trees are frequently buttressed at the base. The bark is, at first, thin and silvery, but eventually becomes furrowed. The head of the subalpine tree is small and open; the branches are frequently gnarled. The leaves are small, coriaceous, rather thick, bright-green, but yellowish in

---

1) Most likely, as now near the Franz Josef glacier, there would be low hills and slopes which would be clad with pre-glacial forest and from this seeds would be carried to the new ground. If this be so, the present podocarp-dicotylous forest of lowland Westland is more or less representative of the first post-glacial forest, and has not replaced a previous *Nothofagus* succession.

the mass and broadly ovate or rhomboid with crenate margins; on the under-surface are fringed domatia. *Nothofagus fusca* (Hook. f.) Oerst., red southern-beech, in the lowlands, is a still larger tree than *N. Menziesii*. It ascends only to the montane and lower subalpine belts. The trunk is frequently 2 m. diam., covered with deeply furrowed bark and furnished at the base with massive plank-buttresses. The leaves are rather thin, 2.5 cm. or more long, broadly ovate, bright-green and deeply serrate. *N. truncata* (Col.) Ckn., hard southern-beech, is a lowland species closely allied to *N. fusca*.

Libocedrus Bidwillii Hook. f. *(Cupress.)*, southern kawaka, is hardly more than 12 m. high in the subalpine belt, but is frequently of smaller dimensions. The trunk, covered with pale chestnut-coloured loose flaking bark, is remarkably straight and often some 54 cm. diam. The upper third of the tree consists of a dense, tapering conical head made up of short branches and leafy twigs forming somewhat horizontal layers. Adult and juvenile shoots are distinct, the latter having a row of flattened acute leaves some 3—4 mm. long on each flank, and an inconspicuous upper and under row of quite minute appressed triangular leaves. The ultimate shoots of the adult are tetragonous, 1.5 mm. diam. and the leaves closely appressed, triangular and minute.

Phyllocladus alpinus Hook. f. *(Podocarp.)*, mountain-toatoa, varies from a small tree some 7 m. high with a trunk about 25 cm. diam., covered with a moderately smooth blackish bark to a shrub 1—2 m. high. The branches are numerous, stout and finally give off many flexible straight opposite branchlets, naked for their lower half or third and then give off cladode-bearing stems. The cladodes, which exactly resemble leaves, are numerous, frequently arranged in threes, moderately close, patent or semi-vertical, pale-green, waxy beneath, thick, coriaceous, oblong to rhomboid in shape and variable in size. The flowers are monoecious.

Hoheria Lyallii Hook. f. and *H. glabrata* Sprague et Summerh. *(Malvac.)*, mountain ribbonwood, are small deciduous trees, 4 to 6 m. high, usually much-branching from near the base and of a rather twiggy habit. The bark is smooth, pale-coloured and may be peeled off in long strips. The leaves are dimorphic; they vary from the lobed, more or less orbicular juvenile of both to the but little hairy cordate adult of *H. glabrata* furnished with a drip-point, and the broadly ovate adult, still faintly lobed, with truncate base and almost tomentose under-surface of *H. Lyallii*. The bright green colour of the leaf of *H. glabrata* resembles that of European deciduous trees; that of *H. Lyallii* is paler. The flowers of both species are large, white, showy, abundant, and not unlike cherry blossoms.

Dracophyllum Traversii Hook. f. *(Epacrid.)*, mountain neinei, is a small tuft-tree varying from about 9 to 5 m. in height, with an erect trunk 60—25 cm. diam. covered with smooth, reddish-brown bark which scales off in papery flakes. At about its upper fourth, the trunk gives off a few very stout branches, which, curving outwards and upwards and branching 3 or 4 times in candelabra-fashion, bear on their ultimate stems great rosettes of reddish leaves which are thick, coriaceous, 30 to 60 cm. long and 5 cm. wide at the base and taper gradually into extremely long, fine points. The inner leaves of the rosette are not fully developed and erect and overlapping, but the outer spread out radially and are strongly recurved, the long points hanging downwards.

Shrubland plants. Shrubby *Compositae* belonging especially to the genera *Olearia* and *Senecio* (Compos.) are a striking both from their multitudes of daisy-like flower-heads, generally rounded habit — branching much after the manner of *Rhododendron ponticum* — and diversity of foliage which is usually coriaceous and tomentose beneath the leaves. With the exception of a few small-leaved species of the divaricating-form these composite shrubs have the same life-form (the "rhododendron-form"). This consists of a generally quite short trunk from which radiate upwards and outwards at about an angle of 45° stout, stiff branches, which branching several times, 2 or 3 branches passing off in close proximity, a close, wide, leafy head is formed. As development proceeds, the lower branches in large measure die and are cast off, so that the shrub is quite open below, but above consists of a rounded mass of dense, short twigs the outer being leaf-bearing. Several species attain the stature of small trees with stout, frequently semi-horizontal trunks, from which hang long strips of papery bark (Fig. 60) e. g. *Olearia ilicifolia, O. arborescens, O. avicenniaefolia, O. lacunosa, O. Colensoi, Senecio elaeagnifolius.* Polymorphic hybrids between several of the species occur in great abundance.

Divaricating-shrubs are common in the high mountains. This life-form (Fig. 31) consists of much-branched, stiff, wiry, sometimes flexuous stems closely pressed together and interlaced, the branching being frequently at, or about, a right angle. There is considerable variation from great rigidity to extreme flexibility. As already seen, the form occurs abundantly in many lowland formations, but it is also characteristic of montane and sub-alpine river-bed and terrace scrub. Taking the whole New Zealand Region there are 51 species of this life-form, including those where it is confined to the juvenile stage. which belong to 16 families and 20 genera.

Species of Hebe[1] *(Scrophular.)* occur in abundance throughout the high-

---

1) CHEESEMAN (1925: 778—783) recognizes 86 species of *Hebe*. On the other hand, COCKAYNE and ALLAN (1926: 11—47), in the light of field taxonomy and garden experience reduce the species to 70, a good many being rejected because they are based on one or more hybrids, but others not recognized by CHEESEMAN are admitted. In addition there are a number of well-marked varieties, numerous unnamed jordanons and many polymorphic hybrid swarms. Also there are certain undescribed valid species and others may certainly be expected; indeed, wide field-observations will be needed, and genetic studies made, before the arrangement of the genus is at all satisfactory.

mountains some of which are confined to the montane belt while others reach the highest altitude at which New Zealand vascular plants have been recorded. Two of their life-forms the "ball-like" and the "cupressoid" are restricted to the high mountains. In the first instance, nearly all the species are built on the same plan. A number of stout straight stems radiate symmetrically upwards and outwards at a narrow angle from a common base. Above, these branch abundantly decussately at an angle of about 45°, the peripheral twigs being green, slender, rather succulent and covered closely with leaves. In many cases, the equality of growth in all directions leads to a remarkable ball-like form, so that bushes perhaps 1.5 m. high and 1 m. trough look as if they had been trimmed by a gardener's hand[1]). The roots do not descend deeply but form a mat near the surface, any stem in contact with the ground readily forming adventitious roots. In the ball-like species the leaves are either patent or sub-imbricating; they are but a few centimetres long, moderately thick, sessile, nearly glabrous, coriaceous and vary both in the species and in individuals in the ratio between length and breadth[2]). In the cupressoid (whipcord) form the leaves are much reduced in size and scale-like, and pressed so closely to the stem as to be almost in the same plane as the bark with which at their bases they coalesce[3]). In outline the shrubs may be more or less rounded, but the absence of spreading leaves makes the growth more open. Juvenile plants have thin, flat, pinnatifid, spreading hygrophytic leaves. The only other class needing mention consists of quite low-growing shrubs with conspicuous thick, more or less imbricating glaucous leaves and prostrate or subprostrate gnarled often black stems. Such play a part in the physiognomy of stony alpine slopes and dry rocks. The number of species is 4 but the forms, most of which apparently result from hybridism, are almost without end. The flowers of hebes are in racemes, or spikes of different lengths or, in the cupressoid forms, in small heads. They are produced in great abundance. White, pure or more or less deeply tinged with lilac, is the prevailing colour but *H. pimeleoides* is blue and *H. Hulkeana* a clear lilac; in some species the flowers are fragrant.

The species of Dracophyllum (Epacrid.) of the high mountains number at least 16, many of which play a distinguished part in the mountain scenery (e. g. *D. uniflorum, D. Traversii, D. longifolium*[3])). On many slopes, 600 m. or more, above the observer, it is easy to pick out, by their brownish

---

1) To this class belong *Hebe leiophylla, H. laevis, H. glaucophylla, H. Traversii, H. subalpina, H. montana, H. Cockayniana* and *H. buxifolia* var. *odora*.

2) Seedling leaves are petiolate more or less deeply-toothed and ciliate; they also occur on reversion shoots.

3) This class includes at least 14 species, some of which cross with the ball-like species and some of the hybrids look like juvenile forms.

colour, those spots which betray the presence of one or other species of the genus, and one conversant with plant-combinations can thus gain a fair idea of what plants occupy that particular station. Subalpine-scrub, too, with *Dracophyllum* present, at once differentiates itself from that where the genus is absent. Some of the species are small trees, others medium-sized shrubs, and one *D. rosmarinifolium*, forms under certain conditions massive cushions (Fig. 53).

The erect shrubby species have a special life-form, the chief characteristics of which are: — stiff, erect stems of a more or less fastigiate habit; branching at a narrow angle and vertical, needle-like leaves, sometimes of considerable length, with sheathing bases. Besides *D. Traversii*, already dealt with, 3 species are of the tuft-tree or tuft-shrub form, the most important being *D. Menziesii* (Fig. 54).

Plants of tussock-grassland, herb-field, fell-field and related formations. *The species of Celmisia (Compos.)* found in the high mountains number about 47. The genus is dominant above all others in New Zealand mountains above the forest-line. Go where you will on subalpine and alpine herb-field or fell-field and their silvery foliage strikes the eye, it may be in stately rosettes of dagger-like leaves, in circular mats trailing over the ground or in dense cushions. Their aromatic fragrance fills the air; from early till late summer some of their white heads of blossom may be seen, while, in due season, gregarious species clothe both wet herb-field and dry, stony slopes with sheets of white. The following characters, with but few exceptions, are common to all the species. They are semi-woody, and the leaves persist throughout the winter. The leaves are stiff, coriaceous, and crowded rosette-fashion, at the extremities of the stems to which they are attached by broad sheaths, the outer enclosing the inner and the whole forming a terete, stem-like mass; or they are arranged spirally along the branches their sheats tightly overlapping; the under-surface is densely clothed with tomentum which varies much in character in different species, being silky, woolly, cobwebbly, kidglove-like &c.; the upper surface is often covered with a silvery pellicle; the sheaths remain attached to the plant long after the blade has decayed, and, as a wet, rotting mass, enclose the stem. The roots are stout, long and cord-like. There are the following types of life-form: — (1.) The leaves are long, the innermost upright and the outer often more or less recurved above making a large semi-erect rosette; the branches of the stem are short so that the rosettes stand closely together and form a circular mass (Fig. 55); (2.) the stems are prostrate, much-branching and put forth adventitious roots at intervals. The leaves are shorter than in the foregoing class, the rosettes, if the leaves are sufficient crowded to form such, less erect and the plant forms wide circular mats, or trails over the face of rocks or banks; (3.) the stem branches so frequently as to bring the rosettes into very close proximity so that a cushion may be formed, especially when abundant peat is produced within the plant from its dead parts.

The giant species of Aciphylla (Umbel.), of which *A. Colensoi* is the most wide-spread, are of special physiognomic importance owing to their life-form which resembles that of certain species of *Yucca* (Fig. 56) rather than one of the *Umbelliferae.* Each individual consists of an upright circular mass of erect, stiff, hard, bayonet-like leaves from the centre of which arises the extremely stout flower-stalk furnished with bracts resembling the leaves in miniature, in the axils of which are small umbels on stout, short, branched pedicels. Average plants have a diameter of about 86 cm. and a height of 50 cm. The leaves are yellowish green, 30 to 60 cm. long and so stiff, thick and rigid as to be almost motionless in a heavy gale; they are pinnate or bi-pinnate with leaflets 6 cm. or more long terminating in sharp, stiff, long spines. The rootstock is quite short and the plant is firmly anchored by means of a very long, flexible, rather fleshy, deeply-descending taproot. *Aciphylla maxima* (Fig. 57) is far larger in all its parts than the above; when in flower it is a most striking object. The flowers of all are dioecious.

The large-leaved species of Ranunculus, all of which have strikingly beautiful flowers of great size, in certain localities occur in sufficient numbers to dominate the landscape. The famous mountain-lily (Fig. 85, 91) is the most noteworthy. It is confined to South Island and is common on the wet mountains from the south of the Spenser Mountains to Stewart Island. The Western district is its headquarters. It forms colonies many square metres in extent to the almost complete exclusion of all other plants. Each individual consists of a very large fleshy, broad, thick rhizome furnished with abundant descending stout, flexible roots. As the plant grows, one end of the rhizome decays while the apex increases in length, the plant thus slowly occupying new ground. From the apical end of the rhizome, long-petioled, peltate leaves are given off, the petioles vertical and the blades horizontal, thus effectively shading the ground beneath. These leaf-blades are smooth, bright-green, flexible, coriaceous and frequently form a concave saucer-like surface which is filled with water after rain. The petioles are stout; they measure 30 cm. or more in length and the blade may be 24 cm., or more, diam. The flowers, borne on tall branched stalks, rise high above the foliage. They are pure white, the petals at times so numerous that the flower looks semi-double, and 30 blossoms to a stalk, each 7 cm. diam. are quite usual. It may easily be seen then, what a glorious spectable is a hillside clothed as far as the eye can see with close colonies of this noble plant! Nor is it when in bloom alone that it is striking, for, when not in flower, the great leaves, almost knee-deep show more plainly their unusual form. In the North-western and Ruahine-Cook districts, *Ranunculus insignis* occupies a similar place to the above. The flowers are golden-yellow, 4 to 5 cm. diam.

and each tall flower-stalk may bear 12 blooms or more. The leaves are dark-green, glossy, somewhat coriaceous, rounded-cordate and about 15 cm. diam.; their petioles are stout and some 16 cm. long. *R. Godleyanus* (W.) is another giant yellow buttercup. *R. Buchanani* differs from any of the above in its smaller much-cut leaves; its abundant large, white flowers are a striking feature of the alpine belt of the Fiord district. It forms many polymorphic hybrids with *R. Lyallii, R. Simpsonii,* and, in a well-marked variety, with *R. Scott-Thomsonii.*

The species of Raoulia (Compos.) are either mat-plants (Fig. 46) — often circular — or cushion-plants, but the life-forms are constructed on the same plan, intermediates between them occur and the difference is merely one of degree. There is a central woody main stem and a deeply-descending chief root. From near the base of the main stem rooting prostrate branches pass off radially. These branch abundantly, the branches tending to grow upwards, while frequent branching and consequent increasing density hinders their horizontal extension. Such closeness of growth, shutting off the light, causes the death of all the interior leaves and many of the stems, the interspaces becoming filled in the case of the thicker cushions with peat from their own decay, and in that of low cushions and patches with wind-blown silt &c. (Fig. 59). According to the relation between horizontal spread and vertical growth, so are mat-plants or cushions produced. The leaves are small, generally more or less imbricating and frequently tomentose. The ultimate shoots are in some species pressed so closely together that they form a hard unyielding surface as in the case of those cushions, the "vegetable-sheep" *(R. eximia, R. Buchanani, R. Goyeni &c.).* In these, large quantities of peat accumulate in the interior and the upper branches put forth adventitious roots by means of which the plant gets most of its water and salts. *Haastia pulvinaris* (NE.) has exactly the same life-form and equals *Raoulia eximia* in size (Fig. 58).

Gentiana corymbifera T. Kirk is a noble plant which lights up the rather desolate montane and subalpine tussock-grassland with its multitudes of large delicate white flowers. From the centre of a rosette of short yellowish green leaves rises up a stout, yellowish, smooth unbranched peduncle, some 40 cm. high, which bears on its summit an umbel or cyme 15 cm. or more diam. of white flowers each 2.2 cm. diam. The species is abundant on the drier mountains of South Island.

The species of Ourisia belong to 2 classes, *the one* with a stout, creeping rhizome, large leaves in rosettes, and erect peduncles bearing whorls of white flowers much in the same fashion as *Primula japonica;* and *the other,* creeping mat-plants with much shorter, more slender peduncles and smaller flowers. *Ourisia macrophylla* (North Island; NW., NE.) is typical of the first class. There is a thick, semi-terete, half-buried rhizome by means of which the plant is capable of rapid vegetative increase and forms wide colonies. At short intervals, leaves are given off from the flanks of the rhizome so closely as to touch. The blades are ovate, crenate, rather thick, vivid green above, but purplish-red beneath and concave, so that they collect water which finds its way to the rhizome by means of the deeply channelled, stout, fleshy leaf-stalk. The flowers are in about 8 whorls on stout peduncles 25 cm. high; each measures about 1.6 × 2.5 cm., they are white on the lips but citron-yellow in the corolla-tube. The two varieties of *O. macrocarpa* (W., F.) and *O. Macphersonii* (F.) are of the same class. The wide-spread *O. caespitosa* represents the second class.

Senecio scorzoneroides Hook. f. *(Compos.)*, in many parts of the wetter high mountains of South Island plays as great a part in the floral physiognomy as *Ranunculus Lyallii* and many hectares of virgin herb-field in the Southern Alps may be white with the abundant blossoms of this plant. The leaves which are summergreen are in erect rosettes from a stout root-stock; they are some 15 cm. long, broadly lanceolate, moderately thick and glandular-pubescent. The flower heads are in corymbs on tall peduncles, each head being 6 to 8 cm. diam. and the ray-florets long. The species crosses freely with the allied, but quite distinct, yellow-flowered *S. Lyallii*, so that hybrids constantly occur with yellow, lemon and cream-coloured flowers.

Chrysobactron Hookeri Col. *(Liliac.)* is a compound species of which all the jordanons are summergreen herbs with short rootstocks and tuberous roots. The leaves are tufted, linear, grass-like, rather fleshy, brownish-green, 30 cm. or so high. The scape is slender 30 to 60 cm. high and bears a raceme some 10 cm. long of bright-yellow hermaphrodite flowers about 8 mm. diam. on short, slender pedicels. The species when in full bloom is very conspicuous through its multitudes of golden-yellow flowers. It has increased greatly with the settlement of the country, since owing to its perennial subterranean tuberous roots it can reappear after the grassland is burned, while also its leaves are not eaten by stock.

Many species of the tussock-form are a most conspicuous feature of the vegetation. Certain lowland grasses (dealt with in Section II) are equally common in the high mountains. At low altitudes, *Festuca novae-zelandiae* dominates but, at higher levels, it is the great tussocks of *Danthonia Raoulii* var. *flavescens* and, especially on low passes, the latter and var. *rubra*, together with their many hybrids. Another tussock, *Schoenus pauciflorus* Hook. f. (Cyperac.) bestows a reddish-brown hue, recognizable from afar. The plant itself consists of a bunched-up mass of close, erect, stiff but slender terete, grooved stems, 30 to 50 cm. high, and more or less purplish-red in colour. *Phormium Colensoi* Hook. f. (Liliac.), really of the tussock-form, must be included here.

Chapter III.
The Autecology of the High-mountain Plants.

1. Life-Forms.

Trees. In this chapter, contrary to the procedure in the first edition of this book, only the 511 species, which never become really members of the truly lowland communities, receive detailed treatment, but brief mention is made of a few other species which play an important part in the high-mountain vegetation.

There are no high-mountain trees which are not at times also shrubs, but 8 may perhaps be considered trees rather than shrubs. Of these "trees" so-called, 7 are evergreen, 1 deciduous and all are low. The following are the life-forms and the number of species to each: — bushy-tree 4 (2 dimorphic), araliad-form 1, tuft-tree 1 and rhododendron-form 2. These trees are not really of much moment in the forests, 3 being rare and local and the remainder absent over wide areas. The really important forest trees are canopy-trees (species of *Nothofagus*), and, to a lesser degree, the low, pyramidal *Libocedrus Bidwillii*, and certain low, small-headed podocarps. The remarkable tuft-tree, *Dracophyllum Traversii* is described in the preceeding chapter. To the tree-composites (rhododendron-form) may be added at least 4 which descend to the lowlands. Generally, these tree-composites have their trunks bent at the base and extending more or less horizontally (Fig. 60). Their outer bark hangs in long strips, as also does that of *Libocedrus Bidwillii* and *Dracophyllum Traversii*.

The leaves of the trees may be characterised as follows: simple 7, compound 1, broad 7, narrow 1, large 5, medium 1, small 2, coriaceous 7, thin 1, glabrous 5, hairy 3 (tomentose 2). The leaves of *Olearia lacunosa* are stiff, thick, narrow and somewhat after the manner of juvenile *Pseudopanax crassifolium* var. *unifoliolatum;* there is a prominent midrib on the under-surface from which pass off at a right angle stout lateral veins, so making sunken interspaces which are filled with rusty tomentum.

Shrubs. Here it must again be pointed out that many of the shrubs under conditions favouring rich growth become trees. Also one and the same species frequently assumes epharmonically different life-forms. Here, however, only what is considered the commonest form of a species is considered.

The number of shrubs is 118 of which 95 are spot-bound, 23 wandering, 14 mesophytes, 104 xerophytes or subxerophytes, 109 evergreen, 9 leafless or nearly so, 80 erect, 9 semi-prostrate, 29 prostrate, 6 tall, 36 of medium height, 60 of low stature and 16 of very low stature of which 11 hug the ground.

The following are their life-forms and the number of species to each: — switch-shrub 3, flat-stemmed leafless 2, divaricating 4 (open cushion 1), creeping and rooting 4 (leafless 3), erect bushy 13 (long needle-leaved 2), tufted 2, cushion 13 (open 2), rooting mat 16, turf-making 1, ball-like 13, straggling 15, cupressoid 15 (bushy 4, mat 1, ball-like 1, straggling 1), semi-cupressoid 3 (straggling 3), rhododendron-form 14.

The branchlets of the shrubs may be roughly divided into more or less stout 22, wiry or stiff 54, slender flexible and twiggy 33 and those much abbreviated of cushions 9. The maximum of rigidity is reached in *Corallospartium*, *Carmichaelia Petriei*, *C. Monroi* and *Hymenanthera alpina*. Wiry, generally slender stems, are characteristic of the divaricating-form; stiff slender stems of the *Dracophyllum*-form; twiggy stems of *Hebe;* and fairly stout, rather brittle stems, covered in the younger parts with tomentum, of the shrub-composites. The only truly spinous shrub is *Discaria toumatou* (also lowland and coastal), but perhaps *Carmichaelia Petriei* may be considered spinous, and both *Hymenanthera alpina* and *Aristotelia fruticosa* produce branchlets spinous at the apex under extreme xerophytic conditions.

Deeply descending roots are a common feature, especially in small, spot-bound shrubs. Obviously, shrubs with creeping rooting stems have less need for long roots. Adventitious roots are readily produced on stems of *Hebe*, *Phyllocladus alpinus*, *Hoheria glabrata* and a few other plants, if they come in contact with the moist ground. The peat-forming cushion-plants give off from their ultimate branchlets many adventitious roots which penetrate the moist peat of the interior of the cushion, so that they are independent of the thick woody, deeply-descending roots for the intaking of water and function mainly in anchoring the plant to the substratum (frequently rock).

The leaves of the high-mountain shrubs may characterized as follows: — simple 109, narrow 20, broad 89, wanting or almost so 9, large 1, medium-sized 8, small 33, very small 67, coriaceous thick or fleshy 100, thin 9, glabrous 73, hairy 36 (tomentose 29). Many species have mesophytic juvenile and more or less xerophytic adult leaves, a distinction strongly marked in the cupressoid form of *Hebe* and *Helichrysum*, in *Carmichaelia* (leafy only when juvenile or in the shade), in *Hymenanthera alpina*, *Aristotelia fruticosa* and the vegetable-sheep.

Lianes, epiphytes and parasites. As these have already been dealt in Section II, Chapter III, along with those of all the altitudinal belts very little need be said here.

Lianes are of little moment in the high-mountain vegetation, *Clematis australis* alone being confined thereto; but *C. marata*, *Rubus schmidelioides* var. *coloratus*, *R. subpauperatus*, *Muehlenbeckia complexa* and *Helichrysum dimorphum* ascend into the montane belt.

As for epiphytes, there are none of the herbaceous or woody classes, excepting a few ferns, but abundance of mosses, liverworts (some store up water) and lichens. The most remarkable fern — also lowland — is

Hymenophyllum Malingii[1]) which is generally found on *Libocedrus Bidwillii* (dead trees or decaying parts of living ones, so not really epiphytic), but occurring occasionally on *Dacrydium intermedium* (PHILLIPS TURNER, 1909 : 3) and perhaps on *Podocarpus Hallii.*

The only strictly high-mountain woody parasite (and this may be merely dwarfed *Korthalsella Lindsayi*) is about 5 cm. high and its branches $\pm$ 2 mm. broad; it has only been recorded, from one locality (E.) where it is confined to certain shrubs of montane debris-scrub. The fair-sized *Elytranthe tetrapetala* and *E. flavida* are abundant on *Nothofagus*, and *Tupeia antarctica* occurs on *Hoheria Lyallii.*

Herbs and semi-woody plants.   These fall into the classes, land-plants and water-plants.

The land-plants number 380 species of which 301 are herbaceous, 79 semi-woody, 13 annuals or biennials, 367 perennials, 39 summergreen, 341 evergreen, 260 spot-bound, 120 wandering, 216 erect, 164 prostrate and 273 xerophytes or subxerophytes, and 107 mesophytes. With regard to stature, 4 are very tall, 6 tall, 42 of medium height, 150 of low stature, 178 of very low stature of which 67 hug the ground. A comparison of the above figures with similar statistics for the lowlands shows some striking differences. For instance, the high mountains have 273 more or less xerophytic species, as compared with but 23 for the lowlands, while species of low and very low stature taken together form 45% of the lowland herbs and semi-woody plants and 86% of those of the high mountains.

Coming now to the life-forms, they fall into the following classes, to each of which is appended the number of species it contains: — (1.) *Annuals and biennials* 13, consisting of: tufted-form 2 (grass 1, herb 1); erect-branching 5; rosette-form 6 (flat low 2, erect 4) — (2.) *Perennials* 367, consisting of: (a.) semi-woody plants 79, made up of, (*a.*) spot-bound 44, the life-forms being: erect tufted 8; erect rosette 15; yucca-form 7; cushion-form 8; flat or low rosette 1; non-rooting mat 2; erect-branching 3; (*β.*) wandering 35, the life-forms being: creeping and rooting mat-form 33; creeping and rooting rosette-form 2 (erect 1, low 1); (b.) herbaceous plant 288, made up of, (*a.*) spot-bound 203, the life-forms being: tufted-form 58 (herb 16,

---

1) The leaves are narrow, more or less pendulous and 3 to 15 cm. long. They are opaque and reddish or silvery in colour owing to their dense covering of stellate hairs. The leaf is without the ordinary lamina. "The vascular bundle (HOLLOWAY, 1923: 597) in all parts of the frond is encircled by thick-walled tissue showing pores in the cell-walls, the outermost limiting layer being thin-walled and protruding as long cylindrical papillae. These latter contain chlorophyll and function as assimilatory tissue. The papillae are separated from one another by narrow air-spaces, into which the water can be drawn in times of rainfall and held during periods of drought." The stellate hairs already mentioned strongly hinder transpiration and evaporation of the water-film, so that the chlorophyll-containing cells are in contact with water except during that most exceptional occurrence — a prolonged drought.

grass 40, fern 2); erect-branching 27 (all herbs); erect rosette-form 40 (all herbs, including 9 *Aciphyllae)*; flat or low rosette-form 19 (all herbs); cushion-form 45 (herb 34, grass 10, moss-like 1); tussock-form 4 (grass 3, rush 1); non-rooting mat 3 (rush 1, herb 2); summergreen tuberous-form 5 (earth-orchid 4, fern 1); straggling-branching 2 (both herbs); (*β.*) wandering 85, their life-forms being: erect creeping and rooting 20 (grass 12, herb 8); creeping and rooting mat-form 39 (grass 6, herb 33); creeping and rooting rosette-form 13 (all herbs); turf-making form 8; straggling and rooting 2 (both herbs), cushion-form 2; creeping and branching herb 1.

The leaves of the high-mountain herbs and semi-woody land plants may be characterized as follows: — compound 63, simple 317, broad 260, narrow 120, very large 8, large 9, of medium size 56, small 125, very small 182, thin 122, coriaceous thick or succulent 258, glabrous 243, hairy 137 of wich 56 are tomentose. This matter of hairiness, though frequently considered epharmonic, is largely a definite characteristic of certain families and genera.

Dark-coloured brownish or even blackish leaves are fairly common amongst high-mountain herbs and in those occasionally lowland, e. g. *Gunnera prorepens,* but I can give no approximate statistics. In some cases, I have proved the colour to be dependant, at any rate in part, on light-intensity, but in other cases this is not the case, for the stain is on the basal sheltered portion of the leaf as in some species of *Cotula.* Dull reddish or purplish leaves belong to the same category. Glaucous leaves occur in some species, e. g. certain *Gramineae, Ranunculaceae, Acaena, Umbelliferae* and *Compositae.*

The subantarctic character of dead vegetative parts remaining attached to the living plant and turning into peat is extremely common amongst New Zealand high-mountain plants and occurs in all degrees of intensity. In the case of *Celmisia* the rotting leaf-sheats, sopping-wet, are more bulky than the living sheaths that they surround. In the peat-filled cushion-plants the water-holding capacity is very great indeed and the plants are independent of soilwater. Many grasses of tussock-grassland have also a sheathing of dead leaf-sheaths, but these are usually of a drier character than those described above. Living leaves with channelled petioles function in conducting water to the roots as may be seen in certain species of *Ourisia* and *Ranunculus.* The great peltate leaves of *Ranunculus Lyallii* are filled with water after rain, but whether such is absorbed is not known. Ecologically they function in strongly shading the partially buried rhizomes and in preventing occupation of the ground by other plants. Filiform leaves with rolled margins are a characteristic of high-mountain grasses in general. Perhaps the most unusual grass-form is that of *Danthonia pungens* (Stewart Island), consisting of large patches of tufted culms from a woody root-stock bearing the somewhat distant sub-imbricating leaves which are not erect but stand out obliquely; they vary in length from more than 30 cm.

to 5 cm. or even less, and are extremely stiff and coriaceous, thick, pale-green marked with brown, and taper gradually to a sharp truly pungent apex; the blade is equitant, striated and waxy on the upper surface; the leaf-sheaths are long, about 9 mm. broad at the base in large examples, and they persist attached to the plant, slowly rotting, and holding much moisture.

Some of the summergreen and semi-summergreen herbs have rhizomes of considerable size, those of *Ranunculus Lyallii* being extremely fleshy and frequently 5 cm. diam.

The roots of the class of plants under consideration are, in no small degree, in harmony with the station. Thus extremely long roots are more common in plants of rock, stony debris, tussock-grassland and fell-field than elsewhere. In some cases, roots do not descend deeply but spread laterally, at times, more or less parallel with the surface. The length of roots of different species growing side by side appear to differ considerably, but I have not sufficient observations to go into this important matter in detail.

The water-plants consist only of the submerged, rush-like *Isoetes alpinus* and the slender stemmed *Schizeilema nitens* with its very small, thin, shining compound leaves, but a number of herbaceous perennials (e. g. several small *Ranunculi, Epilobium macropus, Gunnera dentata, Gnaphalium palu-dosum)* may be considered semi-aquatic, since they occur in shallow streams or where water lies for a long time. Also, many of the lowland aquatics ascend to about 900 m. altitude.

2. Pollination.

As elsewhere in the book the matter of pollination can receive only the most superficial treatment. Though, as in the other belts, there are but few species of butterflies, their individuals are numerous and there is abundance of Diptera, moths and beetles, so that there is no lack of insects for the purpose of pollination. Furthermore, the frequent high winds must carry the pollen long distances. The plants, in general, bear abundance of seed and seedlings of many — not all — species are in plenty. It is true that seeds are more plentiful some years than others, but climate here is the controlling factor with which both abundance of flowers and insects are correlated.

Taking into account only the 511 truly high-mountain species, and putting aside 81 ferns, *Cyperaceae, Gramineae* &c., the number dealt with here is 430 species of which the flowers of 57 per cent may be considered monoclinous and of 43 per cent more or less diclinous but these figures are somewhat misleading, since protandry and protogyny are common.

About 80 per cent of the species here being dealt with possess fairly attractive flowers, but those of the remainder are generally both dull in colour and very small. The attractive flowers are of the following colours: — white &c. 266, yellows 53, purple &c. 13, blue 2, red &c. 7, green 2, black 1. The attractive flowers for the most part belong to the *Ranunculaceae, Pittosporaceae, Onagraceae, Ericaceae, Epacridaceae, Umbelliferae, Gentianaceae, Thymelaeaceae, Borraginaceae, Scrophulariaceae, Stylidiaceae* and *Campanulaceae.* Many flowers are of a size quite disproportionate to the tiny plants that produce them and in some instances they are in such profusion as to hide the foliage, especially in small cushion plants (Figs. 61, 62). Speaking generally, the floral display of the high-mountains is far more striking than that of the lowlands and the coast. A corrie, one sheet of the dazzling white of *Ranunculus Lyallii* or *Senecio scorzoneroides,* the great mats of *Leucogenes Leontopodium* on the Tararuas (Fig. 63) far surpassing the famed Swiss edelweiss, or the Mount Egmont herb-field with *Ourisia macrophylla* and the golden *Ranunculus nivicola* in full bloom are sights not readily forgotten.

Regarding scent, perhaps 20 per cent of the species are sweet-scented. In some cases the flowers may be inconspicuous, e. g. *Stackhousia minima* (also lowland) fills the air with delicious fragrance when in full bloom, its tiny yellow flowers close to the ground not visible a first glance.

3. Dissemination.

Taking only the 511 purely high-mountain species less than one per cent have succulent fruits attractive to birds and still fewer have hooked fruits. But about 30 per cent are specially suited for wind-carriage and many more, though possessing no special apparatus, constantly travel considerable distances by aid of the wind; indeed, powerfully as it functions on the coast and in the lowlands, it is of far greater intensity in the high mountains. Wind, undoubtedly is all-powerful in dissemination not merely in carrying anemochores through the air, but in blowing disseminules over the ground. The conditions of fell-field and pumice-steppe in this regard are not unlike those of a dune-area. Light fruits or seeds, such as those of *Aciphylla, Anisotome* and *Hebe* will be moved by quite a gentle breeze, but there are none that can withstand a gale that hurls small stones and gravel through the air. Run-off water functions strongly as an agent of dissemination and it must play a far greater part than wind in closed associations. Snow-avalanches are of considerable moment. A number of birds (some of them sea-birds) frequent even the highest mountains and they will transport many seeds &c. which though having no special adaptations for that end may adhere to their feet or plumage, in fact it seems probable that the bird is in large measure responsible for the plant-covering of such an isolated mountain as Mt. Egmont.

4. Seasonal changes.

In the upper subalpine and alpine belts the period of blooming for most species cannot extend much longer than 3 months for any locality and within that space of time many species flower and ripen their fruits. On the contrary, if the whole high-mountain flora be considered, and the montane belt be included, blooming commences in October and extends to a limited extent into the middle of March, and even considerably later. There is no rule for the time of flowering of any wide-spread species, all depends on latitude, altitude, and its relation to sun, shade and the melting of the snow-covering. Thus, the seasonal changes in a snow-filled cirque may be identical with or even later than those at a much greater altitude. The relation of seasonal change to insolation is most marked, e. g. the same species will bloom some weeks earlier in the tussock-grasslands on the east of South Island than on the mountains exposed to the western rain-fall a few kilometres distant. In general, the earlier part of the flowering season is marked by the greater blooming of herbs and semi-woody plants and the later part by that of shrubs and the fruiting of the early flowering herbs &c. During the latter half of March, the whole of April, and to some degree in May, the majority of trees and shrubs are ripening or still carrying their fruit. Almost as soon as the snow has melted, or even before it has quite gone, certain species come into bloom, e. g. *Caltha novae-zelandiae*, and *Ranunculus Buchanani*. In what follows, I am indebted for valuable assistance to H. H. ALLAN, C. E. CHRISTENSEN, G. SIMPSON and J. S. THOMSON.

Taking first the montane belt of South Island, flowering commences in October, and about the middle of the month the following, amongst others, will be in bloom in the North-eastern district: — *Hierochloe redolens, Stellaria gracilenta, Clematis afoliata, Ranunculus multiscapus, Geranium sessiliflorum* var. *glabrum, Linum monogynum, Coriaria sarmentosa, Discaria toumatou, Viola Cunninghamii, Pimelea prostrata, Angelica montana, Corokia Cotoneaster, Leucopogon Fraseri, Hebe Raoulii, Wahlenbergia albomarginata, Celmisia longifolia, Helichrysum bellidioides* and *Senecio bellidioides*. In the middle of the same month, *Hoheria Lyallii* is just coming into leaf (lower subalp. belt, NE.) and *Ranunculus Lyallii* is just coming above the ground (upper subalp. belt, F.).

During November, blooming commences at a higher altitude, and the following species may be cited for the North-eastern and Eastern districts: — *Poa intermedia, P. Lindsayi, Astelia Cockaynei, Chrysobactron Hookeri* var. *angustifolia, Nothofagus cliffortioides, Colobanthus acicularis, Ranunculus Haastii, R. Monroi, R. Sinclairii, Notothlaspi rosulatum, Oxalis lactea, Aristotelia fruticosa, Pimelea Traversii, Aciphylla Monroi, Ourisia caespitosa*.

During December in the lower subalpine belt of the Western district the herb-field is a beautiful natural garden, the following being in bloom: *Ranunculus Lyallii, Drapetes Dieffenbachii, Epilobium chloraefolium* var. *verum, Anisotome Haastii, Ourisia macrocarpa* var. *calycina, O. sessiliflora*, several species of *Euphrasia, Forstera sedifolia, Celmisia coriacea, C. Armstrongii, C. petiolata*, and various mat-forming species of the genus.

On the Tararua Mountains during the last week in December ASTON noted as blooming (1910 a: 13 et sep.), — *Phormium Colensoi, Caladenia bifolia, Pimelea Gnidia, Drapetes Dieffenbachii, Aciphylla conspicua, Hebe buxifolia, Forstera Bidwillii, Celmisia hieracifolia, C. spectabilis, Leucogenes Leontopodium* and *Abrotanella pusilla*. On the Waimarino plain (VP.) at the same time as above and earlier, there are blooming in the tussock, tussock-grassland or bog *Herpolirion novae-zelandiae, Stackhousia minima, Aciphylla squarrosa, Wahlenbergia albomarginata* and *Celmisia longifolia*.

During January throughout the Southern Alps — but according to their districts — and the North-western district the greater part of the shrubs are in bloom, some having commenced to flower by the end of December, and the subalpine-scrub and low forest is transformed by the wealth of cherry-like blossoms of *Hoheria glabatra* and the innumerable fragrant snowy flower-heads of the shrub-composites, especially *Olearia ilicifolia, O. lacunosa, O. nummularifolia, O. avicenniaefolia* and *O. moschata* (Mt. Cook southwards). Various species of *Hebe* are in full bloom, e. g. *H. salicifolia* var. *communis, H. subalpina, H. vernicosa, H. Menziesii, H. macrantha*, as also the beautiful *Gaultheria rupestris*. The herbaceous plants still continue their display, including those already mentioned, but supplemented by others, e. g. *Ranunculus Buchanani* (SO., F.), *R. sericophyllus* (W.), *R. Simpsonii* (F.), *R. Godleyanus* (W.), *Gentiana corymbifera, G. bellidifolia, Ourisia glandulosa* (NW., SO., F.), *O. prorepens* (F.), *O. macrocarpa* var. *cordata* (F.), *Forstera sedifolia* var. *oculata* (F.), *Celmisia Dallii* (NW.), *C. laricifolia, C. sessiliflora, C. argentea* (SO.), *C. verbascifolia* (F., SO.), *Leucogenes grandiceps, Cotula pyrethrifolia, Haastia Sinclairii, Senecio scorzoneroides*.

On the Volcanic Plateau, at an altitude of 900 to 1800 m there is a considerable floral display, the following characteristic species being in evidence: — *Pimelea buxifolia, Epacris alpina, Dracophyllum recurvum, Hebe laevis, Veronica Hookeriana, V. spathulata, Ourisia Colensoi* and *Euphrasia tricolor*. At the same season flowering is at its height on other North Island mountains when, according to locality, these important physiognomic plants are in full bloom: *Ranunculus insignis, R. nivicola, Pimelea Gnidia, Gentiana bellidifolia, G. patula, Hebe Astoni, Ourisia macrophylla, Celmisia glandulosa, C. spectabilis, C. hieracifolia, Helichrysum alpinum, Leucogenes Leontopodium* and *Cassinia Vauvilliersii*. By the middle of January, the vegetation of Mt. Anglem (Stewart Island) is quite as much advanced as in the north, owing probably to its lower altitude and the insular character of the climate. Quite 80 per cent of the fl rula is in bloom including *Chrysobactron Gibbsii, Aciphylla Traillii, Dracophyllum rosmarinifolium, Hebe Laingii, Celmisia argentea* and *C. linearis*.

During February, the floral display of the high mountains continues, the shrubs are at their best in the early part of the month, herbs which at a lower altitude are in fruit are blooming in the alpine belt, so that in one place or another examples can be seen of nearly all the alpine flowers. The seeds of a few species are already ripe, e. g. *Uncinia uncinata, Ranunculus Lyallii, R. Buchanani, Epilobium pedunculare, E. glabellum, Anisotome filifolia, Cyathodes empetrifolia, C. Fraseri, Hebe Raoulii, Raoulia glabra* and *Cotula pyrethrifolia.*

Even so late as the first two weeks of March and, indeed, much later, flowers are still to be seen in certain localities especially where the rainfall is excessive and not of necessity at a particularly high altitude. Amongst these late bloomers are *Ranunculus Godleyanus, Carmichaelia grandiflora,* most species of *Ourisia,* several species of *Euphrasia, Hebe buxifolia, H. annulata, Celmisia petiolata, C. verbascifolia* and *Leucogenes grandiceps.* But apart from the above and others not cited, nearly all the species of *Gentiana* are both late and profuse bloomers and their season may extend into April. But long before that, the flowering season is over, and herbs and semi-woody plants have ripened their fruits. The shrubs, however, are mostly laden with fruit and the species of *Coprosma,* thickly covered with translucent drupes of many hues, are objects of great beauty. On river-beds of the Southern Alps, *C. brunnea* forms strings of shining beads in every shade of blue. Throughout April and early May, there are still many berries, drupes &c. to be seen, but by June they have nearly all disappeared, *Hoheria glabrata,* for some time beautiful with autumn colouring, has shed its leaves, dead foliage alone marks the presence of *Ranunculus Lyallii* and *Senecio scorzoneroides,* on river-beds the great cushions of *Raoulia Haastii* are green no longer but of a delightful rich chocolate-brown, and the mats of *R. tenuicaulis* yellowish-green, pale reddish-brown or grey, the shingle-slip species winter beneath the stones, and the alpine vegetation, some of it now reddish or purplish in colour, is at rest until the melting of the snow and the increasing warmth of spring.

5. Epharmonic modifications.

General. The high mountains offer greater opportunities for epharmonic change than do either the coastal or the lowland belts, since not only is there a more diverse variety of growing-places but considerable differences in altitude, and consequently in climate, come into play. Here only a few of the more striking cases, or classes of epharmonic modifications, can be dealt with and examples are generally discussed in which the life-form of the epharmone is identical with the permanent form (if there be such a thing) of some other species; for instance an epharmonic cushion of one species as contrasted with a stable cushion of another species. In some cases it is not possible to say without experiment whether a particular form be epharmonic or not. Thus, judging from field observations alone, the common subalpine *Astelia* of Mount Egmont is apparently *A. Cockaynei,* yet when grown in my garden side by side with a valid example of that species, it clearly revealed itself an epharmone of *A. nervosa* var. *sylvestris.*

Nanism. As is well known, dwarfing is a most common phenomenon in high-mountain species, increase in altitude and xerophytic conditions favouring low growth. Of particular interest is that class in which tall forest-trees become greatly reduced in stature or are transformed to low shrubs. That lofty lowland forest-tree — frequently more than 30 m. high —, *Dacrydium cupressinum,* in the subalpine-scrub of Stewart Island, though still mountaining its trunk and erect habit, is dwarfed to 3.6 or 4.5 m. in height. *Libocedrus Bidwillii,* growing in the subalpine forest of Mount Egmont, is a low tree about 8 m. high, but on the same mountain, when exposed to the westerly wind, it is only 2 m. high, and denuded of all branches on the windward side. The massive *Nothofagus Menziesii* frequently 27 m. high with a buttressed trunk 1.5 m. diam., or more, in the subalpine-scrub of the North-western district has become merely a medium-sized, much branched shrub.

The prostrate-form. As already seen, the prostrate form is eminently characteristic of the high mountains, so that no less than 40 per cent of the 511 true high-mountain species are more or less prostrate. In addition, many shrubs and even some trees are epharmonically prostrate; in fact, it is not feasible in some instances to assign the "normal" either to the prostrate or erect form. Thus, *Podocarpus nivalis* is a spreading shrub 1.8 m. high in the lowland belt of the Western district and it retains this life-form, though its stature decreases, up to an altitude of about 900 m., yet in the far drier North-eastern and Eastern districts, as a member of the fell-field association, it forms far-spreading, *rooting* mats. The small forest-tree, *Dacrydium Bidwillii,* when growing on ancient moraine, spreads for many metres as a more or less circular, prostrate, *rooting* shrub, and this epharmonic-form which epharmonically grades into the tree-form is actually defined in floras as var. *reclinata,* while the tree-form receives no mention.

Perhaps the most striking case of all is that of the ubiquitous *Leptospermum scoparium,* which may be a tree in the lowlands, an erect shrub of subalpine-scrub, or make on boggy ground a veritable turf less than 5 cm. deep, its shoots rooting abundantly. *Dracophyllum rosmarinifolium* (Stew., F.) under one set of conditions is also a turf-maker, under another set a creeping, rooting shrub, and under a third it builds massive, dense cushions (Fig. 53).

Excessive wind, as can be seen further on in regard to the Denniston plateau, is frequently responsible for the prostrate form, but xerophytic conditions, or sour soil under the high-mountain climate, are likewise potent agents.

The cushion-form. This form also is characteristic of the high mountains and occurs in 21 families. It can also be constantly seen in the making, and even the most unlikely forms may become epharmonic cushions. Thus, *Aciphylla Traillii* — an erect plant some 12 cm. high with thick, rigid, spinous, 3-foliolate leaves 5 cm. or more long — on Table Hill (Stew.) where it grew on a piece of moist open ground, all competitors being absent, had its leaves reduced to rosettes, each 5.5 cm. diam. which pressed close together made a fairly large, dense cushion — a form supposed to be constant ("normal") for *Aciphylla Dobsoni*. When *Helichrysum coralloides* — an erect (or sometimes drooping with long shoots) cupressoid-shrub which is confined to rocks — is somewhat more exposed to wind than usual it assumes the cushion-form (Figs. 64, 65). *Dacrydium laxifolium*, a turf-making bog-plant of the Southern Alps, when growing on the dry pumice of the Volcanic Plateau, becomes a close cushion (Fig. 70). But it has already been shown that the hard, dense cushion of *Dracophyllum rosmarinifolium* is an epharmone and the suspicion arises that many other cushions belong to the same category. For example, nothing could seem less plastic than the cushion-species of *Raoulia*, yet *R. bryoides* frequently puts forth shoots with open leaves, and any of these vegetable-sheep, if cultivated in a moist green-house, will put forth mesophytic shoots and eventually assume an open habit.

Divaricating shrubs. This life-form so characteristic of New Zealand, though apparently quite stable, is frequently plastic to an extreme degree. This is well shown by the behaviour of shrubs of this form common in debris-scrub (especially *Pittosporum divaricatum*, *Aristotelia fruticosa* and *Corokia Cotoneaster*) which, when growing in an adjacent forest are open turggy mesophytes which in that life-form flower and fruit. *Hymenanthera alpina* — an extremely rigid divaricating-shrub, so reduced in size as to be an open cushion — behaves similarly, but as a mesophyte it apparently does not flower. Here again come in disturbing features for, in the case of the last-mentioned shrub, one almost of *the mat-form* and one *quite erect* have been oberved growing side by side. Also the divaricating form is strongly fixed in the case of *Nothopanax anomalum* which is hardly altered under forest-conditions.

The occurrence of spines. Spinous plants are very rare in New Zealand. Sharply-pointed rigid leaves are characteristic of *Aciphylla*, and many of the species grow in dry stations. The pungent-leaved grass, *Danthonia pungens*, has been already mentioned. Amongst shrubs, *Discaria*, *toumatou* alone posseses true spines and such are reduced leafy twigs. Increase in xerophytic conditions, however, produces sharp points on the abbreviated branchlets of *Aristotelia fruticosa* and *Hymenanthera alpina*, and in *Carmichaelia Petriei* — which grows in extremely dry stations[1]) — they

---

1) In May 1907 G. Simpson and J. Scott Thomson planted a wild specimen of *Carmichaelia Petriei* in the open ground, keeping it constantly covered with a bell glass, and by the end of November of the same year this entirely leafless, spinous xerophyte

are a usual character. As for the *Discaria,* the leafy juvenile cultivated in moist air remains juvenile, and plants growing wild under strong shade will produce more or less leafy twigs which represent spines. However, a semi-spineless adult of weeping habit has been recorded as growing naturally alongside ordinary spiny bushes fully exposed to strong sunlight and wind (COCKAYNE, L. 1922:208).

Miscellaneous. H. H. ALLAN (1926:75, 76) has shown that a certain form of *Ranunculus,* which may be *R. Monroi* var. *dentatus* is an epharmone, since, according to its growing-place, the leaves alter greatly in (1) *tomentum* — "extremely dense . . . . shaggy, appressed, ferrugineous" — to "reduced to very scant hairs" (2) *texture* — from "coriaceous" to "thin and membraneous" — (3) *size* — from 3 to 5 cm. × 2 to 3.5 cm. to 10 to 13 cm. × 8 to 11 cm.; and the scapes from simple or 1-branched to subcorymbose and much-branched.

The common reduction of leaf-surface through incurving or recurving in grasses and certain *Compositae* is frequently in harmony with dry conditions, for under a moister environment they are flat. This, in the cases of *Olearia cymbifolia* and *Shawia coriacea,* I have proved by experiment. It is possible that *Celmisia longifolia* Hook. f. (*C. gracilenta* Hook. f.) and *C. graminifolia* Hook. f. are epharmones, the one of the other, but in this case the question is complicated through the occurrence of hybrids between *C. longifolia* and one or more broad-leaved species, e. g. *C. Morgani* Cheesem. is a mixture of such hybrids, which CHEESEMAN's (1925 : 953) statement clearly indicates "*C. Morgani* is its *[C. graminifolia]* nearest ally, but can *generally* [italics mine] be recognized by its much larger and broader leaves".

The presence of tomentum on the under-surface of a leaf may be epharmonic as in the shade-leaves (non-tomentose) and sun-leaves (tomentose) of any tree of *Nothofagus cliffortioides.*

Physiognomic of tussock-grassland is the brown or yellowish colour of the tussocks, yet cultivated in a lowland garden in a fairly wet climate the leaves remain green or fairly so all the year round. Relative size and thickness of leaves, as noted for the other belts of vegetation, is also clearly a matter of epharmony, and so too the amount of leafage in so-called leafless woody plants. In the case of cupressoid and phyllocladous shrubs — including *Carmichaelia* and its allies under this term — leafy shoots are put forth under moist conditions especially when accompanied by dim light.

The assumption of the liane-form takes place in certain true highmountain plants. Thus, G. SIMPSON and J. S. THOMSON have noted that when the mat-plants, *Celmisia Bonplandii* and *C. Walkeri* grow beneath the shrubs of dense subalpine-scrub their shoots lengthen greatly and extend

had developed leaves on the old wood. Such a *rapid* change, there being really only 3 months' growing-period, was quite unexpected.

to the roof of the scrub reaching, at times, a length of more than 1.5 m. The carpet-forming grass, *Danthonia australis* also climbs over shrubs to a considerable height.

Chapter IV.
The Plant Communities.

1. Subalpine Forest.

a. General.

The distinction between lowland (in some localities), montane and subalpine forest is arbitrary, all grading into one another and possessing many species in common. Nor can definite altitudinal limits be fixed, for lowland forest in North Island may ascend to over 1000 m. altitude with its composition but little changed, but in the South Otago and Fiord districts at only 150 m. altitude, or even less, it is virtually a lower-subalpine community. So, too, the *Dacrydium intermedium* association of Stewart Island and the south of the Fiord district has a distinct subalpine facies. Nevertheless, in proceeding from the lowlands to the timber-line, well-marked belts of vegetation are encountered, the true lowland species, according to latitude and aspect, going out at certain altitudes and being replaced by those of the high mountains; also, there are purely high-mountain associations.

High-mountain forest, in general, is distinguished by the dominance of medium-sized or small trees which belong to very few species, together with comparatively few kinds of shrubs and ferns as undergrowth, *the former* being more or less of the divaricating-form, together with *Phyllocladus alpinus, Wintera colorata, Myrtus pedunculata, Nothopanax Colensoi, N. simplex, Griselinia littoralis* and *Coprosma foetidisima*, and *the latter* the semi-tree fern *Polystichum vestitum* (in great abundance), *Hymenophyllum multifidum* and *Blechnum procerum* (generally stunted); woody lianes are absent or poorly represented and asteliads and epiphytic shrubs wanting.

The high-mountain forest flora as a whole consists of about 175 species (including hybrid swarms) (pteridophytes 32, conifers 9, monocotyledons 23, dicotyledons 111), as compared with 387 for lowland-montane forest. The species belong to 75 families and 84 genera, the largest of the former being: *Filices* 29, *Compositae* 19, *Rubiaceae* 15 and *Podocarpaceae* and *Araliaceae* 8 each, and of the latter: *Coprosma* and *Olearia* 13 each, *Hymenophyllum* 9 and *Nothopanax* 6.

As in the case of lowland forest, the high-mountain communities are of the rain-forest type (except perhaps bog-forest). Here they are divided into three classes, — (1.) **subantarctic,** (2.) **subtropical** and (3) **bog-forest,** but it is a matter of taste whether the latter should stand alone, or be included in class 1, and when a subtropical association is of a subantarctic facies it is called **conditional subantarctic** and is dealt with in class 1.

Subantarctic high-mountain forest falls into the two divisions, **Nothofagus forest** — this a part of the general *Nothofagus* plant-formation — and **tree-composite forest** (conditional subantarctic) which is a formation in itself. **Subtropical high-mountain forest** is a part of the general subtropical forest plant-formation, and it consists of the two groups, **Libocedrus-podocarp forest** and **low podocarp forest.**

The high-mountain forests have suffered far less from the direct and indirect action of man than have those of the lowland-montane belt, and in localities of high rainfall they still occupy much the same area as in primitive New Zealand. They extend throughout the subalpine belt subject to a high rainfall of both islands and Stewart Island from the East Cape district southwards, but where the climatic-climax is tussock-grassland, sheepfarming has considerably reduced the originally-limited forest area, and the community is now restricted to gullies and, in a lesser degree, to shady slopes. In many parts of the wetter mountains, actual virgin forest still flourishes but, year by year, its area rapidly diminishes through the strong modification of the undergrowth through the action of deer.

The following are the life-forms of high-mountain forest: — trees 47 (medium-sized 7, small 22, tuft-tree 3, tree-composite 15), shrubs 44, herbs and semi-woody plants 25, grasslike plants 12, earth-orchids, 5, lianes 8, parasites 4 (woody 3, earth-orchid 1), saprophytes 1, ferns 29. The number of species of bryophytes is bound up with the rainfall; foliaceous lichens are of large size and abundant.

The special physiognomy of the forest depends upon the numerous, rather slender tree-trunks covered more or less thickly with bryophytes and lichens; the usually rather open undergrowth of shrubs and a few low trees with spreading twiggy branches; the extensive dark-coloured colonies of *Polystichum vestitum*, 1 m. or so high; the floor where open carpeted in places with filmy-ferns and bryophytes. At the upper limit of the forest, the incoming of subalpine shrubs and even herbs lends a new character. But the above is merely general, for the composition of an association in conjunction with its climate and edaphic conditions influences its appearance. In comparison with lowland forest the absence of the characteristic lianes, epiphytes and tree-ferns at a glance distinguishes the two classes.

b. The Nothofagus communities.

a. General.

Any *Nothofagus* community is to be distinghuished by the dominance of one or more species of *Nothofagus,* these in their turn being dependant upon relative altitude and climate. But *N. cliffortioides* (Figs. 66, 67, 68) is frequently dominant, the other species being absent over wide areas.

The distribution of the three mountain species of *Nothofagus* possibly depends upon their relative xerophily, *N. cliffortioides*. the most xereophytic, occupying the driest and loftiest stations. *N. Menziesii* with its leaves thicker and smaller than those of *N. fusca,* and its greater epharmonic plasticity, comes midway in its requirements, so, where the precipitation is excessive, it may form the sole subalpine forest, *N. fusca* dominating at a lower altitude. Where the three species occur in the same locality, *N. fusca,* mixed more or less with *N. Menziesii,* may form the lowest belt and *N. cliffortioides,* either pure or mixed with *N. Menziesii,* the highest. The effect of station on the distribution of these species is sometimes striking. Thus, according to F. G. GIBBS, on one part of Mt. Arthur (NW.) in the subalpine *N. Menziesii* forest at an altitude of 1020 m., *N. fusca* gives out and *N. cliffortioides* appears, but *especially on dry, rocky points.* In the extensive *N. cliffortioides* forest near the sources of the River Poulter (W. eastern part), *N. fusca* and sometimes *M. Menziesii* appear occasionally, but they are *invariably confined to sheltered gullies,* where the conditions are much more mesophytic than on the ridges and slopes.

Nothofagus forest is par excellence *the* high-mountain tree-community of New Zealand. It is absent in the following localities only: — (1.) Mount Egmont and the Pouakai Range (EW.); (2.) from the R. Taramakau almost to the R. Paringa — a distance of about 180 km. (W.); at the sources of the R. Rakaia and most of its tributaries (W., in its eastern part); in Stewart Island.

Evidently this high-mountain part of the great *Nothofagus* formation, which extends for so great a distance, occupies two or more altitudinal belts, and is exposed to considerable extremes of climate especially rainfall (100 to 750 cm.), must be made up of more associations than can be dealt with here but, fortunately these fall into natural groups according to the species dominating and the structure &c. of such groups in relation to rainfall and latitudinal change[1]). Such groups consist, in the first place, of those dominated respectively by *Nothofagus cliffortioides* and *N. Menziesii* and, in the second place, those in which more than one species of *Nothofagus* in concerned. There is also a closely-related well-marked group of associations which are dealt with as a separate formation entitled **mountain bog-forest,** which might, however, come into the conception of the *Nothofagus* formation.

β. The mountain southern-beech *(Nothofagus cliffortioides)* group
of associations.

General. This group may be defined as one in which as a rule all canopy-trees are absent except *Nothofagus cliffortioides,* but if such are

---

1) This does not nearly make such a great difference as might be expected. Thus, taking the whole flora of the *Nothofagus* communities — lowland, montane and subalpine — the total number of species is about 214, of which no less than 78 per cent extend throughout their range, while the small minority — 48 species only — are, for the most part, either quite local or restricted to South Island, where some also are of strictly limited distribution, e. g. local endemics of various districts.

present it must be to a strictly limited extent. It contains nearly all the species of high-mountain forest, but the relative abundance of such and the physiognomy of the forest at any particular point depends on latitude, climate and soil. In South Island, according as the precipitation favours forest or tussock-grassland, so are there two classes of the association here termed respectively **wet** and **dry**.

In North Island, the *Tararua mountains excepted, N. cliffortioides* forest forms the uppermost belt on all mountains where *Nothofagus* is present, but on the Waimarino Plain (VP.) it descends to less than 900 m. altitude, possibly owing to former forest of a different class having been destroyed during a volcanic eruption[1]). Even in South Island a *N. cliffortioides* association rarely descends to much below 600 m. altitude, although, as already explained, it is not an uncommon member of many lowland forests. Where there is an abundant rainfall the association is continuous with montane forest, but where tussock-grassland conditions prevail, it is generally confined to gullies, hollows, or the sheltered side of river-terraces. In such cases, there is no merging of forest and tussock-grassland, but the tree-mass ends abruptly.

Dry Nothofagus cliffortioides forest. This class of forest is distinguished by its paucity of species, the open undergrowth and the few species of bryophytes.

In North Island, the distinct association, described below, occurs on the east of the central volcanoes (VP.) and owes its dry character to the permeable pumice soil rather than to lack of rain. In South Island, dry *N. cliffortioides* forest occurs at various places on the ranges, extending eastwards from the Divide (especially in NE. and E.). There, too, it is not lack of rain alone which stamps the forest, but the violent winds so frequently hot and dry.

The Volcanic Plateau forest, now partly destroyed, consisted of *N. cliffortioides* with straight trunks some 18 m. high and at most 60 cm. diam. bearing a few mats of mosses and many lichens. The undergrowth was fairly open and consisted of shrubs and small trees including the following: *Phyllocladus alpinus* (at times a tree), *Aristotelia fruticosa, Nothopanax Colensoi, N. simplex, Suttonia divaricata, Coprosma parviflora, C. pseudocuneata, C. microcarpa, C. foetidissima* and sapling *N. cliffortioides.* On the floor, were mats of *Hymenophyllum multifidum* and *Lagenophora petiolata,* yellowish-green moss-cushions (with many seedling *Nothofagi), Uncinia uncinata, U. caespitosa, Blechnum penna marina* and *Lycopodium fastigiatum.* Gullies had

---

1) This explanation was first suggested by E. PHILLIPS TURNER (1909: 5) who writes as follows: "The wedges of beech that penetrate the taxad forest may be the result of a volcanic discharge of hot sand or lapilli which have destroyed the original plant covering; the beech having succeeded as being the most suited to withstand the resulting exposed situation, and the (as yet) imperfect soil."

a richer vegetation in which *Polystichum vestitum, Hypolepis Millefolium* and *Astelia Cockaynei* were conspicuous.

The associations of the Eastern and North-eastern districts are chiefly distinguished by the poverty of undergrowth which, frequently, when the substratum is shallow clay on a steep slope, consists merely of young southern-beeches and seedlings with patches of the mosses *Dicranoloma robustum, D. leucomoloides* and *D. setosum*. The trees are slender, about 9 m. high and their branching scanty. In many places, the ground is bare the clay showing through a coating of dead leaves and twigs; where driest the fallen trees are destitute of a mossy covering. Frequently the undergrowth is richer, but large areas may be occupied only by juvenile trees, for *N. cliffortioides* is short-lived, and, as the adult trees fall, light is let in, seedlings grow vigorously and the forest rapidly regenerates (Fig. 68). *Lycopodium fastigiatum* and the summergreen *Hypolepis Millefolium* are generally abundant.

Or in the dampest places, there will be plenty of *Polystichum vestitum*, 60 cm. and more tall and sheets of *Hymenophyllum multifidum*. In some localities *Sphagnum* cushions are present on which may grow *Oxalis lactea*. The undergrowth, may consist of *Coprosma propinqua, C. parviflora, C. linariifolia* (E.), and *Pittosporum divaricatum* (E. and NE.). In the Eastern district, *Hoheria Lyallii, Griselinia littoralis* and various shrubs may grow on the forest's outskirts, but usually the assocation ends abruptly and one steps from the shade of the trees into the open tussock-land.

Wet Nothofagus cliffortioides forest. In this group of associations the undergrowth is denser than in that just described, more species occur, and there is an actual moss-carpet or small moss-cushions. The forest, too, either makes a continuous covering, extending from the montane to the upper subalpine belt, or, succeeding some other forest-association, forms the uppermost belt of tree-vegetation. All the already-mentioned species may be present and, in addition, others according to the geographical position of the forest.

Wet *N. cliffortioides* forest occurs in quantity on the Ruahine Mountains, the Waimarino Plain, to the west and south of Ruapehu and most likely on the highest peaks of the East Cape district. In South Island, it is common on the eastern slopes of the Southern Alps within the area reached by the north-western rain. In the Waimakariri Basin (E.) there is a most extensive area, the mountain slopes being covered with a dark mantle of trees from 600 to 1200 m. altitude (Fig. 39), but in a few localities there are small colonies of *N. fusca* and smaller still of *N. Menziesii*. In many other parts of South Island, the forest under consideration is confined to the uppermost forest-belt.

On the south of Mount Ruapehu a Nothofagus cliffortioides association occupies about the last 100 m. of the forest-mass with its timber-line at about 1300 m. It contains more or less *Libocedrus Bidwillii* (sometimes only 5.2 m. high) and the small podocarps, *Dacrydium Bidwillii* and *D. biforme;* in places *Gleichenia Cunninghamii* forms close masses, and prostrate *Leucopogon fasciculatus* is characteristic. The following, together with those already cited, make up the bulk of the community: — *Polystichum vestitum, Blechnum penna marina, Lycopodium fastigiatum, Podocarpus nivalis, Phyllocladus alpinus, Gahnia pauciflora, Astelia Cockaynei, Libertia pulchella, Griselinia littoralis, Myrtus pedunculata, Nothopanax simplex, N. Colensoi, Suttonia divaricata, Coprosma pseudo-cuneata, C. parviflora* and *C. foetidissima. Hymenophyllum multifidum* is abundant.

The Waimarino Plain's association at 900 m. altitude is much as the last but amongst its important members are some which do not reach the final forest-belt, especially: — *Cordyline indivisa, Rubus australis, Carpodetus serratus, Pittosporum Colensoi, Hebe salicifolia* (one or more jordanon) and *Coprosma tenuifolia.* On the margin of this association there is tall shrubland (an early stage of forest) containing amongst other species *Phormium Colensoi, Cordyline indivisa* (Fig. 69), young *Nothofagus cliffortioides, Nothopanax Colensoi* and *Hebe salicifolia.*

The extensive forest of the Waimakariri (E.), in its lower part, was fairly uniform. In places, young *N. cliffortioides* dominated the undergrowth; in other places, was composed of small *Phyllocladus alpinus, Griselinia littoralis* and the usual *Coprosmae.* The floor was carpeted with several species of *Dicranoloma, Hymenophyllum villosum* and *H. multifidum. Polystichum vestitum* were abundant. Where small rivers pass through the forest there were various subalpine shrubs on its outskirts, particularly in the case of the tributaries of the River Poulter, the chief being *Olearia arborescens, O. lacunosa, O. avicenniaefolia, Senecio elaeagnifolius* and the tuft-tree, *Dracophyllum Traversii.* Near the timber-line, the undergrowth was far denser and species of the subalpine-scrub joined the association. Where there have been land-slips there are bright-green colonies of *Hoheria glabrata.*

On the rocky ground of the Pikikiruna *Range* (NW.), until quite recently, a remarkable *Nothofagus* association occurred, but, since my first visiting it in 1915, it has been almost entirely destroyed by fire and replaced by an open association of shrubs, principally species and hybrids of *Hebe.*

The ground on which this tree-association grew consists of most irregular, much-weathered, honey-combed mounds of hard limestone, full of holes which provide complete drainage and render the habitat, though in a very wet climate, essentially xerophytic.

Nothofagus cliffortioides, about 7.5 m. high was dominant with its slender trunks far apart, so that much light entered the association. In consequence, there was an undergrowth composed of more or less xerophytic (mostly epharmonically) shrubs &c., e. g.: *Asplenium Trichomanes* (rock crevices),

17*

Phyllocladus alpinus, an occasional small *Libocedrus Bidwillii, Uncinia caespitosa, Astelia Cockaynei, Pittosporum divaricatum, Pimelea longifolia, P. Gnidia, Metrosideros lucida, Nothopanax simplex* (mesophytic), *N. anomalum, Corokia Cotoneaster, Suttonia divaricata, Olearia avicenniaefolia, Traversia baccharoides.*

γ. Silver southern-beech *(Nothofagus Menziesii)* group of associations.

General. This group is distinguished by the dominance of *N. Menziesii,* but *N. fusca, N. cliffortioides* and × *N. cliffusca* may be present, as also *Libocedrus Bidwillii, Podocarpus Hallii* and *Weinmannia racemosa.*

In North Island, *N. Menziesii* forest is common on the Dividing Range and the Volcanic Plateau (includes the central volcanoes) and, in South Island, in the North-western, Fiord, and South Otago districts, and portions of the Eastern and southern part of the Western districts. In certain localities, it forms the highest altitudinal belt, but only where *N. cliffortioides* is wanting or in small quantity. Here both montane and subalpine forest are dealt with.

All the species found in any montane-subalpine *Nothofagus* forest are present in some part or other of the community, but more are usually included than in the *N. cliffortioides* communities.

Generally the dominant trees of *N. Menziesii* are low, their crowns scanty, their trunks irregular, buttressed (sometimes to an excessive extent), and frequently moss-clad. The undergrowth consists of an upper tier of shrubs and a lower of floor-plants, these, however, not of equal height. In some localities, certain of the shrubs mentioned below rise as small trees above the average level. The following occur throughout: — *Dacrydium Bidwillii, Phyllocladus alpinus, Wintera colorata, Aristotelia fruticosa, Nothopanax Colensoi, N. simplex, N. anomalum, Griselinia littoralis, Cyathodes acerosa, Suttonia divaricata, Coprosma foetidissima, C. Colensoi, C. pseudocuneata* and *C. parviflora.* Various filmy and creeping ferns, *Enargea parviflora* (creeping amongst moss), *Libertia pulchella* and species of *Uncinia* are common floor-plants, but of far greater physiognomic importance are the taller ferns, especially *Polystichum vestitum,* the silvery masses of *Astelia Cockaynei* and the tall green tussocks of *Gahnia pauciflora* (North Island and parts of NW.) or *G. procera* (South Island).

The upper forest of Mount Te Aroha (Th.). This stands in a class by itself and is excluded from the above general description, since it is a combination of northern montane forest and an *N. Menziesii* association [1]).

---

1) The Northern montane forest is represented by *Astelia trinervia, Phyllocladus glaucus, Ixerba brexioides, Alseuosmia macrophylla, Quintinia serrata, Dracophyllum latifolium* and *Senecio Kirkii,* and the *Nothofagus Menziesii* association by *Phyllocladus alpinus, Libocedrus Bidwillii, Enargea parviflora, Libertia pulchella,* two spp. of *Nothofagus, Griselinia littoralis, Nothopanax Colensoi, Coprosma Colensoi* and *C. foetidissima.*

The mountain is 968 m. high, but only the final 100 m. can be considered subalpine. The forest is low, the floor irregular, the trees more or less gnarled, great clumps of *Gahnia pauciflora* are abundant and bryophytes together with species of *Hymenophyllaceae*[1]), including *Trichomanes reniforme*, clothe the trees with a thick mantle.

Montane forest of East Cape district. At an altitude of about 700 m. on the Huirau Mountains, *Nothofagus Menziesii* and *N. fusca* replace the trees of the subtropical rain-forest and *Nothofagus* forest — possibly with *N. cliffortioides* dominant in the uppermost belt — ascends to the summits. At 900 m. altitude, owing to the *N. Menziesii* trees being far apart, the undergrowth was so dense at the place visited that few, if any, seedlings were present on the forest-floor. This undergrowth was about 3 m. high and composed of *Dicksonia lanata* (the trunkless var.) in great abundance, *Leptopteris superba*, *Astelia nervosa* var. *sylvestris*, *Enargea parviflora*, *Wintera colorata*, *Ixerba brexioides*, *Quintinia serrata*, *Nothopanax Colensoi*, *N. simplex*, *Dracophyllum latifolium*, *Coprosma tenuifolia*, *C. parviflora*, *C. foetidissima*, *Alseuosmia quercifolia* and *Senecio Kirkii*.

Montane-lower subalpine forest of the Volcanic Plateau. *N. Menziesii* is taller and with straighter trunks and larger crowns than is general in forest of this class. *N. fusca* is common, especially at the lower limit of the association. At first, *Coprosma tenuifolia* is plentiful but it gives place higher up to *C. foetidissima*. The fern *Polypodium novae-zelandiae* creeps over fallen trees. *Alseuosmia quercifolia* is plentiful. The other species are those common to *N. Menziesii* forest in general. At the lowest level, *N. Menziesii* is a massive tree with large buttresses; even at 900 m. altitude trees 21 m. high and 90 cm. diam. are not uncommon.

The uppermost belt of the Tararua Mountains. The forest varies much according to exposure to wind; in the gullies, trees are 10 m. high but, on the ridges, they are much smaller. The trunks of *N. Menziesii*, slender, fairly straight much mossed, rise out of the undergrowth, the upper branches short and gnarled. On the floor there is much *Hymenophyllum multifidum*, stunted *Blechnum procerum* and *Astelia Cockaynei*. The belt begins at an altitude of about 600 m. and ends at 900 m. as a minimum. At first, there is plenty of *N. fusca*; *Podocarpus Hallii* and *Weinmannia racemosa* are present throughout, the latter often a shrub. *Pittosporum rigidum* occurs in the uppermost part with other plants of the subalpine-scrub, especially *Senecio elaeagnifolius*. Hybrids in which *Coprosma Colensoi*, *C. Banksii* and *C. foetidissima* all play a part are common.

Upper forest of Mount Stokes (SN.). According to a communication from E. PHILLIPS TURNER, the belt commences at an altitude of about 750 m., there still being some *Nothofagus fusca*, *Metrosideros lucida* and *Podocarpus*

---

1) *Hymenophyllum villosum*, *H. flabellatum*, *H. rufescens*, *H. Malingii*, *H. Armstrongii*, *H. tunbridgense*, *H. multifidum*, *H. scabrum*, *H. dilatatum*, *H. rarum*.

Hallii while common species of the association enter in, especially *Phyllocladus alpinus, Enargea parviflora, Suttonia divaricata, Nothopanax anomalum, N. Colensoi, Pittosporum rigidum* and the various subalpine species of *Coprosma.*

Various associations of the North-western district. From information supplied by F. G. GIBBS the association on Mount Arthur has evidently much in common with the last two, but it differs in the presence of *Pseudopanax lineare* and in the colonies of *Dracophyllum Traversii* of its upper portion. At 1020 m. *N. Menziesii* gives place to *N. cliffortioides.*

The subalpine forest of Mount Rochfort is a combination of *N. Menziesii* and *N. cliffortioides* associations, so far as canopy-trees go, but its contents and habit place it with the former. *Metrosideros lucida, Weinmannia racemosa, Quintinia acutifolia, Pseudopanax lineare* and, near its lower limit, *Metrosideros Parkinsoni*[1]) are common. On the floor are innumerable moss-cushions, of great dimensions which intermingle, and, where a slope descends steeply, form veritable cascades. On the ground, the small shrub *Wintera Traversii*[2]) is fairly common. *Astelia Cockaynei* is abundant and the tuft-tree *Dracophyllum latifolium,* its trunk more slender than that of *D. Traversii* is frequently conspicuous.

Governor's Bush near Mount Cook hermitage (east of W. near junction with E.). In the vicinity of Mount Cook, as is seen more clearly further on, high-mountain species occur at a low level, but this is not particularly in evidence in the small area of *Nothofagus Menziesii* forest.

N. Menziesii is only a slender, small tree, its trunk 45 cm. diam. at most. There is a fairly rich undergrowth as follows: — *Blechnum fluviatile, B. procerum, Polystichum vestitum, Hypolepis Millefolium, Podocarpus nivalis, P. Hallii, Phyllocladus alpinus, Carex Cockayniana, Uncinia uncinata, U. caespitosa, U. leptostachya,* juvenile *N. Menziesii, Pittosporum tenuifolium, P. divaricatum, Rubus schmidelioides* var. *coloratus, Aristotelia fruticosa, Hoheria glabrata* (where much light), *Nothopanax Colensoi, Griselinia littoralis, Coprosma pseudo-cuneata, C. parviflora. C. linariifolia, Lagenophora petiolata* and *Senecio elaeagnifolius.*

Associations of the Fiord and South Otago districts. In these districts pure *Nothofagus Menziesii* forest frequently occurs on the east of the Divide, especially in the south, from sea-level to the final belt (which may be quite

---

1) A low tree of straggling habit with a slender, rigid, leaning trunk covered with warm-brown bark, frequently prostrate and rooting at the base and finally giving off numerous twiggy branches bearing on their flanks the dark-green, rather stiff, ovate leaves, 5 cm. in length.

2) There is a slender, wiry, stiff stem creeping just beneath the surface of the ground and then bending upwards as an erect unbranched stem leafy for the greater part, but with leaf-scars below. The leaves are dull olive-green, waxy beneath, thick, coriaceous, oblong-obovate and 2 cm. long.

narrow) of *N. cliffortioides*. The last-named and *N. fusca* are frequently members of *N. Menziesii* associations. Generally the associations are centered round the Divide and its easterly prolongations, but they occur elsewhere, e. g. Silver Peaks (near Dunedin), Mount Maungatua (bordering the Taieri Plain), the Blue Mountains (near Tapanui), the Takitimu Mountains and the Longwood Range. Where the three species occur, *N. fusca* is confined to the lower altitudinal belts, but *N. Menziesii* ascends almost to the timber-line.

The species are those of *N. Menziesii* forest in general but, on the whole, the conditions on the actual Divide are more favourable for the establishment of subalpine shrubs &c., e. g. *Pseudopanax lineare* and *Archeria Traversii* var. *australis,* and the general bryophyte content of the forest reaches its maximum. As for the undergrowth of the Fiord-South Otago forests it was originally fairly dense the physiognomic species being various divaricating-shrubs, erect bushes of *Wintera colorata* and *Coprosma foetidissima* with far-extending, arching, twiggy branches. The upper forest belt of the Longwood Range was (or is?) remarkable for its great bryophyte cushions.

δ. Montane *Nothofagus Solandri* forest of the Eastern district.

Associations of this type come close to those of dry *N. cliffortioides* forest. They are distinguished by the dominance of *N. Solandri,* but some *N. cliffortioides* may be present and what are probably hybrids between the two; also in some localities a certain amount of *Podocarpus spicatus* was present.

Originally, there was an extensive area near Mount Oxford (E.), whence it extended, but not continuously along the foothills to Mount Grey on the one hand, and, on the other, to the base of Mount Torlesse. The association resembled in its physiognomy that of dry *N. cliffortioides,* but the trees were considerably taller; the undergrowth, too, was similar but richer in individuals and species.

ε. Life-history of *Nothofagus* forest.

Here the *Nothofagus* forest of every altitudinal belt is considered. All the species of the genus are light-demanding and grow far faster than any other tall, indigenous tree. On these properties depend the procession of events both within and without the forest. There are these two aspects of the question, (1.) the establishing of *Nothofagus* forest on a non-forested area and (2.) the turning of subtropical forest into southern-beech forest.

The progress of events within and outside a *Nothofagus* forest can be seen throughout the area of distribution of the formation. Where forest conditions prevail, on any open piece of ground, even in tussock-grassland, seeds of any species of *Nothofagus* will germinate and develope, if not too close, into bushy trees. From such trees seed can be carried into tussock-grassland by the wind for about 8 m., or much further, should there be no obstacle in their path. Also southern-beech seeds germinate well in the shade supplied by open *Leptospermum scoparium* shrubland, the seedlings developing with fair rapidity and when they overtop the *Leptospermum* it is doomed. Still more rapid is the victory of *Nothofagus* when it and *Leptospermum* develope together. Their rate of growth is much the same, but in from 12 to 20 years the *Nothofagus* will be cutting off the light from the strongly light-demanding *Leptospermum* and in a year or two the latter will all be dead. Within certain young forests can be seen the flourishing southern-beech trees and on the floor the dead stems and branches of the original *Leptospermum* shrubs. Once the *Nothofagus* is established, species more shade-tolerating come into the association (e. g. various ferns). Frequently, especially in the lowlands, a considerable number of trees, shrubs &c. commence their existence along with the *Leptospermum* and the *Nothofagus*, e. g. *Cyathea dealbata, Nothopanax arboreum, Cyathodes acerosa, Leucopogon fasciculatus, Coprosma rhamnoides, C. robusta* and *Olearia rani.*

Inside most *Nothofagus* forests, no matter the altitude, the crop of young southern-beeches can be seen ready for filling up the gaps when adult trees fall. Seedlings develop well enough under the ordinary roof canopy, but not beneath the undergrowth, but their growth is slow until exposed to that much greater illumination which comes in when one of the frequently half-rotten[1]) trees crashes to the ground. During heavy gales many trees fall simultaneously in *N. cliffortioides* forest and this accounts for the large, dense stands of saplings in many subalpine localities. As soon as the much brighter light penetrates the forest the saplings and seedlings affected at once respond, no matter to what extent their growth had been restrained.

In some forests where such death of the large trees has taken place over wide areas the sapling community is astonishingly dense. All light sufficient for other plants is cut off, so the ground remains bare, except for dead leaves and twigs, and the lateral branches of the saplings are gradually suppressed. As time goes on, the greater part of the competing young trees are killed, more light reaches the floor, so that conditions allow the most shade-tolerating species of forest-plants to gain a footing and, by degrees, the original forest association is reinstated.

All the species of *Nothofagus* grow at about the same rate which is

---

1) All the species of *Nothofagus* are liable to the attacks of fungi. This is least in *N. Menziesii*, but very severe in the case of the other species. In an apparently vigorous forest of *N. fusca*, frequently more than half the trees are rotten, especially if they are of large size. Sometimes this state of affairs is revealed by striking the trunk which in many instances is hollow. In other cases, the condition of a tree cannot be ascertained until it is felled, though frequently dead branches tell of its being diseased. From the economic standpoint this matter of disease is of great moment and the remedy is probably to fell the trees long before they reach their full development.

in proportion to the amount of light they receive. Saplings *close together* outside the forest, so not under really favourable circumstances as regards light, grow at least three times as quickly as those beneath an average forest-roof canopy. For a well-illuminated sapling an increase in diameter of 25 mm. in 4 years is not out of the common.

Coming now to subtropical rain-forest containing a few trees of *Nothofagus fusca* or *N. Menziesii*, and considering what is likely to be the procession of events, the question is by no means simple. Certainly, if the undergrowth is open, there will be plenty of seedlings or saplings of the southern-beech, but density not openness is the rule for such forest. All the same, under certain conditions *Nothofagus* forest can replace Subtropical rain-forest. Thus, after fire, if both podocarps with their accompanying trees and any species of *Nothofagus* be in close proximity to the bare ground, all is in favour of the new forest being dominated by *Nothofagus*. So, too, in localities of very high rainfall and poor soil in *Nothofagus* - dicotylous - podocarp forest as the podocarps die, there is a fair chance of their being replaced by *Nothofagus*. On the other hand, in lowland rain-forest proper, the rapidity with which a dense undergrowth can be established is entirely hostile to *Nothofagus*. In the Ruahine-Cook and Sounds-Nelson districts with a *Nothofagus* association on the slopes and rain-forest proper in the gullies no southern-beech seedlings have the slightest chance to become established in the latter habitat. The investigation of G. Simpson and J. S. Thomson in the neighbourhood of Dunedin prove conclusively that an original *Nothofagus* forest has been replaced by the present rain-forest proper, while especially hostile to the establishment of *Nothofagus* seedlings are the great colonies of *Blechnum discolor*, and I may add those of tree-ferns also.

c. *Conditional subantarctic forest communities.*

In this type of forest *Nothofagus* is absent and its place taken by species floristically related to those dominating subtropical forest proper, but ecologically suited to subantarctic conditions. The dominant or important trees are of low stature and include small podocarps, *Libocedrus Bidwillii*, *Dracophyllum Traversii* and various arboreal *Compositae*.

Tree-composite forest. This is distinguished by the dominance of several species of *Olearia* and *Senecio*, while the tuft-tree, *Dracophyllum Traversii* is frequently a physiognomic feature of an uncommon kind. The association is closely allied to subalpine-scrub, into which it generally merges.

The following are important members of the association: — *Polystichum vestitum, Blechnum procerum, Hymenophyllum villosum* (epiphytic on trunks and branches), *Libocedrus Bidwillii*, stunted *Podocarpus Hallii, P. nivalis, Dacrydium biforme, Astelia Cockaynei, Phormium Colensoi, Hoheria glabrata, Viola filicaulis, Nothopanax Colensoi, Griselinia littoralis, Dracophyllum longifolium, D. Traversii, Suttonia divaricata, Hebe vernicosa* var. *canter-buriensis, Coprosma pseudo-cuneata, C. parviflora, Olearia ilicifolia, O. lacunosa, O. arborescens, O. avicenniaefolia, O. nummularifolia* (also several hybrid swarms with the first three species of *Olearia* cited above concerned in their parentage) and *Senecio elaeagnifolius*.

The association usually puts in an appearance at about 900 m. altitude and extends upwards for a greater or lesser distance, until, with increase of altitude or exposure to wind, it gradually becomes smaller and is transformed into subalpine-scrub. It probably occurs in the North-western district, but it is extremely abundant in the western part of the Western district where it may form a distinct belt; it also occurs on the eastern side of the Divide, especially near the sources of the Rakaia and some of its tributaries.

As shown by its composition, its most important life-forms are the tree-composite, the tuft-tree, the cupressoid, the araliad, the divaricating and the bushy deciduous tree.

The association is about 4.5 m. high. The trunks of the composites are semi-prostrate (Fig. 60), with long strips of papery bark hanging downwards. Above is a tangle of branches; beneath the forest is more or less open. Seen from without, certain of the trees, either through their colour or form, strongly affect the physiognomy of the forest. Thus *Dracophyllum Traversii* is indicated by its candelabra-like crown and huge reddish-brown leaf-rosettes, *Hoheria* by its bright light-green, *Phyllocladus* by its greyish, green hue, *Griselinia* by its shining green, darker than that of *Hoheria* and the shrub-composites by their rather flat crowns which, in certain species are whitish in the mass. The undergrowth consists of divaricating-shrubs of various genera, species of *Hebe, Astelia Cockaynei* and colonies of *Polystichum vestitum* and *Blechnum procerum. Hymenophyllum villosum* clothes many trunks and branches. A few herb-field species may be present, especially *Ranunculus Lyallii* and *Phormium Colensoi*.

Mountain toa-toa (Phyllocladus alpinus) group of associations. This group is distinguished by the dominance or occasionally subdominance of *Phyllocladus alpinus*, and usually the presence of one or more of the high-mountain podocarps. It is related to bog-forest, but occurs on drier ground where more species can enter in.

The Volcanic Plateau association, here described, occurs on the north of Mount Tongariro, the upper Waimarino Plain, Hauhungatahi and other places. Sometimes, *Podocarpus Hallii* dominates. The trees are about 6 m. high; *Libocedrus Bidwillii* and some, or all, of the podocarps are present. *Hymenophyllum multifidum* and *Astelia Cockaynei* are common floor plants. The following are plentiful: *Wintera colorata, Aristotelia fruticosa, Myrtus pedunculata, Nothopanax Colensoi, N. simplex, Suttonia divaricata, Coprosma tenuifolia, C. microcarpa, C. Colensoi, C. parviflora, C. pseudo-cuneata* and *C. foetidissima*.

On the shaded side of the R. Cass near its mouth (E.) there is low forest where *Phyllocladus alpinus* dominates and grades into river-terrace scrub with which it has many species in common, e. g. *Polystichum vistitum, Blechnum penna marina. Pittosporum divaricatum, Rubus subpauperatus, R. schmide-lioides* var. *coloratus, Coriaria sarmentosa, Discaria toumatou* (3 to 3.6 m. high), *Aristotelia fruticosa, Corokia Cotoneaster, Dracophyllum longifolium* (not in the scrub), *Hebe salicifolia* var. *communis, H. Traversii, Coprosma crassifolia, C. parviflora, C. propinqua* and *Cassinia fulvida* var. *montana.*

On the steep face of the Sealey Range near Mount Cook hermitage (east of W.), the *Nothofagus Menziesii* forest gives place to *Phyllocladus alpinus* forest which, in its turn, gradually merges into subalpine-scrub. The other leading members of the association are *Dacrydium Bidwillii,* small *Podocarpus Hallii, P. nivalis, Griselinia littoralis, Dracophyllum longifolium* and *Senecio elaeagnifolius.*

Mountain ribbonwood (Hoheria glabrata) low forest. *Hoheria glabrata* is essentially a denizen of stony ground. On steep mountains (W., F.), where land-slips have occurred, this association stands out clearly from the dark forest-mass through its bright-green leaves (red and yellow in autumn), wealth of snowy blossoms in due season and deciduous habit in winter.

This association is especially well developed in the Fiord district, where it may form small groves on a substratum of large stones or extend upwards to the timber-line.

H. glabrata, about 7.5 m. high, is dominant, its trunk and far-extending horizontal branches thickly covered with moss, and depending from them an abundance of *Weymouthia Billardieri* (Fig. 71), and perched on them *Asplenium flaccidum. Griselinia littoralis* occurs here and there as a tree; *Polystichum vestitum* dominates the undergrowth; there is more or less *Nothopanax Colensoi* and *Olearia ilicifolia.*

Hoheria Lyallii, a closely-related species, frequently forms small clumps on river-terrace in the North-eastern and Eastern districts.

d. Subtropical rain-forest and allied communities.
a. General.

Subtropical rain-forest is distinguished by the absence of *Nothofagus* and the presence of *Libocedrus Bidwillii, Podocarpus Hallii, Weinmannia racemosa* and *Metrosideros lucida* as dominant tall trees, but all are not of necessity present in an association.

The class of forest under consideration is not nearly so continuously distributed as is subantarctic forest. In North Island, it occurs on some of the mountains of the Volcanic Plateau and Mount Egmont and, in South Island, in the North-western district, the Western from the Taramakau to the Mahitahi, the Eastern on Banks Peninsula, the South Otago near Dunedin and Stewart Island.

β. Associations and groups of such.

Southern kawaka-totara (Libocedrus Bidwillii-Podocarpus Hallii) group. In its various forms this is the most important of the subalpine subtropical forests. It is distinguished by the dominance of either of the species cited above, and the *Libocedrus,* when abundant, with its erect habit and pyramidal head, clearly defines the community even when viewed from afar. In South Island it occurs, but not everywhere, in the subalpine belt of the North-western, Western and Eastern districts and originally to some extent on hills near Dunedin; and, in North Island, on Mount Egmont and the Volcanic Plateau. As will be seen, it is closely related to tree-composite forest, indeed the latter near the source of the Rakaia is a connecting-link between the two groups.

On Mount Hauhungatahi (VP.) there is an association of this class. This mountain is an isolated extinct volcano, 1520 m. high, situated west of Ruapehu. It is forest-clad up to about 1140 m. altitude. At the base of the mountain, and below, at a height of 780 m. or less, the podocarp-forest is replaced by a southern kawaka-totara association. The trees consist of: — *Libocedrus Bidwillii* (first appearing at about 600 m. alt.), *Podocarpus Hallii, P. ferrugineus, Dacrydium cupressinum, D. Colensoi, Weinmannia racemosa* and *Olea Cunninghamii.* The smaller trees and shrubs of the undergrowth are *Phyllocladus alpinus, Wintera colorata, Carpodetus serratus,* juvenile *Elaeocarpus Hookerianus, Aristotelia fruticosa, A. serrata, Melicytus lanceolatus, Myrtus pedunculata, Fuchsia excorticata, Nothopanax simplex, N. Colensoi, N. anomalum, Pseudopanax crassifolium* var. *unifoliolatum, Schefflera digitata, Griselinia littoralis, Suttonia salicina, S. divaricata, Coprosma grandifolia, C. robusta, C. tenuifolia, C. parviflora, C. Colensoi, C. foetidissima* and *Alseuosmia quercifolia. Dicksonia lanata* (trunkless) ascends to 1080 m., or more, and may form much of the undergrowth, its fronds being 1.5 m. long. Many lowland ferns are common and *Leptopteris superba* forms considerable colonies. At an altitude of 960 m. (PHILLIPS TURNER 1909: 3), *Dacrydium cupressinum,* hitherto abundant, becomes much scarcer and the forest is more typically subalpine with *Libocedrus Bidwillii* dominant and *Podocarpus Hallii, Dacrydium Colensoi* and *D. intermedium* abundant. Tussocks of *Gahnia pauciflora* become characteristic and the subalpine-scrub plants of the vicinity enter into the association. In some parts of this association, as where it abuts on the Waimarino Plain, near Horopito and elsewhere, the handsome tuft-tree *Cordyline indivisa* is plentiful. Close to the timber-line the forest becomes very low (scrub-forest), and it is closely related to bog-forest and subalpine-scrub. Its small trees are *Libocedrus, Dacrydium intermedium, D. biforme, Phyllocladus alpinus, Griselinia littoralis* and *Dracophyllum longifolium.*

On Mount Egmont an association containing *Libocedrus* is apparently confined to the vicinity of the North mountain-house and does not extend to the Stratford house. It commences at about an altitude of 900 m. with *Podocarpus Hallii* in abundance, but *Libocedrus Bidwillii* soon becomes dominant, though *Podocarpus Hallii, Weinmannia racemosa* and *Griselinia littoralis* are abundant. This latter is bent, arched and gnarled, while its trunk may be covered by sheets of the dark, curled leaves of *Hymenophyllum villosum* and yellowish-green cushions of *Dicranoloma Billardieri.* The common shrubs of the association are: — *Wintera colorata, Carpodetus serratus, Aristotelia serrata, Melicytus lanceolatus, Fuchsia excorticata, Nothopanax Sinclairii, N. Colensoi, Suttonia divaricata, Hebe salicifolia* (form with narrow leaves), *Coprosma grandifolia, C. tenuifolia* (abundant), *C. parviflora, C. egmontiana* and *Senecio elaeagnifolius* var. *Buchanani.* The sole liane is an occasional plant of *Rubus australis.* Seedlings and young trees are constantly epiphytic, e. g. *Coprosma lucida* var. *angustifolia, Griselinia littoralis, Nothopanax Colensoi* and *N. Sinclairii,* the last two frequently killing and replacing their host. A striking feature is the profusion of bryophytes, especially *Hepaticae,* both on the floor and tree-trunks; the undergrowth is particularly dense and *Wintera colorata* plays a most important part; on the floor is an epharmone of *Astelia nervosa* var. *sylvestris.*

On the west of the Southern Alps, and extending for 160 km. or more southwards from the R. Taramakau, the southern kawaka-totara belt commences at an altitude of about 600 m., *Podocarpus Hallii* being the first tree to arrive. From North Island associations this differs only in certain floristic details. *Weinmannia racemosa* and *Metrosideros lucida* are abundant at first. Tussocks of *Gahnia procera* are a feature of the floor-vegetation. Lianes and tree-ferns are absent. At the upper limit subalpine shrubs by degrees enter in until a distinct belt of forest results.

The highest peaks of Banks Peninsula doubtless originally carried a belt of southern kawaka-totara forest, for there are ample remains, while a small piece in its virgin state still exists on Mount Sinclair. *Weinmannia racemosa, Metrosideros lucida* and other species are absent, but *Cordyline indivisa,* plentiful in Westland, but not occurring elsewhere east of the Divide, is fairly common. Other members of the association are *Wintera colorata, Griselinia littoralis, Fuchsia excorticata* and *Nothopanax arboreum.*

Kamahi (Weinmannia racemosa) group of associations. *Weinmannia racemosa* is frequently dominant in parts of southern kawaka-totara forest, while, in certain localities, it forms an almost pure association — so far as tall trees are concerned.

On Mount Egmont kamahi forest is so striking that it has received the popular and expressive name of "Goblin forest". It occurs as a distinct belt from the neighbourhood of Dawson Falls to the North Egmont house and it probably extends right round the mountain.

At first, there are some comparatively low trees of *Dacrydium cupressinum,* but these soon give out and those of *Weinmannia* decrease in stature and become much-branched, the branches at first more or less erect, but with increase of altitude they extend far horizontally and are gnarled and irregular in shape. Both trunks and branches are covered densely with mosses, liverworts and filmy-ferns *(Hymenophyllum multifidum, H. villosum, H. flabellatum)* which could not be in such profusion but for frequent rain. *Griselinia littoralis* is an important small tree and *Fuchsia excorticata* occurs here and there. The undergrowth consists of ferns and rather freely-branched shrubs, the following being common: — *Blechnum procerum, B. fluviatile, Polystichum vestitum, Uncinia Banksii, Astelia nervosa* (unnamed jordanon), *Wintera colorata, Carpodetus serratus, Coprosma parviflora, C. tenuifolia.* At a higher altitude *Podocarpus Hallii* enters the association which then becomes equivalent to that in which *Libocedrus* is present. At about 1200 m. altitude, the forest gives place to subalpine-scrub.

On Mount Hauhungatahi, in places, there is a *Weinmannia* association more or less of the "Goblin Forest" character.

Southern-rata (Metrosideros lucida) group of associations. Forest of this class is wide-spread in the Western district, it is montane rather than subalpine, but as mountain plants descend so low in that locality, it is here included with subalpine forest.

At above 450 m. altitude in the Western district *M. lucida* becomes dominant and the lowland podocarps gradually decrease in numbers. *Weinmannia racemosa* in places is so plentiful as to dominate. *Quintinia acutifolia* is conspicuous through its somewhat fastigiate habit as a sapling and the yellowish leaves blotched with purple but pale beneath. Many of the lowland shrubs and ferns are present. The undergrowth is dense, especially in gullies. Bryophytes (species of *Gottschea, Schistochila, Aneura, Mniodendron, Plagiochila, Lembophyllum* &c.) and *Hymenophyllaceae* abound. *Leptopteris superba* forms extensive colonies.

At the Franz Josef glacier, the terminal face of which descends to 213 m., the southern-rata association comes on to the ice-worn rocks at a few metres from the ice on either side of the glacier. The forest here, the roof of which has the characteristic billowy appearance, consists principally of the following: *Metrosideros lucida* and *Weinmannia racemosa* (the dominant canopy-trees), *Carpodetus serratus, Coriaria arborea, Aristotelia serrata, Hoheria glabrata, Melicytus ramiflorus, Pseudopanax crassifolium* var. *unifoliolatum, Schefflera digitata, Griselinia littoralis, Hebe salicifolia, Coprosma lucida, Olearia arborescens* and *O. avicenniaefolia.* The pteridophytes include *Hemitelia Smithii* (tree-fern, but here of low stature), several *Hymenophyllaceae, Hypolepis tenuifolia, Histiopteris incisa, Blechnum procerum, B. lanceolatum, Asplenium bulbiferum, A. flaccidum, Polystichum vestitum, Polypodium diversifolium, P. Billardieri* and *Lycopodium volubile.*

At Mount Peel (E.) there is a small southern-rata association with

M. lucida as the sole tree "on rocky knolls and slopes with a westerly aspect (H. H. ALLAN, 1926: 44) which is a succession after *Leptospermum ericoides*"[1]). On the floor there may be much *Blechnum procerum* and *Alsophila Colensoi* and near streams colonies of *Gleichenia Cunninghamii*.

In Stewart Island at an altitude of 300 m. the forest decreases in height, *Metrosideros lucida*, sometimes with prostrate trunks, becomes more abundant, especially on exposed ridges, *Weinmannia* is still plentiful, tall *Leptospermum* may appear, and moss-cushions become far commoner. On the lower hills, so far as is known, at about 270 m., the forest gradually decreases in height until its interior is a tangle of stems from semi-prostrate, slender trunks. On the uneven floor great cushions of *Plagiochila gigantea* and *Dicranoloma robusta* abound (Fig. 72).

e. Montane and subalpine bog-forest.

General. Bog-forest is distinguished by the presence of stunted *Nothofagus* trees, together with nearly always one or more of the small podocarps and *Libocedrus Bidwillii;* more or less sphagnum is generally present.

Taking all the area occupied by this class of forest, the following are important members in one or more of the associations: — *Hymenophyllum multifidum, H. villosum, Blechnum procerum, Hypolepis Millefolium, Gleichenia Cunninghamii, Libocedrus Bidwillii, Dacrydium Colensoi, D. intermedium, D. biforme, D. laxifolium, Podocarpus Hallii, P. acutifolius, Phyllocladus alpinus, Microlaena avenacea, Carex ternaria, C. Gaudichaudiana, Uncinia caespitosa, Gahnia pauciflora, G. procera, Enargea parviflora, Astelia Cockaynei, Libertia pulchella, Nothofagus Menziesii, N. cliffortioides, Nothopanax Colensoi* and var. *montanum, N. simplex, Pseudopanax lineare, Griselinia littoralis, Gentiana Spenceri, Dracophyllum longifolium, Cyathodes acerosa, Leucopogon fasciculatus, Suttonia divaricata, Coprosma pseudocuneata, C. parviflora* and *C. foetidissima*.

Bog-forest occurs in any altitudinal belt of forest from the Volcanic Plateau district southwards to the Fiord and South Otago districts. It is found on ground where the drainage is bad and it is specially common where the rainfall is excessive.

The physiognomy of bog-forest depends upon the low, more or less stunted trees of *Nothofagus,* the cupressoid podocarps, the pyramidal *Libocedrus,* the irregular floor with usually cushions of sphagnum here and there, the rather dense undergrowth which is accentuated by the *Gahnia* tussocks and the divaricating shrubs.

---

1) In no part of this book is a description given of *Leptospermum* forest, such being dealt with as an early succession in the development of kauri or southern-beech forest. It is true that stands of *Leptospermum scoparium* are frequently met with, but it is usually impossible to know whether such are primitive or induced. H. H. ALLAN (1926:44) describes a *Leptospermum ericoides* subassociation for Mount Peel (E.) with the *Leptospermum* 15 m. high and *Suttonia australis* as the chief member of the undergrowth.

High-mountain bog-forest is evidently closely related to lowland bog-forest, or rather, in certain localities, it is its continuation upwards, so it really belongs to the same formation.

Below, examples are given of high-mountain bog-forest in different parts of the Region.

The Volcanic Plateau. *Nothofagus cliffortioides* is dominant and *Dacrydium Colensoi* frequently subdominant. The species not occurring in the associations further south are *Gahnia pauciflora*, *Pittosporum Colensoi* and *Coprosma tenuifolia*. *Astelia Cockaynei* is frequently abundant or, if the soil is particularly wet, *A. nervosa* var. *grandis*. *Gleichenia Cunninghamii* is common; sphagnum may be wanting. Floor and tree-trunks are thickly covered with bryophytes. Both *Leptospermum scoparium* and *L. ericoides* may be present. *Libocedrus Bidwillii* is sometimes abundant.

North-western district. *On the western side of the Tasman Mountains* bog-forest — so far as I have seen it — contains stunted *N. cliffortioides* (about 3.6 m. high), abundance of *Leptospermum scoparium*, also *Dacrydium intermedium, D. biforme, Elaeocarpus Hookerianus, Metrosideros Parkinsonii, Gahnia procera, Blechnum procerum* and *Pseudopanax lineare.*

On the Rahu Saddle (Victoria Range) the association is dominated by very slender *N. cliffortioides* and the following are important members: *Sphagnum* in abundance, mats of *Hymenophyllum multifidum*, low cushions of *Plagiochila, Phyllocladus alpinus* (as a small tree), *Libocedrus Bidwillii, Pittosporum divaricatum, Pseudopanax lineare, Suttonia divaricata, Coprosma parviflora, C. pseudo-cuneata* and *C. foetidissima.* Other communities in the same locality, in addition to the above, contain *Dacrydium Colensoi, D. intermedium, D. biforme* and *Leptospermum scoparium.*

Near Tophouse (eastern part of the district, foot of St. Arnaud Range) the bog-forest contains much *Sphagnum;* both *N. Menziesii* and *N. cliffortioides* are present; *Libocedrus Bidwillii* is common and *Coprosma pseudo-cuneata* the characteristic species of the undergrowth. There is no *Pseudopanax lineare.*

Western district, eastern side of the Divide. On the flat summits of the truncated spurs which are so common a feature of the forest-clad portion of the Waimakariri Basin, there is nearly always bog-forest containing *N. cliffortioides*, abundant *Libocedrus Bidwillii, Dacrydium biforme* and *D. Bidwillii*, but none of the other allied podocarps. Where the light is strong, there are sphagnum cushions carrying prostrate *Leptospermum scoparium* and mats of *Dacrydium laxifolium.*

In the same district and also in the Eastern and wetter parts of the North-eastern districts, and probably elsewhere, there is occasionally bog-forest where *N. cliffortioides* is the sole tree and *Podocarpus nivalis* the only podocarp.

Western district on west side of the Divide in southern part. In this area, as already explained, *Nothofagus* is absent, but in the montane and upper lowland belts there is bog-forest with abundance of *Phyllocladus alpinus, Libocedrus Bidwillii* with epiphytic *Hymenophyllum Malingii, Dacrydium biforme, Podocarpus Hallii, P. acutifolius* and various common species of such communities, including *Gleichenia Cunninghamii, Aristotelia fruticosa, Myrtus pedunculata* and *Nothopanax anomalum.*

2. Shrub Communities.

a. General.

Taken as a whole, the high-mountain shrub communities are made up principally of medium-sized and tall shrubs and dwarfed trees. Other life-forms, however, are important, notably the woody lianoid and parasitic, the large tussock-form, herbaceous perennials, semi-woody plants and ferns especially, *Polystichum vestitum, Danthonia Cunninghamii, D. Raoulii* var. *flavescens, Phormium Colensoi, Aciphylla maxima* and *A. conspicua.* Most of the other grasses, herbs &c. which occur are hardly real members, since some occupy merely the line of tension between shrubland and their proper formation, and others are chance comers.

The number of dwarfed trees, shrubs, woody lianes and parasites is 143 (non-endemic omitting the pteridophytes 3) which belong to 29 families and 43 genera. The largest families and genera are as follows: — (families) *Compositae* 36 species, *Scrophulariaceae* 26, *Epacridaceae* 16, *Rubiaceae* 13; (genera) *Hebe* 25, *Olearia* 17, *Dracophyllum* and *Coprosma* 13 each and *Senecio* 9. Also many hybrid swarms play an important part, especially in *Aristotelia, Dracophyllum, Hebe, Coprosma, Olearia* and *Cassinia.* Taking the species as a whole, 65 per cent are shrubby composites, hebes, epacrids and coprosmas. Many of the species are strongly xerophytic but notwithstanding a considerable number as has been seen, thrive in the forest under hygrophytic conditions, owing in some cases to great plasticity with regard to their life-forms.

When fully developed, all the associations are closed but, in places, a few are open, owing either to being an early stage of succession or to the edaphic conditions not being suitable for some of the leading species. Apparently, the communities fall into two main classes — the one representing **a definite stage of biological or it may be biological-topographical succession,** and the other with a more or less distinct relation to climate may be considered **a climax or subclimax community.**

b. Associations usually of shingly ground.

Discaria thicket. This consists of *Discaria toumatou* either pure or mixed with a few medium-sized shrubs.

As explained, when dealing with lowland low tussock-grassland, the association appears as a succession following the earlier herb &c. stages of river-bed or fan vegetation but, though persisting for a considerable time, it is at best but a migratory community and is replaced by other shrub-associations, tussock-grassland or even *Nothofagus* forest.

Discaria toumatou itself is a semi-divaricating shrub, usually about 1.2 m. high, but varying greatly epharmonically in stature, having spreading, flexible but wiry, slender, dark-coloured branches, at times more or less leafless, furnished at intervals of about 2 cm. with rather long, sharp spines (morphologically reduced shoots).

The shrubs are dotted about on the stony ground or in clumps with spaces between, but eventually they grow into one another. The dark colour of the association shows up from afar, especially in contrast to the adjacent yellow tussock-grassland. Generally, *D. toumatou* is the sole shrub, but one or other of the ball-like species of *Hebe* may be present. As in the lowlands, *Clematis marata* may climb over the *Discaria*. The stony spaces between the shrubs may be bare, but usually there is a sparse growth of tussock and some of its accompanying plants.

The association does not belong specially to the high mountains but is also common in the lowland belt. It is restricted to South Island, but is wanting in the Western and Fiord districts, except east of the Divide, where it occurs on those wide river-beds which extend into the forest-area. In the North Otago district, there is a *Discaria* community at from about 180 m. altitude upwards both in valleys and on slopes of "fertile" mica-schist soil. *Olearia odorata* is common; other species are *Muehlenbeckia complexa, Clematis marata, Carmichaelia Petriei, C. gracilis* (rare) and *Olearia lineata.*

Hebe shrubland. This group of associations is distinguished by the dominance of one or usually more species of *Hebe* and generally some divaricating shrubs are present.

The species number about 50 which belong to 16 families and 24 genera. The following are common members in some part or orther of the group: — *Hypolepis Millefolium, Blechnum penna marina, Muehlenbeckia complexa, Pittosporum divaricatum, Rubus schmidelioides* var. *coloratus, R. sub-pauperatus, Discaria toumatou, Aristotelia fruticosa, Hymenanthera alpina, Corokia Contoneaster,* such species of *Hebe* as belong to the locality, *Coprosma propinqua, C. parviflora,* and one or more of the mountain species of *Cassinia. Hebe* hybrids may be in such profusion that it is difficult or impossible to recognize the species present.

The shrubs are erect. Their principal life-forms are the ball-like and the divaricating. Of less importance are the *Dracophyllum* and shrub-composite forms. The lianes are slender; most not only climb, but form bushes approximating to the divaricating-form. Leaving the ferns out of consideration, 2 of the species are mesophytic and several, at most, sub-xerophytic. Eight have tomentose leaves.

Hebe shrubland is a common feature of the upper montane and subalpine belts. Its presence, generally denotes a tussock-grassland climate, but it demands more shelter from wind than does tussock-grassland, its principal development being in the river-valleys eastward of the South Island Divide. A favourite situation is the sheltered side of river-terrace. It occurs also on the outskirts of the lower subalpine forest; on torrent-fans just where they issue from a gorge, or in the mouth of the latter; on ancient river-bed and on coarse debris at the foot of some disintegrating cliff. The soil that the association affects ranges from stones mixed with fine clay and sand to deep clayey loam. The wind-factor may be extremely powerful on river-bed, but much modified on river-terrace. The soil-water must vary considerably, but even on a steep terrace-face may be fairly abundant. The relation to snow and prolonged frost differs greatly according to aspect, but the richest development is where the sun is least powerful.

The associations vary from a close, bright-green growth of species of *Hebe* to a dense, dark-coloured scrub about 1.8 m. high, of divaricating shrubs[1]) bound together by the various lianes[2]) and relieved in places by the green of *Hebe* or the whitish hue of *Olearia avicenniaefolia; Discaria toumatou* is frequently present. Beneath the shrubs the ground may be bare or occupied more or less closely by certain ferns[3]). On many river-beds, if the rainfall is high, or near streams flowing through tussock-grass-land or fell-field or where water oozes out of the ground are thickets of the glossy-leaved *Hebe buxifolia* var. *odora* round as a cricket-ball. The cupressoid *H. salicornioides* sometimes grows in the North-eastern district in soil saturated with ice-cold water. Coarse rocky debris in the North-eastern and Eastern districts, larger in size than that of "Shingle-slip", is occupied in the first instance by *Hebe* scrub and not tussock-grassland. *Rubus schmidelioides* var. *coloratus* or *R. subpauperatus* and the rigid, open, dark-coloured almost leafless cushions of *Hymenanthera alpina* are often present.

Hebe shrubland follows on as a succession after various mat-plants have occupied the stony substratum and provided a seed-bed. Evidently, it is closely related to *Discaria* thicket, but the latter occupies a more sunny position. In shady situations, it may be an early stage of *Nothofagus* forest.

c. Subalpine-scrub.

α. General.

The term "Subalpine-Scrub" is here applied to that assemblage of stunted trees, — trees no longer, and shrubs of various life-forms, which, on many

---

1) *Aristotelia fruticosa, Pittosporum divaricatum, Coprosma rugosa, C. parviflora, C. propinqua, Corokia Cotoneaster, Olearia virgata,* and *O. odorata.*

2) *Rubus australis, R. schmidelioides* var. *coloratus, R. subpauperatus, Parsonsia capsularis, Muehlenbeckia complexa, M. australis* and *Helichrysum dimorphum* (E., limited to a small part of the R. Waimakariri basin).

3) *Polystichum vestitum, Cystopteris novae-zelandiae, Hypolepis Millefolium, Blechnum penna marina.*

high mountains, either form a belt above the forest-line or make thickets, large or small, on river-beds, in gullies or hollows and even on mountain-slopes.

The species number about 122 which belong to 28 families and 49 genera the largest being: — (families) *Compositae* 26, *Epacridaceae* 14, *Scrophulariaceae* 13, *Rubiaceae* 12, *Filices* 10; (genera) *Hebe* and *Olearia* 13 each and various hybrid swarms, *Coprosma* 12 and *Dracophyllum* 11. But, in addition to the above, various podocarps, *Nothofagi, Hoheria glabrata, Leptospermum scoparium,* araliads, giant *Aciphyllae* and *Suttonia divaricata* are of importance in many localities.

The general ecological conditions that determine the presence of the formation are: *altitude,* (which decreases from north to south or according to edaphic xerophily); *violent wind* (but less than herb and grass communities can tolerate); a *heavy winter snow-fall,* (but not the maximum); and *frequent rain at all seasons.* This last factor leads to the presence of a xerophytic soil rich in peat or raw humus. But, in the dry mountain areas, scrubs likewise occur, some of which are strongly bound up with edaphic conditions, as in the case of the serpentine Mineral Belt and the *Senecio Monroi* scrubs of the North-eastern district.

The subalpine-scrub associations differ from one another in density, floristic composition and physiognomy in different localities, and dissimilar scrubs may occupy contiguous areas. According as shrub-composite, cupressoid-podocarp, *Hoheria glabrata, Phyllocladus alpinus,* stunted *Nothofagus, Dracophyllum* or divaricating-shrubs dominate, so is there a different and distinct facies. In many parts of the Southern Alps *Phyllocladus* and *Hoheria* lend a most distinct appearance to adjacent patches of scrub.

A typical subalpine-scrub of a wet climate consists of a number of rigid or wiry-stemmed shrubs which grow into one another while the main branches of many are parallel to the slope and project downwards. The height may be from 2 to 3 m. and the roof fairly even. The density may be so great that one cannot force a passage through, but must actually walk upon the top! Where there is an actual belt above the forest, it gradually decreases in height as one proceeds upwards and eventually ends in low bushes hugging the ground, herb-field or fell-field cutting gaps into the association. Scrub is taller in gullies than elsewhere and on their shaded side attains its maximum height. In certain cases subalpine-scrub is merely the uppermost belt of forest with its trees stunted to shrubs, and the light-tolerating members of the undergrowth persisting. In other cases, there may be sufficient shelter from wind to allow the shrub-content of grass or herb associations to become dominant, but, on the other hand, an average excessive snow-covering, violent winds, xerophytic edaphic stations, and increase in altitude, favour herb-field at the expense of scrub. The following are the principal classes of subalpine-scrub based on the dominance of distinctive life-forms. Though distinct enough in typical examples, intermediates referable to more than one class are common, and doubtless a number of distinct associations still await discovery and investigation.

β. Shrub-composite scrub.

General. This is distinguished by the dominance, or occasionally sub-dominance only, of shrubby or stunted arboreal species of *Olearia* and *Senecio*, one or both. Various divaricating-shrubs (spp. of *Coprosma*, *Aristotelia fruticosa*, *Pittosporum divaricatum*, *Suttonia divaricata*) will be present. Also one or other of the fastigiate species of *Dracophyllum*, *Phormium Colensoi*, *Cassinia Vauvilliersii*, *Phyllocladus alpinus*, *Nothopanax Colensoi* and one or two species of *Hebe* are frequent members and *Dracophyllum*, *Phyllocladus*, *Cassinia* or even *Hebe* may in places dominate.

The trunks of the tree-composites are generally prostrate and yet tree-like, their horizontal spread exceeding the height of the association. The divaricating-shrub greatly increase the general density. The roof will be fairly level but pierced here and there by *Dracophyllum* (Fig. 73).

Shrub-composite scrub requires a high rainfall for its full development. In North Island, it occurs on Mt. Hikurangi, the Ruahine and Tararua Mts. and Mt. Egmont. In South Island, it is a characteristic feature of the Western district on both sides of the Divide, making, in many places, a broad belt above the forest and partly filling the cirques at the sources of glacial rivers. Similar scrub occurs in the Fiord district; it is also highly developed in Stewart Island. In what follows an attempt is made to give some idea of its chief floristic characteristics in different localities.

Mount Hikurangi (EC.). The scrub occurs (from information generously supplied by W. R. B. OLIVER who recently ascended this mountain on "cliffs and steep rocky slopes, especially on the northern face of the mountain". *Senecio Bidwillii* (50 cm. high) and *Podocarpus nivalis* are dominant and mixed with them are *Dracophyllum recurvum*, *Pimelea buxifolia* and *Hebe tetragona*. Tussocks of *Danthonia Raoulii* var. *flavescens* and *Aciphylla conspicua* (if this can be so termed) are common. *Coprosma pseudo-cuneata* and *Olearia Colensoi* are present. Beneath the scrub are *Hymenophyllum multifidum*, *Lycopodium fastigiatum*, *L. australianum* and *Schizeilema Allanii*.

Mount Egmont (EW.) The scrub commences at about 1140 m. and gives out at about 1240 m. *Senecio elaegnifolius* var. *Buchanani* is generally dominant, but sometimes *Dracophyllum filifolium* rules. The other principal species are *Podocarpus Hallii*, *Carmichaelia australis* var. *egmontiana*, *Nothopanax Colensoi*, *N. Sinclairii*, *Griselinia littoralis*, *Suttonia divaricata*, *Hebe salicifolia* (resembling var. *paludosa* but probably distinct), *Coprosma tenuifolia*, *C. egmontiana*, *C. parviflora*, *Olearia arborescens* and *Cassinia Vauvilliersii*.

Tararua Mountains (RC.). *Olearia Colensoi* is frequently dominant, but *O. arborescens* and *Senecio elaeagnifolius* are often abundant. Other shrubby species are: *Pittosporum rigidum, Nothopanax Colensoi, N. Sinclairii, N. anomalum, Dracophyllum longifolium, D. filifolium, Suttonia divaricata, Hebe salicifolia* var., *Coprosma pseudo-cuneata, C. foetidissima, O. lacunosa, O. arborescens* $\times$ *lacunosa* and *Senecio Bidwillii*.

Southern Alps and mountains of North-western district. *Olearia ilicifolia* or *O. Colensoi* are frequently dominant; *O. arborescens* and its many hybrids with *O. ilicifolia, O. nummularifolia, O. avicenniaefolia, Senecio Bidwillii,* var. *viridis* and *S. elaeagnifolius* are common shrub-composites[1]). The following occur throughout and are often important constituents: — *Phyllocladus alpinus, Dacrydium biforme, D. Bidwillii, Podocarpus Hallii, P. nivalis Phormium Colensoi, Pittosporum divaricatum, Carmichaelia grandiflora, Aristotelia fruticosa, Hoheria glabrata, Nothopanax Colensoi, N. simplex, Pseudopanax lineare, Griselinia littoralis, Gaultheria rupestris, Dracophyllum longifolium, D. Lessonianum*[2]), *Archeria Traversii, Hebe salicifolia, Hebe subalpina, Coprosma serrulata, C. pseudo-cuneata, C. parviflora, C. ciliata. C. foetidissima* and *C. ramulosa*.

In the South Otago and Fiord districts, scrub dominated by *Olearia moschata* is not uncommon. Thus, *in the Lake Harris hanging valley* the combination is *O. moschata* (dominant), *Aristotelia fruticosa, Hebe Cockayniana* (abundant), *H. buxifolia, Coprosma ciliata* and *Senecio revolutus*. A somewhat similar scrub occurs on the Takitimu Mountains.

On Tooth Peaks, at about 900 m. altitude, the dominant shrubby composite is *Senecio cassinioides* and it is accompanied by extensive colonies of *Aciphylla maxima* of the surprising height of 4.5 m. (Fig. 57), together with *Aristotelia fruticosa, Coprosma rugosa, C. parviflora* and *C. rhamnoides*, and as undergrowth, *Hypolepis Millefolium, Poa Cockayniana* and *Acaena Sanguisorbae* var. *sericeinitens*. *On Rough Peaks* (SO.), according to G. SIMPSON and J. S. THOMSON (1926:373) the combination in an allied association is *Senecio cassinioides, Podocarpus nivalis* and *Phyllocladus alpinus*.

Stewart Island. At first the forest-trees, much dwarfed, occur abundantly, but early on *Olearia Colensoi* becomes dominant. *Leptospermum scoparium* and *Dacrydium Bidwillii* are in places plentiful, tussocks of *Gahnia procera* are frequent. Other important species are: *Nothopanax Colensoi, N. simplex, Griselinia littoralis, Dracophyllum Menziesii, D. longi-*

---

1) *O. lacunosa* and *O. arborescens* $\times$ *lacunosa* (*O. excorticata* is merely one of the hybrids) are abundant in the North-western and northern part of the Western districts. *O. moschata* is common from about the latitude of Mount Cook southwards; *O. oleifolia* (but this may be merely one of the swarm, *O. avicenniaefolia* $\times$ *moschata*) occurs in the South Otago and Fiord districts; *O. Crosby-Smithiana*, which is probably *Senecio bifistulosus* is confined to the central and southern parts of the Fiord district.

2) *Dracophyllum Traversii* is abundant in the North-western and Western districts; *D. fiordense* is confined to the central and Southern parts of the Fiord district; *D. Menziesii* is confined to the Fiord and Stewart districts.

folium, D. Pearsoni, D. rosmarinifolium, Suttonia divaricata, species of *Coprosma* as for the Southern Alps and *Senecio elaeagnifolius* (Table Hill).

North-eastern district. Shrub-composite scrub made up of one species only *(Senecio Monroi)* occurs on the Inland Kaikoura Mountains and adjacent parts of the district. The substratum is coarse shingle-slip. Elsewhere the species is a rock-plant, and in this case the stony ground, as it slowly grows, is peopled by plants from the mother-rock.

γ. Other types of subalpine-scrub.

Phyllocladus scrub. *Phyllocladus alpinus* is dominant. The scrub has frequently almost the same composition as the adjacent shrub-composite scrub but dominance of the podocarp lends a distinct facies and colour. The low *Phyllocladus* forest of the Volcanic Plateau already described when of low stature is a scrub, as in certain gullies on Mount Tongariro and the Kaimanawa Mountains. Its other species are: *Dracophyllum montanum* (dominant in places), *Nothopanax Colensoi, N. simplex, N. Sinclairii, Griselinia littoralis, Leptospermum scoparium, Coprosma pseudo-cuneata, C. parviflora, C. foetidissima* and *Olearia nummularifolia.* In South Island *Phyllocladus* scrub is common in the eastern part of the Western district.

The Phyllocladus forest of the Sealey Range (W.), already described, gradually becomes stunted into scrub with *P. alpinus* dominant and the other members of the forest-association present; in its upper part *Danthonia Cunninghamii* comes in, and in places *Hebe macrantha* is common. At an altitude of about 1100 m. facing west, *the scrub on the Mount Earnslaw Spur* (F.) consists of large bushes of *P. alpinus* of great breadth, accompanied by more or less prostrate *Senecio cassinioides, Podocarpus nivalis,* erect *Hebe buxifolia, Coprosma ciliata* and *Aciphylla Colensoi.*

Cupressoid-podocarp scrub. Scrub of this type has usually *Dacrydium Bidwillii* or *D. biforme* dominant. Apparently, it is commonest on boggy, windswept or "poor" stony soil.

Near Mount Cook Hermitage on old moraine, there is a scrub of this character with yellowish-green *D. Bidwillii* dominant, forming bushes ± 1 m. high and 2.5 m. through and with the following more or less common: *Polystichum vestitum, Phyllocladus alpinus, Podocarpus nivalis, Discaria toumatou, Aristotelia fruticosa, Hoheria glabrata, Hymenanthera alpina, Coprosma propinqua, C. parviflora, C. rugosa* and *Senecio cassinioides.*

On boggy ground occupying roches moutonées or truncated spurs in the east of the Western district, open or dense scrub is common with either species of *Dacrydium* dominant.

Cupressoid-podocarp scrub also occurs at the timber-line. For example, Mount Greenland (915 m.), an isolated mountain in the north of the Western district contains near its summit a very distinct form of podocarp-scrub (Fig. 74), which though extremely dense is erect, but on the flat mountain summit becomes low and open. The following is its composition: — *Dacry-dium biforme* (dominant), *Phyllocladus alpinus*, tussocks of *Gahnia procera*, *Pittosporum divaricatum*, *Quintinia acutifolia*, *Weinmannia racemosa*, *Elaeocarpus Hookerianus*, *Nothopanax Colensoi* var. *montanum*, *Pseudopanax lineare*, *Leptospermum scoparium*, *Metrosideros lucida*, *Dracophyllum Traversii*, *D. longifolium*, *Coprosma pseudocuneata*, *Olearia lacunosa*, *O. avicenniaefolia*, *O. Colensoi* and *Senecio elaeagnifolius*. Besides its erect habit a distinct feature of this community is the mixture of forest and scrub species.

Dracophyllum scrub. On dry South Island mountains, *Dracophyllum uniflorum* forms on stony ground a more or less pure rather open scrub of a brownish colour. It burns readily and so is less in evidence than it was in the primitive vegetation. *Hebe Traversii*, species of *Cassinia*, *Olearia cymbifolia*, *Podocarpus nivalis* and *Helichrysum microphyllum* (NE. only) may be associated plants. Such scrub, becoming more and more open, merges into fell-field.

D. Lessonianum or stunted *D. longifolium* are frequently dominant in what otherwise would be classed as shrub-composite scrub but the physiognomy is completely changed. But the change is still greather when *D. Traversii*, a tuft-tree, projects out of subalpine-scrub.

D. Menziesii, of similar life-form to the last named — but in miniature — occasionally dominates in scrubs of the Fiord district. G. SIMPSON and J. S. THOMSON have supplied the following particulars concerning a piece of scrub just above the timber-line on one of the mountains bounding the North Routeburn Vallay (F.). *D. Menziesii* is dominant; the other species are *Podocarpus nivalis*, *Gaultheria rupestris*, *D. Lessonianum*, *Hebe subalpina*, *H. buxifolia*, *H. Cockayniana*, *Olearia moschata*, *Cassinia Vauvilliersii* and *Senecio revolutus*. The remarkable feature is the presence of *Celmisia Walkeri* — usually a mat-plant — as a liane scrambling through the shrubs, its stems exceeding 1 m. in length. In Stewart Island, too, *D. Menziesii* becomes most conspicuous when projecting out of low dense *Olearia Colensoi* scrub (Fig. 73).

Manuka (Leptospermum) scrub. This is chiefly a community of the lower subalpine or montane belts and differs but little from the allied lowland association. In Stewart Island, however, thanks to its tolerance of excessive wind through its epharmonic plasticity, it forms a belt at about 450 m. altitude, or even lower (Fig. 75). At first it is mixed with certain forest-shrubs, but it eventually becomes pure, much reduced in size and with bare stems and small head of twisted branches. Beneath are moss-cushions, carpets of *Lycopodium ramulosum*, tussocks of *Gahnia procera*, low-growing *Cyathodes acerosa* and prostrate *Dacrydium Bidwillii*.

There is closely-related *Leptospermum* scrub on Mount Greenland and probably other mountains in the Western district.

Cassinia scrub. It has been shown, earlier on, that for the lowland-montane belt ericoid-scrub of this kind was almost all, if not all, indigenous-induced. On the contrary, there appear to be high-mountain communities of a primitive nature. For example, on Flagstaff Hill (SO.) there is a *Cassinia* scrub above the timber-line with *Alsophila Colensoi* as an unexpected member. G. SIMPSON and J. S. THOMSON (1926:373) describe a scrub at 900 m. at the foot of Rough Peaks (SO.) with *Cassinia Vauvilliersii* dominant, accompanied by *Dracophyllum uniflorum, Aciphylla maxima* and *Gaultheria rupestris,* and in open spaces amongst rocks various characteristic high-mountain shrubs, including *Olearia moschata* and *Senecio cassinioides.* However, dominance of *Cassinia* generally means there has been burning and the association it governs must always be looked upon with suspicion.

Southern-beech (Nothofagus) scrub. The two subalpine species of *Nothofagus* respond more readily to scrub-conditions than do the trees of other forest-associations, and, in consequence, more than hold their own in competition with subalpine shrubs proper, so that both *N. Menziesii* and *N. cliffortioides* forests are frequently succeeded throughout New Zealand by a scrub in which one or other dominates. Such an association may consist almost altogether of *Nothofagus cliffortioides,* as on many of the drier mountains of the North-western district. Generally where there is an abundant precipitation many of the ordinary subalpine-scrub species accompany the southern-beeches[1]).

3. Mixed communities (shrubs, herbs, semi=woody plants, grasses &c.).

a. Rock vegetation.

a. General.

High-montain rock vegetation consists of a series of associations made up *primarily* of *obligate* and semi-obligate rock-dwelling species and *secondarily* of various plants, principally xerophytes, which belong also to neighbouring associations.

The number of species in the formation is at least 190 which belong to 36 families and 74 genera, the following being the largest: (families)

---

1) On the south of Ruapehu the scrub commences at about 1200 m., *N. cliffortioides* is dominant. *Phyllocladus alpinus, Dacrydium biforme, D. Bidwillii, Nothopanax Colensoi, N. simplex, Myrtus pedunculata, Coprosma foetidissima, C. pseudocuneata* and tussocks of *Gahnia pauciflora* are common. On Mount Rochfort (NW.) the scrub begins at about 890 m. and consists of *N. cliffortioides* (dominant), *N. Menziesii* (both southern-beeches about 2.5 m. high) together with abundance of *Dacrydium biforme* and *Leptospermum scoparium.* Other constituents are: *Phyllocladus alpinus, Pittosporum divaricatum, Pseudopanax lineare, Dracophyllum Lessonianum, Suttonia divaricata, Olearia Colensoi* and *Senecio elaeagnifolius.* The scrub of the Longwood Range (SO.) contains a good deal of *N. Menziesii* but there is much *Dacrydium Bidwillii,* so it may be considered an intermediate type.

Compositae 46 species, *Scrophulariaceae* 21, *Umbelliferae* 18, *Gramineae* 17; (genera) *Hebe* 15, *Celmisia* 10, *Raoulia, Aciphylla* and *Senecio* 8 each, *Anisotome, Myosotis* and *Helichrysum* 7 each. The following species are entirely or almost confined to rock: — *Polypodium pumilum, Gymnogramme rutaefolia* (very rare), *Carex acicularis, Colobanthus acicularis, C. canaliculatus, Hectorella caespitosa, Nasturtium fastigiatum* (genus doubtful, perhaps "new"), *N. latesiliqua, N. Enysii, N. Wallii, Cardamine bilobata, Corallospartium racemosum, Pimelea Traversii, Epilobium gracilipes, E. crassum, E. brevipes, Aciphylla Dobsoni, A. simplex, Anisotome petraea, A. brevistylis, A. Enysii, Myosotis Goyeni, M. macrantha, Hebe rupicola, H. pimeleoides* var. *rupestris, H. ciliolata, H. Hulkeana, H. Lavaudiana, H. Raoulii, H. Bigarii, H. annulata, Pachystegia insignis, Celmisia Monroi, C. bellidioides, C. Thomsoni, Raoulia eximia, R. mammillaris, R. rubra, R. Buchanani, R. bryoides, Ewartia Sinclairii, Helichrysum coralloides, Leucogenes Grahami,* and *Senecio saxifragoides.*

Then there is the class almost confined to or extremely common on rocks but rare elsewhere, e. g. *Asplenium Richardi, Microlaena Colensoi, Trisetum subspicatum, T. Cheesemanii, Agrostis subulata, Poa novae-zelandiae, Carex pyrenaica, Luzula pumila, L. Traversii, Pachycladon novae-zelandiae, Sisymbrium novae-zelandiae, Linum monogynum* (excluding lowland stations), *Aciphylla Monroi, A. similis, A. multisecta, A. Spedeni, Hebe tetrasticha, H. epacridea, Veronica linifolia, Shawia coriacea, Gnaphalium Lyallii, Helichrysum microphyllum, H. Selago, Leucogenes grandiceps, Senecio Haastii* and *S. Monroi.* Taking these classes together it appears that 37 per cent of the rock-flora is composed of plants which are far and away more common on rocks than elsewhere, and *it stands out clearly that the rock-vegetation is an entity quite as distinct as any other type of vegetation and not a mere collection of waifs and strays.*

Taking next all classes of rock-plants together, the following are widespread common species: — *Hymenophyllum multifidum, Blechnum vulcanicum, Pteridium esculentum, Polypodium pumilum, Podocarpus nivalis, Dichelachne crinita, Danthonia Raoulii* var. *flavescens, D. setifolia,* the species of *Koeleria, Poa caespitosa,, P. Colensoi, P. intermedia, Festuca novae-zelandiae, Agropyron scabrum, Schoenus pauciflorus,* various vars. of *Luzula campestris* or allied species, *Phormium Colensoi, Muehlenbeckia axillaris, Scleranthus biflorus, Stellaria gracilenta, Colobanthus acicularis, Tillaea Sieberiana, Acaena Sanguisorbae* (various vars.), *Coriaria sarmentosa, Hymenanthera alpina, Pimelea Traversii, P. prostrata, Epilobium pedunculare, E. glabellum, Leptospermum scoparium, Aciphylla Colensoi, Anisotome aromatica, Angelica montana, Corokia Cotoneaster, Gaultheria rupestris, Suttonia nummularia, Hebe epacridea, Veronica Lyallii, Celmisia bellidioides, Raoulia bryoides, Helichrysum Selago, H. bellidioides, Leucogenes grandiceps* and *Senecio bellidioides* (or one of its near allies).

Rock at various stages of plant-colonization is a common feature of the high mountains, but different ranges vary greatly in this regard. Speaking generally, the alpine belt and the river gorges of the lower levels are the most rocky localities. The volcanoes of North Island furnish much rock with their extensive lava flows (Fig. 76), weathered into fantastic forms, or, as huge blocks, piled one upon another. Mountain rock may form perpendicular cliffs, as in river gorges, much weathered crags, or be worn down to the level of the hillside. In any case, perfectly smooth rocks are rare and there are generally abundant crevices, ledges and hollows where soil can accumulate. The conditions offered for plant-life are most diverse according to the position of the rock with regard to sun, wind and moisture. and they range from intensely xerophytic to distinctly hygrophytic. Even opposite sides of the same gully often contain these opposing classes.

Ecologically, the species fall into obligate and facultative chasmophytes and soil-demanding plants. Such a soil is readily formed in a moist climate both on flat rocks and in depressions and hollows, so that a plant-covering may eventually be established not to be distinguished from that of the adjacent herb-field. At the same time, there are some species that especially affect soil-covered rocks, e. g. *Anisotome pilifera, Dracophyllum Kirkii, Celmisia Walkeri*, and, in general, herb-field species.

The life-forms of the 180 rock-plants are as follows: — trees dwarfed to shrubs 3, shrubs (often dwarfed) 58, grasslike-plants 21, rushlike plants 2. herbaceous and semi-woody plants 88 and ferns 8. Cushion-plants number 15. rosette-plants flattened to the rock 8 and cupressoid-shrubs 5.

As to the physiognomy of rock-vegetation, obviously all depends upon the species present, and this is governed entirely by the variety of rock-habitat they occupy. Frequently, the association is so open that it presents no striking features, but sometimes the shrub-form with some distinct species in excess, or the cushion-form, may be exceedingly striking.

β. The rock-communities.

North Island communities. *On the central volcanoes* there are no obligate rock-plants. The vegetation is scanty and besides mosses and lichens consists only of deep-rooting desert or steppe xerophytes, especially. *Danthonia setifolia, Poa Colensoi, Anisotome aromatica, Gaultheria rupestris* and *Helichrysum alpinum*[1]).

---

1) This name I am giving to the North Island representative of *Helichrysum bellidioides* of South Island *(Xeranthemum bellidioides* Forst. f., based upon material collected at Dusky Sound — F.) By HOOKER the North Island plant was united with *H. prostratum* of the Subantarctic Province, but the figure of this species in the *Flora Antarctica* shows sessile flower-heads, whereas in *H. alpinum*, they terminate the branches, the extremities of which are drawn out into bracteate peduncles. From *H. bellidioides, H. alpinum* is at once separated by its wider, larger leaves which are

On Mount Egmont most of the fell-field and tussock-grassland species occur on rock, but, in addition to the species of the last paragraph, the following need citing: *Coprosma repens, Pentachondra pumila, Drapetes Dieffenbachii, Forstera Bidwillii* and *Celmisia glandulosa* var. *latifolia.*

The Tararua Mountains, though subject to much rain, mist and cloudy skies, possess an extreme xerophyte in *Raoulia rubra*[1]), a typical vegetable-sheep, its cushions *green* however, and about 30 cm. diam. The other rock species are fell-field plants, e. g. *Anisotome aromatica, A. dissecta, Pentachondra pumila*, yellowish-green cushions of *Phyllachne Colensoi, Helichrysum alpinum* and *Leucogenes Leontopodium*. Where rocks are near subalpine-scrub, certain shrubs are present, notably *Senecio Bidwillii* (leaves very thick).

South Island dry mountains communities. In *the North-eastern and Eastern districts*, in their upper montane and subalpine belts, the *Colobanthus acicularis* association occurs with that species dominant. This species is a small, pale-green cushion-plant, made up of very narrow leaves, $\pm$ 12 mm. long with long acicular points. The accompanying plants mostly come from the neighbouring tussock-grassland and are: *Dichelachne crinita, Danthonia setifolia, Poa caespitosa, Festuca novae-zelandiae, Agropyron scabrun, Phormium Colensoi, Muehlenbeckia axillaris, Stellaria gracilenta, Scleranthus biflorus, Tillaea Sieberiana* (semi-obligate rock-plant), *Carmichaelia subulata, Discaria toumatou, Hymenanthera alpina, Leptospermum scoparium, Aciphylla Colensoi, Angelica montana, Cyathodes acerosa, Suttonia nummularia, Wahlenbergia albomarginata, Helichrysum bellidioides, Raoulia australis* and *Senecio bellidioides.*

In *the North Otago district* there is a closely-related association, but *Hebe pimeleoides* var. *rupestris* (twiggy, glaucous, small leaves, blue flowers) is frequently the characteristic true rock-plant, the fern *Cheilanthes Sieberi* is abundant and the *Carmichaelia* is *C. Petriei* or *C. compacta.*

There are other associations, distinguished by the presence — perhaps abundance — of other true rock-plants, e. g. *Epilobium gracilipes* with

---

glabrous above. Grown, side by side, the two species are to be recognized at a glance. In the Lord Auckland Islands, as well as *H. prostratum*, there is *H. bellidioides*, and probably the two cross. In addition to the species cited, *Veronica spathulata* may be present,and on the great lava-flow from Te Mari (part of Mt. Tongariro) various subalpine-scrub shrubs are slowly being established.

1) Apparently this crosses with *Leucogenes Leontopodium*, one of the hybrids being the so-called species *Raoulia Loganii* (Buch.) Cheesem. Elsewhere, similar hybrids occur between *Leucogenes grandiceps and Raoulia bryoides* one of which — described from one individual only — is *Helichrysum pauciflorum* T. Kirk. Probably *Raoulia Gibbsii* Cheesem., *Leucogenes Grahami* Petrie and a Stewart Island plant, originally referred by me to *R. Longanii* (in this case one parent would be *R. Goyeni*) are all edelweiß-vegetable sheep hybrids. H. H. ALLAN and myself have found a number of other forms to which we attribute the same origin, and we have given to all such the provisional hybrid generic name of $\times$ *Leucoraoulia.*

Cardamine bilobata, Anisotome Enysii and perhaps *Myosotis Goyeni* (E.. local in Waimakariri Basin); *Hebe Raoulii* (E. and NE.); *Celmisia Monroi* dominant (NE.); *Ewartia Sinclairii* (NE.); (Fig. 77). *Anisotome brevistylis* with *Senecio southlandicus* (NO., SO.); *Hebe Laraudiana*, a form of *Anisotome Enysii* or an unnamed species and large *Senecio layopus* (Banks Penin.).

The *Pachystegia insignis* association, already described for the lowlands, ascends to upwards of 900 m. with *Senecio Monroi* as a member which — *Pachystegia* absent — dominates dry rock-faces up to 1500 m. or more altitude (NE.).

In the North-eastern district, in the upper subalpine and alpine belts, a strongly xerophytic station is provided by stacks of much-weathered greywacke standing out from vast shingle-slips (screes). A black fruticose lichen may dominate. Pressed as closely to the rock as possible will be numerous, hard, circular greyish cushions of *Raoulia bryoides*, the largest some 30 cm. diam. and 17 cm. deep. But the most striking plant, to which the name of the association must be given, is *Helichrysum coralloides*[1]; many of the following will be present: — *Helichrysum Selago* and *H. microphyllum* (closely related to *H. coralloides*), hybrids between the last-named and *H. selago* — but the 3 species rarely present at the same time, *Hymenanthera alpina*, forming rigid, open cushions of divaricating stem, *Hebe decumbens* or *H. pinguifolia* issuing from crevices, *Pimelea Traversii, Colobanthus acicularis*, perhaps the beautiful white-flowered *Myosotis saxatilis*, hard rosettes of *Epilobium crassum* and probably flattened sheets of *Podocarpus nivalis*. *H. coralloides* is confined to the North-eastern district, but dry alpine rocks in many parts of the Eastern and North Otago districts bear a closely-allied association[2].

The Vegetable-sheep (Raoulia eximia) association (E.) is equally as xerophytic as that just described. The habitat is low greywacke rock at 1200 to 1800 m. altitude weathered so as to stand even with or slightly

---

1) An open shrub some 40 cm. high, if sheltered, but a true cushion, if fully exposed (Figs. 64, 65); its shoots cylindrical, 8 mm. diam. and the small, glossy appressed leaves looking like tubercles, the spaces between being packed with white wool: the leaves are densely tomentose beneath and it is the hairs of adjacent leaves being entangled which makes the white wool, the actual stem being very slender and glabrous and not as stated by CHEESEMAN (1925:986) "¼ in. diam. densely tomentose between the leaves". On shady rocks in gorges of the Inland Kaikoura mountains the shrub may exceed 90 cm. in height with the width equalling this height and the branches spreading and drooping.

2) *Hebe pinguifolia, H. epacridea, H. tetrasticha, Gaultheria rupestris* and stunted *Leucogenes grandiceps* are common in the Eastern district. *Hebe Buchanani*, and near the mountain summits *Pachycladon novae-zelandiae* occur in North Otago. *Nasturtium latesiliquum, N. fastigiatum* and *N. Enysii* with large rosettes clinging to the rock and far-penetrating thick root are local species occurring respectively in the North-western, the North-eastern and Eastern districts, but *N. fastigiatum* also extends into the Western district in the Mount Cook area, and *N. Wallii* (west of So.).

raised above the desert of stony debris (shingle-slip) by which it is partly buried. The great cushions, already described, frequently grow into one another, forming hard, white, amorphous masses 2 m. in length, or more, the woody tap-root penetrating far into the rock. Thanks to the wet raw humus within colonies of *Celmisia spectabilis, C. viscosa, Aciphylla Colensoi* and *Danthonia flavescens* grow as epiphytes on the cushions, being quite independent of the rock. The station, fully exposed to sun and wind and subject to great daily extremes of temperature at all seasons, except when buried beneath the snow in winter, is one of extreme xerophily.

Communities of South Island wet mountains. In the lower subalpine belt the rocks are altogether in the forest-areas, and when in the open occur only on river-beds or the sides of gorges. In such places, various forest-trees and subalpine shrubs are common jutting out from the rocks. In the wettest districts extremely steep cliffs may be actually covered with a close scrub, the rock having become faced with a thick sheet of soil held in position by a network of matted roots. A times, the whole of such a covering slips away for many metres leaving the steep rock-face bare and dripping. Rock of such gorges in the Western district at an early stage may be merely dotted here and there with various herbs and semi-woody plants, e. g. *Veronica linifolia* (the characteristic true rock-plant), *Deyeuxia pilosa, Poa novae-zelandiae, Schoenus pauciflorus, Angelica montana, Epilobium glabellum, Veronica Lyallii, Craspedia uniflora* (one or other of the jordanons), and *Senecio Lyallii;* extensive colonies of *Phormium Colensoi* are characteristic.

In the upper Clinton Valley, on the smooth face of the precipice where water constantly trickles, there is a curious association consisting of a close growth of a species of *Hepaticae* which is hidden by a prostrate grass, its culms pressed closely to the liverwort, pointing downwards and forming an open flat continuous mat. Numerous plants of *Celmisia verbascifolia* grow through the grass their leaves no longer erect but hanging downwards (Fig. 78). Drier cliff is occupied by a combination of *Blechnum procerum, Phormium Colensoi* and *Coriaria angustissima,* to be replaced as more soil accumulates, or where the cliff is less steep, by a shrub-association of *Hoheria glabrata, Aristotelia serrata, Fuchsia excorticata* and *Hebe saliciflora* var. *communis.*

The rocks of old moraine rising out of a subalpine herb-field, or steep buttresses, usually ice-worn, or the irregular, rocky walls of a gully rapidly become covered with peat — where such can rest or cling — made from bryophytes, lichens and early spermophyte settlers. Such stations are colonized by many species from the adjacent herb communities, as well as by a few obligate or semi-obligate rock species and others common on rocks. These include great mats of *Hymenophyllum multifidum* (epharmonic, curled-leaf form), *Microlaena Colensoi,* the prostrate *Dracophyllum Kirkii*

(NW., W.), *Anisotome pilifera, Aciphylla similis, Celmisia Walkeri,* and *Leucogenes grandiceps*; the shrubs *Gaultheria rupestris, Coprosma serrulata* and *Senecio Bidwillii* are common[1]).

The alpine rocks, if steep, have a sparse vegetation which will include some of the following (the extreme altitudes are mostly from A. WALL, (1925: 19, 20): — *Hymenophyllum multifidum, Polypodium pumilum, Microlaena Colensoi, Agrostis subulata, Poa novae-zelandiae* (2100 to 2400 m.), *Carex pyrenaica, C. acicularis* (2400 m.), *Marsippospermum gracile* (2100 m.), *Luzula pumila* (2400 m.), *Colobanthus acicularis* (2400 m.), *Hectorella caespitosa* (2100 m.), *Ranunculus Buchanani* (SO., F., 2100 m.), *R. Grahami* (W. in Mount Cook area, 2700 m. and more), *Nasturtum Enysii* (2400 m.), *N. Wallii* (SO.), *N. fastigiatum, N. latesiliquum* (NW.), *Cardamine depressa, Pachycladon novae-zelandiae,* (SO., F. — 2100 m), *Hymenanthera alpina* (2100 m.), *Epilobium rubromarginatum* (NW., W., 2400 m.), *Schizeilema Haastii* (2100 m.), *S. exiguum* (NO., SO., F.), *Aciphylla Hectori* (NO., F., SO.), *A. Spedenii* (SO., 1800 m.), *A. Dobsoni* (NO., SO — 2100 m.), *A. simplex* (SO., 1800 m.), *Angelica montana, Anisotome imbricata* (SO., F.), *Gentiana* sp., *divisa*[1]), (W., Mount Cook area, 2400 m.), *Myosotis suavis* (east of W., 2100 m.), *M. macrantha, M. pulvinaris* (SO.), *M. concinna* (NW.), *Hebe ciliolata* (2100 m.), *H. epacridea* (2100 m.), *H. Haastii* (over 2700 m.), *Celmisia brevifolia* (SO.), *C. Hectori* (SO., F., W. 1800 m.), *C. ramulosa* (SO. F.), *Raoulia eximia* (2100 m.), *R. Buchanani* (SO., F.), *R. Youngii* (W., SO.), *R. subulata, Leucogenes grandiceps* and *Cotula pyrethrifolia*.

Stewart Island rocks. Almost any of the subalpine plants may be found on rocks. This is not because they are specially adapted for such a station, but because peat is readily formed in the mountain climate on flat rock surfaces or in crevices, and because the frequent rain never allows the rocks to become too dry for bog xerophytes. The following are the only special rock-plants: *Polypodium pumilum, Anisotome flabellata, Raoulia Goyeni* and *Helichrysum grandiceps*.

b. Vegetation of loose, stony debris.

1. The shingle-slip formation and allied communities.

α. General.

The shingle-slip formation consists only of the species mentioned below, but sometimes not more than one or two are present. In any case they

---

1) The following are some of the herb-field species of such rocks: *Pachycladon novae-zelandiae* (SO., F.), *Coriaria angustissima, Acaena Sanguisorbae* vars. *pilosa* and *sericeinitens, Epilobium chloraefolium* var. *verum, E. Matthewsii* (F.), *Anisotome Haastii, A. capillifolia* (SO., F.), *Aciphylla divisa* (W., SO., F.), *A. multisecta* (F.), *Veronica catarractae* (F.), *V. Lyallii, Ourisia caespitosa, O. sessilifolia, O. prorepens* (SO., F.), *O. glandulosa* (F., SO., NW.), the so-called *Plantago Brownii, P. lanigera, Forstera tenella, F. sedifolia* and var. *oculata, Celmisia Gibbsii* (NW.), *C. rupestris* (NW.), *C. Du Rietzii*

grow so far apart — it may be many metres — that the individuals have no relation to one another.

The species fall into two classes — **obligate** (25 species) and **facultative** (8 species) — and taking both classes together the 33 species belong to 14 families and 19 genera. The following list includes all the species: — obligate) *Poa sclerophylla, Stellaria Roughii, Ranunculus Haastii, R. chordorhizus, R. pauciflorus, R. crithmifolius, Notothlaspi rosulatum, Swainsona novae-zelandiae, Epilobium pycnostachyum, Anisotome carnosula, A. diversifolia, Convolvulus fracto-saxosa, Myosotis Colensoi = (M. decora), M. Traversii, M. angustata* — perhaps, *M. Cockayniana, Hebe macrocalyx, Veronica Cheesemanii, Lobelia Roughii, Wahlenbergia cartilaginea, Haastia Sinclairii, H. recurva* and var. *Wallii, Raoulia cinerea, Craspedia alpina* — forms in fell-field may be identical —, *Cotula atrata* and var. *Dendyi* and the hybrids between them, (facultative), *Claytonia australasica, Notothaspi australe, Acaena glabra* — almost obligate, *Anisotome filifolia, Hebe lycopodioides, H. tetrasticha, H. epacridea* and *Haastia pulvinaris.*

The station, as will be seen from what follows, is strongly hostile to plant-life and but few species are so constructed as to be able to gain a footing, or, if such should happen, to thrive and produce offspring.

The much-jointed greywacke and allied rocks, which comprise the greater part of New Zealand mountains, become so rapidly disintegrated that stone-fragments accumulate to such an extent as to cover the slopes for hundreds of metres. Here and there jagged masses of much corroded rock jut out from these stone-fields but hardly break the monotone of the vast, grey even slopes which extend from the lower subalpine-belt to the mountain-tops. Gullies, often with rocky walls seam the mountain sides, their floors occupied by a stream, its source the base of some great stone-field where all on a sudden water bursts forth.

The stones themselves differ in size but the bulk are generally small, perhaps 5 or 6 cm. long by 2 cm. broad, though some may be much larger. Those of the upper layer are quite loose and, as the surface is steep, they are liable to slide downwards, considerable breadths, when disturbed, moving en masse. At 30 cm. or more below the surface, the ground is more stable and there is generally a good deal of finer debris, sand, and even clay mixed with the coarser stones. Although quite dry on the surface, at a few centimetres depth the substratum is damp, and deeper still ample water, but icy-cold, is available for plants. The climatic features of the habitat depend upon extreme exposure to wind; strong radiation of heat from the stones; powerful heating of the stones themselves and, at times, very bright light. Within the space of a few hours the plants are frequently subjected

---

(W., NW. — part of *C. Sinclairii* of the *Manual*), *C. discolor* (= *C. intermedia* Petrie NW., W.), *C. Bonplandii* (SO., F.), *C. petiolata, C. verbascifolia* (SO., F.), *Craspedia uniflora* in a wide sense, *Senecio bellidioides, S. southlandicus* (SO.), *S. scorzoneroides.*

to burning heat and considerable frost, or one hour they may be surrounded by moist air and the next be exposed to a strong, dry wind. Those which are evergreen bear a heavy weight of snow for four months at a time or more. Nor are occasional droughts unknown. It is obvious then that the ecological conditions of shingle-slip are distinctly those of desert, while in addition there is marked instability of surface. This latter character has, in part, led to the occupation of the ground not merely by certain peculiar life-forms but *by 25 distinct species which do not occur in any other formation.*

Near the edges of the shingle-slip, the stones are far less liable to move, and there is stability sufficient for species other than those adapted to the moving debris to settle down, so that, by degrees, the formation is transformed into fell-field. But *towards such change the actual shingle-slip association contributes nothing,* its members are too far apart and too few to supply appreciable humus to the soil. *The formation is indeed distinct in itself and not a phase in the development of fell-field but a definite vegetation-entity* the origin of which is wrapped in obscurity.

Shingle-slip, in its unstable and typical form, is confined to those mountains of South Island with a tussock-grassland-climate and is most strongly in evidence in the North-eastern and Eastern districts. Where there is abundant rain the conditions for accumulation of debris are unfavourable, while its occupation by non-shingle plants is much more easy. Certain scoria-slopes of North Island volcanoes are ecologically similar to true shingle-slip.

Taking both the obligate and facultative species together their life-forms are as follows: — herbs 21 (summergreen 14, evergreen 7), semiwoody plants 4, grass-form 1, shrubs 5, biennials 2. Most of the obligate species have important features in common and several of distinct affinities closely resemble one another. Thus, 17 are much the same colour as the stones; all have thick, fleshy or coriaceous leaves and in 24 species they are in rosettes; underground stems more or less strongly developed occur in 17 species, while in 16 the portion of the plant above the ground is annual; with but 2 exceptions, the shoots lie close to the stones and, although this leads to their being buried, the stems have the faculty of growing upwards, while the leaf-texture in many is such as not to be readily damaged by rolling stones. The roots generally extend, in part, horizontally and then descend more ore less deeply. Twenty one species are glabrous, but, on the other hand, the species of *Haastia* and *Craspedia alpina* are woolly, the latter being very noticeable through its long, snow-white wool.

β. The associations of true shingle-slip and their allies.

Those of the South Island dry greywacke mountains. On the dry greywacke South Island mountains most of the shingle-slip species occur at some point or other, the southern limit being the Takitimu Mountains. In some places, *Stellaria Roughii* is alone present, since it probably can occupy a more unstable position than any other species. But, generally, *Poa sclerophylla, Notothlaspi rosulatum* (Fig. 79), *Ranunculus Haastii, Anisotome Haastii, Epilobium pycnostachyum, Lobelia Roughii, Craspedia alpina* and *Cotula atrata* are present, and such, together with *Hebe epacridea, H. lycopodioides* and *H. tetrasticha* is the usual association of the Eastern district. The plants occur only here and there, indeed one may examine a shingle-slip for hundreds of metres and find no plants or, at best, a solitary example of *Stellaria Roughii*. On the fairly stable ground near the head of the debris-field there may be a few grey mats of the semi-woody *Haastia recurva*.

In the North-eastern district the association is richer than in the Eastern, for, in addition to the species cited above, are *Wahlenbergia cartilaginea* — a rosette-plant somewhat resembling a European crusty saxifrage — *Myosotis Cockayniana, Convolvulus fracto-saxosa, Raoulia cinerea* (rare) and *Swainsona novae-zelandiae* (also E., but apparently rarer).

On fine limestone debris at Castle Hill (E.) there is a distinct association at an altitude of 600 m., and rather more, with the following composition the species marked* being restricted to the above locality. The stony slope is fairly stable and its margin much more so than for shingle-slip in general. On the unstable debris are: *Notothaspi rosulatum*, in great abundance; *Ranunculus pauciflorus**, but confined to one small area; *Myosotis Traversii*; and, where stable, *M. Colensoi**; *Poa acicularifolia*; *Lepidium sisymbrioides*, with extremely long roots; *Oreomyrrhis andicola* var. *rigida; Senecio lautus* var. *montanus* and *Carmichaelia Monroi*.

The Haastia pulvinaris association. This association is characteristic of the alpine belt of the North-eastern district both in its driest part and at its junction with the North-western district. *Haastia pulvinaris* is exactly of the same cushion-form (Fig. 58) as the other great vegetable-sheep, *Raoulia eximia*, already described, but the shoots are much thicker and the leaves larger and more woolly. The great pale-yellow cushions 2 to 3 m. long and 60 cm. or more thick may dot the shingle-slip as far as the eye can reach. The larger examples grow amongst the biggest stones but certainly, in many cases, are not attached to the underlying rock. They do not seem to grow on the finer debris. They are usually much longer than broad, an this is accentuated by de sliding stones from above piling up against them so that the upper surface may be partially buried. Many deaths take place from such burials. Various species are epiphytic on the cushions especially, *Danthonia Raoulii* var. *flavescens, D. setifolia, Celmisia spectabilis* and *C. viscosa*.

The association of the Western district. Nearly all the eastern shingle-slip species are absent and various herb-field or fell-field plants are present especially *Podocarpus nivalis, Acaena Sanguisorbae* var. *sericei-nitens, Epilobium glabellum, Oxalis lactea, Geranium microphyllum, Cotula pyrethrifolia, Celmisia Du Rietzii* and great sheets of *Leucogenes grandiceps*. True shingle-slip plants are represented by *Epilobium pycnostachyum, Hebe macrocalyx, Veronica Cheesemanii* and *Haastia Sinclairii* (distinct from the Fiord var.), while *Epilobium rubro-marginatum* (semi-shingle species) is frequently abundant at the highest altitudes.

Vegetation of scoria slopes (VP., EW.). The scoria slopes of the Volcanic Plateau, and Mount Egmont, present conditions quite as severe as true shingle-slip and, as on certain slopes of Mount Ngauruhoe, may be without plant-life. But there are often a few species distantly dotted about especially: *Luzula Colensoi, Claytonia australasica, Gentiana bellidifolia* (Fig. 80), *Anisotome aromatica*[1]), *Poa Colensoi* and *Gaultheria rupestris*. The most characteristic plant of Ruapehu &c. for this station is *Veronica spathulata*[2]) (Fig. 62).

Vegetation of rock-slides. This association or group of associations is distinguished by the presence of certain species of *Ranunculus* and the local var. of *Haastia Sinclairii*. It was first described by J. SIMPSON and J. S. THOMSON (1926:376—77) who proposed the name "rock-slide".

The station consists of large blocks of stone the spaces between which filled with far smaller particles form the rooting-places for the plants. The large rocks "broken and thrown into vast slopes" are comparatively stable; they cover large areas and are bare of vegetation. *On Rough Peaks* (the area investigated by the above botanists) the association consists of *Haastia Sinclairii*, as above, *Ranunculus Scott-Thomsonii* — a true debris-plant with stout rhizome and rosettes of trifoliolate leathery, grey, glabrous leaves, a polymorphic hybrid swarm between the latter and a var. of *R. Buchanani* growing in its vicinity, and occasional patches of the silvery *Celmisia Hectori.*

On Tooth Peaks (junction SO. and F.), at an altitude of about 1050 m. where the stones are very large, the vegetation is more of a fell-field character and consists of yellowish-brown mats of *Podocarpus nivalis*, almost black open cushions of *Hymenanthera alpina* and plants of *Aciphylla maxima*. There is also some *Acaena Sanguisorbae* var. *pilosa, Muehlenbeckia axillaris, Coprosma propinqua, Senecio cassinioides* and a good deal of *Pimelea prostrata*.

---

1) This may be an epharmone or a distinct variety; at any rate it has a true debris form with coriaceous leaves and an extremely thick, deeply-descending tap-root.

2) A prostrate herb or perhaps semi-woody plant forming a close soft mat upon the ground composed of flexible, decumbent stems branching freely near their extremities; the leaves are small, hairy, soft and thick; the root of extraordinary length giving off many close, more slender secondary roots; and the flowers large, for size of plant, white and produced in the greatest profusion (Fig. 62).

2. Vegetation of river-beds, fans and allied habitats.

River-bed vegetation. This has been described at some length for the lowland and montane belts and the definition given for its class of vegetation applies here also. The physiographical and ecological conditions as there indicated match closely those of subalpine river-bed, as far as the larger rivers are concerned, except just at their glacier sources. Even the species are much the same, as is the procession of events in plant-colonization. Such differences, as there are, arise from climate and the colder water of the substratum, while certain species, according to locality, are present which do not descend to or are rare in the lowland belt. Lowland river-bed of the Western and Fiord districts however approximates closely to that of the mountains.

The South Island species, omitting those of fell-field or herb-field which occur in the upper torrent beds, number about 72 and belong to 23 families and 38 genera. Their life-forms are as follows: — grass-form 11 species, herbs 32, semi-woody plants 18 and shrubs 11. At most 10 species belong to the high-mountain flora but nearly all of these descend to the lower montane belt. As in the lowlands the most important members of the formation are the species of *Epilobium* and *Raoulia*[1]) with *E. melanocaulon* characteristic of the dry areas and *E. glabellum* of the wet. Other common species are the grey, low shrub *Helichrysum depressum* looking half-dead with grey rigid stems and scanty small appressed leaves; great circular mats of *Muehlenbeckia axillaris;* several species of *Acaena; Veronica Lyallii* in the wet and *V. Bidwillii* in the dry areas or stations; *Helichrysum bellidioides, Coriaria lurida; C. angustissima* (where abundant rain) and the vast polymorphic hybrid swarm between them in which *C. sarmentosa* also plays a part, *Veronica catarractae* (F.), *Angelica montana, Discaria toumatou* and *Raoulia glabra. Raoulia Haastii,* which forms large green cushions, is the characteristic plant of many river-beds in the Eastern and Western districts and occasionally occurs in the South Otago district.

Subalpine torrent beds of the wet mountains, especially near their sources, contain more or less of the fell-field and herb-field species and their open plant-covering may resemble that of the adjacent fell-fields. So, too, the old bed of the wider valleys often has a considerable florula and the association may be closed, but it is generally rather tussock-grassland dotted with shrubs than fell-field. *Hebe buxifolia* var. *odora* and *Carmichaelia grandiflora* are common on old river-bed on the wet eastern side of the Divide (W.), *Coprosma brunnea,* its wiry stems hugging the stones, is characteristic, and soft cushions of *Myosotis uniflora* occur occasionally.

---

1) *E. pedunculare, E. glabellum, E. melanocaulon, E. microphyllum, E. rostratum, E. macropus, R. tenuicaulis, R. Haastii, R. glabra, R. australis, R. Parkii* (where the climate is specially dry) and *R. lutescens.*

Vegetation of fans. The vegetation is that of river-bed and commences with the usual species of *Epilobium* and *Raoulia*. Finally tussock-grass-land or *Discaria* scrub may be established but the indermediate stage may possess many circular mats of *Muehlenbeckia axillaris* and extensive colonies of the ferns *Blechnum penna marina* and *Hypolepis Millefolium*.

The association occurs principally in the drier areas and the fans — sometimes of great size — are built up by the stony debris deposited at the mouths of gullies or gorges by their streams as they open out on to the valleys, river-beds or plains. Their vegetation depends upon the supply of stones brought down by the torrent and this again is correlated with the age of the gully and the plant-covering of its walls. Fans may be either active or passive, and every transition between the two can be seen. The stony surface is much steeper than that of river-bed in general. There are water-channels but these are usually dry except during heavy rain, the actual stream running underground. Many of the stones are large and much of the debris coarse and piled up into comparatively high but quite unstable terraces liable during flood to damage or absolute destruction.

Closely related to fan vegetation, on the one hand, and to *Hebe* scrub, on the other, especially in the Eastern district, is an open association with *Hymenanthera alpina* dominant and, growing between the stones, *Hypolepis Millefolium* and *Blechnum penna marina*. The other members of the community are frequently *Geranium microphyllum*, species of *Acaena* and their many hybrids, *Myosotis australis* (yellow flowers, probably not identical with the Australian species), and various shrubs, especially *Discaria*. The substratum consists of large, lichen-covered stones either at the foot of river-terrace or at the base of mountain slopes, particularly in the montane and lower subalpine belts. *Cystopteris novae-zelandiae* is common beneath the shrubs.

c. The fell-field, grassland and herb-field series of communities.

1. General.

This series of communities is considered under one head, though each is probably a distinct formation, for they intergrade so constantly that no hard and fast line can be drawn between them. This resemblance has become greatly intensified during the past 76 years through the sheep-farmer burning the vegetation and through the changes wrought by overstocking with sheep, cattle, rabbits, deer, and other grazing and browsing mammals. At the present time, induced fell-field is to be seen, but little different, if different at all, from virgin fell-field, where, originally, tussock-grassland, or even forest, ruled. Burning tall tussock-grassland rapidly transforms this vegetation-type into low tussock-grassland, whilst burning the latter, year by year, leads to erosion of the substratum and the incoming of fell-field. And still further burning in the presence of sheep &c., brings about a habitat where only obligate or facultative shingle-slip species, or xerophytes of river-bed, can gain a footing.

Though the foregoing state of affairs applies particularly to the dry South Island mountains, it must not be thought that the plant-covering of the wet mountains is altogether virgin. Even there **burning**, on the one hand, and **deer,** on the other, have led to great changes and, lacking a clue, the modified community would unhesitatingly be considered primitive. Nevertheless, there are many wide areas where virgin vegetation can still be seen, especially on certain North Island mountains and in the North-western, Western, Fiord and Stewart districts.

Not only does the presence of exotic species indicate modification of the plant-covering, but even the superabundance of certain indigenous species. For instance, in the lower subalpine belt of dry mountains, profusion of *Celmisia* (Fig. 55), *Chrysobactron*, *Acaena* and *Phormium Colensoi* must be looked at askance. On the other hand, with but few exceptions appar-ently few, if any, indigenous species, absent in a virgin association, seem to have gained a footing in those now modified or induced, nor are the exotic aliens of real importance in forbidding the entry, or spread, of the true members of a community.

2. Fell-field.

α. General.

The fell-field formation consists of a group of associations of an extrem-ely open character made up, for the most part, of low-growing xero-phytes or subxerophytes and with tussock-grasses present only to a quite limited extent.

The number of species, including a few of the commoner hybrid swarms, is 282 which belong to 40 families and 96 genera, of which the most im-portant are: — (families) *Compositae* 62, *Gramineae* 34, *Scrophulariaceae* 30, *Umbelliferae* 17, *Ranunculaceae* 12; (genera) *Celmisia* 26, *Hebe* 16, *Ranun-culus* 12, *Acaena* and *Coprosma* 7 each.

The following are common species of wide range: — *Blechnum penna marina, Lycopodium scariosum, L. fastigiatum, Podocarpus nivalis, Deyeuxia avenoides, Danthonia Raoulii* var. *flavescens* (in a wide sense) and var. *rubra, D. setifolia, Poa Colensoi, P. Lindsayi, Festuca novae-zelandiae, Uncinia compacta, Marsippospermum gracile, Astelia Cockaynei, Exocarpus Bidwillii, Muehlenbeckia axillaris, Claytonia australasica, Stellaria gracilenta, Coloban-thus crassifolius, Geum parviflorum, G. leiospermum, Acaena Sanguisorbae* var. *pilosa, Acaena microphylla, A. inermis* and hybrids between the three species, *Carmichaelia Monroi* (in a wide sense), *Geranium sessiliflorum* var. *glabrum, Coriaria lurida, C. sarmentosa,* ✕ *C. sarlurida, Discaria toumatou, Viola Cunninghamii, Hymenanthera alpina, Pimelea prostrata, P. pseudo-Lyallii, Drapetes Dieffenbachii, Leptospermum scoparium, Epilobium tas-*

manicum (in a very wide sense), *E. Hectori, E. chloraefolium* var. *verum,
E. pedunculare* (in a wide sense), *E. glabellum, E. novae-zelandicae, Hydrocotyle novae-zelandiae* var. *montana, Aciphylla Colensoi, A. squarrosa* (in a very wide sense), *Gaultheria rupestris, Pentachondra pumila, Cyathodes Colensoi, Leucopogon Fraseri, Dracophyllum uniflorum, Suttonia nummularia, Gentiana corymbifera, Myosotis australis* (probably distinct from the Australian species of that name), *Pygmaea pulvinaris, Ourisia caespitosa, Euphrasia revoluta, E. zealandica, Pratia macrodon, Forstera Bidwillii, Phyllachne Colensoi, Celmisia Du Rietzii, C. discolor* (= *C. intermedia* Petrie), *C. spectabilis, C. viscosa, C. laricifolia, C. longifolia, C. Lyallii, C. Haastii, Gnaphalium Traversii, Raoulia grandiflora, Helichrysum bellidioides, Cassinia Vauvilliersii, Craspedia uniflora* (in a very wide sense), *Cotula pectinata, C. pyrethrifolia* and *Senecio bellidioides* ore one of its near allies.

Fell-field vegetation occurs on all the dry mountains but on the wet mountains only on especially stony ground or on a substratum subject to strong erosion; it is absent in Stewart Island. Vertically, it occupies more or less of the subalpine and alpine belts, but, in ascending, the species decrease considerably in number, though a few, absent below, reinforce the depleted community. The greatest development of fell-field is on the dry greywacke mountains of South Island, but it must be emphasized that much is merely induced through the burning of tussock-grassland or *Nothofagus* forest.

Fell-field falls naturally into that of **dry** and of **wet** mountains. Naturally too, the ecological conditions supplied by these opposite classes differ considerably, but the fundamental factors governing the formation as a whole are everywhere similar. *Foremost, stands out the unstable substratum* due to frequent erosion, rapid or insidious, dependant upon rain and wind in relation to a more or less friable surface, or to readily-moved stony debris. Such a substratum is altogether hostile to a closed community, so the excess of bare ground readily becomes the sport of wind and water, while that bearing vegetation is easily undermined. Then, on exposed parts of the mountains, especially ridges, summits and slopes facing the full blast of the wind, only those plants which hug the ground or form dense cushions can exist, even though the substratum be stable enough. Also very shallow soil favours fell-field. The deep snow-covering of the alpine belt leads physically to erosion during its melting and ecologically to a growing-period of short duration and ice-cold water for the roots. The plants are continually exposed to frost, except in winter when lying under their covering of snow, and, though these frosts are not really severe, a plant may be frozen hard during the night and exposed to strong insolation early in the day. In short, *fell-field conditions are less favourable for vascular plants than those of any other high-mountain habitat,* shingle-slip excepted.

Coming now to the life-forms, the main classes and number of species to each class are as follows: — shrubs 60, semi-woody plants 54, herbs 116, grass-form 47, rush-form 1 and ferns 4. With but few exceptions, the plants are of very low stature no less than 144 (51°/$_0$) being prostrate or close to the ground, while hardly any of the remainder exceed 30 cm. in height, the greater part being much less. There are 28 cushion-plants but in the most exposed positions certain species, usually erect, assume that form or else are flattened to the ground. As for the erect species, 10 are tussocks and some 15 have rigid stems or leaves (*Aciphylla*). As for leaves, these are small in 200 species (71°/$_0$), most being very small and 5 species are leafless and several, leafy elsewhere, almost leafless, and in nearly all the species which also occur in other formations their leaves are reduced considerably beyond their average size. With regard to texture, nearly all the species have coriaceous, thick, hard or stiff leaves and those of the grasses are rolled. In fact, *fell-field plants stand in harmony with the violent, frequent, long-persisting winds.*

The life-history of fell-field is fairly clear and on most mountains its beginnings, its prime, and its destruction can be seen. Its earliest stage is the sparse colonization by xerophytes — some coming from neighbouring rocks — of stony debris (e. g. margin of shingle-slip) so soon as this is sufficiently stable. In course of time, should nothing upset this stability, a small amount of humus will accumulate, the stones themselves will break up into fine particles, fairly good seed-beds will be established, particularly through the incoming of mat-plants, and species from tussock-grassland or herb-field can enter the community; indeed, if stability is maintained, as on the flattest ground, well-watered by melting snow, or in hollows where snow lies long, considerable patches of closed vegetation may be established — oases, as it were. On the contrary, the disintegration, ever in progress, renders fell-field far from long-lived it being in a constant state of destruction and renewal. A heavy snowfall is before all else the most powerful factor for damage on a large scale, since it leads to snow avalanches which tear up the surface destroying all vegetation and depositing the remains on the gully-floor hundreds of metres below. Seeds germinate well enough, when they get the chance, and in New Zealand at any rate it is not as WARMING suggests (1909 : 256) a relation between seed-germination and climate that determines the openness of fell-field, but largely the disintegration factor.

At the present time, in lower-subalpine fell-field, though the associations look virgin enough, except for a few exotic species of no real moment, in many cases it is indigenous-induced as already explained. To cite a few examples of induced fell-field there is that of Porter's Pass (E.), Jack's Pass (NE.) and the lower subalpine slopes of Mount Ida (NO.).

β. Fell-field of the dry mountains or a specially dry substratum.

Pumice fell-field. This well-marked community is distinguished by the dominance of *Dracophyllum recurvum*, together with certain shrubs —

usually more or less prostrate — and a few low-growing herbs and semi-woody plants.

The species number about 54 (families 21, genera 37) of which the following are important: — *Gleichenia circinata, Podocarpus nivalis, Dacrydium laxifolium, Danthonia setifolia, Pimelea prostrata, Gaultheria rupestris, Dracophyllum recurvum, Epacris alpina, Pentachondra pumila, Gentiana bellidifolia, Hebe laevis, H. tetragona, Ourisia Colensoi, Euphrasia tricolor, Coprosma depressa, Celmisia spectabilis* and *Raoulia australis* var. *albo-sericea.*

The association under consideration occurs on the Volcanic Plateau and mountains adjacent at an altitude of 1080 to 1350 m. Apparently, there is a related association on the Ruahine Mountains.

The habitat strongly favours xerophytes. The soil to a great depth, is merely pumice, scoria and andesitic lava mixed with sand from their desintegration. Where level, and sufficient plants are present, there is a layer of black sandy humus 2.5 to 5 cm. deep, the dryness of which leads to its being blown away. The water-holding capacity of the soil is of the slightest, and the water-table lies far below the surface, moreover the evaporating action of sun and wind comes strongly into play. On a cloudless summer day, the surface-soil becomes burning hot to the hand and the heat penetrates markedly for at least 7 to 8 cm. The easily-moved soil brings about conditions similar to those of a dune-area, in fact dunes occur. Where there is absence of water and exposure to wind, desert pure and simple results, but with increase of humidity comes a denser plant-covering and the entry of more mesophytic species.

Regarding the life-forms there are 30 shrubs, 15 herbaceous and semi-woody plants, 7 grasslike plants, 1 moss and 1 fern. About 61 per cent are more or less prostrate; the cushion, ball-like, needle-leaved, divaricating, leafless and tussock forms are represented. *Dracophyllum recurvum* itself forms rounded, open cushions or low mats, 60 cm. diam., made up of much-branching, rigid stems bearing apical semi-rosettes of strongly recurved, reddish-orange, stiff leaves each 1.2 to 3.8 cm. long.

The landscape varies from a plantless expanse of scoria, by way of a most open covering of a few species, to one almost closed where the whole florula is present. The abundance of *Dracophyllum recurvum* gives a general reddish or reddish-brown colour to wide areas. Where denser, yellows, pale greens, yellow-greens and silver make a coloured patchwork and, where densest, the species, more or less flattened to the ground, form irregular oases of considerable size, but under unfavourable conditions the black scoria is merely dotted at distant intervals by silvery patches of the *Raoulia,* small straw-coloured tussocks of *Danthonia setifolia,* vivid green semi-cushions of *Pimelea prostrata* and isolated dark rosettes of *Gentiana bellidifolia* (Fig. 80), but the scene is one of desolation. More consolidated ground is occupied by reddish-orange patches of *Dracophyllum recurvum* raised but a few centimetres above the substratum or as higher sand-filled cushions; cushions of the flat-stemmed, leafless *Carmichaelia orbiculata* (Fig. 81) occur occasionally, and epharmonic cushions of *Dacrydium laxifolium* (Fig. 70). On the "oases". the shrubs &c. growing mixed together catch the flying sand and build irregular mounds 30 cm. high with the margins either ragged or hold firmly by a close mat of *D. laxifolium* or *Podocarpus nivalis* which may extend out to the bare ground. Dunes are occupied by the same species as the mounds, but in addition there may be *Coriaria lurida, Muehlenbeckia axillaris, Olearia nummularifolia, Phyllocladus alpinus* and stunted *Notho-fagus cliffortioides.*

The subalpine group of associations of the dry South Island mountains. This group is distinguished by the special physiognomic importance of great sheets of *Podocarpus nivalis* and low, depressed brownish bushes of *Dracophyllum uniflorum,* together with (generally) various species of *Celmisia,* yucca-like *Aciphylla Colensoi* and tussocks of *Danthonia Raoulii* var. *flavescens.*

At a low estimate the number of species and hybrid swarms (9) is 140 (families 33, genera 65). As so many species are mentioned in what follows no list is given here.

The group of associations is found on those mountains situated outside the average line reached by the westerly rainfall and is restricted to the Sounds-Nelson, North-eastern, Eastern and North Otago districts. It extends vertically from about 900 m. to 1200 or even 1500 m. altitude, but the altitude reached and its area are determined by the nature of the substratum, *clay favouring tussock-grassland and rock-debris fell-field.*

Viewed from some lofty ridge in the North-eastern district, grey mountain slopes in all directions meet the eye apparently quite devoid of vegetation, indeed nothing could appear more desert-like (Fig. 67, in *background*). To be sure, not all the dry mountains look so barren but they rarely possess anything like a continuous covering throughout the subalpine and alpine belts, but show, at best, vast debris-fields divided by narrow triangular lines or strips of vegetation which descend towards the brown tussock-grassland or dark forest. Such lines, or patches, raised slightly above the general level stand out from the desert of unstable stones. A closer view shows that there is rarely a continuous plant-covering. Patches of clayey, stony soil, large or small, abound containing sometimes the remains of dying plants, the long roots exposed, but rarely are they to be seen in process of occupation.

The soil is of diverse origin and may be stones merely (e. g. consolidated shingle-slip), loess, glacier-clay or clay from underlying rock. It is obvious, that through unstability of the substratum, these soils must frequently be mixed together. As plant-colonization proceeds, a variable amount of humus accumulates and the soil becomes more or less loamy. The water-content varies much according to the season of the year and to position with regard to the sun, a certain average shade favouring closed vegetation; if the number of rainy days be sufficient.

The physiognomy of the vegetation depends, in the first place, upon the yellowish mats of *Podocarpus nivalis* rooted amongst the stones and extending for several metres and upon the spreading, rather open brown bushes of *Dracophyllum uniflorum*, which give a brownish hue to subalpine slopes conspicuous even from the base of the mountain. Low, dark cushions of the almost leafless divaricating-shrub *Hymenanthera alpina* (Fig. 83, but not subalpine in this figure). small hard-leaved plants of *Gaultheria rupestris* and the glaucous-leaved species of *Hebe*, with prostrate gnarled black stems and imbricating leaves crowded near the ends of the branches are all plants of particularly dry and exposed stations. Other shrubs of dry ground are *Carmichaelia Monroi* (forming open, low cushions of short, broad, leafless, rigid, flat stems) *Cyathodes Colensoi* (dark-coloured mats), prostrate *Dracophyllum pronum*. On certain mountains from the North-eastern to the North Otago district grows the remarkable *Corallospartium crassicaule*, an intensely xerophytic shrub with a stout, rigid, sparingly-branched, yellow grooved stem, 1 m. or more high, and about 2 cm. diam.

The abundance of herbs and semi-woody plants, nearly all prostrate or low-growing, depends upon the degree of moisture in the soil. Where dry, but few species are present, especially *Scleranthus biflorus*, *Muehlenbeckia axillaris*, *Colobanthus crassifolius*, *Acaena inermis*, *A. Buchanani* (Central Otago and Mackenzie Plain), *Geranium sessiliflorum* var. *glabrum*, *Pimelea pseudo-Lyallii*, *P. sericeo-villosa* (NE. to NO.), *Drapetes Dieffenbachii*, *Hydrocotyle novae-zelandiae* var. *montana*, *Pratia macrodon*, *Helichrysum bellidioides*, *Cotula pyrethrifolia*, *C. pectinata* and *Senecio lautus* var. *montanus*.

Moister places, especially where shaded, contain a richer vegetation, which of course may include many of the above. Green mats of *Ourisia caespitosa* (showy when one mass of white blossom) are characteristic. The following may also be present: *Ranunculus Enysii*, *R. Monroi* or var. *dentatus*, *R. multiscapus*, *R. Sinclairii*, *R. gracilipes*, *Geum parviflorum*, *Geranium microphyllum*, *Oxalis lactea*, *Viola Cunninghamii*, *Drapetes Lyallii*, *Epilobium pedunculare* (in a wide sense), *E. chloraefolium*, *Schizeilema hydrocotyloides*, *Oreomyrrhis andicola* (in a wide sense), *Anisotome aromatica*, *Gentiana patula*, *Myosotis pygmaea* (in a very wide sense), *Veronica Lyallii*, *V. Bidwillii*, *Euphrasia Monroi*, *E. revoluta*, *E. zealandica*, *Plantago Brownii*, *Coprosma repens*, *Wahlenbergia albomarginata*, *Forstera sedifolia*, *F. Bidwillii*, *Craspedia* (one or other of the species), *Senecio bellidioides* and *Taraxacum magellanicum*.

The alpine group of associations of the dry South Island mountains. This group is distinguished by the extremely open character of the vegetation and the presence of a few species rarely seen below 1200 m. altitude, or generally higher, e. g. *Luzula pumila*, *L. Cheesemanii* (rare and local) and

Pygmaea pulvinaris. The species number about 65 (families 22, genera 42), but only some 45 belong to fell-field proper. The most important species are mentioned below.

The habitat consists for the greater part of consolidated debris similar to that of shingle-slip but *stable* for the time being. There are, however, hollows where snow lies for a long time, the vegetation of which is closed more or less and far richer in species; indeed, it really stands in a category by itself and is related, on the one side, to herb-field and, on the other, to the carpet-grass association. Indeed, its ecology and species are distinct from those of fell-field, nevertheless it is convenient to deal with it here. For fell-field proper the ecological conditions are much the same as for shingle-slip vegetation, so far as heat, cold, rain, snow and wind are concerned, but the stable substratum puts the plants in a different class so far as their life-forms are concerned.

At a certain altitude, differing considerably according to the ecological conditions, the vegetation of the mountain-side changes all on a sudden, and the alpine-belt commences. The first indication is the incoming of great mats of the stiff-leaved *Celmisia viscosa* and the thin-leaved *C. Haastii* — species which affect ground abundantly watered by melting snow. *Celmisia incana* (west of NW.), accompanied by *Danthonia australis,* occupies a similar position.

As members of the fell-field proper, there will be in some part or other of the community the following: — *Claytonia australasica, Epilobium tasmanicum* (in a wide sense), *Schizeilema hydrocotyloides, S. Roughii* (NE.*),* one or more of the *Aciphylla Monroi* group, *Gaultheria rupestris, Pentachondra pumila, Hebe tetrasticha* (E.), *H. Cheesemanii* (parts of NE.), *H. epacridea, H. decumbens* (SN., NE.), *Pratia macrodon, Wahlenbergia albomarginata* (epharmone, compact and thick-leaved), *Phyllachne Colensoi* (dense, green cushions), *Celmisia laricifolia* (low cushion or mat), *Raoulia grandiflora, R. bryoides* (usually a rock-plant), *Leucogenes grandiceps* (usually a rock-plant), and low, dark-coloured mats of *Cotula pectinata.*

Fell-field dependant on the chemical nature of the substratum. On the Dun Mountain (SN.) at an altitude of less than 900 m., the luxuriant *Nothofagus* forest of the limestone gives place on the magnesian soil almost without any transition to scrub, fell-field and tall tussock-grassland (Fig. 44). Not only does this soil influence the associations but it changes the lifeforms, so that trees beyond its influence are merely shrubs on the belt (the spp. of *Nothofagus, Griselinia littoralis*). Taking the species of shrubland and fell-field together, for these associations merge into one another, the following are the most important: *Phyllocladus alpinus, Poa* sp. (related to *P. acicularifolia*), *Festuca* sp. (unnamed), *Phormium Colensoi, Astelia Cockaynei* var., *Thelymitra longifolia, Libertia ixioides* var., *Exocarpus Bidwillii, Nothofagus fusca, N. cliffortioides, Muehlenbeckia axillaris, Colobanthus mollis, Claytonia australasica, Notothlaspi australis, Pittosporum divaricatum,*

Hymenanthera alpina, Pimelea Suteri, Leptospermum scoparium, Anisotome aromatica, A. filifolia, Griselinia littoralis, Dracophyllum pronum, Cyathodes acerosa, Suttonia divaricata, Myosotis Monroi, Hebe buxifolia, Coprosma propinqua, C. foetidissima, × *C. prorobusta, Helichrysum bellidioides, Cassinia albida* var. *serpentina* and *Olearia serpentina.*

γ. Fell-field of the wet mountains.

This series of associations is distinguished from that of the dry mountains by the presence of plants of a more mesophytic character and by the absence, usually, of the most intense xerophytes.

The species number about 167 (families 30, genera 71); the commoner species are mentioned below. The life-forms are as follows: — shrubs 24, semi-woody plants 34, herbs 77, grass-like plants 28, rush-like plants 1, ferns 3.

The community, in large part, is a thinning out of the species of herb-field and tussock-grassland, together with the addition of certain plants usually rupestral and a few rare except in this class of fell-field.

The Tararua Mountains near the summits of their highest peaks on stony ground bear a fell-field association, consisting principally of the following: — *Danthonia Raoulii* var. *flavescens, Ranunculus insignis, Anisotome dissecta, Hebe Astoni* (cupressoid), green cushions of *Phyllachne Colensoi, Celmisia hieracifolia* (Fig. 83), *C. oblonga, Raoulia grandiflora* (Fig. 84) and *Leucogenes Leontopodium* (Fig. 63).

On Mount Egmont, with increase of altitude, the herb-field becomes more open until on scoria slopes plants are merely dotted about or far distant, the most important being *Poa Colensoi, Luzula Colensoi, Epilobium glabellum* var. *erubescens, Claytonia australasica* and *Anisotome aromatica.*

South Island wet mountain fell-field is virtually restricted to the North-western, Western, South Otago and Fiord districts. The vegetation is richer than that of dry mountain fell-field and the species far more striking. Thus the following *Ranunculi* are particularly noteworthy: — *R. insignis* (NW.), *R. Monroi* (NW. but also on certain dry mountains), *R. sericophyllus* (W.) growing amongst large blocks of stone accompanied by the summer-green fern *Polystichum cystostegia, R. Simpsonii* (F.), *R. Buchanani* (SO., F.) and the hybrid swarm between the last two, *R. Godleyanus* (W. — Rakaia Basin to Mount Cook area), *R. Grahami* (W., Mountain Cook area), *R. pachyrhizus* (SO., F.). Generally the above are much dwarfed as compared with their herb-field form. The following also are characteristic: — *Marsippospermum gracile* (stiff, rush-like tussock, 20 cm. high), *Microlaena Colensoi, Danthonia crassiuscula* (small tussock with stiff somewhat horizontal leaves), *Poa Cockayniana* (large, thick mats), *Uncinia macrolepis, U. fusco-vaginata,* vars. of *Acaena Sanguisorbae* and hybrids between them, *Coriaria angustissima, C. lurida* and hybrids between the two, *Viola Cunninghamii, Carmichaelia grandiflora* (in a wide sense), *Epilobium glabellum, E. rubro-*

marginatum (NW., W.), *E. tasmanicum* (in a wide sense), *Aciphylla divisa,
A. similis, A. crenulata* (W.), *Dracophyllum Kirkii, Suttonia nummularia,
Hebe macrantha, Veronica Lyallii, Ourisia sessiliflora, Phyllachne Colensoi,
P. clavigera, Celmisia sessiliflora, C. laricifolia, C. Du Rietzii, C. brevifolia,
C. Bonplandii, Leucogenes grandiceps, Cotula pyrethrifolia, Raoulia Hectori*
(SO., south of W. on east), *Senecio scorzoneroides* and *S. revolutus* (SO., F.)
forming wide mats.

The community occurs both in the subalpine and alpine belts. The
substratum is too stony to allow the species to grow at all closely e. g.
shingle-slip of large stones fairly stable, which forms a broad path from
the alpine stony ground to river-bed (Fig. 87) or glacier; firm talus below
cliffs; old river-bed at the head of rivers from glaciers; stony ground in the
alpine-belt. Evidently the subalpine area must be the richest in species and
the vegetation merges into tall tussock-grassland and herb-field. Various
mat-forming species of *Celmisia* are frequently physiognomic. In the Western
district, *Leucogenes grandiceps* may form great silvery mats on debris near
the base of alpine cliffs.

In the Fiord district silvery mats of *Celmisia Hectori* may dominate,
accompanied by abundance of the glaucous, cut-leaved *Ranunculus Buchanani*, the pale-lemon flowered *R. Simpsonii* and their many hybrids.
An allied association on Rough Peaks (west of SO.) is thus vividly
described by G. SIMPSON and J. S. THOMSON (1926 : 375). "At about 5000 ft.
altitude large flat stones almost completely cover the ground, and there
Celmisia Hectori formed an almost pure association, the plants growing
in between and over the stones, so that for many acres nothing is to
be seen except *C. Hectori* and the protruding stones, with occasional plants
of *Abrotanella inconspicua, Celmisia Haastii, Marsippospermum gracile,
Phyllachne Colensoi* and *Senecio revolutus*. A few plants of *Aciphylla
similis* and *A. Spedeni* were also present, but very few. This association
is most striking: the great sheets of glistening silvery leaves of the *Celmisia*
framing the grey flat rocks and shining in the sun made a sight never to
be forgotten."

*In the North-western district there is a remarkable montane plant-
association* which occurs on the plateau extending from near the coal-mining
township of Denniston and rising gradually to the slope of Mount Rochfort,
its lowest level being about 600 m. Owing to the plateau being exposed
to the full face of the westerly wind from the Tasman Sea, it is without
forest notwithstanding a very high rainfall. The substratum is flat, rocky
or paved with small stones or, in places sandy or clayey and sticky; rocks
of various sizes stand out here and there. The plants are all flattened closely
to the ground: they root in the chinks between the stones; there is far
more bare ground than vegetation. Where large rocks supply shelter,
shrubs prostrate in the open grow more or less erect and even *Nothofagus*

cliffortioides is present i. e. *but for the wind there would be forest.* The following, nearly all high-mountain plants, are the principal species: — *Leptospermum scoparium* (dominant), *Dacrydium Bidwillii, D. laxifolium, Danthonia Raoulii* var. *rubra?, D. australis* (a species of the alpine belt), *Carpha alpina, Hypolaena lateriflora, Pimelea Gnidia* or an unnamed species, a species of *Dracophyllum* allied to *D. uniflorum, Cyathodes empetrifolia, Gentiana Townsoni, Donatia novae-zelandiae, Celmisia dubia* and *Senecio bellidioides.*

3. Grassland.

a. General.

Grassland is that great group of associations — the largest of the high mountains — in which grasses dominate but where, at times, herbs play an almost equal part. The group falls into the distinct ecological classes of **tussock-grassland** and **mat-grassland** — this quite limited in extent —, and the former naturally divided into **tall** and **low,** as already defined for the lowlands.

It is not feasible to supply definite figures regarding the floristic composition of grassland since, as will be seen, it is a matter of opinion whether to include tall tussock-herb field in herb-field or in the group under consideration. Therefore, I am giving a rather low estimate which omits those species belonging mainly to associations grasslike in physiognomy but herb-field in content.

The number of species admitted here is 249 which belong to 39 families and 104 genera, the largest being as follows: — (families) *Gramineae* 56 species, *Compositae* 43; (genera) *Danthonia* 14, *Celmisia* 12, *Poa* 11, *Acaena* 10, *Ranunculus* 9 and *Carex* and *Epilobium* 8 each — 4 hybrid swarms are included, but many are omitted, and rather more than half the species belong also to fell-field. The most common species are cited when dealing with the communities.

Grassland, though abundant on all mountains, is most in evidence where forest is wanting. At the present time, its area is much smaller than in primitive New Zealand, since burning and grazing have eradicated the grasses and left behind the non-inflammable and unpalatable species most of which belong to fell-field, the latter, as an indigenous-induced community having replaced the grassland. Where forest is absent, high-mountain grassland is merely a continuation upwards of that of the lowland-lower montane belt, the main difference in the former being the dropping-out of a few lowland species and the gradual, or sometimes sudden, incoming of true high-mountain plants. Grassland requires a more "fertile" soil than fell-field and attains its greatest development on a clay substratum with more or less humus on the surface; indeed, the tussocks themselves gradually supply the latter.

The amount of rain required is different for different classes of grassland but, *the higher the rainfall the larger the herb-field content.* The ordinary high winds of the mountains and extreme insolation are not antagonistic, but snow lying to a great depth and for a long period is more or less harmful.

The life-history of grassland has been discussed when dealing with lowland-lower montane grassland. In the upper subalpine and alpine belts, grassland appears to be a succession following fell-field dependant on increase in the average stability of the substratum. Once tussocks are established, their humus-making power comes into play and species other than those tolerating fell-field conditions can gain a footing. On the other hand, the tussocks themselves will increase in quantity and eventually make the entrance of further species difficult, the tussock-form being a dominant climax-form just as is the tree-form in forest. Nor are these two forms, so distinct in themselves, as ecologically different as one might think since both are strongly hostile to light-demanding species.

β. Low tussock-grassland.

General. This is a group of associations distinguished by the complete dominance of small tussock grasses $\pm$ 30 cm. high, together with a number of herbs and semi-woody plants and a few shrubs and ferns. There are several distinct groups of associations dependant upon the dominant tussock-grass.

The number of species &c. of low tussock-grassland is about 215 which is not far from the estimate for grassland in general, but it must be remembered that in that estimate many herb-field species were excluded.

The Festuca novae-zelandiae group of associations. This group, or possibly association, is distinguished by the dominance (usually) of *Festuca novae-zelandiae* and the presence — but not always — of *Poa caespitosa*, frequently in very limited amount, though occasionally dominant under special circumstances. This class of vegetation up to $\pm$ 600 m. altitude has been already dealt with.

High-mountain low tussock-grassland occurs to a limited extent in North Island on the Volcanic Plateau at an altitude ranging round 900 m., but, in South Island, it is abundant throughout on mountains where forest is absent and ascends upwards to 1200 m. or considerably higher. At the highest altitude it occupies sunny positions, those more shaded being clothed with tall tussock-grassland. At there is little, if any, of the community that is not more or less modified, the chief change being great reduction both in size and number of the tussocks, so that their ecological influence is lessened as well as there being far more open ground. In primitive New Zealand the tussocks would touch one another, much light being cut off from the ground and the light-demanding members of the association

— a considerable majority — be wanting or present only in small quantity[1]).

Coming now[2]) to the general structure and composition of the community, as the high-mountain plants enter in, the facies changes to some extent. Specially important in this regard is the abundance of yucca-like *Aciphylla Colensoi* with its yellowish, erect bayonet-like leaves, 40 cm. high, now dotted here and there, but in the primitive grassland forming dense thickets impassible to even horsemen. The low, green circular cushions of *Celmisia spectabilis,* 90 cm. diam. and the mats of *C. novae-zealandiae* or its allies also lend a distinct aspect to the plant-covering. Most of the species enumerated for the lowland lower-montane belt are present and, in addition, the following may be cited: — *Blechnum penna marina, Hypolepis Mille-folium* (where stony), *Lycopodium fastigiatum, Ophioglossum coriaceum, Koeleria novo-zelandica, Danthonia Buchanani* (E., NO., SO., F.), *Poa Kirkii, P. Lindsayi, Uncinia rubra, Phormium Colensoi* (moist or shaded places). *Chrysobactron Hookeri* vars., *Muehlenbeckia axillaris, Stellaria gracilenta, Ranunculus Monroi* var. *dentatus, Acaena inermis, A. microphylla* and their hybrids, *Viola Cunninghamii, Drapetes Dieffenbachii, Epilobium elegans, Hydrocotyle novae-zelandiae, Anisotome aromatica, Gaultheria depressa, Pernettya nana, Dracophyllum uniflorum* (dotted here and there and rising above the tussock), *Cyathodes Colensoi, Gentiana corymbifera* (abundant in many places and very handsome with its numerous large white flowers raised on a stout peduncle 30 cm. high), *Myosotis australis* (flowers yellow), *Hebe pimelioides* var. *minor* (stony ground, flowers blue), *Coprosma Petriei* (forming a continuous turf several sq. metres in area), *Celmisia novae-zelandiae* or its allies, *Helichrysum bellidioides, Raoulia glabra, R. subsericea, Cassinia Vauvilliersii, C. fulvida* var. *montana, C. albida* (SN., NE.), the hybrids between the last three (these shrubs in patches or dotted about) and *Taraxacum magellanicum.*

Poa Colensoi or P. intermedia (blue-tussock) grassland. On the mica-schist mountains of North and South Otago, at an altitude varying according to rainfall, the *Festuca* tussocks give place to the smaller ones of *Poa Colensoi* or *P. intermedia,* as the case may be, and probably to a con-siderable amount of those of *Agropyron scabrum.* In parts of the North Otago district such grassland ascends to the summits of mountains of medium

---

1) Experiments and observations have clearly shown that the number of species increases when the tussock is burnt and heavy grazing permitted, and that the species decrease when stock is excluded and burning no longer practised. The great masses of dead leaves which surround mature, unburnt tussocks are also hostile to the presence of other species. Further, various turf-making species — e. g. *Raoulia subsericea, Coprosma Petriei* — greatly extend their areas of occupation.

2) What is said regarding the ecological conditions of low tussock-grassland in the lowland-montane belt applies here also, so nothing is said here on this head.

height. On the flat-topped mountains of the South Otago district at about 1200 m. altitude a belt of *Poa Colensoi* gradually appears replacing the tall tussock-grassland and continuing upwards to the herb-moor of the narrow plateau above. But *Poa Colensoi* grassland, in certain cases is really indigenous-induced[1]).

Danthonia setifolia grassland. How wide-spread this class of grassland may be I do not know, nor whether it is primitive or induced. On Horseshoe Hill (comparatively wet part of NE.) at an altitude of 1200 m. *Danthonia setifolia* is dominant; amongst other common species present are: some *Danthonia Raoulii* var. *flavescens*, *Pimelea pseudo-Lyallii*, *Drapetes Dieffenbachii*, *Dracophyllum uniflorum*, *Coprosma ramulosa* and *Cassinia albida*.

γ. Tall tussock-grassland.

General. Tall tussock-grassland is distinguished by the strong dominance of one or more of the varieties — several unnamed — of *Danthonia Raoulii*, together with a more or less large content — in one class very large — of shrubs, semi-woody plants and herbs.

As tall tussock-grassland grades, on the one hand, into low tussock-grassland and fell-field — yet still with one or other tall tussock dominant — and as, on the other hand, if strong dominance of a life-form counts, the transition to herb-field — rich in species of that formation — must be included, it becomes evident that tall tussock-grassland must contain pretty well *all* the species belonging to the three communities. Any statistics then regarding the species as a whole would be superfluous and misleading.

Tall tussock-grassland is common in both islands and there is a hint of its presence in Stewart Island. At the present time it is most common in the subalpine belts, but originally in the drier mountains it descended to the montane belt, nor can it really be separated from lowland tall tussock-grassland, except for convenience. Burning, however, quickly spells the death of the large tussocks and the smaller tussocks take their place, such replacement making a community not to be distinguished from one that is virgin, except by its history.

The tall, dense tussocks, sometimes 1.4 m. high, when growing close together, as they do in certain virgin associations, forbid the presence of a large majority of the species, so that an association may consist of little else than the tall tussocks (Fig. 86). But, as the substratum is never homogeneous, many spots — too dry, too stony &c. — are unfavourable for the tussocks but favour various

---

1) At Cass (E.) adjacent to the Biologic Station burning has led to the dominance of *Poa Colensoi*. H. H. ALLAN (1926 : 81) describes as indigenous-induced a *P. Colensoi* association for Mount Peel (E.) caused by burning the *Celmisia Lyallii* fell-field of 1500 m. altitude, the *Celmisia* not being able to regenerate after burning, but the *Poa* regenerating quite well. On the other hand, *Celmisia spectabilis* — well able to tolerate burning — becomes more plentiful.

herbs, semi-woody plants and shrubs. Then, erosion is always playing its part and making places suitable for species to enter in from neighbouring rocks, fell-field, herb-field &c. Some of such species will be robust enough to defy the tall tussocks, others of creeping habit can rapidly occupy the ground and others may assume a lianoid form (e. g. certain species of *Acaena)* and so become firmly established.

The community falls into two distinct classes dominated respectively by *Danthonia Raoulii* var. *rubra* or var. *flavescens* or its close allies. *D. crassiuscula,* a smaller tussock than the above, may also be dominant, but its presence usually denotes herb-field rather than tussock-grassland.

Red tussock (Danthonia Raoulii var. rubra) grassland. This is distinguished by the dominance of *Danthonia Raoulii* var. *rubra.* It occurs in bot North and South Island. The lowland associations of South Otago and Stewart Island really come here.

The association of the Volcanic Plateau is no longer virgin, wild horses and cattle having grazed on it for many years and more recently rabbits having become common, nevertheless, the facies is still primitive. The association covers much of the pumice-scoria soil on the Volcanic Plateau at an altitude of from 900 to 1200 m. The tussocks some 75 cm. high usually a metre or more apart, but in places they touch. *Celmisia longifolia* (silvery in colour) is abundant. Erect, dark-coloured, fastigiate bushes of *Dracophyllum subulatum,* 45 cm. high, affect the general physiognomy. The following also play a notable part: — *Lycopodium fastigiatum, Hierochloe redolens, Danthonia setifolia, Poa anceps, P. caespitosa* (probably distinct from the common South Island variety), *Gaultheria depressa, G. perplexa,* many *Gaultheria* hybrids, *Leucopogon Fraseri, Cyathodes Colensoi, Pentachondra pumila* (bluish-green patches), circular mats of *Coprosma depressa,* the beautiful *Euphrasia tricolor,* low cushions 90 cm. diam. of *Celmisia spectabilis* and rooting, flat glaucous mats of *Celmisia glandulosa* var. *rera.*

In South Island red-tussock grassland apparently denotes a colder or wetter situation than low tussock-grassland. It appears to be essentially a community of the montane or, at most, the lower subalpine belt. In the South Otago district it may be continuous with the tall grassland of the lowland and montane belts but further north it rarely descends much below 900 m. altitude. *Sphagnum* bog is readily occupied by red-tussock and so converted into grassland. In South Otago, *Herpolirion novae-zelandiae* and *Oreostylidium subulatum* are abundant.

Snowgrass (Danthonia Raoulii var. flavescens) grassland. The snowgrass group of associations is distinguished by the complete dominance of one or more of the jordanons[1]) of *Danthonia Raoulii* var. *flavescens,* all of which have broader leaves than var. *rubra.* When the herb-content

---

1) To accurately ascertain the status of the many forms of *D. Raoulii* var. *flavescens* will be a matter of considerable difficulty involving much close study both in

becomes excessive *there is a community which is intermediate in character between tall tussock-grassland and herb-field,* but, for reasons given later, it is dealt with as a part of the herb-field community.

Snow-grass associations are extremely common in South Island and appear in their greatest strength on the wet mountains, but much vegetation where *D. Raoulii* var. *flavescens* is abundant must be referred to herb-field. On the dry mountains, and even in the fairly wet South Otago district, much of the community has been wiped out by fire and the large area remaining is usually greatly modified. Originally in South Otago, as may be seen even yet, snowgrass associations descended nearly to sea-level.

On the driest mountains (NO., NE.) the community is not wanting. For instance, on *the Inland Kaikouras,* there is more or less grassland of this type which is specially distinguished by the presence of numerous, immense, globose, bright-green plants of an unnamed variety of *Aciphylla squarrosa* together with low cushions of *Celmisia spectabilis* and some patches of *Carmichaelia Monroi.*

On Mount Peel (E.), a rather wet mountain, H. H. ALLAN (1926: 76—79), shows that in the lower part, the community contains much the same species as low tussock-grassland but, on the shaded slopes, it is more or less of a herb-field character.

On the range separating the Ahuriri Valley from Lake Hawea (E.) the snowgrass association commences at about 1200 m. altitude and has almost the same character up to 1540 m., when *Celmisia Lyallii* puts in an appearance.

In the South Otago district the association of the sunny side of *Walter Peak* remains purely grassland up to the final rocks, the herb-field element being virtually absent. But, on the shaded side of the mountain, where, from the end of April till the beginning of August, there is no direct sunshine, at less than 900 m. altitude the association contains a good many high-mountain shrubs and herbs. On *the Hector Mountains* the snow-grass belt begins at about 1100 m. altitude. In many places the tussock is extremely dense. *Aciphylla Colensoi* is abundant; there is a good deal of *Chrysobactron Hookeri.* The following species are fairly common: *Deyeuxia*

---

the field and the garden. In the first place, comes in the question of epharmony. For instance, is the form of the Mineral Belt an epharmone or a jordanon? A plant I have had for many years in cultivation supports the belief that it is a jordanon, but this only genetic experiment can solve. And this case is one of comparative simplicity. Frequently one or more of the jordanons of var. *flavescens* grow side by side with var. *rubra* and, in such cases, there is a swarm of hybrids. Possibly, too, *D. Cunninghamii* is involved in such hybridity. Moreover, jordanons of *D. flavescens* may be ecologically different, especially in the degree of rolling of the leaf; i. e., the associations cannot all be similar in their ecological demands. Cases of this kind occur again and again in the New Zealand flora, and probably in all floras, and, in no few examples **it is hopeless to really understand the ecology of a community until the taxonomic status of its members is understood** — a matter almost invariably neglected by plant ecologists.

avenoides, Agrostis Dyeri, Poa intermedia, Festuca novae-zelandiae, Gentiana corymbifera (or an allied species), *Wahlenbergia albomarginata, Celmisia coriacea, C. Lyallii* and *Helichrysum bellidioides.* At its upper limit (1320 m.) the *Danthonia* tussocks become scattered and *Poa intermedia* becomes dominant while many of the higher-mountain plants put in an appearance.

δ. Mat-grassland.

Mat-grassland as dealt with here includes **mat-grassland proper** where the grass extends in a continuous thick sheet over the ground together with those communities where tussock is absent and the grass either forms a continuous close turf or consists of flat mats or low cushions growing either closely or at some distance apart.

The carpet-grass (Danthonia australis) association. This remarkable association, together with its subassociations, is distinguished by the dominance of *Danthonia australis*[1]) which forms a continuous carpet upon which grow (exist) many fell-field species. It is confined to the alpine belt of the North-western, the wettest part of the North-eastern and the north of the Eastern districts, where it may cover the mountain-side for many square metres, or, where the conditions are unfavourable, make merely small patches.

The mats form a slightly hummocky dense covering to the ground quite slippery when walked on. The closeness of growth is antagonistic to other plants, the most important of which are flat cushions of *Celmisia spectabilis, C. viscosa, C. Lyallii* and *C. discolor* according to the locality. Here and there are a few tussocks of *Danthonia Raoulii* var. *flavescens* or in some localities *D. crassiuscula.* Where there is a little open ground a few fell-field species, e. g. *Pimelea pseudo-Lyallii, Hebe lycopodioides, Aciphylla Monroi, Celmisia laricifolia* &c., may be present, but are not really a part of the association.

Probably the association is a climax succession following fell-field, and the species in part are gradually dying out, *Danthonia australis* being most aggressive and changing its form and becoming lianoid can kill and replace vigorous shrubs.

Needle-grass (Poa acicularifolia) association. This is distinguished by the strong dominance of *Poa acicularifolia.* This grass is much-branching and forms close flat leafy cushions, or patches, 7 to 15 cm. diam. The leaves are 4 to 8 mm. long, involute, curved, rigid, smooth and sharp-pointed.

---

1) *D. australis* forms dense mats 5 cm. or so thick. Beneath, there is much dead grass. The culms and leaves are flattened to the ground and all point one way. The stem is creeping, slender, much-branched and covered thickly for 10 cm. or more with old leaf-sheaths. At this point, quite short roots are put forth, but longer ones, 6 to 7 cm. in length pass off from the base of the leafy stems. The leaves have broad sheaths almost equalling the blades. The latter are stiff, hard, wiry, closely involute, smooth, polished, 2.5 to 5 cm. long, needle-like, and rather sharp-pointed.

So far this grass is known only from the Mount Arthur Plateau (NW.), certain lowland limestone hills in the east of the North-eastern district, the Mineral Belt of the Dun Mountain (SN.) and the Trelissick Basin of the Upper Waimakariri (E.), where it occurs between 600 and 900 m. altitude. The following description refers only to the area last mentioned.

The closeness of the grass cushions is in proportion to the stoniness of the ground, the plant itself thriving on consolidated stony debris. Near shingle-slip various species of that association, as described earlier on, are present. In the body of the association many species of low tussock-grassland occur; *Plantago spathulata* is very plentiful.

Mountain-twitch (Triodia exigua) association and its allies. *Triodia exigua* is a small turf-making grass spreading extensively by means of its long, much-branching rhizomes which form a matted tangle. The leaves are narrow, about 2.5 cm long, filiform, stiff and almost pungent. Where dominant, there is a close, dense sward looking almost as if regularly mown. Certain other quite low-growing plants, mostly with creeping underground stems, grow in this association, e. g. *Triodia pumila, T. Thomsoni, Carmichaelia nana, C. Enysii, C. uniflora* (one or other of the three), *Leucopogon Fraseri, Wahlenbergia albomarginata* and *Gentiana serotina* (E.). The association occurs from the North-eastern to the North Otago districts.

Where the ground is particularly stony one or other of the species of *Carmichaelia* noted above is dominant forming broad patches of short, flat, erect green stems. There may also be present more or less prostrate *Discaria toumatou*, round mats of reddish-leaved *Acaena inermis*, the close turf-making semi-woody *Coprosma Petriei, Wahlenbergia albomarginata*, low, flat, silvery cushions of *Raoulia lutescens* and probably *R. australis*. The association occurs throughout the drier mountain valleys of South Island.

In the upper Clutha Valley (NO.) there is an ecologically-similar community with the usually coastal *Zoysia pungens* dominant, but this is at best a lower montane association.

Poa pusilla in certain parts of the montane and lower subalpine Western district forms a fairly close turf with which may be associated *Carmichaelia uniflora, Asperula perpusilla, Mazus repens* and *Pratia angulata*.

4. Herb-field.

a. General.

Herb-field is distinguished by the abundance of various large herbs — some more or less mesophytic — which may either dominate, or tall tussock grasses or even shrubs be the dominant life-form. The vegetation is generally closed but open herb-field must also be distinguished where the composition is virtually that of closed herb-field.

Though it seems best to treat the community as defined above, it should certainly be divided into the following classes: (1.) **Herb-field proper,** where herbs and low shrubs dominate and tall tussock, if present, plays a very small part, (2.) **shrub herb-field,** where tall shrubs are conspicuous, the herbs filling the spaces between; (3.) **tall tussock herb-field,** where *Danthonia Raoulii* var. *flavescens* dominates but the var. *rubra* may be important and probably *D. ovata;* and (4.) *D. crassiuscula herb-field,* where that grass is dominant. To give an account of herb-field on these lines, though most desirable, simplifying, as it would, herb-field ecology, is impossible at the present time, no one having made the necessary studies. All the same these terms are used to some extent.

The number of species for the whole formation including only 10 hybrid swarms (there are at least 35) is 308, about 68 per cent of which belong also to fell-field. They belong to 35 families and 91 genera of which the following are the largest: — (families) *Compositae* 68, *Gramineae* 30, *Umbelliferae* 29, *Scrophulariaceae* 27; (genera) *Celmisia* 43, *Aciphylla* 15. *Ranunculus* and *Gentiana* 13 each, *Ourisia* 9, *Danthonia, Epilobium, Hebe* and *Euphrasia* 8 each and *Coprosma* and *Anisotome* 7 each. As so many of the species are cited in what follows, to supply a list of the most common would be superfluous.

Herb-field occurs in North Island on the Dividing Range from the highest peaks of the East Cape mountains southwards (EC., RC.), on Mount Egmont (EW.) and, to a limited extent, on the central volcanoes; and, in South Island, on both sides of the Southern Alps subject to the western downpour together with the mountains of the North-western district, many in the South Otago district, and those of Stewart Island, but there, for the most part, the community is that form I am calling „Herb-moor". The vegetation of snow-patches in fell-field is closely related to herb-field.

The formation, especially herb-field proper, occurs on all the high mountains subject to frequent rain and where the climate is less favourable for disintegration of rock and soil than on the dry mountains but much more favourable for the making and accumulation of raw humus or peat. Where fully developed, the members grow closely and there is no bare ground, but there is every transition in this regard from herb-field to fell-field.

As already explained, it is not easy to draw the line between herb-field and tall tussock-grassland, for the two imperceptibly merge one into the other. The degree of steepness of the ground is apparently the main selecting-factor the steeper the slope, so long as humus can accumulate, the more favourable is it for tall grassland while, the more level the ground, the richer the herbaceous content. Thus, herb-field proper attains its richest development on the various passes of the Southern Alps (900 to 1500 m. altitude); on the basin-like beds of corries; on the floors and lower slopes of glacier-cirques and on the more gentle mountain slopes in general. Throughout the greater part of the western Southern Alps the vegetation was until recently absolutely virgin, but it is now modified, in places, by the destructive action of red-deer, chamois, and other introduced mammals.

The special soil-conditon, a surface of peaty humus, has been brought about by the plants themselves. Although frequently sopping wet, cessation from rain for a few days, a quite common occurrence, brings about a striking change for the plants, which is met by their xerophytic „adaptations".

The composition and structure of the associations stand in close relation to the average amount of water in the soil, so that the formation may be divided into **wet** and **dry,** the soil conditions being reflected in the composition of the vegetation. Altitude is not really a factor of much moment, for the formation at its most luxuriant development, so far as the size of the herbaceous plants is concerned, occurs at sea-level in the Lord Auckland Islands and herb-moor at the same altitude in Stewart Island. Also, at 300 m. altitude or thereabouts, well developed herb-field is present in some of the narrow valleys of the Fiord district where forest has either been destroyed or never been established owing to the frequent snow avalanches.

Coming now to the life-forms of the 308 species they include 47 shrubs, 147 herbs, 65 semi-woody plants, 40 grass-like, 3 rush-like and 6 ferns. Cushion-plants number 12, summergreen plants 17, herbs &c. grasses and ferns above 15 cm. high 54, plants with very hairy or tomentose leaves 59, and *nearly all the shrubs are prostrate* though the usual form may be erect. Speaking generally, the formation, especially dry herb-field, is less xerophytic than fell-field, indeed certain characteristic members are mesophytes distinguished by large, glabrous leaves *(Ranunculus Lyallii, Ourisia macrocarpa* vars. *calycina* and *cordata, Anisotome Haastii).* Even species with marked xerophytic structure can develope leaves of considerable size *(Celmisia coriacea,* 60 × 7 cm. *Aciphylla maxima,* 2 m. long).

The physiognomy of herb-field is dominated over considerable areas by the extraordinary abundance of various species of *Celmisia* (Figs. 83, 89, 90), their more or less silvery leaves and, in season, multitudes of white flowers (Fig. 89) being very conspicuous. Then come the great *Ranunculi,* sometimes covering several hectares at a time, or areas — equally large — may be a glory with the great snowy flowers of *Senecio scorzoneroides.* Many shrubs add to the variety with beauty of form and numerous blossoms while, ecologically, they provide shade and shelter; in short, the most beautiful herbaceous plants of the New Zealand Region, those of the Subantarctic province excepted, occur in subalpine herb-field, and, when in full bloom, the highest slopes of the Ruahine-Tararua Mountains, or the virgin passes of the Southern Alps, where alpine and subalpine vegetation mingle, are lovely natural gardens (Figs. 84, 89 and 91).

Dominance in a minor degree is exhibited by the presence in quantity of other plants. Thus, one or other of the large species of *Astelia* may by abundant (Fig. 91) or giant *Aciphyllae* the striking feature. Altitude brings in one or more of the cushion-forms of *Celmisia*, (Fig. 92) and the species as a whole become smaller, while the special grass will be *Danthonia crassiuscula*. In fact, the large mesophytes are replaced by xerophytes or subxerophytes of the same genera, e. g. *Ourisia macrocarpa* by *O. sessilifolia*. But little can be said here concerning the smaller modifications of the plant-covering, all that is attempted being a quite general account of the communities, laying stress, as far as my knowledge goes, on what seems fundamental.

β. *North Island dry herb-field.*

Mounts Te Moehau (T.) and Hikurangi (EC.) Te Moehau (725 m. alt.) is the northern outpost of the true high-mountain plants (ADAMS, J., 1889: 39—40) where upon the open ground appears to be a small patch of herb-field, containing *Danthonia setifolia?*, *Oreobulus pectinatus*, *Carpha alpina*, *Pentachondra pumila*, *Cyathodes empetrifolia*, *Ourisia macrophylla* and *Celmisia incana*. No more herb-field vegetation occurs until the summit of Hikurangi (1704 m.) 297 km. distant southwards is reached where some 44 high-mountain species occur, including — of a herb-field character — all the above and the special high-mountain species *Ranunculus insignis*, *Gentiana bellidifolia*, *Celmisia spectabilis*, *Raoulia grandiflora* and *Leucogenes Leontopodium*.

The Tararua Mountains. The striking feature of the association is the enormous quantity of *Astelia Cockaynei* in clumps 2.5 m. × 1.9 m. or as a continuous covering for many square metres at a time. In many places *Danthonia Raoulii* vars. *rubra* and *flavescens* dominate. The circular clumps of *Ranunculus insignis* are characteristic, the green leaves contrasting with the equally abundant silvery sheets of *Leucogenes Leontopodium* (Fig. 63). The following conspicuous plants are abundant: *Chrysobactron Hookeri*, *Aciphylla conspicua*, *Anisotome dissecta* (rather tall with finely-cut leaves), flat circular cushions of *Celmisia spectabilis* and *C. hieracifolia* (Fig. 83) and the much smaller *C. oblonga*. Certain subalpine shrubs appear in the association, e. g. *Coprosma pseudo-cuneata*, *Hebe evenosa* but the following belong properly to the community: *Gaultheria depressa*, *Dracophyllum pronum*, the cupressoid *Hebe Astoni*, prostrate or rounded bushes 30 to 60 cm. high of the *Daphne*-like *Pimelea Gnidia* (Fig. 89). The following are the more or less important of the smaller grasses and herbs: *Poa anceps*, *Triodia australis* (sometimes turf-forming), *Ranunculus geraniifolius*, *Oxalis lactea*, *Epilobium chloraefolium* var. *verum*, *E. Cockaynianum*, *Veronica diffusa*, *Ourisia caespitosa* and *Euphrasia tricolor*. The association consists of about 53 species.

Mount Egmont. Herb-field[1]) proper passing into fell-field at the highest

---

1) By the settlers the association is called the *Moss*, so that one might think flowering-plants were absent. Certainly large white hummocks — greyish when dry —

altitude, occurs in gullies above the scrub-line where snow lies much longer than on the slopes which are occupied by the tall tussock herb-field described further on.

The substratum consists of scoria with a certain amount of humus on the surface. The rainfall is excessive and but for the porous soil the surface would be boggy.

The association consists of a more or less close turf dotted here and there with taller plants either singly or in small groups. The turf consists principally of the following: — *Hymenophyllum multifidum, Lycopodium fastigiatum, Danthonia setifolia, Anisotome aromatica, Gaultheria depressa, Coprosma repens, C. ramulosa, Forstera Bidwillii* var. *densifolia, Celmisia major, C. glandulosa* var. *latifolia, Raoulia glabra, Helichrysum alpinum* and *Cotula squalida*. Everywhere in January the plant-covering is marked by the beautiful white whorls of flowers of *Ourisia macrophylla*, 27 cm. or so high rising from the rosettes of pale-green, purple-edged leaves. *Ranunculus nivicola*, with its large golden blossoms, also gives special stamp to the association. In places, silvery patches of *Celmisia major*, its narrow leaves recurved, are physiognomic. Other species rising out of the general groundwork are brown bushes of *Dracophyllum filifolium* (12 cm. high), flat-topped yellowish cushions of *Cassinia Vauvilliersii* (17 cm. high, 27 cm. diam.) and bright-green *Hebe buxifolia*.

γ. South Island dry herb-field.

It would be impossible, even were anything approaching exact details available, to set forth in a short general account of South Island dry herb-field proper the relation of this or that dominant or subdominant species to topographical conditions, soil, climate &c. Suffice it to say that in some places shrubs dominate, in other places tussocks dotted about or in small groups are subdominant, in others *Phormium Colensoi* or again in others species of *Celmisia* or *Ranunculus* are physiognomic.

The most important, as a rule, are those species of *Celmisia*[1]) with large, erect rosettes and great white heads of blossom, which, constantly form colonies more or less pure. Above all *C. coriacea* in one or other of its

of *Rhacomitrium pruinosum* on certain parts of the mountain give a characteristic aspect to the scene. These hummocks are dense, rounded and average about 70 cm. in length to 60 cm. in breadth and 10 to 12 cm. in depth. Frequently, two or more of the cushions, becoming united, form amorphous masses, or they may be separated by the general carpet of herbs. Within a cushion the moss is dead and full of moisture. Usually various herbs grow on them, but eventually such are buried. On some parts of the mountain the cushions are absent, their representative being patches of a yellowish-green moss.

1) They fall into the two categories of those with **wide leaves** which in one class (*C. petiolata, C. Traversii* &c.) arch downwards at their extremities, and those with **narrow stiff leaves** (*C. Armstrongii, C. Petriei, C. Lyallii*).

form is the dominant species throughout South Island herb-field, its broad silvery leaves, frequently 30 cm. long, and the great white flower-heads, 8 to 12 cm. diam. render it extremely conspicuous (Fig. 87). Various parts of the Southern Alps possess particular species, and as these are by no means closely alike, they affect the physiognomy to some extent[1]).

The giant buttercups lighting up a hillside with their white or yellow blossoms are both of extreme physiognomic importance and beauty. *Ranunculus Lyallii* (Fig. 91) is the most striking and wide-spread occurring, as it does, from the Spenser Mountains to Stewart Island; it is particularly plentiful in the Western district. The great glossy, green peltate leaves spreading horizontally, raised on stout stalks 30 to 45 cm. high, cut off the light from the ground and forbid other plants to gain a footing, this being also assisted by the broad, half-buried, thick rhizomes paving the substratum. In the North-western district, *R. insignis* occupies a similar station, and in the Fiord district *R. Buchanani* makes wide colonies, but usually only at a higher altitude and on more stony ground, a happening in accord with its more xerophytic structure. When growing in company with *R. Lyallii* great variety is lent to the community by the polymorphic hybrid swarm between the two, one individual of which has been described as a species under the name *R. Matthewsii*.

Astelia Cockaynei and *A. Petriei* are common in many places. The latter with its shorter, more rigid, glossy leaf ascends to the alpine belt and completely fills snow-patch hollows; the smaller *A. monticola* (W., F.) behaves similarly.

The large species of *Ourisia (O. macrophylla* — NW., *O. macrocarpa* var. *calycina* — W. and var. *cordata* (F.) and *O. Macphersonii* (F.), in places, are almost as striking as the *Ranunculi*. The fern-like tufted leaves of *Anisotome Haastii* are a feature in herb-field generally. Various moderately tall species of *Aciphylla* are of moment at times, e. g. *A. crenulata* (W.), *A. Hookeri* (NW.), *A. Lyallii* (F.). The mat and cushion (Fig. 90) species of *Celmisia* are of prime importance. One or more of the *C. discolor* group and *C. sessiliflora* (silvery cushion) occur throughout, but the following are of more or less restricted range: — *C. dubia* (NW.), *C. Dallii* (NW.)[2]),

1) *C. Traversii* with rusty tomentum and leaf margined with same is characteristic of the northern part of the Southern Alps and neighbouring mountains; it occurs also in the Fiord district. *C. Armstrongii*, its leaves stiff, sword-like and with a conspicuous orange midrib is the plant par excellence of the Westland side of the Southern Alps. *C. verbascifolia, C. holosericea* and *C. Petriei* are characteristic of the Fiord district, the latter has dagger-like leaves, green and glossy above, satiny-tomentose beneath and the margins much recurved. *C. lanceolata*, intermediate between *C. coriacea* and *C. Armstrongii* is common in the southern half of the Fiord district.

2) To J. A. MCPHERSON (Director of Parks and Reserves, Invercargill) I am greatly indebted for particulars regarding the herb-field proper of Mount Glasgow (1400 m. — coastal range of NW.) — hitherto undescribed. As about 1100 m. altitude the subalpine-

C. Walkeri (W., SO., F. — Fig. 88), *C. rupestris* (NW.), *C. Gibbsii* (NW.), *C. Hectori* (south of W., SO., F.), *C. ramulosa* (SO., F.). When in flower, species of *Gentiana* are conspicuous, e. g. *G. bellidifolia, G. patula, G. divisa, G. Townsoni* (NW.), *G. montana* (NW., W., F. — but records doubtful), and other badly-understood species of one or two localities. Some of the mat-forming spcies of *Ourisia* are striking when in flower, e. g. *O. caespitosa, O. Cockayniana* (W., F.), *O. prorepens* (SO., F.) and *O. glandulosa* (NW., SO., F.),

Certain subalpine shrubs invade the associations especially: — *Phyllocladus alpinus, Hebe subalpina, Coprosma pseudocuneata, C. ramulosa, C. crenulata, C. serrulata* (spreading widely by means of underground stems), *Gaultheria depressa, G. rupestris, Pentachondra pumila* and *Carmichaelia grandiflora.*

The following — equally important — are of restricted distribution: — *Dracophyllum Menziesii* (F.), *Hebe vernicosa* in various forms (SN., NW., north of NE.), *Coprosma depressa* (NW., north of W. — but records untrustworthy) and *Senecio revulutus* (SO., F.).

The general carpet of the community is a striking feature where the taller plants are not greatly in evidence or it may merely fill intermediate spaces between them. The following are more or less common throughout: — *Hymenophyllum multifidum, Blechnum penna marina, Lycopodium fastigiatum, Hierochloe Fraseri, Agrostis Dyeri, Deyeuxia pilosa, Deschampsia Chapmani, Trisetum antarcticum, Danthonia setifolia, Poa pusilla, P. caespitosa, P. Cockayniana, P. Colensoi, Festuca* one or more unnamed species, *Uncinia compacta, U. fusco-vaginata, Carex dissita* var. *monticola, Astelia linearis, Ranunculus foliosus, Caltha novae-zelandiae, Geum parviflorum, G. uniflorum, Acaena Sanguisorbae* vars., *Geranium microphyllum, Coriaria lurida, C. angustissima, Viola Cunninghamii, Epilobium chloraefolium* var. *rerum, Aciphylla similis, Anisotome aromatica, Angelica decipiens, Euphrasia revoluta, E. zelandica* and other species according to locality, *Plantago Brownii, Wahlenbergia albomarginata, Phyllachne Colensoi, P. clavigera, Forstera sedifolia, F. Bidwillii, Gnaphalium Traversii, Helichrysum bellidioides, Craspedia* various allied species, *Cotula pyrethrifolia, Senecio bellidioides, S. Lyallii* and *Taraxacum magellanicum.*

---

forest gives out and is fringed by an extremely narrow belt of *Olearia Colensoi* and *Phormium Colensoi* passing through which the herb-field at once commences. This association is distinguished by the extreme abundance of *Celmisia Dallii* accompanied by large quantities of *C. Armstrongii* and *Anisotome Haastii* and various species of *Hebe* are dotted about, e. g. *H. vernicosa, H. buxifolia* and their hybrids. There are many forms of *Senecio Bidwillii* including some with toothed leaves (perhaps hybrids with *Olearia arborescens*). Other species of moment are *Aciphylla Hookeri, A. polita, Gentiana montana, Ourisia. macrophylla, Hebe macrantha* and a cross or crosses with *H. buxifolia* and *Celmisia incana.*

δ. Tall tussock-herb field.

The North Island communities. Most likely there is herb-field of this character on all the high mountains, but its nature on Mount Egmont is beyond dispute. *Danthonia Raoulii* var. *rubra* is dominant and between the tussocks there is a close turf made up of all the neighbouring herb-field species, as already described. The association occupies the slopes, whereas herb-field proper is confined to the gullies, the controlling factor as to which kind of herb-field shall occupy the ground being the length of time this is covered by the winter snow, the tall tussock-herb-field being far longer free from snow.

The South Island communities. In South Island the communities are dominated by *Danthonia Raoulii* var. *flavescens,* or it may be *D. crassiuscula* or perhaps *D. ovata* (F.). Most, if not all the herb-field proper species of the particular locality will be represented. Seen from a distance the community resembles tall tussock-grassland proper, but a close view reveals the large herbaceous plants *(Ranunculus, Celmisia)* and various shrubs. Though the group of associations stands intermediate between herb-field proper and tall tussock-grassland it is no mere tension-belt (ecotone), for it occupies wide areas where herb-field proper may be absent.

Tall tussock-herb field occurs principally in those South Island mountains exposed to an excessive rainfall. A familiar instance is the community near Mount Cook (east of W.) which may be the sole representative of herb-field from the scrub-line to the alpine fell-field. As for its ecological distribution I have not nearly sufficient data, but suggest *it may require a gentler slope and rather moister soil than tall tussock-grassland proper.* The presence of the herbaceous, semi-woody and shrubby members depends upon open spaces between the tussocks, but, those shade-tolerating as young plants, are the most likely to become established. The tall rosettes of various species of *Celmisia (C. Lyallii* and the great *C. Petriei* themselves tussocks), the shade-casting leaves of *Ranunculus Lyallii* and its massive rhizomes covering the ground, and the dense habit of certain shrubs, enable all such to bid defiance to the tussocks which, nevertheless, remain dominant.

As for the history of the community it may be a succession following herb-field proper and the herbaceous-shrubby element be the remains of the original association. On the other hand, tall tussocks and the herb-field element may have developed side by side as they increased the humus content of a preceding fell-field association, or, in places, the community may replace bog by way of wet herb-field. In any case, no matter its origin, at the present time tall tussock-herb field is apparently a climatic climax-community, as in the hanging valley near Lake Harris (F.).

There is no need to give details regarding the species making this class of herb-field, for all those of herb-field proper are concerned. In the alpine belt, however, the dominant grass is *Danthonia crassiuscula,* which is much smaller than *D. Raoulii* var. *flavescens*, and which may take on to some extent a mat-grass habit and so be somewhat hostile to other species. In this group of associations the alpine herb-field species are strongly represented.

ε. South Island wet herb-field.

This group of associations is usually distinguished by the dominance of the summergreen *Senecio scorzoneroides*, together with abundance of *Schoenus pauciflorus*, and *Hebe buxifolia* var. *pauciramosa* (a compound var.) is characteristic.

The presence of wet herb-field depends upon a more badly-drained substratum than that of dry herb-field and a more peaty soil. Frequently, streams carry away more or less of the surplus water so that the soil — though always wet — is not so "sour" as that of herb-moor. The flat ground of subalpine passes and the most depressed parts of corries and hanging vallies are the special site of wet herb-field.

In some localities, the *Senecio*, when in full bloom, renders extensive areas white as a snow-field or, if the yellow *S. Lyallii* is present the many hybrids will add flecks of yellow, pale-yellow and cream. Where there is no stagnation *Ranunculus Lyallii* and many dry herb-field species may be present, including the species of *Astelia*. The following may receive special mention: — *Dacrydium laxifolium, Hierochloe redolens, H. Fraseri, Deyeuxia pilosa, Danthonia crassiuscula, Carex ternaria, Chrysobactron Hookeri, Ranunculus foliosus, Geranium microphyllum, Viola Cunninghamii, Epilobium chloraefolium* var. *verum, Aciphylla Lyallii, Oreomyrrhis andicola* in a wide sense, *Dracophyllum prostratum, Gaultheria depressa*, various species of *Gentiana*, various species of *Euphrasia, Ourisia macrocarpa* vars. *calycina* and *cordata, O. caespitosa, Coprosma repens, Donatia novae-zelandiae* (Fig. 92), *Celmisia petiolata, C. verbascifolia, C. glandulosa* var. *vera, Craspedia uniflora* in a wide sense, *Helichrysum bellidioides, Microseris Forsteri* and *Taraxacum magellanicum*.

5. Herb-moor.

General. Herb-moor is here considered as a formation intermediate in character between wet herb-field and bog, which is distinguished by the absence of tall plants and the abundance of turf-forming, prostrate and cushion species.

The species number about 141 which belong to 31 families and 20 genera, the largest being: (families) *Gramineae* and *Cyperaceae* 14 species each, *Scrophulariaceae* 12, *Umbelliferae* 11; (genera) *Celmisia* 11, *Carex* 8 and *Raoulia* 6.

The following are characteristic South Island species: — *Carpha alpina, Oreobolus pectinatus, O. strictus, Astelia linearis, Hectorella caespitosa,*

Ranunculus Berggreni, certain small *Aciphyllae, Anisotome imbricata, A. la-nuginosa, Dracophyllum muscoides, Myosotis pulvinaris, Hebe dasyphylla, Ourisia glandulosa, Plantago lanigera, Coprosma repens, Phyllachne clavigera, P. rubra, Donatia novae-zelandiae, Celmisia ramulosa, C. brevifolia, C. sessili-flora, C. argentea* and the hybrids between the last two, *Raoulia Hectori* and *Cotula sericea.*

Herb-moor is confined to the flat-topped mountains of the South Otago District, the Dunstan Mountains (NO.) and Stewart Island. The association called by me **winter-bog** (1908: 30) of the Volcanic Plateau is similar in various ways, and small areas of wet herb-field in the high mountains generally come into the ecological conception of the formation.

Herb-moor attains its full development on badly-drained, peaty ground exposed to a subantarctic climate. Thus, the long, narrow, plateau-like summits of those South Otago mountains at over 1800 m. altitude, lying within the full influence of the south-west wind, and where the winter snow lies until November or even the end of December, are ideal. Deep, wet peat; frequent driving rain or snow; violent gales; cloudy skies but, at times strong insolation, are the leading ecological factors. The base of the plants being clothed with a wet, sticky, peaty covering reaches about its maximum development. Evidently, the formation is closely related to bog, but certain bog species, including *Sphagnum*, are wanting and the substratum, though wet enough, in no part lies under water unless for a brief period.

The life-forms of the community are as follows: — shrubs 17 species, herbs 67, semi-woody plants 23, grass-like plants 30, rush-like plants 2, ferns 2. Cushion-plants number 30, but many other species at times assume this form and there are 48 mat-plants and, again others epharmonically become mats.

South Island herb-moor. On the narrow summit-plateau of the Old Man Range (SO.), herb-moor has attained its richest development. The winter snow lies well on into the summer and in shady places may not melt. The plateau is frequently swept by the full fury of a south-west gale, rain-bringing or snow-bringing, but always piercingly cold. The peaty surface of the ground is cut into hummocks separated by basin-like depressions about 38 cm. deep and 60 cm. across which are united into a more or less connected trench, so that the plateau is traversed by parallel trenches. The hummocks are rounded, about 90 cm. across, and usually densely covered with vegetation which extends, though different in composition, into the basins.

The following species, all of low stature, are common on the hummocks: — *Hierochloe Fraseri, Agrostis subulata, Poa Colensoi* (greatly dwarfed), *Ranunculus novae-zelandiae, Drapetes Lyallii, Anisotome lanuginosa, Dracophyllum muscoides, Hebe Hectori* (cupressoid), *H. dasyphylla, Phyllachne Colensoi, P. rubra, Celmisia argentea, C. brevifolia, C. viscosa, Craspedia uniflora* in a wide sense, *Raoulia grandiflora, R. Hectori* and *Senecio belli-dioides* (glabrous form).

In the basins the small rosettes of *Ranunculus Berggreni* are abundant; in shallow hollows where snow lies long are wide breadths of *Celmisia viscosa* and *C. Haastii,* as on South Island mountains in general. Where the wind attains its maximum velocity all vegetation may be destroyed except the rigid low, dense cushions of *Dracophyllum muscoides.* Natural monoliths of micaschist stand up here and there — one the "Old Man" from which the range gets its name — and on the debris near them there are many silvery cushions of *Raoulia Hectori,* also small cushions of *Myosotis pulvinaris* and dense *Dracophyllum muscoides.*

On *Mount Pisa Plateau* (1500 to 2100 m. alt.) much the same community as just described is present, except, that over wide areas the prostrate reddish-brown branches of a species of *Dracophyllum* (possibly prostrate epharmonic *D. uniflorum*) are physiognomie. Much of the ground is covered with pebbles of quartz; also, it is traversed by streams and much better drained than the Old Man Range.

The *Rock and Pillar Range* (1430 m. alt., SO.) has much the same association as described above, but the species are rather fewer, and in full strength occur on consolidated flattish stony debris. Maungatica (SO., 900 m. alt.) likewise carries herb-moor but with fewer species. *Celmisia coriacea* var. *stricta, C. argentea* and *Dracophyllum prostratum* are common.

Stewart Island herb-moor. This association is distinguished by the abundance of *Danthonia pungens, Dracophyllum rosmarinifolium* (its mat and turf forms), *Astelia linearis, Carpha alpina* and *Microlaena Thomsoni,* and is made up of two subassociations — the one confined to granite and the other to mica - schist, but the relative steepness of the slope may be an important factor.

Various species of the lowland belt, though eminently suitable, are absent, e. g. *Blechnum penna marina, Gleichenia alpina, Schizaea fistulosa* var. *australis* (probably an epharmone), *Dacrydium laxifolium, Danthonia Raoulii* var. *rubra, Gaultheria perplexa, Ourisia Crosbyi* and *O. modesta.* Further, the following are wanting in South Island herb-moor: *Lycopodium ramulosum, Microlaena Thomsoni, Danthonia pungens, Chrysobactron Gibbsii, Aciphylla Traillii, Dracophyllum rosmarinifolium* (but in herb-field, F.), *D. Pearsonii, Gentiana lineata, Hebe Laingii* and *Raoulia Goyeni.*

As for the subassociations on the granite (Mount Anglem), the following are wanting: *Euphrasia Dyeri,* the so-called *Celmisia linearis* (most likely *C. argentea* × *longifolia*) and *Senecio scorzoneroides* and *Celmisia argentea* is rare, and the following are restricted thereto: *Schoenus pauciflorus, Uncinia compacta* var. *caespiciformis, Hebe Laingii, Ourisia prorepens, O. sessilifolia* and a mat - *Celmisia* of the *C. discolor* group (originally referred to *C. Sinclairii*).

Seen from a distance, the association presents a most even and smooth surface, but when walking on it, one soon perceives it to consist of a dense mass of plants, springy to the tread, into which the feet sink but which rise back into their former position. There is a ground-work of the dominant species, others entering in according to changes in water-content, and perhaps becoming dominant. The above ground-work consists of pale-green grass-like tufts of *Carpha alpina*, the leaf apices straw-coloured dead and twisted; close low cushions of *Donatia novae-zelandiae*, the green leaves with bright orange tips pressed together into small rosettes; *Dracophyllum rosmarini-folium* much like *Donatia* but the leaves more open and tipped with red; green cushions of *Oreobolus pectinatus* and growing through and mixed with the whole almost everywhere *Astelia linearis* and the small, flat, glaucous leaves of the grass *Microlaena Thomsoni*. Although the covering is made up so largely of cushions, that form is masked through their growing into one another. At less than one metre apart, everywhere project obliquely above the general mass tufts of the rigid, sharp-pointed pale-green or reddish-brown leaves of the ungrass-like grass *Danthonia pungens*. On the schist mountains, where the soil becomes drier, the general carpet is everywhere dotted with the silvery, dense cushions of *Celmisia argentea* (Fig. 90) and those more open, and greenish-tinged of *C. linearis* (its status already explained), which, growing together form conspicuous, silvery cushions.

Stewart Island herb-moor is in harmony with that semi-subantarctic climate which permits the establishment of what are really high-mountain communities at sea-level; indeed it commences at less than 600 m. altitude and barely reaches 900 m. at its highest point.

a. Bog, water, and swamp communities.

a. Bog associations.

General. Bog associations are those in which more or less *Sphagnum* is nearly always present, together with some of the following species: — *Gleichenia alpina*, *Dacrydium laxifolium*, *Carpha alpina*, *Oreobolus pectinatus*, *Carex Gaudichaudiana*, *Hypolaena lateriflora*, species of *Gaimardia*, species of *Drosera*, *Halorrhagis micrantha*, *Liparophyllum Gunnii*, *Phyllachne Colensoi*, *Donatia novae-zelandiae*, *Celmisia glandulosa* and *Olearia virgata*.

The number of species for the high-mountain part of the formation is 138 which belong to 33 families and 77 genera, the largest being *Compositae* 15 species, *Cyperaceae* 14 and *Epacridaceae* 9, but none of the genera exceed 6 species. Those most widely spread appear in what follows.

Bog occurs throughout all the high-mountain area wherever the drainage is insufficient to forbid water lying on the ground for considerable periods. Usually the bogs are small. The habitat is not governed principally by rainfall, for bogs occur both where such is heavy and where it is quite light; in the latter case, the bog-community depends upon the substratum being in close proximity to streams liable to flood or springs and shallow lakes. On the contrary, in an extremely wet climate, bogs may originate on flat ground distant from streams &c., for, in such a climate *Sphagnum* can become established on steep slopes and other unlikely places.

As to the life-forms of the species, there are 29 shrubs, 11 semi-woody plants, 65 herbs, 23 grass-like plants, 6 rush-like and 4 ferns. Most of the shrubs are prostrate, even if usually erect elsewhere, e. g. *Leptospermum scoparium, Olearia virgata;* cushion-plants number 11.

A bog-association may be a succession following swamp; primary succession culminating in tussock-grassland, herb-field or shrubland; or even a climax where the water-supply is constant. In some measure the permanency of the bog depends upon the ability of the sphagnum by its upward growth to bury the invading grasses, herbs and shrubs. If these grow faster than the moss the latter is doomed. Thus, small bogs are rapidly transformed into grass, herb and shrub associations, as in the case of one I have known on Jack's Pass (NE.) for some 25 years and noted getting gradually smaller and smaller. A closed community of *Celmisia coriacea* and *Astelia Cockaynei* in its vicinity was originally *Sphagnum* bog.

Various classes of bog. *Schoenus pauciflorus* bog is extremely common in the montane-subalpine tussock-grassland area of South Island and it occurs also to some extent on the Volcanic Plateau but as a somewhat different association. The reddish colour of the *Schoenus* tussocks in contradistinction to the brown grass renders the community conspicuous even from a distance. Although the bog is traversed by running streams, the ground is sopping wet. *Sphagnum* cushions are usually plentiful and growing on them certain small herbs, e. g. *Blechnum penna marina,* species of *Drosera, Geranium microphyllum, Viola Cunninghamii, Haloorhagis micrantha, Nertera Balfouriana* and *Celmisia longifolia* var., *Chrysobactron Hookeri* var. *angustifolia* is usually abundant.

Closely related to the above in the North-eastern and Eastern districts are associations where *Schoenus* is not dominant, *Sphagnum* abundant and shrubs present e. g. *Leptospermum scoparium, Gaultheria depressa, G. rupestris, Dracophyllum uniflorum, Hebe buxifolia* and *Cassinia Vauvilliersii. Astelia Cockaynei* is frequently an important feature. In many places, *Carex Gaudichaudiana* dominates. More or less *Danthonia Raoulii* var. *rubra* is usually present, but rather as an invader from the grassland surrounding the bog than as a true member of the community.

Cushion-bog, so-called from the abundance of that life-form in the associations, is common in the subalpine-belt of many South Island wet mountains. The bog is usually quite small. The special feature is the numerous, small, dense cushions of the following, some of which may be absent: — *Oreobolus pectinatus, O. strictus, Gaimardia ciliata, G. setacea, Phyllachne clavigera, P. Colensoi* and *Donatia novae-zelandiae.* In addition most of the following will be present: — *Gleichenia alpina, Dacrydium laxifolium,* bushes of *D. Bidwillii, Carex Berggreni* (SO.), *Carpha alpina, Hypolaena lateriflora, Astelia linearis,* species of *Drosera,* prostrate *Leptospermum scoparium, Plantago triandra, Pentachondra pumila, Cyathodes empetrifolia, C. pumila,* one or more species of *Gentiana, Celmisia longifolia* var. *alpina* or other variety and *C. glandulosa* var. *vera.* The extensive bog association of the upper Routeburn hanging valley is of this class and, in addition to most of the above are *Hebe buxifolia* var. *paucibrachiata* and *H. Hectori.* Sometimes, especially in open places in subalpine forest, prostrate *Dacrydium Bidwillii,* is physiognomic and stunted *Nothofagus cliffortioides* common.

Hypolaena-Gleichenia bog is another type. It greatly resembles the allied bog of the lowland belt. Bog of this character occurs on the Volcanic Plateau, but its presence depends upon abundant ground-water supplied by melting of the winter snow since the drainage is good. *Gleichenia alpina* is plentiful but the grass-like *Carpha alpina* with its pale-green leaves, withered and twisted at their extremities, is the special feature. There is abundance of *Oreobolus pectinatus, Viola Cunninghamii* and *Celmisia glandulosa* var. *vera.* In the same district where the conditions are truly boggy the dark-green *Liparophyllum Gunnii* forms close mats together with a turf of *Carpha alpina, Scirpus crassiusculus, S. aucklandicus* and *Gnaphalium paludosum. Utricularia monanthos* is extremely plentiful. Other common plants are *Danthonia Raoulii* var. *rubra, Juncus antarcticus, Carex ternaria, Aciphylla squarrosa, Gentiana bellidifolia, Coprosma repens* and *Craspedia minor.* Better drained, though constantly wet ground, brings *Hypolaena* and *Gleichenia* into dominance accompanied by abundance of *Hebe buxifolia,* and many fell-field shrubs much dwarfed are raised on drier mounds above the general level.

Hypolaena bog on the Rahu Saddle (NW.) is distinguished by the abundance of bushes of *Dracophyllum palustre* forming patches about 90 cm. long and about 60 cm. high of its slender twiggy, blackish, crowded branches; also *Cyathodes empetrifolia* is conspicuous and there is some *Phormium Colensoi.*

β. Water and swamp vegetation.

Water communities. Here are included associations of running and still water and those of flushes, but excluding those of dripping rocks. The species number only 11 which belong to 8 families and 9 genera.

Where streams flow swiftly *Algae* and perhaps a moss or two are present, their presence on stones marking the usual depth of the water. In winter many glacial streams, even of considerable size, are quite dry and during the rest of the year the amount of water varies greatly. *Montia fontana* is characteristic and often fills small, shallow brooks, even floating upon the surface or forming close cushions in shallow, fairly rapid streams (Fig. 93). *Epilobium macropus* — physiognomic from its large flowers — is common in shallow parts of slow-flowing subalpine brooks and in shallow water of depressions in river-beds, a station likewise of *Claytonia australasica.*

21*

In South Island on margins of montane streams reddish tussocks of *Carex Buchanani* grow in abundance. On the Volcanic Plateau and in many parts of South Island, *Gunnera dentata* forms close masses in very shallow water and on the wet ground adjacent, its leaves flattened to the soil excluding other plants. In many South Island shallow streams, there is a close growth of the slender-stemmed *Schizeilema nitens* distinguished by its small shining, trifoliolate, thin, glabrous leaves. *Ranunculus Cheesemanii* (NE., north of E.) is frequent in places where water generally lies. *R. depressus*, in a very wide sense, is common on still water. In fairly rapid brooks there is often a dense growth of *Myriophyllum propinquum*, the leaves submerged and on the water-surface float the brown leaves of *Potamogeton Cheesemanii*.

Soil constantly wet through water rising out of the ground is occupied by many species, of which the following are particularly common: — *Carex Gaudichaudiana, C. Petriei, Scirpus aucklandicus, Juncus novae-zelandiae, Claytonia australasica, Acaena saccaticupula, Oxalis lactea, Viola Cunninghamii, Epilobium pedunculare* var. *brunnescens, Pratia angulata, Plantago triandra, Gnaphalium Mackayi* and species of *Craspedia*.

The associations of lakes and tarns differ but little from those of the lowland belt. The same species of *Potamogeton* and *Myriophyllum* are present and the swamp species of the margin are lowland plants, e. g. *Heleocharis Cunninghamii, Cladium glomeratum, Carex subdola, C. secta, C. pseudocyperus. Isoetes alpinus* and *Pilularia novae-zelandiae*, submerged aquatics, are probably common.

Swamp. There is but little swamp of a high-mountain character. *Carex virgata* or *C. secta* are frequently dominant and *C. ternaria, C. stellulata* and *C. diandra* are extremely common. In the deepest water *Typha angustifolia* var. *Muelleri* dominates. On the margin, there is usually a bog community in which *Plantago triandra* and *Isotoma fluviatilis* may be plentiful.

Swamp occurs principally in montane valleys of South Island and owes its origin to frequent flooding of the ground by rivers and streams, especially during rapid melting of snow; where the drainage is insufficient, it is readily transformed into bog.

Section IV.
The Vegetation of the Outlying Islands.
Chapter I.
The Vegetation of the Kermadec Islands.

General. The species of vascular plants number 117 (pteridophytes 38, monocotyledons 25, dicotyledons 54) and they belong to 42 families and 89 genera, the largest families being *Filices* 33 species, *Gramineae* 13, *Com-*

positae 9 and *Cyperaceae* 6, but only 6 genera contain 3 species or more, the largest being *Asplenium* with 5 species. The elements of the flora are as follows: — *Endemic* 15 species (9 closely related to New Zealand species and 3 of Polynesian, 2 of Norfolk Island and 1 of New Hebrides affinity); New Zealand 89 of which 65 are Australian, but 45 of such occur in Norfolk Island or Lord Howe Island — one or both; and Australian (other than those also *New Zealand*) 7. With regard to the destribution of the New Zealand element in New Zealand proper it is as follows: North Island 89 (14 Auckland districts only), South Island 74 (17 extending but little beyond lat. 42), Stewart Island 41.

The general ecological conditions of the Kermadecs are a combination of a humid subtropical[1]) climate, coastal conditions, and an extremely porous soil. According to W. R. B. OLIVER (1910:125) from records from Feb. to March 1908 inclusive calm days are rare and the wind often blows with violence. The rainfall is fairly evenly distributed throughout the year (171 cm. spread over 176 days — Feb. to Oct.). In cloudy and rainy weather a mist continually hangs about the hilltops and its influence is reflected by a distinct forest association. The temperature ranged between a minimum (Aug. 4th) of 8.1° C. and a maximum (Feb. 1st) of 29.4° C. but there were hotter days in January.

The physiognomic plants. The following are the leading physiognomic plants all of which are either identical with, or closely related to mainland species: — *Metrosideros villosa* (Myrtac.), *Rhopalostylis Cheesemanii* (Palmae), *Suttonia kermadecensis* (Myrsinac.), *Ascarina lanceolata* (Chloranthac.), *Myoporum laetum* (Myoporac.), *Corynocarpus laevigata* (Corynocarp.), *Melicytus ramiflorus* (Violac.), *Coprosma petiolata* (Rubiac.), *Macropiper excelsum* var. *major* (Piperac.), the endemic tree-ferns *(Cyathea Milnei, C. kermadecensis)* and a number of herbaceous ferns, especially, *Pteris comans, Polystichum aristatum, Nephrolepis exaltata, Dryopteris glabella* and *Blechnum norfolkianum.* Only a few of the above need notice.

Metrosideros villosa is either a lofty tree 20 m. high or its massive trunk is more or less prostrate. In the latter case, trunk-like branches pass upwards from the prostrate trunk. Aereal roots are extremely abundant in both forms; in the erect tree they grow downwards, pass into the soil and function as stems; in the prostrate form they make entangled masses at the base of the trunk. The leaves are similar to those of *M. tomentosa* but smaller.

Rhopalostylis Cheesemanii closely resembles *R. sapida* but is taller. *Suttonia kermadecensis* is a low tree, 6 to 10 m. high with smooth reddish-gray bark and dense foliage of elliptic-oblong, dark-green, coriaceous leaves

---

1) According to CHEESEMAN (1888: 156—7) various tropical fruits or vegetables were grown by the settler who then, with his family lived on the main island, e. g. the banana, sugar-cane, pineapple, guava, custard-apple, mango, oranges, shaddocks and citrons.

and with the small flowers produced, in part, on the naked twigs. *Ascarina lanceolata* is a tall shrub or low tree much resembling *A. lucida,* already described. *Coprosma petiolata* is very similar to *C. chathamica;* it varies from a quite prostrate shrub to a low tree 6 m. high. *Cyathea Milnei* has a stout trunk 2 to 8 m. high, clothed above with persistent withered leaves while *C. kermadecensis* is a magnificent plant, 15 to 20 m. high with a naked trunk 1 to 2 m. diam. at the base which consists chiefly of aereal roots.

Autecology of the plants. *The trees and shrubs* number 22 (trees 17, shrubs 5), but many of the former epharmonically become shrubs. In stature 6 trees are of medium height (sometimes almost tall) and 11 of low stature and 2 shrubs are tall and 3 of medium height. Two trees of but low stature in New Zealand proper in the Kermadecs reach a height of 20 m. or more.

The life-forms and the number of species to each are as fellows: — canopy-tree 4, tuft-tree 3 (ferns 2, palm 1), bushy-tree 10, brushy-shrub 4, straggly shrub 1. One species may perhaps be considered xerophytic and 1 subxerophytic. Aereal roots are abundant in *Metrosideros villosa* and are in harmony with a moist atmosphere. In wet forest, states OLIVER, "the prostrate trunk of a tree may be a metre above ground supported by hundreds of large and small root-props, and sending up large branches like distinct trees".

As for the leaves of the trees and shrubs they may be defined as follows: — compound 5, simple 17, very large 5, large 2, medium-sized 11, small 4, coriaceous, thick &c. 9, thin 13, glabrous 19 and hairy 3.

Herbs and semi-woody plants number 78 of which 13 are annuals or biennials, 4 summergreen, 11 semi-woody, 67 herbaceous, 15 xerophytes or subxerophytes and 63 mesophytes. Regarding height 23 are very tall, 20 tall, 27 of medium stature and 8 of low stature.

The life-forms and number of species to each are as follows: — (1) *Annuals and biennials* 13, consisting of creeping and rooting grass 1, erect tufted grass 2, turf-making grass 1, erect-branching herb 8 and erect-branching semi-woody plant 1. (2) *Perennials* 65, consisting of: (a) semi-woody plants 10, which include (α) wandering 3, consisting of erect-creeping lycopod 1, straggling and branching plant 1 and mat-herb 1; and (β) spot-bound 7, consisting of: prostrate-straggling plant 1, large succulent mat-plant 1, erect-branching plant 3 and straggling plant 2; and (b) herbaceous perennials 55, which include (α) wandering 19 consisting of: erect-creeping 12 (ferns 9, rush 1, grass 2), mat-herbs 6 and tufted rush 1; and (β) spot-bound 36 consisting of: erect unbranched fern 1 (summergreen), tufted 20 (ferns 12, grass-form 8) tussocks 6 (grass 2, rush 3, sedge 1), earth-orchids 2, erect-branching herbs 5, semi-erect branching 1 and trailing herb 1.

The leaves of the plants here dealt with may be characterized as follows: compound 25, simple 49, very large 16, large 6, of medium size 18, small 25, very small 9, coriaceous &c. 26, thin 48, glabrous 65 and hairy 9; 4 species are leafless.

Lianes and epiphytes, similar to those so characteristic of forests in New Zealand proper, are wanting, but there are a few lianoid coastal herbs and in wet forest many bryophytes and some of the usual New Zealand epiphytic pteridophytes.

The plant communities. Here I am not usually giving names to the associations based on the nature of the vegetation, or the dominance of species, but am mostly distinguishing them by the nature of their edaphic position.

Coastal-rock vegetation is related to that of the North Auckland district. The following are important members: — *Asplenium obtusatum, Poa polyphylla* (endemic), *Mariscus ustulatus, Scirpus nodosus, Parietaria debilis, Mesembryanthemum australe, Tetragonia expansa, Apium prostratum, Samolus repens* var. *strictus, Lobelia anceps* and *Coprosma petiolata* (endemic).

Talus slopes at the base of sea-cliffs, named by OLIVER **Mariscus slopes,** are covered with close-growing tussocks of *Mariscus ustulatus* together with the ferns *Pteris comans* and *Hypolepis tenuifolia*, tussocks of *Scirpus nodosus* and *Carex kermadecensis* and shrubby *Myoporum laetum.*

Coastal scrub forms a belt, where there is sufficent space, between the foot of the cliffs and high-water mark; it also constitutes most of vegetation of Meyer Island. Shrubby *Myoporum laetum* is dominant. Other members of the association are: *Pteris comans, Mariscus, Carex kermadecensis, Macropiper excelsum* var. *major, Canavalia obtusifolia* and *Sicyos australis* (tendril liane).

Sand dunes are extremely limited in area. *Ipomaea pes caprae* is easily dominant and extends for some metres on to the beach, other species are *Imperata Cheesemanii* (endemic), *Scirpus nodosus* and *Apium prostratum.*

Gravel flat vegetation occurs in Denham Bay and extends for 2 km. with an average width of 74 m. It is subject to frequent drenching by sea-spray. The leading species are: — *Ipomaea pes caprae* (dominant), *Imperata Cheesemanii, Deyeuxia Forsteri, Mariscus ustulatus* (sub-dominant), *Scirpus nodosus* (abundant), *Tetragonia expansa, Myoporum laetum, Calystegia Soldanella, Scaevola gracilis* and *Erechtites prenanthoides.*

Inland rocks bear the following important rock-plants: — *Cyclophorus serpens, Asplenium Shuttleworthianum, Psilotum triquetrum, Carex kermadecensis, Peperomia Urvilleana, Mesembryanthemum australe, Hydrocotyle moschata, Hebe breviracemosa* (endemic), *Scaevola gracilis* (endemic) and *Lagenophora pumila.*

Warm ground near steam vents where the steam strikes directly contains no plants, but at a short distance away there is a close growth of *Nephrolepis axaltata, Polypodium diversifolium, Lycopodium cernuum, Psilotum triquetrum* and *Paspalum scrobiculatum.*

Swamps are few and small. *Typha angustifolia* var. *Brownii* is dominant. Other species are: — *Blechnum procerum, Histiopteris incisa* and *Juncus polyanthemos*; where the water is shallow near the margin of the Green Lake there is a belt of *Typha*.

Forest occupies virtually the whole of Sunday Island and is a distinct expression of climate. The forest may be divided into the two classes, **wet** and **dry**.

Wet forest is composed of a mixture of trees none of which is dominant. The principal trees are: — *Ascarina lanceolata, Melicytus ramiflorus, Notho-panax kermadecensis, Suttonia kermadecensis* and *Coprosma acutifolia. Metrosideros villosa*, of huge dimensions, occurs in places. Frequently one species or another makes a pure stand. *Rhopalostylis Cheesemanii* and *Cyathea kermadecensis* are generally extremely plentiful. The abundance of epiphytic ferns is a striking feature.

Dry forest is characterised by the dominance of *Metrosideros villosa*. The average height is some 20 m. There are 3 tiers of vegetation. Different combinations form the lowermost tier in different localities, e. g. *Polystichum aristatum* 1 to 2 m. high may make an impenetrable mass, or *Nephrolepis cordifolia* occupy wide areas, its matted roots spreading over the ground or fallen logs or climbing the tree-fern trunks, or again the undergrowth be merely a dense mass of the stems of *Macropiper*. The second tier may consist of small trees, the palm and tree-ferns, the first-named being especially *Melicope ternata, Boehmeria dealbata, Coriaria arborea, Corynocarpus laevi-gata, Melicytus ramiflorus, Suttonia kermadecensis, Myoporum laetum* and *Coprosma acutifolia*. In places the palm forms colonies (Fig. 94). The third tier consists of the *Metrosideros* together with *Corynocarpus* and *Myo-porum* which almost equal it in stature. The forest-roof is fairly dense.

Chapter II.
The Vegetation of the Chatham Islands.

General. The species of vascular plants number at least 257 (pterido-phytes 53, monocotyledons 71 and dicotyledons 133) which belong to 57 families and 152 genera, the largest being the following: — (families) *Filices* 48, *Cyperaceae* and *Compositae* each 24, *Gramineae* 17, *Onagraceae* 13 and *Umbelliferae* 9; (genera) *Epilobium* 13, *Hymenophyllum* and *Carex* each 7, *Blechnum, Scirpus* and *Coprosma* each 6 and *Asplenium* and *Lyco-podium* each 5. The florula consists of 36 endemic species, 218 of New Zealand proper (Endemic 85, Australian 121 of which 31 are cosmopolitan or sub-cosmopolitan, subantarctic South American 3, Polynesian, Lord Howe or Norfolk Island 9), 2 of Subantarctic New Zealand and 1 Australian but confined to the Chathams.

With regard to the distribution of the 218 species, 161 occur in all three islands of New Zealand proper, 49 in North Island and South Island, 1 in South Island and Stewart Island, 1 in South Island, 5 in North Island only and 1 in Stewart Island only. Nearly all belong to the lowland flora only 30 at most ascending in New Zealand proper to above the forest line. As for the relationship of the Chathams' flora to that of the other outlying islands 38 species occur in the Kermadecs and 52 in the Subantarctic Islands.

The endemic element is remarkable as containing 23 species so closely related to one or more species of New Zealand proper that most of them, at first glance, seem hardly distinct. On the other hand, the endemic genera *Coxella* and *Myosotidium* stand out plainly, as also *Agropyron Coxii*, *Geranium Traversii*, the 3 shrubby species of *Hebe*, *Olearia Traversii* and *Cotula Featherstonii* and the related *C. Renwickii*. Endemics of this class seem to be either representatives of an earlier flora, or relics of one belonging originally to New Zealand proper.

The ecological conditions of the Chatham Islands consist of a combination of coastal, rain-forest and subantarctic factors. Thus, at no part of the main island are you distant 7 km. from the sea and the extreme is, in most places less than one half of this. High winds are frequent, especially those from the north-west and south-west. The rainfall (about 90 cm.) is not great, but the number of wet days is excessive, while cloudy skies are frequent. Frost is trivial, snow falls but rarely and a maximum temperature of 24° C. is seldom reached. The soil in many parts consists of peat frequently more than 6 m. in depth and always in a complete state of saturation. There is however, in certain localities, a much more "fertile" soil known locally as "red clay", the outcome from weathering of the volcanic rock. The main features of the vegetation, as will be seen further on, clearly reflect the climatic and edaphic conditions.

The leading physiognomic plants and their life-forms. Certain physiognomic plants common in New Zealand proper need no further description here, e. g. the various tree-ferns, *Pteridium esculentum*, *Desmoschoenus frondosus*, *Rhopalostylis sapida*, *Leptocarpus simplex*, *Rhipogonum scandens*, *Phormium tenax*, *Muehlenbeckia australis*, *Mesembryanthemum australe* and *Corynocarpus laevigata*. Other species again have life-forms identical with those of their mainland relatives and differ merely in taxonomic characters, e. g. *Carex sectoides*, *Pseudopanax chathamicum*, *Cyathodes robusta*, the species of *Dracophyllum*, the large-headed species of *Olearia* and *Cotula potentillina*. There remain then only the following all endemic except the first, to be considered: — *Suttonia chathamica*, *Myosotidium Hortensia*, *Hebe gigantea*, *Coprosma chathamica*, *Olearia Traversii* and *Senecio Huntii*.

Suttonia chathamica (F. v. Muell.) Mez *(Myrsin.)* varies from a dense shrub 1 to 2 m. high to a round-headed low tree 4 to 8 m. high with the trunk 30 to 60 cm. diam. covered with dark bark. The leaves vary in size according to the degree of exposure, but in the forest they are obovate, about 5.5 cm. × 3 cm., dull-green, rather thick, coriaceous and crowded at the ends of the twigs.

Myosotidium Hortensia Hook. *(Borrag.)*, the Chatham Island lily, has a stout rootstock 5 cm. diam. which creeps just below the ground-surface, or is partly unburied. The rhubarb-like leaves are in erect rosettes; they are ovate-cordate to reniform, 25 cm. long × 30 cm broad, more or less, and consists of comparatively thin blade strengthened beneath by a stout framework of midrib and veins; the leaf-stalks are stout and about 12 cm. long. The flowers, each about 1.2 cm. diam., the central half of the corolla bright blue and the outer half white, are arranged in dense, sub-globose, many-flowered cymes borne on stout stalks 60 cm. high.

Hebe gigantea Ckn. *(Scroph.)* is a low forest-tree 4.5 to 9 m. high with an erect trunk 30 cm. or more diam. and a close, rounded crown. The leaves are lanceolate, 7 to 9 cm. long, soft, dull-green and moderately thick. Though like a gigantic form of *H. salicifolia*, the seedling-form is altogether distinct in its early leaves deeply and coarsely toothed, their margins ciliated, with hooked, white hairs and the strongly pubescent soft, purple stem. Succeeding leaves are larger than the adult; and, although entire are still ciliated and also pubescent on the midrib. Plants more than 60 cm. high maintain the juvenile character.

Coprosma chathamica Ckn. *(Rubiac.)* is a tree 4.5 to 15 m. high with a trunk 30 to 60 cm. diam. covered with light-brown bark and having oblong, obovate-oblong or lanceolate leaves about 4.9 cm. long, dark-green, somewhat thin, shining above but pale beneath, flat or with the margins more or less incurved. The species is closely related to the Kermadec *C. petiolata*, to which it was referred by HOOKER.

Olearia Traversii (F. v. Muell.) Hook. f. *(Compos.)*, the Chatham akeake, is a small tree 4.5 to 9 m. high with a trunk 30 to 60 cm. diam. covered with pale, rough bark and a rather dense crown. The leaves are opposite, oblong to ovate, 5.5 cm. long or shorter, soft, rather thick, bright-shining-green above and beneath clothed with dense, silky tomentum, as are also the branchlets, inflorescence-branches and involucres.

Senecio Huntii F. v. Muell. *(Compos.)*, the rautini, is a small tree 3 to 6 m. high of the tree-composite type, with a trunk 30 cm., or so, diam. and 20 to 24 leaves in rosettes at the ends of the branchlets. The leaves are lanceolate, sessile about 12 cm. × 3.5 cm., pale but shining green above and greyish-green beneath with short, glandular hairs. The flower-heads, each some 1.3 cm. diam., with 15 to 20 yellow ray-florets, are in massive much-branched, terminal panicles, 10 cm. or more in breadth. Flower-stalks, involucres, branchlets etc. are densely covered with glandular hairs which fill the air with a powerful aromatic scent. Seen from a distance, the foliage forms a bluish semi-spherical mass, but when in full blossom, fully one half is concealed by the brilliant yellow.

The autecology of the plants. The trees number 20. All are of low stature but at times a few may reach a height of 15 m., or more, 18 are mesophytes and 2 xerophytes or subxerophytes. Their life-forms are as follows: — canopy-tree 9, tuft-tree 6 (ferns 5, palm 1), bamboo-like 1, bushy-tree 2, araliad-form 1, semi-globose 1. Several readily assume the shrub-form, especially *Suttonia chathamica* and *Olearia Traversii*. *Dracophyllum arboreum* is not of the usual fastigiate habit of its congeners and is here included amongst the canopy-trees. It usually commences life as an epiphyte.

Generally, the tree-crowns are denser than those of the same species of the mainland and resemble those of coastal trees. The trunks are slender and straight. The appearance of *Hebe* and *Coprosma,* as true forest-trees, must be specially noted. *In species or genera strongly hetero-morphic in New Zealand proper the phenomenon is exhibited to a very trifling extent,* e. g. the juvenile form of *Plagianthus chathamicus* gives but a hint of the divaricating form. In *Pseudopanax chathamicum,* the juvenile stage has leaves of an adult type, but it nevertheless persists for some time with straight unbranched stem. *Dracophyllum arboreum* has broad spreading leaves as juvenile after the manner of *D. latifolium* but not nearly so broad or large and absolutely distinct from the narrow grass-like leaves of the adult. They appear on reversion-shoots in the crown of the tree and this stage may blossom. The juvenile *Coprosma chathamica* has much more mesophytic leaves than the adult and they are frequently much larger. The heterophylly &c. of *Hebe gigantea* has been already described.

Coming now to the leaves of the trees, they may be characterized as follows: compound 7, simple 13, broad 19, narrow 1, very large 6, large 2, medium 12, coriaceous &c. 13, thin 7, glabrous 17, hairy 3; all are evergreen except *Plagianthus chathamicus* and most likely *Edwardsia chathamica*.

The flowers of 5 species are more or less showy, 9 have succulent fruits and those of 2 (both endemic) are specially suited for wind-carriage.

The shrubs number 23, of which 12 are xerophytes or sub-xerophytes 11 mesophytes, 3 wandering, 20 spot-bound, 2 very tall, 8 tall, 8 of medium height and 5 of low stature. The life-forms and the number of species to each are as follows: Wandering 3, consisting of: bamboo-form 1 *(Sporo-danthus),* open rooting cushion 1, erect twiggy 1; spot-bound 20, consisting of: tuft-shrub 2 (both ferns), bushy-shrub 5, ericoid-form 4 (fastigiate 2, bushy 1, ball-like 1), twiggy 1 (tall, open, spreading), rhododendron-form 2, divaricating and semi-divaricating 3, straggling 2 (prostrate 1, semi-prostrate 1), open cushion 1 (divaricating).

The leaves of the shrubs (1 species leafless) may be characterized as follows: compound 2, simple 20, broad 14, narrow 8 (needle-like 2), very large 2, large 2, of medium size 7, small 5, very small 6.

The flowers of 11 species are more or less showy, *Olearia semidentata*, with its multitudes of brilliant purple ray-florets, being of extraordinary beauty. Eight species have fleshy fruits and in 2 species they are suitable for wind-transit.

Lianes are only represented by 7 common species of New Zealand proper and include 2 root climbers (ferns), 1 scrambler and winder (combined), 2 high-climbing, woody and 2 semi-woody winding-lianes.

Epiphytes number 13 all of which are common also in New Zealand proper. They comprise creeping filmy ferns 5, erect herbaceous ferns 3 (creeping 1, tufted 2), drooping herbaceous fern 1, low creeping fleshy fern 1, low creeping unbranched semi-woody plant 1 and thick-leaved orchids 2.

Parasites are represented only by *Gastrodia*.

Herbaceous and semi-woody plants number 187 of which 13 are annuals, 174 perennials (herbaceous 137 semi-woody 37), 114 spot-bound, 73 wandering, 145 mesophytes, 4 hygrophytes, 38 xerophytes, 24 very tall, 23 tall, 37 of medium stature, 58 of low stature, and 45 of very low stature.

The life-forms and number of species to each are as follows: — (1) Annuals or biennials 13 consisting of: (a) semi-woody 4, all erect-branching, and (b) herbaceous 9 made up of tufted 2, mat-form 1, erect-branching 5 and prostrate-spreading 1. (2) Perennials 174, consisting of: (a) semi-woody 37 which include (α) spot-bound 26, made up of erect rosette-form 1, erect tufted 3, erect straggling 2, prostrate trailing 1, cushion-form 1, erect-branching 17 and straggling 1; (β) wandering 11 made up of, erect-creeping 4, prostrate-creeping 1 mat-form 6; and (b) herbaceous perennials 137, which include (α) spot-bound 75 made up of, erect-tufted 24 (ferns 4, grass-form 15, and rush-form 5), prostrate-tufted 3, erect unbranched summergreen 2 (ferns), tussocks 12 (grass-form 7, rush-form 3, iris-form 2), earth-orchid form 11, mat-form 4, erect-branching 8, yucca-form 2, straggling 1, rosette-form 8; and (β) wandering 62, made up of carpet-plants 4, turf-makers 1, mat-form 23, erect-creeping 25, sand-binder 1, wiry interlaced 1, iris-form 1, erect rosette-form 2, straggling 2, tufted-grass 1 and *Utricularia*-form 1.

Coming now to the leaves, they may be characterized as follows: very large 14, large 14, of medium size 30, small 63, very small 58, simple 141, compound 38, broad 106, narrow 73, coriaceous &c. 77, thin 102, glabrous 137 and hairy 42.

Water-plants number 4 and consist of submerged filiform 1, floating-leaved 2 and partly submerged 1.

The plant-communities. Here, again, in certain cases an account is given of the vegetation of habitats rather than of definite associations.

On sandy shores the leading species were *Ranunculus acaulis, Calystegia Soldanella* and *Myosotidium Hortensia*. The last-named was originally extremely abundant and formed a belt just above high-water mark, but it is now extinct in this association.

On stony shores beyond the reach of sheep, pigs &c. *Myosotidium* forms broad colonies which are a remarkable and beautiful spectacle. Its huge leaves have the laminae bent so as to be funnel-shaped and thus an ample supply of water is conducted to the roots during light showers. Other common plants of the association are: — *Urtica australis* forming thickets 30 to 35 cm. high, its stems 1.5 cm. diam. and leaves 15 × 10 cm.; *Rumex neglectus, Ranunculus acaulis, Lilaeopsis novae-zelandiae, Selliera radicans* and *Cotula potentillina.*

Sand-dunes of great extent occupy a large part of the coast-line. The unstable and partly stable dunes bear a plant-covering similar to that of New Zealand proper with *Desmoschoenus spiralis* and its accompanying species except that *Spinifex hirsutus* is absent. Where the sand is stable comes dune-forest made up, near the sea, entirely of *Olearia Traversii* and *Suttonia chathamica,* but further inland other trees occur.

Where sand has blown on to flat rock-ledges the endemic *Sonchus grandifolius*[1]) grows luxuriantly (Fig. 95), and, in its company the endemic creeping, xerophytic grass *Festuca Coxii, Salicornia australis, Apium prostratum* and *Samolus repens* var. *procumbens.*

Cliff and rock vegetation is by no means uniform. A rocky coast-line is a frequent feature especially in the south of the main island and of the smaller islands. Where there is a maximum of spray the endemic *Hebe chathamica*[2]) is characteristic and its companion-plants are: *Mesembryanthemum australe, Apium prostratum, Salicornia australis, Senecio radiolatus,* stunted *Olearia Traversii* and the endemic *Geranium Traversii,* but this last seems hardly so tolerant of salt. *Blechnum durum,* in less halophytic stations, forms extensive pure colonies.

Near the margin of the great lagoon, there are limestone cliffs, in many places, which, at one time purely littoral, now bear a covering partly coastal, partly inland. The most striking plants is the endemic *Hebe Dieffenbachii*[3]). Other common plants of this association are *Adiantum affine* (in hollows), *Phormium tenax, Geranium Traversii, Linum monogynum* var. *chathamicum, Acaena novae-zelandiae, Leucopogon parviflorus* and *Senecio lautus.*

---

1) A summergreen herb with thick, juicy rhizome, great oblong, pinnatifid or pinnate, palegreen, thick, hard leaves, 30 to 60 cm. long and large flower-head protected by extremely thick, spinous bracts.

2) A shrub with rather thick main-stem, issuing from a crevice, and giving off numerous, supple, slender branches which trail over the rock or hang downwards. The elliptic leaves, 2 to 2.5 cm. long, are rather thick, fleshy, pale-green and more or less downy. There are many distinct forms of the "species" and such are probably jordanons, crosses between such, and crosses with *Hebe Dieffenbachii* and *H. Dorrien-Smithii.*

3) A shrub with far-extending branches, much stouter than those of *chathamica,* but nevertheless extremely pliant, and thick, fleshy, pale-green, linear-oblong leaves, 6.5 cm. long, confined to the ends of the branches.

Frequently, flattish rock is covered with peat which varies considerably in depth. In such places, a turf may be formed out of a close growth of *Scirpus cernuus, Triglochin striata* var. *filifolia, Ranunculus acaulis, Lilaeopsis novae-zelandiae, Pratia arenaria* and *Selliera radicans*. *Mesembryanthemum australe* is characteristic and may form pure stands. In the north of the island, where the peat is deep, and where certain petrels burrow, are extensive colonies of the endemic *Cotula Featherstonii*[1]) while alongside are great clumps of *Myosotidium* and sheets of *Mesembryanthemum*. In the south of Chatham Island, and on Pitt Island, there is still to be seen the one time more widely-spread endemic *Coxella Dieffenbachii*[2]).

Salt-meadows were originally common. Unfortunately, I saw none in their virgin state but W. MARTIN has very kindly given me details from his field notes which show that the association is dominated by *Juncus maritimus* var. *australiensis* and all the halophytes of the florula are present.

In shallow water containing more or less salt there is abundance of *Ruppia maritima*.

Inland rocks support no special rock plants. The most frequent are the following: — *Cyclophorus serpens, Polypodium diversifolium, Asplenium flaccidum, Phormium tenax, Earina autumnalis, Linum monogynum* var. *chathamicum, Cyathodes robusta, Hebe Dorrien-Smithii, Coprosma* sp. (probably undescribed) and *Olearia chathamica*.

Lakes and sluggish streams of dark peaty water are abundant but water-plants are few and scarce. The sole species recorded so far are *Potamogeton Cheesemanii, Polygonum serrulatum, Ranunculus rivularis, Callitriche Muelleri* and *Myriophyllum elatinoides*.

Swamp is essentially a lowland association. In the deepest water, but rarely mixed, grow *Leptocarpus simplex* and *Carex sectoides*. Where shallower there is a scrub of *Coprosma propinqua* (dominant), *Blechnum procerum, Arundo conspicua, Deschampsia caespitosa, Carex Darwinii* var. *urolepis, Phormium, Astelia nervosa* (or undescribed species), *Epilobium pallidiflorum* and *E. chionanthum*. As the swamp becomes drier small trees or shrubs appear, especially, *Coriaria arborea*, the endemic *Pseudopanax chathamicum, Dracophyllum arboreum* and *Suttonia Coxii*. *Hymenanthera chathamica* and *Senecio Huntii* also occur to some extent.

Bog associations are specially characteristic both in the lowlands and on the tableland. Lowland bog has been greatly modified by draining, burning and trampling of stock, so that as it was doubtless similar to the

---

1) A biennial herb, 15 to 30 cm. high, greyish in colour and rather like *Matthiola incana*. Its leaves are oblong- or obovate-spathulate 2 to 5 cm. long, soft, pubescent and in rosettes at the ends of the naked branches. The life-form is quite different from that of the genus in general.

2) A perennial herb of the *Aciphylla*-form, but the leaves are neither rigid nor spinous. The flower-stalk measures 60 cm. or more and the leaves are 30 to 60 cm. long.

tableland association, the latter, being quite virgin in places is alone dealt with here.

The soil consists of peat which varies considerably in its water-content, this latter being reflected in the vegetation, so that the following successive subassociations can be defined: Sphagnum bog; *Sporodanthus-Olearia* bog; *Olearia-Dracophyllum* bog, and pure *Dracophyllum* bog, which is a connecting-link between bog and forest.

In sphagnum bog a species of *Sphagnum* forms large, rounded cushions on which grow in the wettest portions, *Scirpus inundatus* and a species of *Carex.* Where a little drier are *Hierochloe redolens, Pratia arenaria* and the endemic grass *Poa chathamica.* Other associated plants are, *Gleichenia circinata, Heleocharis Cunninghamii, Drosera binata, Myriophyllum pedunculatum,* and *Utricularia monanthos.* The *Gleichenia* frequently establishes wide colonies. Tiny plants of *Dracophyllum paludosum* and *Olearia semidentata* in bloom are not uncommon.

Sporodanthus-Olearia bog follows *Sphagnum* bog, as soon as the ground becomes a shade drier. The soil is peat, fully saturated with water, and, at a depth of 20 cm. it has the consistency of porridge. The vegetation is extremely dense and consists of *Sporodanthus Traversii* mixed and entangled with *Olearia semidentata* and *Dracophyllum paludosum,* the mass averaging perhaps 1.5 m. in depth.

Olearia-Dracophyllum bog which succeeds the last-described as the ground becomes drier, consists of about an equal amount of the *Olearia* and *Dracophyllum* while in open places are *Poa chathamica, Drosera binata, Gentiana chathamica* and the *Utricularia.* Possibly *Phormium* was originally an important constituent. When the *Olearia* is in full bloom the brilliant purple flower-heads, in hundreds on every bush, render these Chatham Islands' bogs wonderfully beautiful.

Pure *Dracophyllum* bog is a distinct transition between bog and forest. *D. paludosum* 1.6 m. high is dominant and mixed with it here and there are trees of *D. arboreum* while juvenile plants of this latter equalling *D. paludosum* in height, but distinguished by their broad leaves, are abundant. On the ground grow *Poa chathamica, Gentiana chathamica, Pratia arenaria* and seedlings of both species of *Dracophyllum.*

Forest occupies a very considerable part of both the main island and Pitt Island. There are two classes, **lowland** and **upland**; the first distinguished by the dominance of *Corynocarpus* and an abundance of *Rhopalostylis* and *Olearia Traversii,* and the second by the absence of the above three trees, the dominance of *Senecio Huntii* and *Dracophyllum arboreum,* and, with the exception of *Hebe gigantea,* a smaller percentage of the other forest-trees than in the lowlands. Shrubs, such a common feature of New Zealand forest, are absent, and the tallest undergrowth consists mainly of the 5 tree-ferns (*Dicksonia squarrosa, D. fibrosa, Cyathea Cunning-*

hamii and *C. medullaris)* and, in some localities, the stems of the liane *Rhipogonum.* On the floor are many of the usual New Zealand forest-ferns. The lowland forest is generally quite low (6 to 13 m.) and the roof is flat and close, the trees being of equal height. So too, is upland forest low and flat-roofed, except where *Dracophyllum arboreum* raises its foliage above the general level. Filmy-ferns occur in both classes of forest, but they are most abundant in the upland and may carpet the ground or hide the tree-trunks or fern-trunks. Forest on limestone near the large lagoon is distinguished by the presence of *Edwardsia chathamica,* but according to W. MARTIN it occurs elsewhere to some extent. Lowland forest is really ecologically equivalent to *Corynocarpus* coastal forest of New Zealand proper.

Heath, in exposed situations, succeeds bog as the ground finally becomes fairly dry, or, in many places at the present time, it has seized the ground from which forest has been eradicated by fire. There are two classes of the formation, **shrub-heath** and **bracken-heath.**

Shrub-heath is distinguished by the presence of the eriocoid shrubs *Leucopogon parviflorus* and *Cyathodes robusta,* the rush-like *Scirpus nodosus* and the *Iris-*like *Libertia peregrina. Pimelea arenaria,* or an allied species, is common in some places. Other species are *Pteridium esculentum, Deyeuxia Forsteri, Danthonia semiannularis, Acaena novae-zelandiae, Pratia arenaria, Helichrysum filicaule* and *Gnaphalium japonicum.*

Bracken-heath frequently consists of pure *Pteridium esculentum* forming dense thickets and attaining a height of more than 1.5 m.; where more open, various plants of shrub-heath gain a footing and there are transitions between the two associations.

Chapter III.
The Vegetation of the Subantarctic Islands.

General. The florula consists of 193 species (pteridophytes 38, monocotyledons 63, dicotyledons 92) which belong to 37 families and 91 genera. The largest families and the number of species to each are as follows: — (families) *Filices* 34, *Gramineae* 26, *Compositae* 22, *Orchidaceae* 12, *Cyperaceae* 11, *Juncaceae* 9, *Rubiaceae* 7, *Umbelliferae* 6 and *Caryophyllaceae, Cruciferae, Ranunculaceae* and *Onagraceae* 5 each; (genera) *Hymenophyllum* 10, *Poa* 8, *Blechnum* and *Coprosma* 6 each, and *Polypodium, Corysanthes* and *Ranunculus* 5 each. The following 60 species (31 % of the flora, but 3 are doubtfully endemic here considered endemic: — *Hierochloe Brunonis*† (A. C.), *Deyeuxia Forsteri* var. *micranthera** (C. Ant.), *Danthonia antarctica** (A. C.), *Deschampsia gracillima*† (A.), *D. penicillata*† (M.), *Poa ramosissima*† (A. C.), *P. Hamiltoni*† (M.), *P. aucklandica*† (A. C.), *P. incrassata*† (A. C.), *P. imbecilla* var. *breviglumis** (A.C.), *Atropis antipoda** (Ant.), *Triodia macquariensis*† (M.), *Uncinia Hookeri** (A.C. Ant. M.), *Luzula crinita* (A. C. Ant. M.), *Chrysobactron*

*Rossii** (A. C.), *Urtica aucklandica** (A.), *Stellaria decipiens**, (A. C. M),
S. decipiens var. *angustata** (Ant.), *Colobanthus muscoides*† (S. A. C. Ant. M.),
*C. Hookeri** (A. C.), *C. glacialis* var. *subcarnosa** (A. C.), *Ranunculus pinguis**
(A. C.), *R. subscaposus*† (C.), *R. aucklandicus*† (A.), *Cardamine stellata**
(A. C. S.), *Geum albiflorum** (A.), *Acaena sanguisorbae* var. *minor** (A. C.
Ant. M.), *Epilobium confertifolium** (A. C. Ant.), *E.* sp. aff. *E. alsinoides**
(Ant.), *Stilbocarpa polaris** (A. C. Ant. M.), *S. robusta** (S.), *Schizeilema
reniforme** (A. C.), *Anisotome latifolia*† (A. C.), *A. antipoda*† (A. C. Ant.).
*A. acutifolia** (S.), *Dracophyllum subantarcticum** (C.), *D. scoparium** (C.),
*Gentiana cerina** (A. C.), *G. concinna** (A.), *G. antarctica** (C.), *G. anti-
poda** (Ant.), *Myosotis antarctica** (C.) *M. capitata** A. C.), *Hebe Ben-
thami*† (A. C.), *Plantago aucklandica* (A.), *P. subantarctica*²)† (A. C.), *Co-
prosma antipoda* (A.), *C. myrtillifolia*³) (A. C.), *C. cuneata*⁴) (A. C.), *Olearia
Lyallii** (S. A.), *Pleurophyllum speciosum*† (A. C.), *P. criniferum*† (A. C.
Ant.), *P. Hookeri*† (A. C. M.), *Celmisia vernicosa*† (A. C.), *C. campbellensis*†
(A. C.), *Cotula lanata*† (A. C.), *Helichrysum prostratum** (A. C. Ant.), *Abrotanella spathulata** (A. C.), *A. rosulata** (C.) and *Senecio antipodus*† (Ant.).
The genus *Pleurophyllum* is endemic. It is closely related to *Celmisia* and
Olearia from which it differs chiefly in habit. The two species of *Celmisia*
belong to the subgenus *Ionopsis*. *Stilbocarpa* is common in Stewart Island,
but it occurs in the South Island only in south-west Otago to a most
limited degree.

The florula is thus distributed in the different groups of islands: —
Snares 23 species, Lord Auckland Islands 159, Campbells 115, Antipodes 57
and Macquaries 34. Its elements fall into the following classes: — (1.) *En-
demic* 60, as above; (2.) *New Zealand proper* 122 (63%), of which 65 are
endemic and 36 Australian (omitting the cosmopolitan &c.); (3.) *Subantarctic
American* (including Kerguelen Land &c. and the species extending also to
Australia) 28, of which the following do not extend to New Zealand proper: —
Festuca erecta, Carex Darwinii var. *urolepis* (also in Chathams), *Rostkoria*

---

1) The marks or letters attached, to the specific names signify: — * = closely
related to New Zealand mainland species, † = not closely related to mainland species:
S. A. C. Ant. M. = Snares, Lord Auckland, Campbell, Antipodes and Macquarie Islands
respectively.

2) *Plantago subantarctica* Ckn. sp. nov. = *P. carnosa* R. Br. ex Hook. f. in *Flora
Antarctica*, I, (1844) 65, t. 43.

3) According to observations made recently by G. E. Du Rietz, it seems not unlikely
that *C. myrtillifolia* and *C. ciliata* are one and the same, but, until further evidence is
available. I am treating the former as endemic and distinct from *C. parviflora* Hook. f.,
and maintaining the latter.

4) Du Rietz has pointed out to me the resemblance between *C. cuneata* and
C. Astoni Petrie of New Zealand proper; indeed, they may be same. The series of
jordanons of South and North Islands, hitherto merged into *C. cuneata*, are dealt with
in this book as *C. pseudocuneata*, excepting the Mount Egmont jordanon *(C. egmontiana)*.

magellanica, Juncus scheuchzerioides (perhaps not the South American species), *Ranunculus crassipes, Cardamine corymbosa, Acaena adscendens, Callitriche antarctica, Azorella Selago* and *Cotula plumosa.*

The distribution of the New Zealand proper element within the Botanical Region is as follows: — Kermadecs 12 (pteridophytes 7), Chathams 45 (pteridophytes 22), North Island 98 (pteridophytes 35), South Island 115 (pteridophytes 36) and Stewart Island 110 (pteridophytes 37) or about 22 per cent of its florula. With regard to their vertical distribution about 60 species extend above the timber-line in New Zealand proper.

The ecological conditions. The climate is uniform; there is but little difference between the means or extremes of winter and summer temperature. The sky is generally cloudy; showers are frequent (annual rainfall $\pm$ 140 cm.); the atmosphere is saturated; periods of sunshine are brief; there is a general average low temperature (maximum $\pm$ 10° C. and minimum about $-$ 1° C.) with but slight winter frosts at sea-level; cold and violent winds accompanied by showers of sleet, hail or even snow are of constant occurrence. The wind indeed is a master-factor. Its mark is on the vegetation everywhere both in regard to form and distribution. The moisture-laden air and lack of sunshine favour the formation of peat. Dead stems and leaves of the herbaceous plants, slowly rotting, remain attached to the living plants. Bryophytes on the ground, on tree-trunks and on rock-faces build thick layers and cushions of peat, the outer shoots alone alive. The ferns, *Blechnum durum* and *Asplenium obtusatum* of the coastal cliffs form, from their dead rhizomes, masses of peat, 30 cm. or more in depth, which completely cover the flatter rocks. The soil of all the islands, indeed, for a depth of 9 m., or more, is made up altogether of plant-remains. Such a soil becomes saturated; pools lie on the surface and holes, masked by vegetation, full of water, are frequent on the open hillsides. The indigenous birds, especially penguins and albatrosses, and the seals, in some of the small islands, play an important part in plant-distribution and where numerous bring about a regular "rotation of crops" (Fig. 96). The foregoing conditions, that dependant on birds &c. being excepted, have brought about a vegetation akin, at all levels, to that of the subalpine belt of Stewart Island and of the flat-topped Mountains of the South Otago district.

The leading physiognomic plants and their life-forms. With the exception of the *Metrosideros,* the following physiognomic plants, common to the Subantarctic province and the main islands of New Zealand, need no description: — The herbaceous and filmy ferns, especially *Polystichum vestitum; Nothopanax simplex; Dracophyllum longifolium; Suttonia divaricata* and *Coprosma foetidissima.*

Metrosideros lucida (Myrtac.), the southern-rata, in New Zealand proper, is an erect, evergreen tree with a maximum height of some 18 m., but in Lord Auckland Islands the trunk is nearly always more or less prostrate (Fig. 97), irregular in shape, bent or arched and far-spreading either on, or just above the forest-floor. The branches, too, are gnarled and twisted and extend at first more or less horizontally, but finally put forth erect branches which ultimately terminate in numerous twigs bearing abundant leaves and so forming flattened or rounded masses. The bark is of a dull reddish-brown and frequently hangs in long strips from the horizontal branches. The leaves are lanceolate to elliptic-lanceolate, 3 to 5 cm. long, thick, stiff, coriaceous and bright-green with a yellow midrib. The flowers are very numerous, bright-crimson and arranged in short terminal cymes.

The purely, or almost, endemic species of physiognomic importance are: — *Danthonia antarctica, Poa foliosa, P. litorosa, Chrysobactron Rossii,* 2 species of *Anisotome, Stilbocarpa polaris, Dracophyllum subantarcticum, Olearia Lyallii* and 3 species of *Pleurophyllum.*

The 3 grasses of the above list are of the tussock-form, all build trunks, those of *Poa litorosa* at times attaining a height of 1.5 m. The leaves of all are coriaceous, those of *P. foliosa* are bright-green and flat, of *P. litorosa* rather stiff, filiform and involute and of *Danthonia antarctica* thick, involute and pale or yellowish green.

Chrysobactron Rossii Hook. f. *(Liliac.)* is a summer-green herb with a short, stout rootstock and numerous leaves arranged somewhat like those of a garden hyacinth. They are linear, 29 cm. $\times$ 9 cm., bright green and fleshy, the outer ones curved outwards and the inner erect but all so bent as to make channels down which water is conducted to the long, thick, fleshy roots. The flowers are dioecious, orange coloured, in dense racemes terminating stout scapes 30 cm. or more long. The male inflorescence may be 10 cm. $\times$ 5 cm. and is larger and more showy than the female.

Anisotome latifolia and *A. antipoda (Umbell.)* are stately herbs of the *Aciphylla*-form with large, erect, thick, coriaceous, long-petioled, dark-green, pinnate leaves, about 60 cm. long, and an inflorescence of compound umbels forming a head of great size, borne on a stout stalk 73 cm. high. The two species are closely related but *A. antipoda* has the leaves much more finely divided.

Stilbocarpa polaris (Araliac.) much resembles *S. Lyallii* already described for Stewart Island but there are no stolons, the bright-green, fleshy, coriaceous, orbicular-reniform leaves, 20 cm. long $\times$ 30 cm. broad, are bristly on both surfaces and there is a stout, branching rhizome, $\pm$ 6 cm. diam., which creeps along the surface of the ground.

Dracophyllum subantarcticum (Epacrid.) is of the ordinary erect *Dracophyllum*-form and about 1.5 m. high; it needs no special description; it crosses freely with *D. longifolium.*

Pleurophyllum speciosum (Compos.) is a semi-summergreen herb of great size with 4 to 5 bright-green, fleshy, coriaceous, ovate leaves, some 57 cm. long $\times$ 39 cm. broad, arranged in a cup-like rosette. On the under-surface

22*

of the leaf, at about 10 mm. apart, are a number of stout, almost parallel ribs which are connected by a network of raised, stout veins forming lacunae in the deep furrows between the ridges filled with loosely entangled cobwebby hairs. The upper surface of the leaf above the veins is sunken and the intermediate parallel spaces are raised, so giving a corrugated appearance to the leaf. The flower-stems are 80 cm., or more, high and terminated by a raceme of 15 flower-heads, each some 5 cm. diam., the disc dark-purple but the rays paler.

P. criniferum Hook. f. is a summer-green herb with thinner, much more erect leaves than the above, 30 to 90 cm. long by 15 to 30 cm. broad, the under-surface loosely tomentose and strengthened by stout flexible ribs. The massive flower-stalk, often more than 1 m. high, bears a raceme of globose heads, each 4 cm. in diam., the florets purplish-brown (Fig. 98).

P. Hookeri Buch. is a semi-summergreen herb with rosettes of obovate, rather thin leaves, about 29 cm. long by 6.5 cm. broad covered on both surfaces with silvery adpressed silky hairs which render the plant conspicuous.

The autecology of the plants. Since not only are the special subantarctic species considered in what follows but the whole florula, so far as spermophytes and pteridophytes are concerned, and as many species are of restricted distribution, the statistics given are somewhat misleading and do not truly reflect the relation between the life-forms and the ecological conditions. The abundance of the following life-forms must therefore be specially emphasized, although some of them are represented by very few species: — 1. *The tussock with a trunk.* 2. *The prostrate tree.* 3. *The divaricating-shrub.* 4. *The cushion.* 5. *The rosette.* 6. *The creeping and rooting herb.* Further, *the number of large-leaved herbs is much greater than in any formations of the open elsewhere in the region.*

The trees number 6, all of which, leaving the tree-fern on one side, are frequently shrubs also. The growth-forms are: — Canopy-tree 1; rhododendron-form 2; araliad-form 1; low bushy-tree 1.

The trunks of 3 species are more or less prostrate for half their length or more, curved, arching or irregularly twisted. In the case of *Olearia Lyallii* and *Senecio Stewartiae* their trunks are firmly anchored to the substratum by adventitious roots. As for *Metrosideros lucida*, roots are abundantly produced from the trunk, but they rarely enter the substratum. However, they branch abundantly in the liverwort covering of the trunk and are thus in a position to take up a good deal of rain-water. *Nothopanax simplex* and *Dracophyllum longifolium* have generally quite short trunks, more or less erect, which are branched from the base; they both stand on the border-land between trees and shrubs.

The leaves of all the trees are thick, coriaceous, stiff (except in the *Senecio*) and glossy (except in *D. longifolium*). Those of *O. Lyallii* and *S. Stewartiae* are large, measuring 20 × 13 cm. and 18 × 4.5 cm. respectively.

The otherwise excessive transpiration from the wide surface is checked by a thick tomentum on the under-surface. *Metrosideros lucida, Nothopanax simplex* and *Dracophyllum longifolium* have leaves respectively measuring 5×1.9 cm.; from 6.5×2 cm. to 10.5×3 cm. and 16 cm. $\times 3$ mm. The last-named has long, narrow leaves tapering to a fine point, sheathing at the base, concave on the upper surface and bunched together 15 or more at the apices of the ultimate branchlets, after the manner of a tufted grass. They are also vertical or thereabouts and the inner leaves are sheltered by those which have withered. This xerophytic form leads to the tree occupying the most exposed station in the forest or, as a shrub, growing on wind-swept slopes.

The tree-fern *(Hemitelia Smithii)* may be considered a tree, though in the Aucklands it does not exceed 2 m. in height; its leaves are thin.

Shrubs number 15 (including the rare *Fuchsia excorticata* which quite likely is a tree, as in New Zealand proper). As most vary epharmonically to an extreme degree[1]) it is not easy to decide as to the common form. Arbitrarily they may be classified as follows: — Mesophytes 8, xerophytes or sub-xerophytes 7, tall shrubs 4, of medium height 7, of low stature 3 and very low 1. As for their life-forms, 5 are bushy, 5 divaricating, 2 of the fastigiate *Dracophyllum*-form, 1 low straggling, 1 a flat open cushion or mat, and 1 a tuft-shrub (fern).

The branches of the shrubs are generally slender; those of the divaricating-form are stiff and more or less rigid; those of the two species of *Dracophyllum* fastigiate and dense, those of *Hebe elliptica* and *Fuchsia* fairly stout, and *Cyathodes empetrifolia* has very slender, flexible, wiry branches.

The leaves may be characterized as follows: — compound 1, simple 14, broad 11, narrow 4, very large 1, of medium size 1, small 4, very small 9, coriaceous &c. 11, thin 4, glabrous 13, hairy 2.

Herbs and semi-woody plants number 160 of which 21 are semi-woody, 139 herbaceous, 4 hygrophytes, 103 mesophytes, 53 xerophytes, annuals 7, perennials 153 (semi-woody 20, herbaceous 133), wandering 52, spot-bound 108, summer-green 25 (including 7 annuals), evergreen 135, very tall 18, tall 10, of medium stature 13, of low stature 67, of very low stature 52.

The life-forms and the number of species to each are as follows: — (1) *Annuals* or *biennials* 7 consisting of: (a) semi-woody erect-branching 1, and (b) herbaceous 6 made up of: tufted 3 (erect 1, spreading 1) and erect-branching 3. (2) *Perennials* 153 consisting of: — (a) semi-woody 20 made

---

1) For instance *Hebe elliptica, Coprosma foetidissima* and *probably Fuchsia excorticata* may be trees with a distinct trunk. The tall spreading *C. foetidissima* in the wind-swept open becomes prostrate, lying on the ground beneath the tussocks. So, too, the stiff-stemmed, tomentose-leaved *Cassinia Vauvilliersii* and the divaricating *Coprosma myrtillifolia* form mats upon the ground. Likewise all the forest-trees are dwarfed to dense shrubs in the mountain scrub, or in shallow gullies on the hillside.

up of: (α) wandering 9 consisting of: mat-form 6, cushion-form 2 and erect-creeping 1; and (β) spot-bound 11, made up of: erect-tufted 5 (ferns 5), erect-branching 4, cushion-form 1 and rosette-form 1; and (b) herbaceous 133, made up of: (α) wandering 43 consisting of mat-form 27, erect-creeping 11, turf-making 2 and rosette-form 3, and (β) spot-bound 90 consisting of: tussock-form 9 (grass 7, rush 2), erect-tufted 38, earth orchid form 12, cushion-form 6, erect-branching form 1, rosette-form 23, and straggling 1.

Coming now to the leaves they may be characterized as follows: — simple 130, compound 29, broad 125, narrow 34, very large 14, large 13, of medium size 22, small 46, very small 64, thin 64, coriaceous &c. 95, hairy 27, and glabrous 132. The most striking fact regarding the leaves of the Subantarctic plants is the occurrence amongst the *endemic* species of leaves not only of great size but of distinctly mesophytic character. This luxuriance is truly remarkable when the wind-factor is considered, but it must be pointed out that the leaves of several are only summergreen or semi-summergreen and that those of some are strengthened by stout veins. Moreover, *special luxuriance* is in harmony with considerable shelter and *decrease in size combined with flattening to the ground* comes on in pro-portion to increase in exposure.

There is no need to give any details regarding special plants since this is done to some extent both when dealing with the physiognomic species and with the associations. The species of *Anisotome*, with their very large leaves in erect rosettes, are semi-mesophytic or sub-xerophytic representatives of the intensely xerophytic *Aciphyllae* of subalpine and alpine New Zealand. At the same time, their leaf-anatomy, shows various xerophytic features[1]).

Speaking of some of the endemic species, the roots of *Chrysobactron*, *Anisotome* and *Pleurophyllum* are long, numerous, thick and fleshy. Several, including the cushion-plants (some of these non-endemic) have deeply des-cending tap-roots. In the case of creeping stems the roots are comparatively short.

The flowers of some of the endemic species show colours almost if not quite unknown in New Zealand proper; examples are: — *Pleurophyllum speciosum* (disc dark-purple, rays whitish-purple); *Myosotis capitata* (brilliant dark-blue), a most beautiful flower; *M. antarctica* (blue); *Hebe Benthami* (blue); *Epilobium confertifolium* (pink); *Gentiana cerina* (white to brilliant crimson); *Celmisia vernicosa* and *C. campbellensis* (disc purple, rays white); *Anisotome latifolia* (pale lilac, rosy-lilac, rosy-purple), *A. antipoda* (bright purple) and *Chrysobactron Rossii* (yellowish-orange). This far-greater pro-

---

1) *Anisotome latifolia* has thick, wrinkled cuticle; very thick-walled epidermal cells; strong development of stereome below epidermis, continuing through leaf to the vascular bundles; stereome at margin of leaf; 4-layered palisade. *A. antipoda* has cuticle and epidermal cells as above; stomata on both surfaces; subepidermal stereome; dense palisade and pneumatic tissue in centre of leaf.

portion of brilliant coloration[1]) than elsewhere in New Zealand and the occurrence, too, of brilliantly colored vicarious species, represented by dull-colored in New Zealand, is certainly not due to a greater proportion of insects, but quite the contrary, and even were the insects present, they could do little in the face of constant gales.

Epiphytes, including the lianoid *Polypodium diversifolium*, number 12. They consist of 7 erect, spreading or pendent creeping filmy-ferns, 3 low tufted ferns, 1 creeping rather large-leaved fern, and 1 erect or drooping semi-woody plant. But, with so many prostrate, bryophyte-clad trunks at their disposal, virtually all the forest species may be epiphytes, and so this term be to some degree a misnomer.

With regard to seasonal changes the vegetation is comparatively quiescent from May to the end of September. As the forest and scrub are evergreen the winter and summer aspect are the same, except that the leaves of *Histiopteris* are for the most part dead. In the open, where tussock does not dominate, the aspect is considerably changed, for the great leaves of *Pleurophyllum criniferum* and those of *Chrysobactron Rossii* are absent and those of other species of *Pleurophyllum* much reduced in size. *Hypolepis Millefolium* and *Polystichum cystostegia* are likewise summer-green. At about the middle of November, the herbs of Lord Auckland Islands are just coming into flower the first to appear being those of *Ranunculus pinguis, R. aucklandicus, Myosotis capitata, Hebe Benthami* and *Phyllachne clavigera*. By December, many more species are blooming freely and during that month the herb-field is full of colour, though *Pleurophyllum speciosum* is at its best in early January. Then too, *Metrosideros lucida* transforms the forest roof into a blaze of crimson. More or less of a floral display extends till March but most likely it is chiefly a few species flowering out of season that persist so long, e. g., *Stilbocarpa, Anisotome antipoda, Pleurophyllum criniferum, Chrysobactron* &c.

The plant-communities. *Dune* occurs only on Enderby Island (Lord Aucklands). True sand-binding plants are absent, the wet climate alone keeps the sand fixed and non-dune species form the association, especially *Tillaea moschata, Ranunculus acaulis, Epilobium confertifolium, Pratia arenaria* and a moss of dense habit. Where the dunes are moving inland, through disturbance by cattle, there are pure colonies of *Rumex neglectus* which may have been absent in the virgin association.

Rock and cliff vegetation (including stony shore) is made up of the following species: — *Blechnum durum* and *Asplenium obtusatum*, which finally form deep masses of peat; *Poa foliosa*, but generally where there is

---

1) Brilliance, however, is judged only from the human standpoint, and from that standpoint, too, *the colour visible from the greatest distance is white*, which also has the double advantage of being visible in dim light.

considerable depth of peat; *Scirpus aucklandicus;* green cushions of *Colobanthus muscoides*[1]) growing on solid rock and offering a station for small, shallow-rooting plants; *Callitriche antarctica,* where the rock is wet; *Tillaea moschata* and *Hebe elliptica,* this latter frequently on the summit of cliffs. In addition to the above, there occur on the Lord Auckland group and Campbell Island *Poa ramosissima* (Fig. 99), hanging on the cliff in thick, broad, pale, bluish-green sheets; *Montia fontana,* where water drips; and the two subantarctic species of *Cotula,* of which *C. plumosa* is also coastal on Antipodes Island. Certain species occur only on one or other of the islands, e. g.: — *Myosotis albida* (Snares); *Plantago subantarctica*[2]) (Aucklands); *Apium prostratum* (Antipodes).

Forest includes two distinct associations — **the southern-rata** and **the Olearia Lyallii,** the latter occurring on the Snares and the Lord Aucklands, the former on the Lord Aucklands only. The remaining groups are without forest.

The southern-rata forest occurs on the Lord Auckland Islands as a belt along the shore wherever there is sufficient shelter and extends to a varying altitude up the hills, where it is succeeded by scrub. The trees vary in size according to the degree of exposure, so that, in many places, the forest is little more than scrub. The florula numbers about 49 species of spermophyta and pteridophyta and more than 50 species of bryophyta. There are 3 trees, 6 shrubs, 8 herbs and 29 ferns including 1 tree-fern, 1 semi-tree-fern and 10 filmy ferns. Seen from a distance, the forest appears as a close, homogeneous dull-coloured mass of shrubs, rather than trees, with a slightly undulating roof of extreme density. Within, the view is truly remarkable. Everywhere are the massive prostrate and semi-prostrate trunks of *Metrosideros lucida* some-times pressed to the ground, at other times forming great arches, or at others again bridging the deep depressions of the forest-floor. From the trunks branches pass off bent and twisted, in every way conceivable, and forming frequently a rigid tangle. Ultimately, branches arise which pass upwards, branch several times and terminate in close masses of leafy twigs (Fig. 100). Without there will be a boisterous gale, but within the forest all is calm and intense hygrophytic conditions prevail. Thus there is a wonderful wealth of filmy-ferns, mosses and liverworts, the first-named forming sheets of delicate green on tree-trunks and floor and the bryophytes forming great cushions or continuous masses on the ground or trunks, or covering these with a thick mantle through which

---

1) An extremely dense and solid autosaprophytic cushion-plant measuring, at times, 54 cm. diam. The leaves are linear, glabrous, fleshy and about 6 mm. long. Within the cushion, there is a mass of yellow, sticky peat into which the peripheral shoots send roots. The cushion absorbs water like a sponge.

2) This grows on solid rock, the fleshy, rather stiff, bright-green almost glabrous leaves forming symmetrical hard, flat rosettes, each about 4 cm. diam.

the rhizomes of the filmy ferns ramify. Especially are the yellowish *Dicrano-loma Billardieri* and *Plagiochila ramosissima* and *Mastigobryum involutum* conspicuous cushion-builders, while the dark-green *Aneura multifida* makes wide patches on the forest-floor. In many places, there is a close under-growth which may consist chiefly of the semi-tree-fern, *Polystichum vestitum,* or of *Suttonia divaricata,* the 3 divaricating species of *Coprosma* and *C. foetidissima.* Here and there, the small trees *Dracophyllum longifolium* and *Nothopanax simplex* are abundant.

Olearia Lyallii forest is closely related to *O. Colensoi* coastal-scrub of Stewart Island. On the Snares, it occupies the gullies and more sheltered slopes covering much of the main island. *O. Lyallii* is usually pure, but in places there is a little *Senecio Stewartiae* of similar life-form. The dominant tree has generally a fairly thick trunk which lies prostrate for half its length, or more. A few more or less horizontal branches are given off, which, branching sparingly, finally bend upwards and branching several times into twos or threes bear, on the ultimate, stiff, white, tomentose branchlets, rosettes of dark-green, very thick, coriaceous, large leaves white beneath with dense tomentum and so close as to touch (Fig. 101). Seen from within, the forest is about 4.5 m. high; trunks sprawl over the ground, sometimes for a distance of 9 m. and everywhere there is a rigid tangle of stiff, grey branches, while above are naked stems and a close roof of white foliage. Generally, the floor is bare and undergrowth wanting, except at the bottom of a gully, where there will be a few plants of *Blechnum durum* and *Polystichum vestitum.*

The association on Ewing Island (Lord Aucklands) is similar, except that the trees are taller (6 to 9 m.), less prostrate and *Senecio Stewartiae* is absent.

Scrub occurs where wind or altitude are antagonistic to forest. The community differs on the different islands and is absent on the Snares, unless the lowest *Olearia* forest be so designated, and the Macquaries.

Lord Auckland Island's scrub consists of the forest-trees and shrubs, the former now shrubs merely (Fig. 102), with the addition of *Cassinia Vauvilliersii.* Though, in exposed positions, it occurs at sea-level, as on Enderby Island, its greatest development is as a continuous belt on the hills, united to the upper forest, at an average altitude of perhaps 150 m. At its upper limit, it merges into tussock-moor, but in the shelter of the gullies it still continues for some time. The chief peculiarity of the asso-ciation is its astonishing density. The shrubs are so rigid, much-branching and interlaced, that it is frequently impossible to force a passage through, or even to crawl beneath them; the only feasible mode of progression is to roll over their top. *Suttonia divaricata* is dominant and to this especially is the extreme density due. The scrub varies from 1 to 2 m. in height, according to exposure; its surface is uneven.

Dracophyllum scrub is the common scrub association of Campbell Island. The members are *Dracophyllum subantarcticum* (dominant), *D. longifolium*, the hybrids between these two, *D. scoparium* (rare), *Suttonia divaricata*, *Coprosma parviflora*, *C. ciliata*, *C. cuneata* together with *Blechnum procerum* and *Polystichum vestitum* when sufficiently open. Seen from a distance, the association presents an even surface and recalls *Leptospermum* shrubland of New Zealand proper. The relative proportion of *Dracophyllum* or divaricating-shrubs varies in different localities and situations, so that either life-form may dominate and, in the latter case, the association, according to LAING (1909 : 488) might be called **Coprosma scrub**. *Dracophyllum* scrub is equally as dense as the allied association of the Lord Aucklands from which it differs only in the greater abundance of the *Dracophyllum*-form, in the absence of *Metrosideros*, *Cassinia* and *Coprosma foetidissima* and in the presence of the two species of *Dracophyllum*. *Antipodes Island scrub* occurs in the sheltered gullies, descending in long dark lines down the hillsides. It consists of *Coprosma ciliata?* and *C. antipoda* and attains a height of about 1.5 m. Probably *Polystichum vestitum* is an associated species.

Moor embraces the two distinct types of vegetation — **tussock-moor** and **herb-moor**. Tussock-moor includes 3 subassociations which are dominated respectively by *Poa foliosa*, *P. litorosa* and *Danthonia antarctica*.

Poa foliosa moor occurs on the Snares and Aucklands and probably, to some extent, on the Campbells and Antipodes. The subassociation stands out conspicuously through its bright-green colour. The tussocks are about 50 cm. high and 55 cm. diam.; they grow closely and the broad leaves droop somewhat. On the Snares, *Stilbocarpa robusta* is a companion plant, either singly or in broad patches, its darker-green, great orbicular leaves constrasting with the paler grass. *Asplenium obtusatum* and *Blechnum durum* occur sparingly. On Lord Auckland Islands, apart from a frequent belt of varying breadth near the shore, where *Carex trifida* is a member, *Poa foliosa* moor exists merely as small pure patches in the *Poa litorosa* subassociation.

Poa litorosa moor is common on the Snares, Campbells, Antipodes and Disappointment Island (Lord Auckland's), but of limited extent on Lord Auckland Island itself. Frequently the tussocks are on trunks 57 cm. high, but, on the coastal slopes of Antipodes Island, the are 1.5 m. high, and grow so closely, that progress can be alone made by stepping from tussock to tussock. On Campbell Island, the association clothes the hillsides, where scrub is absent, to a height of perhaps 150 m. The most important companion-plant, at the present time[1]), is *Chrysobactron Rossii*, which has increased greatly through burning the tussock. Various other species are

---

1) *Chrysobactron* was conspicuous in the virgin vegetation, for HOOKER write (1847 : 73) "It covered the swampy sides of the hills in such profusion as to be distinctly visible at a full mile from the shore". The Campbell Islands vegetation was by the

present, e. g. *Polystichum vestitum,* dwarf shrubs, especially *Dracophyllum scoparium, Pleurophyllum speciosum* and *Hebe Benthami*[1]). On Disappointment Island, many of the herbs of the herb-moor are present, almost black clumps of *Polystichum vestitum* abundant and, in places, hidden by the tussock, is a close growth of the scrub-shrubs, *Metrosideros* excepted. *Acaena Sanguisorbae* var. *minor* scrambles over the tussocks. On Antipodes Island, where the tussocks are tall, as already described, there is little else, but, where lower, there are masses of *Polytichum vestitum* and, in the shelter this and the tussock afford a rich vegetation made up chiefly of the following: — many lichens (chiefly species of *Sticta* and *Cladonia*[2]), various bryophytes[3]), *Blechnum penna marina, B. procerum, Asplenium bulbiferum, Hypolepis Millefolium, Hymenophyllum multifidum, Lycopodium fastigiatum, L. varium* var. *polaris, Luzula crinita, Stellaria decipiens* var. *angustata, Epilobium linnaeoides* and an undescribed species[4]), *Coprosma repens,* stunted *Coprosma antipoda, Pratia arenaria* and *Helichrysum prostratum or bellidioides.* The soil of the association throughout its range is deep peat always extremely wet.

Danthonia antarctica moor is a well-marked subassociation which forms the next altitudinal belt after scrub in the Lord Aucklands and after *Poa litorosa* moor in Campbell Island giving, at a distance, a brown colour and smooth appearence to the hillsides. But such smoothness is quite illusory, the ground being most uneven while water lies in hollows of the peat which throughout is extremely wet. The dominant *Danthonia,* of tussock-form, is usually raised above the ground-surface on irregular-shaped, peaty trunks. Shrubs[5]) are dotted about, and, in the shelter they and the tussocks afford, are mats of *Coprosma repens* together with *Nertera depressa, Epilobium linnaeoides, E. confertifolium, Helichrysum prostratum*[6]) and *H. bellidioides,*

---

grazing of sheep &c., so much changed so far as the lower hill-slopes were concerned, before any ecological observations were made, that only a guess can be made as to the primitive tussock-associations. As for Antipodes Island, it has been examined only in a superficial manner.

1) An erect, loosely-branched shrub, 20 to 40 cm. high with naked, terete, very flexible branches marked with old leaf-scars, branching near their extremities into short branchlets covered with close-set, coriaceous, thick, bright-green leaves, 2.5 cm. long × 1 cm. broad and bearing short racemes of violet-blue flowers each 8 mm. diam.

2) *Sticta Freycinetii, S. orygmaea, S. filicina, Cladonia aggregata, C. verticillata, C. pycnoclada, C. gracilis* var. *campbelliana, Stereocaulon argodes* and *Usnea articulata.*

3) *Pallavicinia connivens, Lepidolaena Menziesii, Lophocolea pallida, Metzgeria glaberrima, Tylimanthus homomallus* and *Leptostomum inclinans.*

4) Perhaps *Epilobium antipodum* Petrie.

5) On the Aucklands, *Cassinia Vauvilliersii, Dracophyllum longifolium* and prostrate *Coprosma foetidissima;* on the Campbells the divaricating species of *Coprosma.*

6) Both *H. prostratum* and *H. bellidioides* occur in the subantarctic Islands but in all previous publications only the former has been recorded, though probably the rarer of the two, but no accurate information is available.

In places, tussocks without trunks grow so close together that their leaves mingle and the yellow mass, waving in the breeze, looks like a field of ripe corn. But such an apparently pure association, on the Lord Aucklands, may conceal a wiry undergrowth of *Coprosma foetidissima, C. cuneata* and *C. myrtillifolia*. When the ground becomes wetter, the tussocks are further apart and there is space for other smaller grasses and many of the smaller herbaceous plants, especially *Carpha alpina* (Lord Aucklands), the endemic *Hierochloe Brunonis, Deschampsia Chapmani, Agrostis magellanica, Deyeuxia setifolia, Chrysobactron Rossii, Ranunculus pinguis,* the endemic *Acaena, Epilobium confertifolium, E. linnaeoides, Celmisia vernicosa*[1]) and *Helichrysum prostratum* or *H. bellidioides.*

Herb-moor is distinguished by the dominance of herbaceous life-forms other than tussock which if not absent, plays quite a subordinate part. It may be divided into the subassociations **tall** and **low** herb-moor, the former lowland and not strongly xerophytic, but the latter subalpine xerophytic.

Tall herb-moor, formerly named by me "*Pleurophyllum* meadow", occurs in its greatest luxuriance on the shores of Carnley Harbour, and, in a more modified form, on the slopes of Disappointment Island and in Campbell Island. All its members are present in one or other of the communities, but here the array of stately herbs with immense leaves and, in some cases, masses of showy flowers, are gathered together, so that the glory of nearly all the magnificent endemic species can be seen at a glance. "Fairchild's garden", on the gentle slope of Adams Island near the Western Channel, is the most striking example of the subassociation and to that charming spot the following specially refers.

The plant-covering presents an irregular surface of varied greens. Near the shore are dark-green masses of *Anisotome latifolia,* knee-deep and deeper, the huge purple inflorescence more than 70 cm. high. The great pale-green, corrugated leaves of *Pleurophyllum speciosum,* in loose rosettes, are everywhere, so that one has most regretfully to trample them underfoot. The bright green leaves of *Poa foliosa* and the brown ones of *Carex appressa* are scattered through the whole. Colonies of *Stilborcarpa polaris* abound (Fig. 103), their fine, round, bristly leaves are vivid green, *Pleurophyllum criniferum* raises up its immense leaves on all sides and its flowering stems 1.5 m. high (Fig. 98). Blackish patches of *Polystichum vestitum* add a contrast to the prevailing greens. Masses of the orange blossoms of *Chrysobactron Rossii* are everywhere. In places, the beautiful *Gentiana cerina* is plentiful, its delicate flowers varying from white to crimson. Here and there is the briliant dark-blue *Myosotis capitata.* Other smaller herbs are

---

1) A semi-woody plant forming loose cushions 93 cm., or so, diam. with rosettes 4 to 6 cm. diam. of excessively glossy, dark-green, stiff, coriaceous linear leaves about 8 cm. long and bearing nomerous flower-heads 3 to 4 cm. diam., their disc purple and ray-florets white.

plentiful where there is space, e. g. *Epilobium confertifolium, E. linnaeoides, Acaena Sanguisorbae* var. *minor, Coprosma repens, Nertera depressa, Cotula plumosa* and the species of *Helichrysum.* Finally, the association shades off into *Danthonia antarctica* moor.

On Antipodes Island, in certain sheltered places there is a distinct subassociation made up of the large-leaved *Urtica australis, Poa foliosa, Polystichum vestitum,* extensive colonies of *Stilbocarpa polaris* and flat-topped bushes of *Coprosma ciliata?.*

Ground on Antipodes Islands, manured by the giant petrel *(Ossifraga gigantea)* is occupied by the endemic *Senecio antipodus,* a branching perennial herb, 20 to 60 cm. high, but, according to W. R. B. OLIVER and G. E. DU RIETZ, it also occurs on unmanured ground.

Low herb-moor occurs on sopping wet ground near the summits of the hills of the Lord Auckland Island. The association consists of species which, almost all, grow close to the surface of the ground. *Pleurophyllum Hookeri* is dominant, the rosettes being frequently so close together that many square metres glisten with the silvery covering. Sometimes the glossy-green cushions or mats of *Celmisia vernicosa* are in vast numbers. *Carpha alpina* and *Astelia linearis* generally form the groundwork of the association. The following are also common: — *Hymenophyllum multifidum, Agrostis magellanica, Luzula crinita, Chrysobactron Rossii, Ranunculus pinguis, Gentiana cerina, Myosotis capitata, Hebe Benthami* and *Coprosma repens.*

Where the ground is wettest actual bog-conditions prevail and a subassociation occurs in which the cushion-plants *Phyllachne clavigera* (Fig. 61) and *Oreobolus pectinatus* are dominant and sub-dominant respectively and the following are common: — cushions of *Gaimardia ciliata,* the rush-like *Schizaea fistulosa* var. *australis,* a turf of *Astelia subulata, A. linearis* and *Coprosma repens. Celmisia vernicosa* and *Chrysobactron* will also be present. Similar bogs exist on Campbell Island, but these in addition contain *Sphagnum antarcticum.*

On Antipodes Island there are numerous bogs in flat depressions, but their composition as follows is more nearly related to tall herb-moor: — *Marchantia cephaloscypha* forms broad, flat patches, *Hymenophyllum multifidum, Uncinia Hookeri, Carex ternaria, Luzula crinita, Anisotome antipoda, Stilbocarpa polaris, Coprosma repens, C. antipoda* (stunted) and *Pleurophyllum criniferum.*

Rock and debris communities are only to be found near the shore and on the summits of the hills in Lord Auckland Islands and Campbell Island. The special rock plants are: — *Polypodium pumilum, Colobanthus subulatus. Geum albiflorum* (Lord Aucklands only), *Schizeilema reniforme* and *Abrotanella rosulata* (Campbells only). Two associations occur to be called respectively. *the subalpine rock-association and the subalpine rock-debris association.*

In the latter association on the bare rock is the suffruticose lichen, *Stereocaulon ramulosum;* also there are black patches of several species of *Andreaea.* Peat very readily accumulates on ledges and in hollows, so that there is hardly any species belonging to neighbouring communities that does not occur on rock; in fact, *Anisotome antipoda, Phyllachne clavigera, Coprosma repens* and other species may form a virtually closed association.

The principal crevice-plants are *Cardamine depressa, Colobanthus subulatus, Geum albiflorum* (Lord Aucklands), *Schizeilema reniforme* and *Abrotanella rosulata* (Campbells). Very characteristic are *Polypodium pumilum* and *Hymenophyllum multifidum* either forming lines in crevices or great sheets on the rock-face. Certain mosses are common on wet rocks, especially *Braunia Humboldtii, Lophiodon strictus, Conostomum australe* and *Hypnum hispidum.*

Beneath the cliffs which form the actual summit of many of the hills of both the Aucklands and Campbells there are debris-fields. But although he primitive substratum would be rock-fragments merely, it would take little time in the wet subantarctic climate before plants settled down, many of which from their peculiar nature would rapidly cause peat to be formed. The substratum is always as wet as possible, so that there is here no need for shingle-slip "adaptations", rather is wind-resisting power the chief desideratum, here attained by lowness of stature, the cushion-form, the ground-rosette form and the prostrate-habit. The plants peculiar to the association, or nearly so are: — *Polystichum cystostegia* (growing where debris is largest), *Marsippospermum gracile, Cardamine glacialis* var. *subcarnosa* and *Plantago aucklandica* (Lord Aucklands). The other species are the same as for low moor but the relative percentage is different. *Myosotis capitata, Ranunculus pinguis* and *Celmisia vernicosa* are abundant. *Pleurophyllum Hookeri* still plays a most important part.

Swamp is essentially a lowland community. It is especially distinguished by plants of the trunks-tussock form. *Carex trifida, C. appressa* and *Poa litorosa* are abundant, the former on the Lord Aucklands and Campbells, and the two latter on Antipodes as well, where *Polystichum vestitum,* in dense masses is very characteristic. *Blechnum procerum* is also a frequent swamp-plant.

The vegetation of Macquarie Island forms, for the most part, a class by itself. Not having visited the island myself, what follows is taken from the writings of SCOTT (1883), A. HAMILTON (1895) and CHEESEMAN (1919), the latter based largely on specimens collected by H. HAMILTON, but containing little new regarding the vegetation.

On cliffs subject to drenching with salt-water, there are, in abundance, cushions of *Colobanthus muscoides,* together with the small endemic grass *Triodia macquariensis* and *Tillaea moschata.* The last named and *Cotula plumosa* occur on the beach. The subantarctic American - Kerguelen *Festuca erecta* is also a plant of coastal rocks. Where the ground is swampy there is a close growth of *Poa foliosa* with tall trunks. If tussock is absent, there is *Cardamine corymbosa, Montia fontana* and *Callitriche antarctica.*

The hill-slopes are occupied by a tall growth of *Poa foliosa* tussock, *Stilbocarpa polaris* and the silvery rosettes of *Pleurophyllum Hookeri;* here too is *Acaena adscendens* and *A. Sanguisorbae* var. *minor.* This association is of considerable extent. All the above is distinctly New Zealand-Subantarctic, but on the exposed hill-tops the scene changes; the wind has here the mastery and the formation is allied to the "wind-desert" of Kerguelen Land. Here is A. HAMILTON'S vivid account: "At about 300 feet you gain a plateau so swept by the antarctic gales that the vegetation is reduced to compact closely growing mosses, small *Uncinias* and the conspicuous cushion-like masses of *Azorella Selago.* In the hollows of the uplands are countless little tarns or lakes, some of considerable extent. Round the tops of the hills the wind has cut out wonderful terraces from a few inches to a foot or two in height, with completely bare rock, much disintegrated by the weather on the top. In some of the more sheltered places or gullies stunted plants of *Stilbocarpa* and *Pleurophyllum* cover the ground."

Ligneous plants are absent, the representative of the subantarctic divaricating species of *Coprosma* being the mat-forming herbaceous or semi-woody *C. repens.*

The Bounty Islands consist of a small group of islets and rocks, the largest about 1 km. in length and 90 m. high. Their sole rock is granite worn smooth as glass by the polishing action for ages of the feet of thousands of penguins and many seals. Large quantities of guano are deposited during the breeding season, but it is washed away by the rains of winter. Except for one species of fresh-water alga, vegetation is absent on the land-surface, but at the shore-line is, in places, abundance of *Durvillea antarctica.* Doubtless the islands were once a part of "Greater New Zealand" and possessed a fairly rich flora.

Part III.

The Effect of Settlement upon the Plant-Covering of New Zealand.

Chapter I.

The Introduced Plants growing wild without Cultivation.

General. In Part II of this book an attempt has been made to paint a picture both of the primeval plant-covering of New Zealand and of those communities as yet but slightly modified by the action of man. This Part, on the contrary, treats of that greatly altered or entirely new vegetation which now occupies so much of the land, and some of the fundamental causes are discussed which have operated in bringing about the vast difference between the New Zealand of 1769 and that of 1927. On this intensely interesting subject a good deal has been written, some of which contains important data and luminous suggestions, but, on the other hand, erroneous statements are even yet widely accepted as truths and misconceptions have crept into authoritative scientific writings[1]).

At the time of COOK'S first visit in 1769—70, except for changes wrought by the aborigenes, *the vegetation almost everywhere was intact* and *the flora contained no aliens.* To what extent the native race had altered primitive New Zealand it is not possible to determine, but although the neolithic population may have reached 200 000, its power to damage the vegetation was slight[2]). Here and there clearings for cultivations were made in forest, shrubland or heath, and though new combinations of indig-

---

1) A. R. WALLACE (1889 : 15) writes, "In New Zealand there are more than 250 species of naturalised European plants, more than 100 species of which have spread widely over the country often displacing the native vegetation", and on pp. 28, 29 it is stated that, "in New Zealand (white clover) is exterminating many native species, including even the native flax (*Phormium tenax*)."

2) Even now, in the Urewera Country (EC.), where the Maori still maintains, in large measure, his primitive conditions, except in the wider valleys and, to a limited extent on the adjacent slopes, the vegetation is primeval, notwithstanding the people possessing agricultural implements and their grassing areas by means of "bush-burns" and subsequent grazing, as described further on.

enous species would arise on the cleared ground when abandoned, they would hardly be of a permanent character but in time would revert to the original plant-covering[1]). On the other hand, judging from the observations of BANKS[2]), the Maori appears to have made considerable use of fire for clearing forest &c., so that certain apparently primeval areas, especially *Pteridium* heath and *Leptospermum* shrubland, and perhaps tussock-grassland, to some extent, may have originated from ancient forest-fires. But when the vast areas of undoubtedly virgin forest that only *now* are vanishing are borne in mind, it is clear that *the aborigenes brought about no changes of moment,* and that *it was a truly, primeval scene that met the gaze of the early botanists.* How great the difference in much of the present plant-covering is clearly brought home from the statistics in Chapter III and the details in Chapter II of this section, which show that the greater part of the lowlands has now a plant-covering resembling that of Europe rather than New Zealand, and that there are also wild associations of *recent* origin composed altogether of indigenous plants. Thus, as well as an indigenous flora, there is a second composed of introduced species, some of which are so well suited to their new environment that they flourish side by side with indigenous plants making new associations, or, in other cases, the exotics, thanks generally to the direct influence of man, have formed pure communities. In other words, a new vegetation and flora are being evolved and various stages of the process everywhere afford invaluable material for research.

Statistical. The number of exotic species more or less firmly established in New Zealand is about 514[3]), including 3 confined to the Ker-

---

1) RUTLAND (1901: 324—326) shows how *Podocarpus totara* invaded abandoned Maori clearings and considers that much of the forest on the shores of Pelorus Sound is a regrowth. Various varieties of *Phormium tenax* were frequently cultivated, and COLENSO describes (1881: 19) how he has seen remains of old plantations miles away from any Maori dwelling. Also some of the groves of *Corynocarpus laevigata* are due to the planting of that tree.

2) "Here we saw many great smokes, some near the beach, others between the hills, some very far within land, which we looked upon as great indications of a populous country"... "At night we were off Hawke's Bay and saw two monstrous fires inland on the hills. We are now inclined to think that these, and most if not all the great fires that we have seen, are made for the convenience of clearing the land for tillage" (BANKS 1896: 183, 189). But such fires would be in heath or shrubland, since it is impossible to burn standing virgin subtropical rain-forest.

3) This is 62 species less than in the second edition of *The Manual of the New Zealand Flora,* notwithstanding it includes 31 species admitted as indigenous by CHEESEMAN some of which, however, he considered exotic. The species rejected include those recorded by T. KIRK (1896: 501—507) as occuring on a ballast-heap which have not since been reported from any other locality, together with plants which linger in deserted gardens, species once recorded which have died out, and a few others of doubtful occurence. In fact, my estimate might be somewhat reduced with advantage.

madecs, (pteridophytes 2, gymnosperms 3 — but probably more — mono-cotyledons 113 and dicotyledons 396). They belong to 75 families and 279 genera, the largest, together with the number of species for each, being the following: — (families) *Gramineae* 83, *Compositae* 75, *Leguminosae* 46, *Cruciferae* 30, *Caryophyllaceae* 23, *Labiatae* 16, *Scrophulariaceae* 15, *Rosaceae* 14, and *Polygoonaceae* and *Ranunculaceae* 13 each; (genera) *Trifolium* 16, *Ranunculus* 13, and *Bromus* 9. Twenty-four families and 219 genera are represented which are wanting in the New Zealand flora.

As for the source of this large exotic element, by far the greater part is European, no fewer than 375 species belonging to this class. Australian species number only 30, notwithstanding constant traffic for many years with Australia, including the importation of living plants, seeds and agricultural products. North American, South American and South African species number respectively 26, 18 and 12, and the remainder are mostly subtropical or tropical of wide distribution.

Coming now to the distribution within the Region of the exotic species, it is as follows: — North Island 469 species, 111 of which are restricted thereto; South Island 399, 41 of which are restricted thereto; Stewart Island 72 (COCKAYNE: 1909) but there must be more, yet so much of the vegetation is virgin and introduced plants cannot become established; Kermadec Islands, as recorded by OLIVER, 51; Chatham Islands (list supplied by E. N. NORTHCROFT) 128; and Subantarctic Islands 25, including *Phormium tenax* and *Acaena Sanguisorbae* var. *pusilla* of New Zealand proper, but Campbell Island having now being used for sheep-farming for more than 24 years, there should be considerably more. With regard to their vertical distribution, the greater part is confined to the lowland belt of which some 12 are restricted to the coast-line or there abouts, and 50, at a high estimate, gain the montane and lower subalpine belts, but none are restricted to the high mountains.

Regarding the lifeforms of the species, 23 are trees, 28 shrubs, 94 of the grass-form, 5 of the rush-form, 344 herbs or semi-woody plants, 6 lianes, 3 parasites and 10 water-plants.

The species themselves differ greatly in their relative abundance and there is a gradual decrease from those of the widest distribution, and with ample individuals, to such as are only recorded so far from one or two localities, where they just hold their own. Certain species, too, are present in abundance, but they are confined to a definite habitat of perhaps limited extent and others, again, are restricted to the cultivated areas, or to waste ground, and really have little or nothing to do with modifying the primitive associations. It is then a most difficult matter to decide as to relative abundance or importance, and I am far from satisfied with the figures about to be given, since species belonging to different categories are counted as equal. About 64 species may be considered very common, 65 common, 105 neither common nor rare, 54 of local occurrence, 100 rare and 126 — the largest class — very rare; indeed, if the species of the last two classes be taken together, it is plain that 44 per cent of the exotic flora is of no moment. On the other hand, the very common and common species form together only 22 per cent of the aliens. Perhaps, on the whole, the most wide-spread species are *Rumex Acetosella* and *Hypochoeris radicata* followed, but at some distance, by *Holcus lanatus* and *Trifolium repens*. But none of the commonest species, nor, indeed, any of the exotic plants, are nearly so "aggressive" or have spread so widely on *uncultivated ground*, under the influence of settlement, as the indigenous *Pteridium esculentum, Leptospermum scoparium* or perhaps *Danthonia pilosa* in its various forms. Various forms of *Acaena novae-zelandiae* and *A. Sanguisorbae* are also almost the equal in aggression of any exotic alien. Without giving a full list of the 64 very common species the following may give some idea of their character: — *Anthoxanthum odoratum, Agrostis alba, Holcus lanatus, Dactylis glomerata, Poa pratensis, Ranunculus repens, Rubus fruticosus* (in a wide sense), *Ulex europaeus, Cytisus scoparius, Trifolium repens, Centaurium umbellatum, Prunella vulgaris, Plantago major, Erigeron canadensis, Chrysanthemum Leucanthemum, Cirsium arvense* and *Hypochoeris radicata*.

In order to ascertain more fully the position the exotic plants hold in the vegetation it is necessary to consider their habitats. In this regard **waste ground** far and away stands first in the number of its alien plant-population. Ground of this character is always in process of being made in a "new" country and it occurs on roadsides, on little used roads themselves, on railway embankments and the like, on cuttings, on unused sections in towns and villages and, indeed, on bare ground generally. Obviously, far more than one habitat is included under the above term, but in all there is *open ground* ready for occupation by plants and all kinds of growing-places are provided in regard to sun, shade and moisture-content of the soil. If to the species of waste ground be added those of cultivated land in its widest sense, by far the greater part of the exotic flora is accounted for and 100 *species is perhaps too wide an estimate for those aliens which really come into competition with the indigenous species.*

Chapter II.
The New Vegetation.

1. General.

Although the communities dealt with in this chapter fall conveniently under certain heads, there are not only intermediates between them, but it is sometimes difficult to decide into what class any particular combination of plants should go. Even in a series apparently so clear-cut as *artificial farmlands,* nature, simultaneously, or very soon, may alter the intention of the farmer by bringing in more or less **weeds;** indeed, the latter element not infrequently dominates the community which is then transformed from one purely artificial to one of an exotic-induced or even an indigenous-induced character. As for the communities themselves, they are generally successions, the progress of which cannot usually be predicted for certain. Even when an apparently well-established association is present over wide areas (e. g. *Ulex* thicket, manuka shrubland, water-cress tangle), there is no telling what may eventually come abut through the establishment of some species new to the community, or some disease attacking the dominant member. Here, then, the term **community** is generally used rather than one speaking more definitely of fixity.

In considering this new vegetation as a whole, one fact — surprising enough to those taught to believe in the extreme aggressiveness of what HOOKER termed the "Scandinavian flora" — is *the far-greater aggressiveness — in their native land — of the New Zealand indigenous species themselves.* Where rain-forest or swamp-forest has been felled and artificially replaced by European pasture-plants, as detailed further on, were it not for the constant presence of the farmers grazing and browsing animals, there would be rapid reversion to forest. *In fact, it is hardly going too far to declare, that were such animals entirely removed from North Island, the whole of the present "permanent pastures" would in one hundred years, or less, be well on the road once more towards dense rain-forest!*

Certainly the effect of the grazing and browsing mammal cannot be overestimated. Even, a plant-covering, so persistent as *Pteridium* heath (as detailed further on), is constantly being turned into grassland by aid of overstocking and burning, but alone the latter does nothing. If all stock, including rabbits, were removed from the induced steppe of semi-arid Central Otago, experiment has clearly demonstrated (COCKAYNE, L., 1922 a : 142) that, in certain habitats, there would be a gradual return to something not unlike the primitive plant-covering, while, if suitable seeds were sown, the process of pasture-establishment would be rapid, except in the most arid and stony places (lo. cit.: 142—43).

2. Modified communities.

General. A plant-community, without destroying its primitive stamp and general ecological and floristic structure, may readily be altered by a change in the relative abundance of its constituents, which may be so marked that a species of but little moment in the primitive community, or one introduced, may become common, or even dominant, and in this way a new community come into being.

The modifications may consist merely of increase or decrease in abundance of certain species already present but, on the other hand, species either indigenous, exotic, or both together, frequently join the community and become definite members.

The causes of modifications are principally grazing or burning, or these combined; also seeds may be purposely sown or plants planted, or land may be drained, forest felled; indeed, there are many causes. In associations containing edible species, the **relative palatability**[1]) of their various members plays the foremost part and it can be seen how grazing, or browsing, may rapidly alter the composition of certain classes of pasture or forest without affecting their general physiognomy.

Fire, of course, functions in proportion to the damage done and in regard to the species destroyed or damaged. In forest, regeneration after fire will depend, to some extent, on the ability of certain species to put forth suckers or the contrary. It also destroys the humus of the forest-floor, so bringing about new conditions for the germination of seeds.

Both grazing &c. and burning lead to increase in light-demanding species, both in number and variety, but such additions as arrive come, almost invariably, from plants very near the scene of damage. On the other hand, stock brought from some distant locality frequently bring in seeds of plants not present in the association (e. g. species of *Acaena, Danthonia pilosa, Carex resectans, Poa maniototo*).

Some examples of modified communities. The student of New Zealand vegetation has usually not far to go in order to see examples of modified associations; indeed, much of the apparently primitive vegetation he will encounter in the lowland and montane belts will be more or less modified. This has already been stressed in Part II, where various modified communities are dealt with. As the changes in these partly-modified communities become, by degrees, intensified, according as exotic or indigenous species rule, so will the communities become **exotic-induced** or **indigenous-induced**. All the same, so long as the ecological conditions governing a modified community remain constant, it will bear the stamp of fixity. Thus, many modified types of vegetation may well be expected to remain for a long time much as they now are provided no new disturbing factor appears. Virtually all lowland and montane tussock-grassland, swamp and salt-meadow, and a good deal of forest, dune and even rock-vegetation come into this class.

In some of the modified communities certain conspicuous species play

---

1) *Thus, in the great tussock-grassland formation*, nearly all the exotic edible plants possess palatability to a far higher degree than do the indigenous grasses. For example, *Festuca rubra* var. *fallax*, a species of low palatability, ranks in this respect with *Agropyron scabrum* and *Poa intermedia* — species of great repute among high-country sheep-farmers — and the dominant *Festuca novae-zelandiae*, also regarded as of high palatability, is not eaten at all by sheep except the young leaves produced by burning. I have also observed sheep grazing in semi-coastal forest and how various shrubs &c. were eaten, some with avidity (e. g. *Brachyglottis repanda)* and others rejected.

such an important part as to affect their physiognomy. In what follows
various examples are given.

Where *Ammophila arenaria* has been purposely planted to arrest dune-
movement, self-sown plants may invade the adjacent primitive sand-grass
association and grow side by side with *Desmoschoenus spiralis* or *Spinifex
hirsutus;* so, too, *Oenothera odorata* may become abundant and make many
parts of more stable dunes, and also sandy river-beds, gay with its yellow
blossoms. *Sambucus nigra* has become firmly established in certain lowland
forests (its seeds brought by birds) where the light has been greatly increased through destruction of undergrowth; also, in the different environment
of shady gullies in semi-arid Central Otago, it is now a common member
of certain patches of scrub. *Hypericum Androsaemum* is frequent both in
damaged lowland forest and in montane *Pteridium* heath (SO.), it having
come in after fire. *Senecio mikanioides* climbs over small trees and shrubs
on the outskirts of semi-coastal forest in some parts of North Island and
north of South Island. *Verbascum Thapsus* is now a most important member
of the plant-covering of river-terraces and other stony or dry spots in the
montane belt of South Island; by the settlers it is well known by the name
"tobaco-plant". *Plantago Coronopus* forms extensive colonies in certain salt-
meadows looking exactly as if indigenous; so, too, *Lepturus incurvatus*
is equally at home in the same formation; in places, some of the indigenous
species of salt-meadow are the hosts of the parasitic *Cuscuta Epithymum.*
Pinus radiata has extended into the *Leptospermum* shrubland (VP.) and,
in course of time, should make a distinct association. *Hypochoeris radicata*
is a most important constituent of low tussock-grassland.

3. Exotic-induced communities.

a. Tree associations.

At Waitati, near Dunedin (SO.) the burning of dense *Leptospermum*
shrubland — really an indigenous-induced community — allowed the seeds
from a mature tree of *Eucalyptus radiata* to germinate and there is now
Eucalyptus forest, thanks to the destruction of the *Leptospermum* having
provided the essential light-requirement for the development of the seedling
eucalypts, and their far more rapid growth than those of *L. scoparium.* In
Kawarau Gorge (NO.), *Eucalyptus globulus* has formed dense groups on
rocky ground, the seeds having come from neighbouring trees, originally
planted by gold-diggers in the "sixties" of last century.

Salix fragilis and *S. babylonica,* to a lesser extent, planted in the first
instance on banks of rivers, thanks to the rooting-power of their broken
twigs, line the margins of many streams.

Albizzia lophantha is self-established in many parts of the North Auckland district. *Acacia dealbata,* both on dunes and the northern gumlands,

in places, spread from parent trees by means of creeping, underground stems and forms dense groves. So, too, *Robinia Pseud-acacia* makes close thickets in the neigbourhood of old mission stations in several localities in North Island.

Near Lake Wanaka (NO.) windborne seed of *Cupressus macrocarpa* and *Pinus radiata,* trapped by bushes of *Discaria toumatou,* germinates freely and many trees have been so established in the presence of stock and rabbits thanks to their natural spiny protector.

b. Shrub communities.

Leguminous shrubs. *Ulex europaeus*[1]), *Cytisus scoparius, Lupinus arboreus* and, to a much lesser degree, *Cytisus candicans,* invade open communities replacing and displacing the primitive vegetation. The seeds germinate in the neighbourhood of the parent shrub, and, in the first instance, the invasion was from shrubs purposely planted. At the present time, there are vast impenetrable thickets of pure (usually) *Ulex* and *Cytisus* on stony river-bed and hillsides where forest has been burnt, which when in blossom are a glorious spectacle; both species, too, are abundant on lowland and lower-montane tussock-grassland and fixed dune. At an altitude of 700 m. in South Island, the above are no longer aggressive.

Lupinus arboreus[2]) is confined to dry sandy or stony stations forming close thickets, 1.8 m. or more high, on fixed or semi-stable dunes; it also occurs to some extent on river-bed.

Rosaceous shrubs. *Rosa Eglanteria, R. canina*[3]), *Rubus fruticosus* in a wide sense and *R. laciniatus* form individually, in many localities, extensive thickets which differ from those of the *Leguminosae* inasmuch as they owe their distribution to birds[4]) while climate restricts *Rubus* to wet and *Rosa* to dry areas.

Rubus thicket is especially aggressive in forest-clearings. Though occurring abundantly in many places, it attains the greatest luxuriance in the East Cape, Western, North-western, Egmont-Wanganui and Chatham districts.

Rosa thicket becomes readily established on low tussock-grassland, making eventually a pure association. It is specially abundant in the North-

---

1) This was early on introduced as a hedge-plant and is still extensively used for that purpose. Where left uncut, the hedge exceeds 3.6 m. in height and seed is shed in profusion. Burning has no effect in eradicating the plants; seedlings also are produced in millions and grow with great rapidity.

2) This shrub was purposely planted, or sown, on dunes, in the first instance, in order to check drifting sand. This it is unable to do, but it forbids all sand-movement on the ground it occupies.

3) Of local occurrence.

4) Some forms of *Rubus fruticosus* spread vegetatively by means of natural "layering".

eastern district, ascending to about 900 m. In Central Otago it is common on old mining tailings and in gullies in company with *Discaria* and the introduced *Sambucus nigra*.

Australian shrubs[1]). *Hakea acicularis*[2]), *Epacris purpurascens*, *E. microphylla* and *E. pulchella* form extensive colonies in *Leptospermum* shrubland (NA., SA.) the first-named being wide-spread, but the three epacrids being confined to one locality. The *Hakea* spreads after *Leptospermum scoparium* is burnt and probably the other shrubs first made their appearance after fire.

c. Communities of herbaceous and semi-woody plants.

Thymus vulgaris (garden thyme) association. This well-known culinary plant occurs sporadically in several parts of semi-arid Central Otago, but on old mining tailings at Ophir several hectares are occupied by a close growth of this species, the flowers of which vary from crimson to white in many intermediate shades.

Some herbaceous plant communities. *Centranthus ruber* occasionally forms a distinct and beautiful association with its abundant red, white and pink blossoms. The substratum preferred appears to be rocky slopes. One area under my close observation for some years is gradually extending its limits.

Where forest has been burned in areas subject to a considerable rainfall *Digitalis purpurea* forms colonies which extend for several kilometres at a time over hillsides. It also occurs in profusion on old mining tailings (SO.), rocky ground (NA.), and in the Taieri Gorge (SO.) along with *Leptospermum scoparium* until the average limit of the south-westerly downpour is reached, when the community halts, all on a sudden, and the induced steppe country begins.

The case of *Eschscholzia californica* is of particular interest. On the old flood-plain of the R. Clutha near Clyde and Cromwell, (NO.) on sandy ground, there is virtually a pure association of the above extending for several hectares at a time which, when in full bloom, is a dazzling spectacle with the yellows, brilliant oranges and creams of the fully opened

---

1) It is obvious that Australian trees and shrubs, if sufficiently hardy, are much better suited to New Zealand conditions than was supposed. The belief that Australian plants would not thrive was partly based on the statement that BIDWILL was in the habit of habitually scattering Australian seeds during his travels through the country. But we have no evidence as to the nature of the seeds, their age, the soil or locality where scattered or the time of year; indeed the experience of BIDWILL, even if a fact, proves nothing.

2) *Hakea saligna* forms close thickets on sour, boggy soil near Collingwood (NW.), but F. G. GIBBS has informed me that they have originated from seed purposely sown, but probably seedlings are being established from seed shed from this *artificial* shrubland. Where *H. acicularis* equals the *Leptospermum* in quantity, or is not so abundant, the community will be merely a modified one.

flowers under the cloudless sky. The species also extends on to the depleted land. It is common, too, on the Marlborough Plain (NE.), but there the plants are more scattered. Also, it is greatly in evidence on certain stony river-beds (E.). The following species, growing under similar conditions to *Eschscholzia* — which also may be present — on more or less depleted grassland, and forming a distinct association are: — *Echium vulgare, Acaena ovina* and *Gilia squarrosa* and with is frequently *Marrubium vulgare* which, also, along with *Urtica urens* and *Hordeum murinum* makes a community on ground where sheep camp.

4. Indigenous-induced communities.

a. General.

There is nothing out of the way in indigenous and exotic species coming together and forming communities, or in the exotics making pure associations or colonies of themselves alone; on the contrary, these are what all phyto-geographers must expect. But, that the indigenous plants, in the presence of hundreds of exotics dispersing far and wide highly viable seeds, can construct pure association, must cause some surprise, which cannot fail to be intensified on learning that *communities, purely New Zealand in facies, but unknown in the primitive vegetation, have come into being when in competition with well established exotic-induced associations.* In other words, contrary to what has been authoritatively reiterated *ad nauseam,* the indigenous species, *in their own domain,* are far more aggressive than the exotics. That is, that after all, *the life-forms and "adaptations" of New Zealand plants are better suited for New Zealand conditions than those of any class of plants as yet introduced into the Region.* In fact, *wherever any part of New Zealand is in its primitive condition and uninter-fered with by man or the animals he has introduced, none of the exotics have gained a foothold, their great powers of dissemination notwithstanding, although the virgin area may be pierced in all directions by ground occupied by man where there are introduced species in plenty.* On the other hand, where man has separately, or collectively, brought into play fire, draining, cultivation and introduced domestic or feral animals, he has created a new environment where indigenous and introduced species, if the latter be present, alike go to the wall and new associations arise, or are purposely produced, made up, it may be, purely of exotics, though frequently of such and native plants. Thus, *there are two distinct areas, the one dominated by primitive New Zealand conditions and the other by such as approximate to those of Europe, while between these extremes is a gradual series of intermediates.*

Between the modified, exotic-induced and indigenous-induced communities there are frequently many transitions or successions leading from one to the other, so that it sometimes becomes a matter of personal opinion into which class this or that community should go.

At a low estimate, *there are probably 50 more or less well-marked indigenous-induced associations or successions,* some of trifling extend and others occupying wide areas; here only a limited number can be dealt with.

b. The grassland series.

Danthonia pilosa grassland. As far as can be judged from an examination of the present modified tussock-grassland, and from and estimate of its primitive structure, it seems certain that *Danthonia pilosa*[1]) was rare, or, in places, absent in the unmodified formation. It would, however, be fairly common on rocks, open stony ground, between low shrubs, and like places. Certainly there was no primitive grassland where it was dominant.

With the introduction in 1850 of sheep-farming in South Island a powerful factor — the grazing mammal —, unknown in primitive New Zealand, came into play. Nor was this all, for the dominant tussock-grasses (*Festuca novae-zelandiae* and *Poa caespitosa* were quite neglected by the sheep. So, in order to add to the scanty food supply, the tussocks were set ablaze, many hectares at a time being burnt in order that there would be plenty of young green leaves for food. The vigorous tussocks of the early days would tolerate a good deal of burning, but as this was carried out to excess, and at all seasons, the tussocks became smaller, and some after repeated burning, followed by grazing, died outright and so, as the years sped, more and more bare ground became available for recolonization. Such was quickly occupied, light-demanding species, previously kept in check by the tussocks, seized on the new ground, so that the modified tussock-grassland of the present day came into being, with its mat-plants, low cushion-plants, small grasses and dwarf shrubs. Perhaps, first of all, it was in the dry North-eastern district that *Danthonia pilosa,* so well-equipped for occupying the open spaces, gradually increased in amount. Nor was this because it was inedible, but quite the contrary, since it is liked by sheep especially in spring, while the grazing is really beneficial since it encourages the turf-making habit. Nor is this all, for "danthonia" — as it is now called — not only tolerates burning, but the young leaves are specially palatable. It is no wonder, then, that the North-eastern pastures, over wide areas are now devoid of tussocks and present a more or less continuous turf of *Danthonia pilosa*[3]), not merely the association but the formation being changed.

------

1) The species is compound but, so far only one jordanon has been recognized (var. *racemosa*). Probably it crosses with *D. semiannularis,* another compound species. *D. pilosa* is low-growing, of a more or less turf-forming habit (intensified by grazing), with leaves narrow, flat or involute (according to circumstances). *D. semiannularis* is of tufled habit. Both are light-demanding, hence their absence in dense tussock-grassland.

2) Many areas were not grazed until 1860, or later.

3) On stony or wet ground, exotics and indigenous enter in, but occupy little space compared with the dominant danthonia. Exotic grasses, of highly aggressive power

Induced steppe. In particularly dry parts of South Island the primitive tussock-grassland has been entirely replaced by a desert-like formation, here designated **steppe.** How this striking transformation has come about, and of what the new vegetation consists, are matters of far more than local interest.

The maximum steppe development occurs in the most arid part of New Zealand, the upper basin of the R. Clutha[2]). Originally the valleys and hillsides, now virtually desert up to about an average altitude of 900 m., at the advent of sheep-farming were clothed with low tussock-grassland which must have existed for a very long period. Its subsequent disappearance, then, is due to those new conditions imposed by the sheep-farmers. But such were identical with those experienced for the same period by the formation elsewhere, yet, with the exceptions noted, steppe is unknown. Plainly, then, the sole difference for the North Otago grassland lay in its special rainless climate[3]) and its temperature, except in winter, easily the highest in New Zealand.

In order to render their harsh feed palatable for sheep, as already explained, the tussocks were burnt at all seasons. Now, in an arid climate or situation, a tussock can badly tolerate burning even during a moist period and the subsequent eating of the young leaves is highly detrimental to its well-being, while burning during hot dry weather may cause death. Also, consider the effect of a second burning on tussock not recovered from the previous attack, and then think of the result of indiscriminate burning coupled with heavy grazing year in and year out!

The lower mountain slopes and valleys, now depleted to the full (Figs. 82, 104), are where the flocks have been wintered yearly, nor can they be driven on to the high pastures until late in spring, hence this lower country has been heavily overstocked. So, what with burning and overstocking, more and more bare ground would gradually appear, the palatable plants would be eaten out, and the tussocks become weaker and weaker until they too vanished. In some parts of the valleys blown sand and gravel would be an additional agent of destruction (Fig. 59).

---

elsewhere, are powerless, e. g. *Anthoxanthum odoratum, Agrostis tenuis, A. alba* (in a wide sense) and *Poa pratensis.*

1) An account of such sowing is given when dealing with artificial communities near the end of this chapter.

2) Steppe in the making can be observed in other parts of the great lowland-montane tussock-grassland formation, notably on the Mackenzie basin-plain (north of NO.) and near the source of the R. Awatere (NE.), so that the factors concerned stand out clearly enough.

3) That the special climate, above all else, is the deciding factor with regard to depletion, is further supported by the facts that in the arid area itself, as the rainfall gradually increases, so does depletion gradually decrease and that the sunny slope of even a shallow gully may be fully depleted while on the sunless slope there will be more or less tussock.

Finally, in the early "eighties" of last century, or a little earlier, the rabbit arrived from the lowlands and, as the supply of food[1]) became insufficient, crept higher and higher up the mountains. With an eminently favourable climate, abundant food, soil suitable for burrowing or rocks in plenty for their homes, these rodents increased almost beyond belief, so that with them and the many thousands of sheep the runs became greatly overstocked. Every plant at all palatable was eaten to the ground, the depleted area ascended higher and higher, those plants alone surviving which increased rapidly from seed, spread by means of underground stems, or were not eaten at all or only to a negligible extent.

At the present time, the scene is changed beyond recognition, instead of a close array of tussocks there is a man-made desert. With the tussocks are likewise gone nearly all the members of the ancient grassland. Viewed from a distance the mountains resemble giant sand-dunes. A close view shows, however, on sunny slopes a surface of bare, hard soil in places destitute of all visible plant-life except on rocks (Fig. 82), but usually having growing upon it at from about 60 cm. to several metres apart, that hall-mark of the rabbit, the hard, flat, silvery cushions or mats of *Raoulia lutescens* which are frequently 1 m. or more diam. (Figs. 59, 104). There are also more or less of the thinner, open mats of *R. apice-nigra* and occasionally those of *R. australis,* while, in the extreme west of the depleted area, the soft, silvery mats of *R. Parkii* abound and with them are many dull whitish-green, soft, hairy cushions of *Pimelea sericeo-villosa* and glaucous-green mats of *Acaena Buchanani.* Various plants, usually inconspicuous, are dotted about the otherwise bare ground, or epiphytic on the *Raoulia* cushions, especially the little tufted grass, *Poa maniototo.* the tiny hard brown cushions of *Colobanthus brevisepalus* and small brown mats of *Stellaria gracilenta.* Where conditions are slightly more favourable perhaps a dozen more species are present[2]), most of which along with

---

1) Besides the indigenous edible species, various exotic plants of higher palatability had become wide-spread, easily the most important of which are *Rumex Acetosella* and *Erodium cicutarium.* In spring, there are various annual grasses, especially *Hordeum murinum* and species of *Bromus,* several of which including *Hordeum* are eaten when young. *Carduus pycnocephalus* is extremely abundant but, as certain palatability experiments conducted by me proved, it is only eaten when there is little else; so, too with the abundant *Cirsium arvense.* Where a few tussocks of *Poa caespitosa* still persist on badly-depleted ground, notwithstanding the host of rabbits they remain untouched.

2) The slender *Poa Lindsayi,* a small tufted jordanon or epharmone of *Danthonia semiannularis* (eaten to the ground by rabbits but never killed), green mats of *Carex resectans,* occasional small rosettes of *Geranium sessiliflorum* var. *glabrum,* tufts of stunted *Oxalis corniculata* (in a wide sense), the small erect *Hypericum gramineum,* rigid open cushions of *Hymenanthera alpina* (Fig. 83) and the erect slender reddish *Epilobium Hectori.* On sunless slopes are many colonies of the exotic *Cnicus arvensis. Rumex Acetosella* is at home almost anywhere. On sunny slopes are *Reseda luteola* and

those already cited would be either absent or rare in the original grassland. Where there is rather more rain various species of *Acaena*, particularly *A. Sanguisorbae* var. *pilosa*, are in extreme abundance and, on the less sunny slopes, *Chrysobactron Hookeri* in great quantity — an unexpected species which usually grows in very wet ground.

Near the upper limit of depletion (800 to 1100 m.), dead and moribund tussocks of *Agropyron scabrum* and *Poa intermedia* appear in quantity and, *crossing* this belt, the almost primitive subalpine tussock-grassland is entered with *P. intermedia* dominant.

In addition to the perennial steppe-vegetation there is a remarkable vernal florula[1]) made up largely of annuals, many of the members growing close together, so that for a brief period many more or less shady hillsides and the valley floors, desert-like for most of the year, become green and look quite fertile. If the flora of the whole Central Otago induced-steppe be considered, it consists of about 91 species (54 indigenous, 37 exotic) about 40 of which are common, but not more than 30, indigenous and exotic, occur on the worst depleted ground.

The evolution of induced steppe can be seen near the source of the R. Awatere (NE.). There, on the rather loose soil of gentle slopes (altitude about 800 m.), where there are thickets of *Discaria toumatou*, are many rabbit warrens, in the vicinity of which every stage[2]) of grassland depletion is present.

Celmisia spectabilis and Chrysobactron colonies. These two communities, though of different life-forms, are taken together because both

---

Verbascum Blattaria both of which increase greatly from seed when protected from sheep and rabbits. Where there are loose stones the endemic far-creeping *Urtica aspera* is frequently abundant.

1) The following are the principal species: — (exotic) *Bromus hordeaceus, B. sterilis, Hordeum murinum, Aira caryophyllea, Festuca Myurus, Urtica urens, Gypsophila tubulosa, Cerastium vulgatum, Ranunculus falcatus, Draba verna, Erodium cicutarium* (but of far wider distribution), *Anagallis arvensis, Gilia squarrosa, Myosotis arvensis, Carduus pycnocephalus* (but of far wider distribution); (indigenous) *Chrysobactron Hookeri, Myosurus novae-zelandiae, Claytonia australasica* (summergreen, grows on shady slopes in ground wet in winter), *Myosotis pygmaea* (a linneon, with 2 or more distinct jordanons amongst the Central Otago forms, and apparantly many hybrids between them).

2) Where the tussocks are close, bare earth dotted with rabbit-holes stands out conspicuously. By degrees, the tussocks are killed, not merely because they are eaten more or less, but *through the intolerable manuring they undergo*, and this area with holes and moribund tussocks is encircled by a wall of healthy tussocks. A little later, a temporary association occupies the new ground consisting of: abundant *Rumex Acetosella, Geranium sessiliflorum* var. *glabrum* (inedible), *Epilobium novae-zelandiae* (inedible), more or less circular mats of *Acaena Sanguisorbae* var. *pilosa* and *A. inermis, Carex breciculmis, Cnicus lanceolatus, Stellaria media* and *Cerastium vulgatum*. In course of time, as the depleted ground increases, there are established on its oldest part, their "seeds" windborne, low circular cushions of *Raoulia lutescens* between which, where fully exposed to the wind, no other plants are present, the dry soil having been blown away.

arise from repeated burning and the reason for their remarkable increase is similar.

Celmisia spectabilis is common in dry mountain fell-field (SN., NE., E.) and forms large, green, circular semi-cushions made up of erect leaf rosettes crowded together. It also descends into the montane low tussock-grassland and was probably a rare member of the virgin community in its upper part. At the present time, in many places (E.) the tussocks are much stunted or eradicated and *Celmisia* cushions are everywhere at a distance apart of 2 m. or more.

On examining an area where burning has taken place it looks as if the *Celmisia* — its leaves burnt to their bases — was irretrievably damaged, but not so. The leaf-buds are closely surrounded by the woolly sheaths of the burnt leaves and they also contain much moisture. Thus, the young leaves altogether escape damage from the fire and, in due course develope, so that, in quite a short time, the plant is a vigorous as ever. Further, as burning the tussocks leads gradually to extension of the bare ground, in which the *Celmisia* seed can germinate, in a comparatively short time *C. spectabilis* becomes the physiognomic plant, the open cushion replacing the tussock-form. Other species behave similarly, e. g. *C. coriacea* var. *stricta* in the Takitimu Mountains (SO.), Fig. 55, and *C. dubia* on the coastal mountains of the North-western district; indeed, wherever great fields of *Celmisiae* occur burning may be suspected and the community be indigenous-induced.

In the case of *Chrysobactron* it is its summergreen habit and tuberous roots which save it from grassland fires. Though naturally a plant of wet ground, as already noted, it can also grow under dry conditions, so it, too, can readily occupy bare ground resulting from burning tussock-grassland, while its rapid spread is furthered by its absolute unpalatability for any kind of grazing animal.

The incoming and spreading of Lycopodium fastigiatum. At the head of Lake Wanaka (east of W.) there is a good deal of *Lolium perenne* pasture which has been laid down by aid of the plough. On what I feel pretty certain was an area of this kind — though unfortunately my notes say nothing on this head — *Lycopodium fastigiatum* has come into the community by means of its spores and, by its unexpected centrifugal growth, has formed huge "fairy rings" many metres diam. (Fig. 105). The plant extends by means of underground stems which branching put forth orange-coloured erect shoots 10 to 12.5 cm. high. Within the rings there is much dead *Lycopodium;* evidently as it exhausts the supply of nutritive salts in the soil, the stolons extend outwards into the new ground, the ring gradually increasing in diameter at the expense of the artificial pasture.

Between the rings there is a close turf of small indigenous plants common in tussock-grassland, together with others of more restricted distribution, e. g. *Herpolirion novae-zelandiae, Triodia australis, Poa pusilla, Acaena Buchanani, Stackhousia minima, Gentiana Grisebachii, Plantago triandra* and *Cotula squalida.*

Invading shrub associations. Montane low tussock-grassland, montane ploughed land (neglected), lowland artificial meadow and some other classes of vegetation are frequently invaded by different species of *Cassinia.* These are heath-like, erect composite shrubs $\pm$ 2 m. high, but much lower when growing closely. The seed is carried by the wind and by the sheep. As the species are of rapid growth, and seedlings abound, a closed association rapidly comes into being.

c. The swamp series.

The Phormium association. In various parts of New Zealand proper, especially where large rivers overflow their banks, there are extensive areas of ground, perhaps extremely wet in winter, but quite dry in summer, occupied by bushes (tussocks) growing closely of *Phormium tenax*, making apparently a pure association. To one unversed in this class of vegetation, it would unhesitatingly be considered primitive, yet *it has arisen solely through draining the Typha-Phormium swamp described in Part II.* Nor would it remain stable for long were it not that various invading species are removed by the owners of the "flax-swamps", so called.

A has been explained, the primitive swamp is wet at all seasons and contains abundance of *Typha angustifolia* (in a wide sense) but, as for *Phormium,* it only occurs — more or less stunted — here and there in the shallowest water or fringing the margin of the swamp. Into the latter deep drains are cut, its water-content is greatly lessened, and in a year or two the *Phormium* present attains its full dimensions; then the subtratum becoming drier the *Typha* will languish and young *Phormium* plants come up thickly all over the drained ground. In a few years such attain their full dimensions and the former swamp will be almost pure *Phormium* 3.6 m. high, under favourable conditions.

In North Island "flax-swamps" the bushes about every 4 years are cut to within some 65 cm. of the base and as much as 100 quintals of leaves per acre are secured.

Generally, there are spaces between the bushes or groups of bushes; also paths are cut through the jungle of *Phormium* to enable the crop to be conveyed to the mill. Such spaces and paths are soon occupied by invaders — some exotic, some indigenous — but the latter, which form colonies of purely indigenous species, are alone considered here, though, it must be pointed out they have to compete with the eminently aggressive exotics, *Festuca gigantea* and *Rubus fruticosus* which form dense colonies. The indigenous invaders consist chiefly of the following: — (ferns) *Blechnum procerum, Hypolepis rugosula;* (grass-form) *Arundo conspicua, Carex secta;* (herbaceous perennials) *Epilobium junceum* (in a wide sense), *E. erectum, Hydrocotyle pterocarpa;* (shrubs) *Hebe salicifolia* var. *palustris* (R. C.), *Coprosma propinqua, C. robusta,* × *C. prorobusta, C. tenuicaulis, Olearia virgata;* (trees) *Cordyline australis, Hoheria angustifolia;* (lianes) *Muehlenbeckia australis, M. complexa* var. *trilobata, Rubus schmidelioides* var. *coloratus, Parsonsia capsularis, Convolvulus Sepium.* The *Cordyline* is frequently dotted about the "swamp" and the woody lianes soon form dense masses; where the ground remains specially wet *Typha* abounds. Cattle pastured in the "swamp" help to dry and consolidate the soil, destroy the undergowth and make space for invading species.

Successions following felling of swamp-forest. *Cabbage-tree (Cordyline australis) swamp* is now a common feature of lowland plains and valleys of North Island and has a truly pimitive aspect. But it has come into being after the destruction of *Podocarpus dacrydioides* forest and it is distinguished by the close groves of the *Cordyline* which are frequently accompanied by more or less *Typha.*

Divaricating-shrub swamp is a succession after the felling of.*Podocarpus dacrydioides* forest when more or less of the undergrowth is left standing. Its composition depends, in addition, upon the incoming of species not present in the original forest, especially the following: — *Arundo conspicua* (wetter parts), *Cordyline australis, Edwardsia microphylla, Hoheria angustifolia, Hebe salicifolia* var. *palustris* (RC.). The lianes — *Muehlenbeckia australis, M. complexa* var. *trilobata, Rubus schmidelioides* and *Parsonsia heterophylla* — become most luxuriant and bind the shrubs together, so that the thicket becomes impenetrable. Great tussocks, 3 m. high, of *Gahnia xanthocarpa* are abundant. Various divaricating-shrubs represent the original undergrowth especially: *Paratrophis microphylla, Melicope simplex, Nothopanax anomalum, Coprosma rigida, C. tenuicaulis, C. propinqua, C. robusta* (bushy shrub) and × *C. prorobusta.*

An induced Phormium association has arisen at the base of Ruapehu (VP.) after the burning of the *Dacrydium Colensoi* association, and, without good evidence, such an origin could never be suspected. On the drier ground is *Leptospermum scoparium* thicket.

d. Some high-mountain indigenous-induced communities.

Hebe shrubland. Probably a considerable percentage of *Hebe* shrubland has originated from the burning of various kinds of vegetation, notably, *Nothofagus* forest, tall tussock-grassland and subalpine-scrub. On Mount Dick (SO.) at 830 m. altitude *Hebe buxifolia* (in a wide sense) shrubland has originated most likely from burning forest and, at 1200 m., that of *H. propinqua* var. *major* from burning tall tussock-grassland. On Arthur's Pass (W.), at 900 m. altitude in 1908, several species of *Hebe*, with *H. subalpina* dominant, had taken possession of ground where scrub containing no *Hebe* had been burned some 18 years previously. On Mount Egmont there is now pure *H. salicifolia* var. *egmontiana* (may = var. *paludosa)* where at about 1100 m. altitude the vegetation had been removed.

Induced fell-field or herb-field. On Mount Miromiro near Hanmer (NE.), at an altitude of 840 m. on stony, wind-swept ground the *Nothofagus* forest on which had been burnt perhaps 20 years before, there is, as the dominant *Helichrysum bellidioides*, dwarfed to the ground and of raoulia form and other common members of the induced fell-field are *Poa Colensoi, Holcus lanatus* (exotic), *Uncinia divaricata*, dwarf *Leptospermum scoparium, Leucopogon Fraseri, Cyathodes acerosa, Pentachondra pumila, Helichrysum microphyllum* and *Raoulia glabra.*

By degrees as the altitude increases more typically high-mountain species come in, e. g. *Geum parviflorum, Celmisia Traversii* (a little), and a plant or two of the alpine *C. sessiliflora.* At a little higher altitude still, the following characteristic high-mountain species appear: *Pratia macrodon, Raoulia bryoides, Celmisia Du Rietzii, C. viscosa, Helichrysum Mackayi* and *Leucogenes grandiceps.*

At only 980 m. altitude, the species grow so closely that, in a sheltered gully, true herb-field appears. Here *Celmisia Traversii* — at at least 300 m. below its usually lowest range — is so plentiful as alone to catch the eye at first glance; there is also some *C. coriacea* and naturally some of the hybrid swarm between the two species of *Celmisia* — × *C. Morrisonii.* Also the following are more or less common members of this remarkable community: — *Lycopodium fastigiatum, Hypolepis Millefolium, Danthonia setifolia, Luzula picta, Astelia Cockaynei, Geum parviflorum, Acaena Sanguisorbae* var. *pilosa, Oxalis lactea, Epilobium chloraefolium* var. *verum, Drapetes Dieffenbachii, Anisotome aromatica, Hebe buxifolia* (prostrate), *Ourisia macrophylla, Coprosma ramulosa, C. parviflora, Brachycome sp.* of the *B. Thomsonii* group, *Hypochoeris radicata* (exotic) and *Traversia baccharoides. Here and there stand the burnt tree-stumps of the former forest.* The gully lies right in the track of the north-west wind.

Induced cushion-plant association. On the flat saddle connecting Mount Judah with Stone Peak (Richardson Mountains, SO.), at 1300 m. altitude, the primitive tall tussock-grassland has been burnt and is now replaced by a most unexpected type of vegetation. Dotted about everywhere are yellowish-green cushions of *Phyllachne Colensoi*, warm-brown mats of *Dracophyllum uniflorum* and straw-coloured tussocks of *Festuca novaezelandiae.* In the adjacent unburnt grassland there is no trace of the *Phyllachne* but, in places, the *Dracophyllum* grows through it. The other principal species of the new association are: *Lycopodium fastigiatum, Drapetes Dieffenbachii, Gaultheria depressa, Hebe dasyphylla, Ourisia caespitosa* and *Celmisia Lyallii.*

5. Artificial communities

(by A. H. COCKAYNE, Director of the Fields Division, New Zealand
Department of Agriculture).

Under this head come all those communities — some designed to last
for a considerable time, and others of brief duration — which are **produced
directly by man** in his ordinary agricultural, forestry (so far as direct
planting, sowing and thinning goes) and horticultural operations. Here only
two classes of communities are dealt with, but they are of particular interest
representing, as they do, agricultural proceedings unknown in Europe.

**Displacement of rain-forest and swamp-forest by burning and
replacement by artificial pasture without ploughing the ground.** Some
4050000 hectares of forest, the greater part in North Island, have been
converted into grassland. For many years from 400 to 800 sq. km. were
dealt with in each year but, at the present time, only a comparatively
small area is available and the *bush-buru* days of New Zealand's pastoral
development have almost passed away. The process of conversion of forest
to grassland is of extreme phytogeographical interest since *in one year's
time a formation, apparently attuned to a special habitat, is replaced by
another ordinarily supposed to depend upon altogether different conditions*,
which more-over remains permanent so long, at any rate, as it is kept
fully grazed. The processes involved consist of the following phases: —
1. Felling the forest. 2. Burning the fallen timber. 3. Sowing the seed.
4. Stocking the ground. 5. Burning the old and fallen logs. 6. Stumping
the ground.

The forest is felled in winter and early spring. First, the undergrowth
is cut and allowed to lie where it falls in order to provide the actual
kindling-wood for the succeeding fire. Next, all the trees having a smaller
diam. than 90 cm. are cut down, the others being left untouched, but the
procedure differs in different localities and larger trees are felled on level
ground than on steep slopes. Burning usually takes place after Christmas
but the date depends entirely upon the state of the weather, since on a
successful "burn" depends the future success of the subsequent operations.
The following conditions are essential: — 1. the fallen trees &c. must be
dry enough; 2. the weather must keep fine during the burn; 3. the wind
must be favourable both in direction and intensity. The lighting of the
fires takes place along as long a line as possible at right angles to the
wind. Rapidity of burning is essential, so that on a large "block" as many
as forty men may be required. Almost before the ashes have stopped
smoking, and certainly before they are cold, the seed is sown.

The sowers sow by hand, carrying the seed[1] in bags, which are ordinary

---

1) The seed is packed on horses to the ground from the nearest road in bags
containing 35 kg., all the different seeds having been previously mixed together, and
they are placed in position over the burnt area.

sacks cut half-way down, a flap being thus formed in front, while a hole is cut in the back through which the man thrusts his head. The seed is scattered right and left, both hands being used, and, at each step, a handful of seed is thrown. The sowers form a diagonal line, so that one man slightly overlaps the work of another. The sight of the line of sowers crossing a log-strewn area where walking unburdened is no easy matter for a novice, each carrying a heavy bag and scattering the seed without cessation is not easily forgotten.

The amount sown per hectare varies from 20 to 30 kg. The following are the chief species sown: — *Dactylis glomerata, Lolium perenne, L. italicum, Phleum pratense, Alopecurus pratensis, Poa pratensis, Cynosurus cristatus, Agrostis alba, Trifolium pratense, T. repens, T. hybridum*. Other grasses (*Festuca* spp.) and clovers are occasionally used but the bulk of the seed consists of the first two species in the list[1]). A certain amount of rape, mustard and soft turnip is included in the mixture, so as to provide food for such stock as are turned on to the land within a few months of sowing.

Within 12 months from felling the forest the land is fully stocked and the trampling of the animals consolidates the ground and greatly assists in forming a sward. Where hilly, sheep are generally pastured, but the richer bottomlands are used for cattle.

By slow degrees, in process of time, the unburned logs decay or are burnt, the standing trees fall, and, if the ground is to be cropped, the stumps are extracted. At present, every stage of the conversion of forest into meadow is to be seen, but there are many areas where no vestige remains of the original plant-covering.

If the grass does not entirely cover the ground, certain indigenous shrubs may become abundant, especially *Aristotelia serrata*. Also, plants not present in the original vegetation may appear, especially *Leptospermum scoparium* and, near the coast, *Cassinia leptophylla*. *Pteridium* heath may also enter in, and were it not for abundant "stocking" would become permanent.

Replacement of Pteridium heath by artificial grassland ("Fern crushing"). As seen in Part II, there are large areas of *Pteridium* heath, much of which is certainly not primitive. Many areas of such heath have been converted into artificial grassland by a somewhat similar process as that just described. The heath ("fern" of the settlers) was burnt in the early autumn and from 15 to 30 kg. per hectare of grass and clover seeds, similar to those used on bush-burns, sown by hand. Between autumn and early spring little, if any, growth of *Pteridium* occurs, so time is afforded

---

1) In many localities *Lolium perenne* does not persist, so that eventually *Dactylis* and *Trifolium repens* dominate, and such form the basis of many meadows on soils which are "fertile", but over wide areas where lower "fertility" conditions were present the dominant species of grasses consist of such as were not originally sown but species of *Danthonia* (indigenous) or *Agrostis* (exotic).

for the germination and establishment of sown grasses and clovers. But
in the spring new leaves of *Pteridium* are rapidly developed and, if left
to grow, the whole area during the one growing-season would be dominated
by the fern and *Pteridium* heath hold sway, the young grass and clover
being killed. However, as soon as the young leaves commence to develop
and before their circinate tips have unrolled, large numbers of either dry
sheep or dry cattle, depending upon which was the more economic, are
grazed on the area and, to use the expressive term of the farmer, the fern
is "crushed out". The stock are kept on the ground until lack of feed
leads to considerable loss in body weight at which time they are removed.
The area is then spelled until the fern begins to put forth fresh leaves
when the crushing is repeated. According to the topography of the area,
and the number of the stock available for crushing, in a variable time the
fern becomes so weakened that the artificial grassland obtains the mastery
and can be continuously grazed with that number of animals that, without
falling away in condition, can make use of the feed produced. Not in
every case, by any means, is success attained after a single burning of
the fern, but, in certain instances, several resowing of grass seeds become
necessary and in many cases this leads to the incoming of *Leptospermum*
shrubland.

The underlying principle of fern-crushing is one that also is regularly
adopted on bush-burn areas which are developing into induced *Pteridium*
heath. In most cases, the application of crushing is successful, but in certain
localities heavy crushing of such heath is followed by the establishment of
circular sheets or deep mats of *Paesia scaberula,* and if the crushing is
continued, the *Paesia* patches will coalesce and unpalatable species become
dominant.

Chapter III.
Agriculture and Horticulture in New Zealand.

Agriculture. New Zealand with its various soils and climates is admirably adapted for different branches of agriculture and horticulture. A
full account of how the virgin plant-associations have been displaced or
utilized during the past 77 years, and made to yield in exports of agricultural produce a sum of nearly fifty-two million pounds sterling in 1925
would be full of interest, but lie far beyond the limits or scope of this
chapter.

Primarily, the phytogeographical character of the region has directed
the progress of agriculture, forcing it into certain channels. Thus, first of
all, apart from the work of the early missionaries and their converts in
the far north, the extensive tussock-grassland provided fair pasturage, so
that the first progress was in the direction of sheep-farming upon a large

scale without in the least attempting to "improve" the land. On the other hand, the dense rain-forests, quite unlike anything the European settler had been accustomed to, seemed to offer an insuperable barrier to agricultural advance. But, with increase of population, and, before all else, with the practical application of certain scientific discoveries, the markets of the world have been brought, as it were, to the very door of the most distant lands, so that from the end of the eighties, in the case of New Zealand, agriculture has advanced by leaps and bounds. No longer did wool form the mainstay of the industry but the production, first of meat, and later of dairy produce, became of prime importance. The small farm, which hitherto had provided a scanty livelihood, became a paying concern, the demand for land increased and still increases, so that such thought to be of no value, or impossible to "reclaim", now yields an abundant harvest, while, above all, certain forest-lands have been converted, at but little cost, into the richest of dairy-farms. Another formation that possessed great agricultural capabilities was the swamp, and this early on through drainage, ploughing and sowing with meadow-grasses, was transformed into pasture quite foreign to the soil. Some swamps, however, were too vast for private enterprise to deal with, but even these are now being subdued by aid of the State. So it comes about that only certain plant-formations remain comparatively undisturbed, especially *Leptospermum* shrubland, forests too far distant or in too wet a climate for profitable occupation, much of the dune-area and the herb-field of the Southern Alps. But all these, too, are being slowly occupied, so that the time is not far distant when the whole of New Zealand, save the most inhospitable and rugged portions, together with the National Parks and similar reserves, will be subservient to the will of man.

The land now being utilized for agricultural purposes falls into the two main classes of arable and grassland, while this latter must be subdivided into natural and artificial pasture. The last-named, again, comes into the two categories of ploughed and unploughed. Speaking phytogeographically, the arable land and ploughed grassland represent primeval swamp, alluvial tussock-grassland, lowland hillside, tussock-grassland and to some extent forest and even heath, while the unploughed grassland is, for the greater part, rain-forest converted into pasture by the method described in the last chapter. The natural pastures represent the original tussock-grassland at all altitudes now considerably modified by the reduction in number of certain indigenous species, the increase of others and the presence of many foreign plants especially *Hypochoeris radicata* and *Rumex Acetosella*.

The following statistics for the season 1926—27, most kindly supplied for this book by Mr. MALCOLM FRASER (Government Statistician), give at a glance detailed information as to the crops grown the area occupied and their yield when such can be stated.

Crop	Area		Yield		Yield in Quintals
	Acres	Hectares	Unit of Quantity	Quantity	
Wheat	220 083	89 063	Bushel	7 952 442	2 164 282
Oats	117 326	47 479	„	4 997 535	906 730
Barley	29 886	12 094	„	1 243 333	281 981
Maize	10 249	4 148	„	491 468	133 755
Peas & beans	15 495	6 271	„	454 722	123 754
Rye-grass	42 082	17 030	lb.	18 083 120	82 023
Cocksfoot	9 820	3 974	lb.	1 358 082	6 160
Chewings fescue	9 634	3 899	lb.	2 177 125	9 875
Red clover and cow-grass .	8 540	3 456	lb.	1 935 328	8 778
White clover	4 029	1 630	lb.	671 828	3 047
Potatoes	24 616	9 962	tons	116 771	118 644 tonnes
Turnips	462 360	187 108			
Mangolds	11 870	4 804			
Pasture grasses	16 680 348	6 750 198			
Area under cultivation . .	18 830 436	7 620 295			
Area in occupation . . .	43 587 698	17 639 057			

Production figures for butter and cheese are not available for the year 1926-27. For the previous year the production of butter was 1 544 722 cwt. (78 475 tonnes), and of cheese 1 520 169 cwt. (77 228 tonnes).

Live-stock figures for 1927 (sheep at 30th April, others at 31st January), are: — horses 303 713; cattle 3 257 729; dairy cows 1 303 225; sheep 25 649 016 pigs 520 143.

Horticulture. Horticulture, since it deals with an unlimited number of species, reflects climatic and edaphic conditions to no small degree. Speaking generally, the lowland climate throughout the region, with the exception of the Subantarctic province, permits the cultivation, in the open air, of many species not hardy in Europe, except in the South or in districts possessing a mild insular climate e. g., — *Eucalyptus globulus, Acacia me-lanoxylon, Agave americana, Hakea saligna, Pelargonium zonale.* But the greater part of the plants, whether of flower-garden, kitchen-garden or orchard, are those most commonly cultivated in Great Britain and the garden-fashions of that country, so different ecologically, are for the most part slavishly followed. At the same, almost from the foundation of the Colony, there have been enthusiastic amateur gardeners in the different centres and by them, at one time or another, have been introduced an immense and heterogeneous collection of plants hardy in their several localities. Further, there are a number of semi-botanic gardens where considerable collections of trees, shrubs and herbs have been brought together.

In proceeding from north to south, leaving on one side the rank and file of the garden species, a considerable change takes place in the flower gardens more especially, which is most marked on entering the Southern botanical province on its eastern side and thus quite in accordance with the distribution of the indigenous vegetation.

Various species absent in most parts of South Island, and in many other parts of North Island, give a special character to the gardens of Auckland. Amongst such are: — *Schinus Molle* which attains a great size especially at Thames, *Bougainvillea glabra, Datura cornigera, Euphorbia pulcherrima*, the indigenous *Meryta Sinclairii, Phytolacca dioica*, species of *Hibiscus, Tibouchina semidecandra, Ficus macrophylla, Erythrina crista-galli*, lemons and oranges. The two latter are grown for commerce, especially near Whangarei. Many species of *Eucalyptus*, not tolerant of frost, grow excellently in lowland Auckland. Napier possesses a climate much the same as Auckland so far as temperature goes, but drier. Virtually the same subtropical species grow excellently. The horticultural feature of the town is the splendid row of *Araucaria excelsa* along the esplanade. Taranaki has been named the garden of New Zealand. Here, too, many subtropical plants grow luxuriantly. But the chief horticultural features are the splendid vigour of shrubs of all kinds and the rich abundance of Camellias and Chinese Azaleas. By the time the city of Wellington is reached, the subtropical element has weakened somewhat, still it is plainly to be seen in the presence of *Ficus macrophylla*, many garden forms of *Fuchsia, Lagerstroemia indica, Clethra arborea*, species of *Bouvardia, Araucaria Bidwillii, Boronia megastigma*, many species of *Abutilon, Eucalyptus ficifolia* and Himalayan rhododendrons in great variety.

Turning now to South Island, various subtropical species can be grown in Nelson and along the coast on the west to as far south as Hokitika and probably much further, but on the east, as soon as the Canterbury Plain is gained, the winter frost forbids the presence of all plants which will not tolerate above — 10° C. Thus the trees and shrubs consist largely of species belonging to Europe, as a whole, California and Japan. Many Tasmanian species, too, are quite hardy. Alpine plants are somewhat difficult to cultivate on account of the north-west wind, and this too forbids the use of treeferns which can be grown with success in the open in the wetter parts of North Island. Lowland Banks Peninsula does not answer to the above description, for the climate is much warmer in winter. In Otago, near the coast, the climate of winter is milder than in Canterbury, but the summers are cooler and there is less sunshine. In consequence, more tender species can be grown in many localities e. g. *Leucadendron argenteum*. The climate especially favours alpine plants so that not only can European, North American and Himalayan species be grown with the greatest ease, but the indigenous plants, much more difficult to cultivate, thrive amazingly. Further south still, at Invercargill, the great subantarctic herbs grow most vigorously.

Fruit-growing is becoming a thriving industry. Apples, pears, plums and small fruits are grown in many parts of the lowland and montane belts, but commercial orchards are situated for the most part in the following districts, to each of which the area in orchards (including private orchards) is appended: — South Auckland (2800 hectares), East Cape (1070), Sounds-Nelson (2700), North Otago (1800). The best class of gumlands' soil of Auckland is suitable for all classes of fruit, including grapes and citrous fruits. Apple orchards are a special feature of Sounds-Nelson. Central Otago (NO.), with its climate far drier and more sunny than any other part of the region, is particularly suitable for the growing of stone-fruits, notwithstanding its winter cold, greater than elsewhere. But, in order to secure the best results, irrigation is essential.

A most important branch of horticulture or agriculture is *forestry*. From the earliest days of settlement, first the missionaries, and later the colonists, planted the trees of the homeland, and, with the settlement of South Island, plantations and shelter-belts were established on its treeless plains (the Canterbury Plain more particularly) consisting principally of *Eucalyptus globulus* (frequently sown *in situ*), *Pinus radiata* and *Cupressus macrocarpa*. But these were far from being the only species planted; indeed all the forest-trees used in Great Britain for economic and ornamental purposes (European, Californian, Japanese &c.) have been introduced, together with many species of *Eucalyptus* and some of *Acacia* — the result being that there are now many noble plantations of mixed exotic trees.

Since the subtropical rain-forest regenerates very slowly after its timber trees have been removed, it became necessary as the natural forests were being rapidly destroyed — partly for their timber (wood being almost the sole building material) but far more to make place for grassland — to establish extensive plantations of exotic trees. To begin with, this essential national work was placed in the hands of the Department of Lands and Survey and for a number of years it was ably directed by the late H. T. MATTHEWS whose name must ever stand high as a pioneer of New Zealand commercial forestry; nor must it be forgotten that much of his work was necessarily of an experimental nature.

In 1919 a separate Forestry Department was formed which in 1920 was reorganized as The State Forest Service with Mr. L. M. ELLIS as Director of Forestry. MATTHEWS had early on selected a portion of the Volcanic Plateau near Rotorua with its fern heath and low shrubland as the principal scene of his operations, the general climatic conditions from 300 to 600 m. altitude or more, and the easily worked pumice soil being ideal for tree-planting on a large scale, the value of the land also being very low. Various areas, too, in South Island were planted with trees, including part of the tussock-clad Hanmer Plains (NE.). But, under the State Forest Service, planting is being carried out on a far more extensive scale, and the introduction of new methods, and labour-saving devices, have greatly reduced the cost of planting, while also it has been definitely ascertained what species are best fitted for the artificial forests, and trees of inferior value are no longer used. Indeed, the future of New Zealand forestry is full of high promise and in no part of the British Empire does the same class of forestry stand on a higher plane.

Since the inauguration of State afforestation in 1896, nearly 32000 hectares of plantations have been established (*N. Z. Official year-Book*, 1927:494), and during the season 1926—27 about 8000 hectares was added to this area (*Ann. Rep. State Forest Service*, 1927:2).

At the present time, the following are the principal species which are being planted: — *Pinus radiata* (matures in from 30 to 40 years, i. e. 3 crops in 100 years), *P. ponderosa*, *P. laricio* and *Pseudotsuga taxifolia*. But there are many more species in the various plantations, amongst which are *Larix decidua* (grows splendidly at Hanmer), *Sequoia sempervirens* (grows rapidly in certain localities), *Pinus austriaca*, *Picea excelsa*, and various species of *Eucalyptus* (many not hardy throughout North and South Island). A rather curious circumstance is the fact that *Pinus sylvestris* is useless for afforestation in any part of the region; the trees becoming diseased at an early age.

Part IV.
The Flora of New Zealand, its Distribution and Composition.

Chapter I.
The Botanical Subdivisions of the Region.

1. General.

So many areas throughout New Zealand are insufficiently explored, or altogether unknown, so far as plant-distribution is concerned, that few statements as to the range of any species are absolutely valid. Therefore, the classification proposed in this chapter is advanced with some hesitation and offered merely as a more or less provisional attempt to deal with a subject that will be treated with much greater precision by some investigator in the future. Nevertheless the classification, first put forth by me in 1914, has been tested in the field and modified from time to time, so that is may be asserted with some confidence that the botanical divisions are fairly natural and appear to serve the purpose for which they were first designed. Yet in most cases it is not feasible to fix definite boundaries to the various botanical districts; for not only is there a gradual merging of one into another, but local climates, or edaphic conditions, occur which permit the occasional presence of species, or indeed associations, not in keeping with the general florula, or vegetation.

The major divisions of the region are here designated **botanical provinces.** These are based largely upon climatic change depending on latitude. As before six **provinces** are admitted. Except with regard to junction of the Central and Southern provinces all seems simple. For fixing the southern boundary of the Central Province two courses are open. The first is to restrict the Province to North Island — a very convenient arrangement, admirable for most purposes but hardly natural. The second course is to extend the Province to South Island and to let it *include* that portion of South Island lowland vegetation which comes very close to that of the southern part of the Ruahine-Cook district, but to *exclude* all the adjacent high-mountain vegetation. This course is followed here, the main objections to its adoption are, (1.) that the three northern botanical districts of South Island will each belong to parts of two provinces and (2.) that the southern boundary must be ill-defined and its position always open to discussion. But these objections seem slight in comparison with separating into two classes vegetation and florulas so similar as those of the forest and coast-line of all the Sounds-Nelson, much of the North-western and part of the North-eastern districts.

In dealing with the botanical provinces — subject to the explanation in the last paragraph — the ground is fairly secure for their basis is the stable one of gradual change in species in proceeding from north to south. But, when the question of smaller subdivisions, here called **botanical districts,** comes in, the ground is much less stable, not merely because new discoveries of species, or of distribution, may become disturbing factors, but because facts of various kinds — floristic, ecological and geological — have to be considered.

2. The Botanical Provinces.

1. The Kermadec Province. This embraces all the Kermadec group of small islands. Their flora and vegetation are dealt with at some length in Part II, and no more need be said here.

2. The Northern Mainland Province. This includes that part of North Island, as shown on the phytogeographical map, to the north of a line passing from a little south of Tauranga to the River Mokau but extending to the north of the Mamaku Plateau. Certainly this boundary is to some extent artificial and a more natural line of demarcation is lat. 38⁰, but it takes in far too much characteristic Volcanic Plateau vegetation and so also becomes in part an unnatural boundary-line. In the north the Three Kings Islands are included and to the east all the other small islands.

The Central Mainland Province. This includes the remainder of North Island, Kapiti Island, and that part of South Island, already defined, up to a line passing from Greymouth to Amuri Bluff.

The Southern Mainland Province. This includes all the remainder of South Island, together with Stewart Island. It is certainly a natural division and its flora, with but a few exceptions, is that of South Island proper.

The Chatham Province. This includes all the Chatham group as already described.

The Subantarctic Province. This includes all the New Zealand Sub-antarctic Islands as already dealt with. The sole critical matter is whether the Snares Islands should be excluded and added to the Stewart district. But the character of the vegetation is essentially subantarctic, and the presence of *Colobanthus muscoides* and *Stilbocarpa robusta* — allied to *S. polaris* and not to the Stewart *S. Lyallii* — turn the scale in favour of the position here assigned.

3. The Botanical Districts.

The Kermadec District. This corresponds with the Kermadec Province. The flora and vegetation have been dealt with in Part II.

The Three Kings District. This includes the Three Kings group of small islands[1]). The *flora* numbers 143 species of which the following are locally endemic: — *Davallia Tasmani, Paratrophis Smithii* (near to *P. opaca*), *Pittosporum Fairchildii* (near *P. crassifolium*), *Alectryon grandis* (a species based only on leaf-form and may be an epharmone of *A. excelsum*, large leaves being characteristic of the small, outlying islands), *Hebe insularis* (closely allied to *H. diosmaefolia*, and *Coprosma macrocarpa* (close to *C. grandifolia* but fruit twice as large). Both the flora and vegetation are of a coastal character. On the West King there is a remarkable low forest of tropical aspect with *Meryta Sinclairii* (an araliad with immense, thick, glossy, smooth, entire, shining, oblong leaves in great rosettes) dominant, together with *Cordyline australis* and, as the principal constituent of the undergrowth, *Macropiper excelsum* var. *major*.

The North Auckland District. This includes all the North Auckland Peninsula lying to the north of latitude 36°. Much of the district consists of comparatively flat gumlands but there is also a good deal of hilly country rising in places to nearly 800 m. altitude. The extreme north consists of a narrow, much-dissected tableland about 300 m. high, which, at one period, was an island, but is now united to the mainland by a narrow spit about 80 km. long consisting of dune both recent and consolidated.

The *climate* is the warmest and least disturbed of the main islands. In summer, easterly breezes of a subtropical nature prevail. Winter is decidedly the rainy season. The mean temperature is 16.2° C., the mean of the absolute maxima of the year being 4.5° above this and the mean of the minima 4.1° below, showing a mean daily range during the year of only 8.6°. Frosts occur inland at times but are extremely light. Snow is altogether absent. The mean annual rainfall is about 151 cm. its distribution being as follows: spring 22.7%, summer 13.1%, autumn 29.4%, winter 34.6%. In the west, from Herekino Harbour southwards the rainfall is higher.

The *flora* numbers about 630 species of which the following rather large number are locally endemic (but not of necessity so for the region): — *Diplazium japonicum, Todaea barbara, Microlaena Carsei,* (very close to *M. avenacea), Cladium complanatum, Lepidosperma filiforme, Hydatella*

---

1) They lie nearly 53 km. W. N. W. of Cape Maria van Diemen in 34° 6′ south lat. The Great King, easily the largest island, measures only 2.8 km. long by 1.2 km. broad. The islands rise abruptly from the sea and — leaving the smallest out of consideration — are from 106 to 303 m. high. All that is known of their flora and vegetation is derived from two brief visits of CHEESEMAN (1888, 1891), but only the two largest islands were visited, so probably the flora is larger than is given here.

inconspicua, Xeronema Callistemon (Poor Knights only), *Thelymitra inter-media, T. Matthewsii, Pterostylis Matthewsii, Chiloglottis formicifera, Corysanthes Matthewsii, C. Carsei, Cassytha paniculata, Phrygilanthus Raoulii, Tillaea pusilla, Pittosporum pimeleoides, P. reflexum,* (and a great hybrid swarm with *P. pimeleoides), Ackama rosaefolia, Hibiscus diversifolius, Halorrhagis cartilaginea* (North Cape only, may be an epharmone of *H. erecta), H. incana, Pseudopanax Gilliesii,* (probably *P. Lessonii × Nothopanax arboreum), Leucopogon parviflorus* (also in Chathams), *Hebe brevifolia, H. Bollonsii* (Poor Knights, Hen and Chickens), *H. ligustrifolia, Coprosma neglecta, Olearia albida, Celmisia Adamsii* var. *rugulosa,* and *Cassinia amoena* (North Cape Peninsula only).

The most striking features of the *vegetation* are kauri-dicotylous forest with *Beilschmiedia taraire* frequently dominant; extensive areas of *Leptospermum* shrubland and *Gleichenia* bog; *Avicennia* salt-forest or salt-scrub; and in places *Vitex lucens* forest.

Farming is making considerable headway, so that in alluvial valleys there are many fine dairy farms with meadows of *Paspalum dilatatum. Lotus hispidus* is also used for pastures, but it spreads naturally. Certain exotic species are restricted, so far, to the district, e. g. *Kyllinga brevifolia, Carex Brownii, Panicum Lindheimeri, Polygala virgata, Lantana camara, Helenium quadridentatum, Erechtites valerianaefolia* and *E. atkinsoniae.*

The South Auckland District. This is divided, more or less naturally, into the 3 subdistricts dealt with below. The district, as a whole, is distinguished by, (1.) the extension southwards of those species peculiar to the Northern Province which reach their southern limit in the neighbourhood of lat. 38°, or a little beyond, (2.) the coming in, especially from the south, of a number of species absent in the North Auckland district, (3.) the continuation of kauri-dicotylous forest, but with *Beilschmiedia tawa* and *Weinmannia sylvicola* dominant and not *B. taraire,* and (4.) the presence of a number of local endemics.

The Kaipara Subdistrict embraces that portion of the North Auckland Peninsula lying between lat. 36° and 37°, but excluding Little and Great Barrier Islands and the Coromandel Peninsula. Generally, the surface is flat but there are some isolated low hills and, in the south-west, the Waitakerei Hills which exceed a height of 400 m.

The average rainfall is 110 cm., or more in some parts, distributed over 184 days. Cumulus clouds frequently gather on summer afternoons, but generally disperse without rain. The climate is mild, frost is almost unknown. The maximum temperature rarely exceeds 26° C.; the absolute daily range is 7.4°; the mean of the maxima is 3.6° above and the mean of the minima 3.8° below the average mean of 15.2° C. The rainfall is distributed as follows: — spring 24.1%, summer 18.7%, autumn 24.9%, and winter 32.3%.

The *flora* numbers about 580 species of which the following are locally endemic: — *Thelymitra aemula, T. caesia* and *Hebe obtusata*. The following species, which also belong to the North Auckland district, reach their southern limit: *Ranunculus Urvilleanus, Mida myrtifolia, Alseuosmia Banksii, A. linarifolia*.

A large part of the district is occupied by the gumlands' communities. Originally there were considerable kauri-dicotylous forests, especially on the Waitakerei Hills. *Farming* is similar to that described for the preceding district.

The Thames Subdistrict includes the area between the River Thames and the east coast, together with the Coromandel Peninsula, the two Barrier Islands and the small islands lying off the coast. On the south, it is bounded by the line shown on the map, which commences a little to the South of Tauranga and obviously is a quite arbitrary boundary. The *land-surface* is much broken and mountainous, some of peaks rising to 900 m. and more. There is a considerable coast-line much of which is rocky.

The mountainous character of the district leads to various local climatic differences. In summer, the temperature frequently exceeds 27° C. and in winter — 4° C. is not uncommon. The mean annual rainfall is 165 cm. which is distributed as follows: — spring 24.2%, summer 18.2%, autumn 26.8% and winter 30.7%.

The subdistrict is distinguished by, (1.) the presence originally of extensive kauri-tawa forests, (2.) the commencement of the real high-mountain flora, (3.) the first appearance in the northern province of true *Nothofagus* forests — not merely isolated trees or small groups, (4.) the possession of a distinct locally-endemic element, and (5.) the large number of southern species which extend no further north, a matter, in part, depending upon the fairly high continuous mountain range.

The *flora* numbers about 652 species of which the following are locally endemic: — *Elytranthe Adamsii, Pittosporum virgatum* var. *Matthewsii, P. Huttonianum, Pomaderris rugosa, Hebe pubescens, Veronica irrigans, Olearia Allomii* (Great Barrier Island only), and *Senecio myrianthos*. The following are specially important members of the group of species having its northern limit in the subdistrict: — *Hymenophyllum pulcherrimum, H. peltatum, Blechnum penna marina, Polystichum vestitum, Lycopodium fastigiatum, L. varium, Libocedrus Bidwillii* (and with it *Hymenophyllum Malingii*), *Dacrydium Bidwillii, Podocarpus nivalis, Phyllocladus alpinus, Carpha alpina, Oreobolus pectinatus, Uncinia ferruginea, Cordyline indivisa, Arthropodium candidum, Enargea parviflora, Nothofagus Menziesii, N. fusca* (the species further north is *N. truncata*), *Metrosideros Parkinsonii* (at a high elevation on Great Barrier Island, 697 km. distant from its nearest station in the south!), *Nothopanax Sinclairii, N. simplex, N. Colensoi, Gaultheria rupestris, Pentachondra pumila, Cyathodes empetrifolia, Dracophyllum strictum, Pimelea*

Gnidia, Ourisia macrophylla, Coprosma linariifolia, C. foetidissima, C. Colensoi, Olearia virgata, Celmisia incana and *Raoulia tenuicaulis.*

The *Waikato subdistrict* extends from lat. 37° on the north to the southern boundary of the Northern Province and eastwards to the River Thames. Much of the district is quite flat and situated but little above sea-level. To the north, west, and south there are low hills, while out of the low country rise the extinct volcanic cones Mount Pirongia (860 m.) and, near the west coast, the rather lower Mount Karioi.

The range of temperature, especially in summer, is greater than in the Kaipara subdistrict, January having a mean maximum of 24.1° and a mean minimum of 11.4°. The mean annual temperature is 14.2°, the mean maximum 19.7°, the mean minimum 8.7° and the mean daily range 11°. The winds are possibly more steady and less variable in direction than in the surrounding districts. Dews are heavy but frosts are not often experienced until at the junction with the Volcanic Plateau. The average annual rainfall is 140 cm. distributed as follows: — spring 26.8%, summer 17.1%, autumn 25.2%, winter 30.9%.

The main feature of the *vegetation* was originally the far-spreading swamps grading, in places, into sphagnum bog, an important feature of such swamp the abundance of *Sporodanthus Traversii* (also characteristic of the Chathams). Kauri-dicotylous forest occupied only a small area but the hills were forest-clad with an association containing the usual northern species. On Mount Pirongia, however, in its uppermost belt the following southern species appear: — the trunkless var. of *Dicksonia lanata, Polypodium novae-zelandiae* (central North Island, not southern), *Cordyline indivisa, Enargea parviflora, Weinmannia racemosa, Nothopanax Sinclairii, N. Colensoi, Coprosma tenuifolia* (another central species), *C. foetidissima* and *C. Colensoi.* The *flora* numbers about 600 species. There are apparently no local endemics. *Asplenium Trichomanes* and *Discaria taumatou* reach their northern limit in the subdistrict.

Dairy farming is easily the leading branch of *agriculture.* Some of the highest butter-fat producing farms are in this subdistrict in which is represented the highest grade of grass-farming of the region. Also, the area of supplementary crops grown annually is becoming less and less through the application on all grassland of phosphate top-dressing which reaches as high as 52.6 quintals per year for each cow pastured.

The Volcanic Plateau District. This district occupies the central portion of North Island. Its boundaries as shown on the map are still provisional. Much of the area is at an altitude of more than 600 m., and consists partly of a tableland and partly of high mountains (the central volcanoes with Ruapehu ice-clad near the summit and the Kaimanawa Mountains). In certain localities, (White Island to Mount Tongariro) many hot springs and fumaroles occur. There is a short strip of sea-coast extending from the mouth of the R. Rangitaiki nearly to Tauranga, and White Island is included. The soil consists of pumice and other kinds of volcanic ash.

The *climate* of the district is far from uniform extending, as it does, from the coast to the summits of the central volcanoes. Taking Rotorua (altitude 282 m.), as typical of a good deal of the eastern part of the district, the temperature in summer frequently exceeds 26° C. and, in winter, — 6° C. is frequently registered, while in the subalpine and alpine belts frost occurs at all seasons. At from 900 to 1200 m. altitude snow lies in winter, on an average, for a few days only, but near the glaciers of Ruapehu there are perpetual snow-fields. The mean annual rainfall, combining the stations at Rotorua and Taupo, is 137 cm. distributed as follows: — spring 24.9%, summer 22%, autumn 25% and winter 28.1%. To the west and south of the volcanoes, the climate is far wetter; in fact, it is a rain-forest climate as compared with the tussock-grassland climate of the eastern highlands.

The *flora* consists of about 726 species, the following of which are locally endemic: — *Nephrolepis cordifolia* (also Kermadecs, a "hot-water" fern), *Gleichenia linearis* (another "hot-water" fern), *Scirpus crassiusculus*, *Bagnisia Hillii*, *Pittosporum Turneri* (very close to *P. patulum* but with a different juvenile form), *Utricularia Mairii* (originally on L. Rotomahana and not seen elsewhere, but the lake was destroyed by the Tarawera volcanic outburst in 1886), *Logania depressa*, *Veronica Hookeriana* and *V. spathulata* (the last two extend somewhat beyond the eastern boundary). Though not absolutely confined to the district, the following are characteristic: — *Polypodium novae-zelandiae*, *Dracophyllum subulatum*, *Gaultheria oppositifolia*, *Hebe laevis*, *H. tetragona*, *Ourisia Colensoi*, *Coprosma tenuifolia* (but of fairly wide range) and *Raoulia australis* var. *albosericea*.

Much of the area below 600 m. altitude is occupied by *Leptospermum* shrubland. There are also fine forests, those of the lower montane belt containing much *Podocarpus totara* but at lower levels *Beilschmiedia tawa* dominates *and Dacrydium cupressinum* attains a great size. In the subalpine belt the forests consist of *Nothofagus*, all the species being present in some part or other. Where the climate is drier, or for edaphic reasons, there is tussock-grassland — tall and low. Above the forest-line, or lower in the drier climate, there is subalpine-scrub, fell-field, shrub-steppe and actual desert.

From the *agricultural standpoint* the most notable feature is the splendid plantations of exotic forest-trees with which the State has already replaced 27,442 hectares of *Leptospermum* shrubland and *Pteridium* heath, which was mostly unsuitable for ordinary farming purposes. All the same much pumice land, considered worthless a few years ago, is now well clothed with grass and a good many dairy farms have been established. But in order that really profitable live-stock farming can be carried out it is essential that heavy applications of phosphate fertilizers be made in order to

produce permanent pastures. Also transportation over the area is difficult
and, lacking this, intensive farming is impossible.

The East Cape District. This includes all the area lying to the east
of the R. Rangitaiki and thence southwards to the northern extremity of
the Ruahine Mountains, together with the East Cape semi-peninsula, and
bounded on the south by a line extending from a little to the south of
Cape Kidnappers to about Kuripaponga. As far as my experience goes the
Rangitaiki boundary seems fairly satisfactory, but the extension of the
western boundary and the position of the southern boundary is little better
than a guess; probably the latter should extend further to the south.

Much of the country, especially inland, is mountainous and much broken
with narrow valleys containing torrents in time of flood, but near the coast
the land, though hilly presents gentler contours. Both pumice and calcareous
soils are common.

Owing to the hilly nature of the area there is much modification of
climate. The littoral is sheltered from westerly winds. East to south-
easterly winds in cyclones bring heavy rains and occasional floods, 20 cm.
falling in a day. In the southern part of the district, even on the coast-
line, there is frequently a temperature over 26° C. from December to the
end of March. Frost is only of moment in the upper montane and subalpine belts. The mean annual rainfall is 108 cm. and, except in the
forested parts, the number of rainy days is small ($\pm$ 110). The rain is
distributed as follows: — spring 20.1%, summer 19.5%, autumn 29.7% and
winter 30.7%.

The *flora* consists of about 734 species, the following of which are locally
endemic: — *Lemna gibba* (possibly a mistake), *Danthonia nuda*, *Peperomia
tetraphylla*, *Edwardsia tetraptera* (in a narrow sense), *Myosotis saxosa*,
M. amabilis, *Hebe Darwiniana* (perhaps hybrid), *H. macroura*, *H. Cookiana*,
Jovellana Sinclairii, *Utricularia Colensoi*, *Olearia pachyphylla* and *Senecio
perdicioides*. *Pittosporum Ralphii*, which extends into the Egmont-Wanganui
district, is a characteristic species. Many common northern species reach
their southern limit in the north of the district so that the forests of the
East Cape semi-peninsula have a facies very similar to those of the Auckland districts.

Extensive forests still remain in the northern part of the district, both
dicotylous-podocarp and *Nothofagus*, the former the southern limit for a number
of northern trees. Further south, forest is rare, except on the mountains,
the climate being dry, so that possibly the present covering of *Pteridium*
and *Leptospermum* is more or less primitive. On the coast, there is a
characteristic rock association of *Phormium Colensoi* and *Hebe macroura*,
and, on river-banks, in the south-east of the district, *Pittosporum Ralphii*,
Edwardsia tetraptera and *Hoheria sexstylosa* are greatly in evidence. On
the high mountains there is a small subalpine flora.

With regard to *farming* sheep are depastured in large numbers on the rich artificial pastures the raising of fat lambs being all-important. Thus, in the Gisborne land district, a part of the area, the number of sheep in 1926 was 1,582,521 and in Hawkes Bay, which however extends for a considerable distance into the Ruahine-Cook district the number was 1,660,556, the total for all New Zealand being nearly 25,000,000. The greater part of the undulating hilly "bush-burn" pastures have danthonia dominant and they support far more sheep per hectare than apparently similar grassland elsewhere. The southern limit of maize, as a grain crop, is in the neighbourhood of Gisborne, though as a forage crop it is grown as far south as Cook Strait. In gardens, subtropical plants, similar to those cultivated in the Auckland districts grow vigorously. The vine is grown to some extent for wine making.

The Egmont-Wanganui District. Here again the boundaries, as shown on the map are quite uncertain, but this district embraces the Wanganui coastal plain which gradually rises, on proceeding inland, from sea-level to 600 m. altitude. The plain is everywhere deeply cut by rivers and streams which flow far below the apparently hilly suface of the plain. Mount Egmont and the adjacent Pouakai Range come into the district.

The district has a westerly aspect and most of the rain comes from that quarter. The rainfall on Mount Egmont must be very high, judging from the rich bryophyte content of its forests. Taking a number of stations there is a mean maximum temperature of 18.4° C., a mean minimum of of 8.3° and a mean daily range of 10.1°. In New Plymouth during summer the maximum rarely exceeds 24° and in winter the actual freezing point is rarely reached. The mean annual rainfall is 118 cm., and its distribution is as follows: — spring 25.1%, summer 21.8%, autumn 25.4% and winter 27.7%. In certain localities the westerly winds carry salt spray inland for several kilometres and are hurtful to many species of plants.

The *flora* consists of about 630 species of which the following are locally endemic: — *Carmichaelia australis* var. *egmontiana*, *Hebe salicifolia* var. *longeracemosa*, and perhaps the narrow-leaved var. of Egmont, *Coprosma egmontiana*, *Plantago Masonae*, *Celmisia glandulosa* var. *latifolia*, *Olearia Thomsonii* (perhaps hybrid), and *Senecio Turneri*.

Excepting where the full force of the sea wind struck, *Beilschmiedia tawa* forest originally occupied most of the district. Cliff vegetation of the deep, river-gorges is a characteristic feature with *Blechnum procerum*, *Cladium Sinclairii*, *Senecio latifolius* &c., as described in Part II. About 100 high-mountain species occur on Mount Egmont and *Ourisia macrophylla* and *Ranunculus nivicola* are particularly abundant.

As *farmland* the soil grows grass amazingly, so that certain areas produce almost unbelievable amounts of butter and cheese. At the Kaupokanui cheese factory, for example, 2,730,000 kg. of cheese are produced yearly from the milk of 10,000 cows grazed on 10,000 hectares. Near the coast, the immense hedges of *Lycium horridum* are physiognomic.

The Ruahine-Cook District. This comprises the area east of the Rangitikei and Hautapu Rivers and bounded on the north by a line passing from a little to the north of Taihape to a little to the south of Cape Kidnappers. The other boundaries are Cook Strait and the Pacific Ocean. The Ruahine-Tararua-Rimutaka wall of mountains extends through the district, dividing it into eastern and western parts. Much of the area is flat, embracing, as it does, the Manawatu and Wairarapa fluviatile plains. There is an extensive coast-line, rocky on the east and south but with extensive dunes on the west.

Much of this district has a *climate* considerably affected by the proximity of Cook Strait. The predominant weather is westerly but not infrequently it falls under the influence of subtropical disturbances. The Dividing Range distinctly influences the climate, so that the plain on its eastern side receives less rain than other parts of the district. It is also hotter in the summer and colder in the winter. The mean maximum temperature for the lowland belt is 17.5° C., the mean minimum 8.9° and the mean daily rang 8. 6°. In the subalpine belt snow lies from about 1 to 3 months and frosts occur at all seasons. The mean annual rainfall for the lowlands is 108 cm. and distributed as follows: — spring 22.6%, summer 23.3%, autumn 26% and winter 26.7%. Much more rain, however, falls on the mountains and in certain lowland localities. High winds of considerable duration are characteristic.

The *flora* consist of about 805 species of which the following are locally endemic: — *Gahnia robusta, Carmichaelia odorata* (according to T. Kirk, it extends to the Marlborough sounds), *Epilobium Cockaynianum, Aciphylla intermedia, Anisotome dissecta, Myosotis Astoni* (close to, if not identical with *M. petiolata*), *Hebe evenosa, H. Astoni, H. elliptica* var. *crassifolia, Coprosma Buchanani* (possibly a hybrid), *Craspedia maritima, Abrotanella pusilla, Senecio Greyii* and *S. compactus*.

Originally, there were extensive dicotylous-podocarp forests, and there are still wide breadths of *Nothofagus* forest which ascend to the timber-line. The high-mountain flora contains nearly all the North Island species of that class and, excluding local endemics, the following have not been recorded elsewhere for North Island: — *Deschampsia tenella, Triodia australis, Uncinia fuscovaginata, Drosera stenopetala, Aciphylla conspicua* (unless it is the *A. Colensoi* of various records), *Dracophyllum pronum* (if it be that species which — *D. rosmarinifolium* of CHEESEMAN &c., but not *D. rosmarinifolium* (Forst. f.) R. Br. of which *D. politum* (T. Kirk) Ckn. is a syn.), *Epilobium pernitidum, Olearia lacunosa, Celmisia hieracifolia, C. oblonga* and *Senecio Adamsii*.

Dairy farming and grazing are the chief *agricultural pursuits,* the Manawatu area being particularly fertile. There are *permanent* pastures of *Lolium perenne* on rich soil — an uncommon occurrence elsewhere in New Zealand. Large areas of indigenous-induced *Phormium tenax* are an important feature. *Danthonia* grassland replacing artificial meadow of European grasses is a common occurrence.

The Sounds-Nelson District. This includes that portion of the north of south Island bounded by the river Wairau on the south up to nearly Tophouse and, on the west, by a line denoting the average limit of the westerly rainfall. Most of the district is hilly and part is occupied by lofty mountains. In the Marlborough Sounds portion of the district the sea fills many valleys, but others are well sheltered and their floors rich alluvial soil. The celebrated **Mineral Belt** extends as a narrow band throughout much of the district, ascending in some places to the subalpine belt and in others descending to sea-level.

Near the coast and in the valleys the *climate* is mild, as testified by the numerous subtropical garden plants. The average annual rainfall for the city of Nelson (coastal) is nearly 95 cm. with 122 rainy days, but it is higher in the Marlborough Sounds. As for the temperature, the average maximum for January (the hottest month) at Nelson is nearly 24° C. and the average minimum for the same mouth is 12.2°. For July (the coldest month) the average maxima and minima are respectively 12.5° C. and 3.1°. In the high-mountains there are frequent frosts and much snow.

The *flora* consists of at least 700 species of which the following are locally-endemic: — an unnamed var. of *Poa acicularifolia* (Mineral Belt), an unnamed *Festuca* (probably confined to the Mineral Belt), *Notothlaspi australe* var. *stellata, Poranthera microphylla* (also Australian), *Pimelea Suteri, Myosotis Monroi, Scutellaria novae-zelandiae, Hebe rigidula, H. divaricata, H. Gibbsii, Olearia serpentina, Celmisia Rutlandii, C. cordatifolia, C. Macmahoni* and *Cassinia Vauvilliersii* var. *serpentina.*

In the Sounds portion of the district there was originally much forest, both *Beilschmiedia tawa* - podocarp and *Nothofagus*, the two almost identical with those of the southern part of the Ruahine-Cook district. Southwards and westwards from Nelson City, the lowland-montane belt is occupied in large part by *Leptospermum* shrubland — mostly induced. On coastal rocks, *Astelia Solandri, Phormium Colensoi, Arthropodium cirrhatum* (absent in Ruahine-Cook) and *Griselinia lucida* are characteristic. On tidal mud-flats, there are wide carpets of *Salicornia australis* (Fig. 2).

The mountainous nature of so much of the land-surface militates strongly against *agriculture*. Sheep grazing and dairy forming, especially in the Sounds area, are of considerable importance. In the Nelson area, fruit-growing is of prime moment, much *Leptospermum* - clad country having been utilized for that purpose, the area in orchards — 2700 hectares —

being the largest in New Zealand. Hops are cultivated to a small extent (241 hectares), the district being the only one in the region where they are grown commercially.

The North-eastern District. This is bounded on the east by the coast-line from the mouth of the River Wairau on the north to the mouth of the River Hurunui on the south; on the west by a line, denoting the average limit of the western rain, extending northwards from near the source of the Hurunui till near Tophouse where it joins the boundary between this and the Sounds-Nelson district; and, on the south, by the R. Hurunui.

The *surface* consists largely of lofty mountains composed chiefly of greywacke, but in places limestone is abundant. As for flat ground, there is the gravel-plain south of the R. Weirau, the small Kaikoura gravel-plain and the basin-plains of Hanmer (montane in character) and Culverden.

The *climate* throughout much of the district is dry and of a semi-continental character, particularly in the Awatere and Clarence basins, but, in the west and south, it is wetter. Rain frequently comes from the east, hence the forest-climate of the Seaward Kaikoura Mountains. The mean annual rainfall in the lower part of the Awatere Valley, at 7.2 km. from the coast, is 61.7 cm., and for the upper part of the valley, 72 km. from the coast, 72.7 cm., but in the extreme western part of the district more than 100 cm. is not uncommon. Droughts of considerable duration occur, as in 1914 at Seddon, when only 6.2 cm. of rain fell from July to December inclusive, and the drought extending into 1915, some trees of *Eucalyptus globulus* were killed thereby.

With regard to temperature, 33° C., or rather more, are occasionally registered during summer and at Hanmer (alt. 368 m.) — 8° C. is not unusual in winter while frosts (mostly light) may occur on 106 days distributed over 7 months. High winds are frequent, and the north-west is much as that described in Part I.

The *species* number about 715 of which the following large number are locally endemic: — *Ranunculus lobulatus* (perhaps a jordanon close to R. *Monroi*), *Geum divergens, Carmichaelia Monroi* (the type), C. *juncea* var. or undescribed species (COCKAYNE, L., 1918 b: 165), *Notospartium Carmichaeliae, N. glabrescens* (perhaps identical with N. *Carmichaliae*), *Chordospartium Stevensoni, Epilobium Wilsoni* (probably identical with the next, but upheld by CHEESEMAN — 1925: 608), *E. chloraefolium* var. *kaikourense*, E. *rostratum* var. *pubens,* E. *brevipes, Schizeilema Roughii, Gentiana Astoni, Convolvulus fracto-saxosa, Myosotis Laingii,* M. *Cockayniana,* M. *saxatilis, Hebe rupicola, H. Hulkeana, Wahlenbergia Matthewsii,* W. *cartilaginea, Pachystegia insignis. Shawia coriacea, Celmisia Monroi, Haastia pulvinaris* (may extend for a short distance into SN. and NW.), H. *pulvinaris* var. *minor,* H. *recurva* var. *Wallii, Gnaphalium nitidulum, Ewartia Sinclairii, Raoulia cinerea, Heli-*

chrysum coralloides, Cassinia albida vars. *typica* and *canescens, Abrotanella Christensenii, Senecio Monroi,* and *S. lapidosus.*

Ranunculus crithmifolius, Gunnera densiflora and *Angelica trifoliolata* have each been noted only in one locality in this district and in one locality far to the south in the Eastern district.

By far the greater part of the *vegetation* is low tussock-grassland, forest usually being present — but much has been destroyed in the west — in gullies or shady slopes. There was, however, luxuriant dicotylous-podocarp forest on the east side of the Seaward Kaikouras, thanks to the moisture-laden east winds. This forest was of a semi-North Island character, as shown by the presence of *Rhopalostylis sapida, Melicope ternata* — absent further south but here meeting *M. simplex* and $\times$ *M. tersimplex* appears — and *Coprosma grandifolia.* Also, *Corynocarpus* forest follows the coast-line. A striking feature of the district is the cliff vegetation with *Pachystegia insignis* dominant and abundance of *Hebe Hulkeana* and *Senecio Monroi.* Fell-field, usually most open, and shingle-slip, reach their maximum development for the Region.

With regard to *agriculture,* sheep are depastured on all the open mountains and the most extensive sheep-runs of New Zealand occur in this district. In the districts already dealt with the dominant breeds of sheep have been of the coarse long-wool types but from the R. Wairau southwards throughout South Island it is the fine-wool sheep, largely of merino origin which are used. It is also rapidly becoming of peculiar agricultural importance through the ease with which *Medicago sativa* (lucerne, alfalfa) can be established, there being no need for special effort to keep down weeds. Malting barley is grown to a fair extent. Oat chaff is harvested earlier than in any other part of New Zealand. Indigenous-induced danthonia pasture is common but unlike that of the wetter districts it is not suitable for ewes with lambs. *Eschscholzia californica* is common on the Blenheim gravel-plain.

The North-western District. This lies to the west of the Sounds-Nelson and North-eastern districts extending to the coast-line of the Tasman Sea and is bounded on the north by the sea-coast from near Motueka to C. Farewell and on the south by the R. Taramakau. It is essentially an area of high mountains, but it contains a number of rather broad valleys. The extensive coast-line is frequently rocky or precipitous.

The *climate* is humid and the rainfall heavy. Much of the district is fully open to the prevailing westerly winds, those from the north-west bringing rain but, as they change to the south, the weather rapidly clears. Both frost and snow are abundant on the mountains in winter, but they occur to some extent at all seasons. Near the coast, the climate is mild and almost frostless and there the mean annual rainfall is 188 cm. and it is distributed as follows: — spring 25 %, summer 22.6 %, autumn 26.2 %

and winter 26.2%. On the mountains the rainfall must be considerably higher, and near the southern boundary it is much higher than as given above.

The *flora* consists of at least 905 species of which the following are locally endemic: — *Carex trachycarpa*, *Gaimardia minima*, *Townsonia deflexa*, *Colobanthus caniculatus*, *Ranunculus verticillatus*, *Nasturtium Gibbsii*, *Wintera Traversii*, *Tillaea Helmsii*, *Pittosporum Dallii*, *Carmichaelia Fieldii*, *Poranthera alpina*, *Aciphylla Hookeri*, *A. indurata*, *A. trifoliolata*, *A. Townsoni*, *Anisotome diversifolia*, *Dracophyllum Townsoni*, *D. pubescens*, *D. palustre*, *Myosotis angustata*, *M. concinna*, *Mitrasacme montana* var. *Helmsii*, *Gentiana filipes*, *G. gracilifolia*, *G. vernicosa*, *G. Townsoni*, *G. Spenceri* (perhaps in F.), *Hebe salicifolia* var. *serrulata*, *H. coarctata*, *Euphrasia Cheesemanii*, *Celmisia rupestris*, *C. Gibbsii*, *C. Dallii*, *C. semicordata*, *C. parva*, *C. dubia*, *C. lateralis*, *Senecio glaucophyllus* and *S. Hectori*.

A number of North Island species reach their southern limit in this district, many of which are absent in other parts of northern South Island: the following is a list, those coming into the last class are marked with an asterisk: — *Adiantum aethiopicum**, *Pteris macilenta*, *Blechnum Fraseri**, *Athyrium australe*, *Lycopodium cernuum**, *Bromus arenarius**, *Eleocharis neo-zelandica**, *Cladium capillaceum**, *Astelia Banksii**, *Pterostylis puberula**, *Acianthus Sinclairii*, *Calochilus paludosus**, *Corysanthes Cheesemanii**, *Nothofagus truncata*, *Peperomia Urvilleana**, *Ranunculus insignis**, *Laurelia novae-zelandiae*, *Lepidium flexicaule**, *Hibiscus trionum** (only recorded from specimens collected by LYALL and may be an error), *Pimelea longifolia*, *Metrosideros Parkinsonii**, (only elsewhere on Great Barrier Island), *M. Colensoi*, *M. robusta*, *Myriophyllum robustum*, *Epacris pauciflora**, *Dracophyllum latifolium**, *Suttonia salicina*, *Coprosma retusa*, *Nertera Cunninghamii**, *Gnaphalium subrigidum** and *Brachyglottis repanda*. Also some southern plants attain their northern limit in this district e. g. *Blechnum durum*, *Microlaena Thomsoni* (absent from lat. 46° to lat. 42°), *Pittosporum patulum* (absent from lat. 44° to lat. 42° 30'), *Ranunculus Lyallii*, *Gunnera albocarpa*, *Pseudopanax lineare*, *Anisotome Haastii*, *Actinotus suffocata* (absent form lat. 46° to lat. 42°), *Gentiana montana* (this may not be identical with the Dusky Sound plant), *G. saxosa* (extends a short distance into W.), *Coprosma serrulata*, *Celmisia Traversii* (absent from lat. 46° to lat. 42° 30') and *Senecio rotundifolius* (absent from lat. 44° to lat. 42°).

Forest, except on the particularly sour soils and certain broad, windswept valleys, covered the surface up to the timber-line, at the higher levels *Nothofagus*, and, at lower levels all grades from pure dicotylous-podocarp forest with a North Island facies of *Nothofagus* forest *(N. fusca, N. truncata, N. Menziesii* and *N. cliffortioides* and many hybrids). The high-mountain flora is rich and herb-field abundant. Notwithstanding the wet climate tussock-grassland occurs in the broad, montane valleys.

There is some dairying in the lowland alluvial valleys and flats and a good deal of *Nothofagus* forest has been felled and burned and the land grassed, but frequently with disappointing results. Various exotic species of *Rubus,* belonging to the great linneon, *R. fruticosus,* form far-extending thickets on cleared land.

The Eastern District. This is the continuation southwards of the North-eastern district. On the west, it extends to the average line reached by the westerly rain (Fig. 39) and on the south is bounded by the irregular line, shown on the map, which commences at the mouth of the River Waitaki. It is naturally subdivided into two subdistricts as below.

The Banks Subdistrict is restricted to Banks Peninsula. This is an oval-shaped mass of hills of volcanic origin about 56 km. long by 32 km. broad, pierced by several narrow arms of the sea and with the highest peaks from 600 to 900 m. altitude. The soil is partly loess partly volcanic, the latter particularly "fertile".

The *climate* is equable, the difference in average winter and summer temperature being possibly about 8° C., and the maximum 33° C. or rather more. Compared with the Canterbury Plain the amount of frost in winter is trifling, so that early crops of potatoes and tomatoes can be readily grown. On the summits, snow may lie in winter for a few days at a time. The annual rainfall apparently ranges from about 75 to 112 cm. The winds are similar to those of the Canterbury Plain, but the contour of the land leads to far more shelter.

The *flora* consists of about 472 species of which the following are possibly locally endemic: — *Anisotome* sp. (LAING and WALL consider this identical with *A. Enysii* — 1924: 442 to 43 — but I still look upon it as distinct), *Myosotis australis* var. *lytteltonensis* (almost certainly none of the New Zealand plants belong to *M. australis* R. Br.), *Hebe leiophylla* var. *strictissima, H. Laraudiana, Celmisia Mackaui, Cotula Haastii* and *Senecio saxifragoides.* There is also a distinct form of *Anisotome* belonging to the linneon *A. aromatica,* and possibly *Gunnera monoica* does not occur else-where in New Zealand proper. The following species have their southern limit on Banks Peninsula: — *Adiantum fulvum, Polypodium dictyopteris, Pteris tremula, Mariscus ustulatus, Rhopalostylis sapida, Spiranthes australis, Macropiper excelsum, Rhagodia nutans, Corynocarpus laevigata, Dodonaea riscosa, Alectryon excelsum, Tetrapathaea tetrandra* (also in swamp forest on Canterbury Plain), *Angelica rosaefolia, Griselinia lucida,* (am not sure of the identification). *Coprosma grandifolia.*

Originally the area, except where struck fairly by the strongest or most persistent wind, was forest-clad with dicotylous-podocarp forest containing a good deal of *Podocarpus totara* and *P. spicatus* in the lowland-lower montane belt and at a higher altitude *P. Hallii* and higher still some *Libocedrus Bidwillii.* At the present time, there is a good deal of tussock-grassland but how far primitive or induced it is no longer possible to say.

So, too, with the colonies of high-mountain plants, few of which, however belong to the true high-mountain element.

Early on in the history of settlement, the rich lands and excellent climate attracted the settler and by degrees almost all the forests were replaced by artificial meadows in which *Dactylis glomerata* is dominant. This grass is harvested yearly for its seed, but less land is used for this purpose than formerly. Long before cheese and butter factories were established Banks Peninsula was celebrated for its dairy produce. At the present time dairying and grazing are the main branches of agriculture.

The Canterbury subdistrict, with the exception of the Canterbury Plain, is exceedingly mountainous including as it does the eastern extension of the Southern Alps, many of the peaks exceeding 1800 m. in altitude, while some reach more than 2400 m. Shingly river-beds formed of greywacke debris, 1 km. wide or considerably more with their accompanying high terraces, and fans terminating torrents, are features of the land-surface rather more fully developed than in South Island generally, the rivers coming from glaciers of great extent. There is a long coast-line which is mostly sandy or shingly.

The climate is semi-continental in character. The hot wind, described in Part I, though not peculiar to the district, is an important climatic feature. Extremes of climate constantly occur and a sudden change of the wind to a southerly direction brings a rapid decline in temperature. These southwesters are frequently accountable for heavy showers and thunderstorms, but the most generous rains are from the south-east. The mean temperature is 11.5° C., the mean maximum 16.4° — but over 32° is occasionally recorded in summer — and the mean minimum 6.6°. Frost is frequent and more severe close to the coast than in any other part of the Region at that altitude and a temperature of —8° C. is not unusual.

The rainfall differs greatly in different parts. At about sea-level in the extreme east of the subdistrict it averages 62.5 cm., near the base of Mount Torlesse — 70 km. inland, altitude 360 m. — 100 cm. and, in the extreme west, at least 130 cm., while for the subalpine belt 170 cm. is not an unreasonable estimate. Near the coast the rainfall is distributed as follows: — spring 24.3 %, summer 23.8 %, autumn 25.9 % and winter 26 %.

The flora numbers about 750 species of which the following are locally-endemic: — *Botrychium lunaria, Carex cirrhosa, Korthalsella clavata* (may be epharmonic *K. Lindsayi*), *Ranunculus Enysii* (apparently a linneon made up of 2 species and their hybrids), *R. Monroi* var. *dentatus, R. pauciflorus* (may be *R. chordorhizus*), *R. rivularis* var. *glareosus, Carmichaelia robusta, Hoheria Allanii, Pimelea Haastii* (may be *P. aridula*), *Epilobium gracilipes, Dracophyllum acicularifolium, Gentiana serotina, Myosotis Colensoi* (= *M. decora* T. Kirk), *Hebe anomala* (a doubtful species), *H. amplexicaulis* and its vars., *H. Allanii, H. Armstrongii* and *Brachycome pinnata*.

The subdistrict, as a whole, even from sea-level, up to the lower sub-alpine belt was occupied by tussock-grassland (mostly low). There is a rich development of fell-field vegetation, much of which is modified. Shingle-slip vegetation is a common feature.

Agriculture on ploughable land consists principally of rotation farming, in which wheat plays an important part. A conspicuous feature of distribution, from the agricultural standpoint, is the inability to successfully grow hard turnips every season owing to the ravages of aphides and diamond-backed moths, and consequently winter catch-cropping with forage oats is practised. Dairying is practised on the drained swamp-land of the plain. The mountainous area and the stony parts of the plain are devoted to sheep-farming, the raising of fat lambs being an important branch. Owing to the somewhat severe winters many garden plants, which succeed well further south, cannot be successfully grown.

The Western District. This extends from the R. Taramakau on the north to a line passing from the north end of Big Bay and extending by way of the sources of the Rivers Cascade and Arawhata in a straight line to a spot on the east of the Divide marking the average limit of the westerly rainfall and from this point northwards to the junction with the North-eastern district the line denoting the above limit forms the eastern boundary.

The district includes the highest part of the Southern Alps, their western foothills and the narrow Westland coastal plain. Really it should be sub-divided into 2 subdistricts, **eastern** and **western,** distinguished in part by the preponderance of rain-forest proper on the west of the Divide and of *Nothofagus* forest on the east, but I have quite insufficient data for such treatment. The district is superabundantly watered and is traversed by numerous, torrential glacial rivers. The Franz Josef and Fox glaciers descend far into the lowland belt (211 m. and 204 m., respectively).

The *climate* is extremely wet, the average rainfall at Hokitika — *almost the driest locality in the west* — being 290 cm. on 187 days; and, on the east, that of the old Bealey township — only just in the district — being 257 cm. on 174 days. In the mountains the rainfall cannot be less than 500 cm. On the high mountains the snow-fall is extremely heavy, as demonstrated by the size, number and low altitudes of the glaciers. Not-withstanding the excessive precipitation there is an average of 1,915 hours of sunshine per year at Hokitika. The temperature is without extremes the average maximum and minimum at Hokitika being respectively 16^0 and 8^0, while the frost is not sufficiently severe to kill garden plants which cannot be grown for any length of time at Christchurch (E.).

The *flora* consists of about 755 species of wich the following are locally endemic: — *Deyeuxia Youngii, Danthonia oreophila* var. *elata, Colobanthus monticola, Ranunculus Godleyanus, R. Grahami, R. sericophyllus, Epilobium westlandicum, Aciphylla divisa, A. similis, Myosotis explanata, M. suavis,*

Nertera ciliata, Hebe macrocalyx, H. Treadwellii, Brachycome polita and *Leucogenes Grahami* (possibly *L. grandiceps* × *Helichrysum Selago*).

The lowlands and mountains up to a height of about 1200 m. are densely clothed with forest (dicotylous-podocarp except in the south and *Nothofagus* in the east). Characteristic of lowland forest are *Quintinia acutifolia, Ascarina lucida* and the lianes, *Freycinetia Banksii, Metrosideros scandens* and *M. perforata*. Subalpine-scrub is richly developed and allied shrub associations descend to the lowlands. The most extreme xerophytes are absent; herb-field of a luxuriant character is characteristic with *Ranunculus Lyallii, Ourisia macrocarpa* var. *calycina, Celmisia Armstrongii, C. coriacea* and *Senecio scorzonerioides* in abundance.

Owing to the wet climate it is difficult to bring in pasture by bush-burning. Some dairying is carried out on the river-flats and sheep are pastured on river-beds. Saw-milling is an important industry and will remain so for a considerable time. After the forest is felled, in many places indigenous-induced bog is the first succession.

The North Otago District. This forms the southernmost portion of the South Island tussock-grassland area and it includes the driest and hottest part of New Zealand proper. It is bounded on the west by the line marking the average limit of the north-westerly rain, and, on the south by a much more irregular line which denotes the average limit reached by the south-westerly downpour. Also, the northern boundary is most irregular in shape and designed to separate the dry area from one rather wetter, but its position is merely a guess based on insufficient investigation. Really the district is not quite as shown upon the map, for it extends along the Kaiwarau valley almost to L. Wakatipu, but the upper parts of Mount Cardrona, Mount Pisa, the Carrick Range and the Dunstan Mountains clearly belong to the South Otago district. The northern part of the district includes part of the Mackenzie plain — a basin-plain — but most of the remainder is occupied by a series of straight, wall-like block mountains, up to 2100 m. high, composed of mica-schist or greywacke, as the case may be, while between these mountains are large or small depressions, the former forming basins or plateaux (Maniototo Plain, Ida Valley, Manuherikia Valley &c.).

The district is far and away the driest in the Region. The effect of such drought is plainly visible in the rocks weathered into fantastic shapes by blown sand &c. and by the present state of the vegetation described in Part III. The average annual rainfall in the upper Clutha Valley is some 35 cm., and the average number of days on which rain — usually very hight — falls is 60. Droughts extending over 2 months or more are common. Even at Kurow on the Waitaki, far from the centre of aridity, the rainfall in 1907—8 (March to Febr., inclusive) was only 36 cm. and the rainy days 94. On the Maniototo Plain, too, as low as 32.2 cm. in 94 days has been registered, and that area is far from having reached anything like the depletion *(Raoulia lutescens* colonies) of the upper Clutha. The low rainfall of the district is accompanied by a clear sky day after day, a burning sun, and frequent violent hot winds with clouds of dust or sand. In summer and early autumn, a shade temperature of 32° C. is common and even over 38° not infrequent. The cold of winter, too, attains its maximum for New Zealand and — 16.6° has been registered for the Maniototo Plain.

The *flora* numbers about 480 species of which the following are locally endemic: — *Carex decurtata, Colobanthus Buchanani, Lepidium Kirkii, L. Kawarau* (possibly an epharmone of the next), *L. matau, Acaena Buchanani* (apparently extends some distance into the South Otago district), *Carmichaelia Petriei, C. compacta, C. curta, Pimelea aridula, Halorrhagis depressa* var. *spicata, Myosotis pygmaea* var. *imbricata, M. albosericea, Limosella Curdieana, Hebe pimeleoides* var. *rupestris.*

Except in a few gullies in the east, forest is wanting. *Olearia lineata* is dominant in certain scrubs. Tussock-grassland with *Poa Colensoi* or *P. intermedia* dominant, and *Agropyrum scabrum* abundant, may ascend to the summits of the mountains but in the driest part of the district it has been replaced by induced-steppe with *Raoulia lutescens* dominant. Near Lake Wanaka the soft cushions of *Pimelea sericeo-villosa* are conspicuous in such steppe and green or silvery mats of *Acaena Buchanani* and *Raoulia Parkii* respectively are abundant. *Carmichaelia Petriei* (an extreme xerophyte. now eaten to the ground by sheep and rabbits) was originally abundant throughout. On dry rocks, *Hebe pimeleoides* var. *rupestris* and, in sandy ground, *Pimelea aridula* are characteristic.

For many years the district has been the site of numerous sheep stations, the carrying-capacity of which has been greatly reduced by burning and, at one time, overstocking, and the presence of rabbits in vast numbers. In the valley of the Clutha and some of tributaries fruit-growing is a flourishing industry, thanks to irrigation, stone fruits growing better than elsewhere in the region. Irrigation, too, has led to the cultivation of lucerne (alfalfa) so that now the district comes next to the North-eastern in regard to the area occupied by this important crop. Malting barley is grown to some extent, especially near L. Wakatipu. With irrigation, both vegetables and flowers are grown to perfection, the latter, roses for instance, being most brilliant in hue.

The South Otago District. This occupies the south and south-eastern portion of South Island subject to frequent south-west gales accompanied by rain. The boundary between this district and the Fiord district is, as yet, problematical, since it is not easy to fix an average limit for the north-western downpour for, in the south, continuous forest extends from the one district to the other, and most of the transitional area is unexplored. The irregular northern boundary is shown on the map. This boundary is well-defined in many places by the sudden incoming of induced steppe. At Deep

Creek, in the Taieri Gorge, *Leptospermum scoparium* and the exotic *Digitalis purpurea* suddenly give out, thus marking *exactly* the junction of the two districts. So, too, at a certain point on the Palmerston South-Central Otago road, *Carmichaelia Petriei* comes to a stand, and another species (perhaps unnamed) takes its place eastwards — a remarkable fact first noticed by J. S. THOMSON and G. SIMPSON.

In general, the surface is hilly and, in the west, mountainous, but there is the flat Taieri Plain — liable to floods — and the extensive Southland Plain. Rivers and streams are abundant. The long coast-line is frequently rocky and precipitous, but along Foveaux Strait there are sand-dunes of considerable extent.

Owing to the frequency of cold rain-bearing south-west winds with cloudy sky and the comparatively low summer temperature, the climate of this area approximates more to the subantarctic type than that of any other of the mainland botanical divisions, except the Fiord district. At the same time, it must be pointed out that there is much bright sunshine and cloudless sky, as in New Zealand generally. The north-west wind is dry. its moisture having been lost in passing over the Fiord district. High winds are frequent and at times sweep over the open Southland Plain with fury. In the mountains the winter snow is of comparatively long duration. Snow at sea-level too is not uncommon, but frosts at that altitude are less severe than on the coast of the Eastern district. At Tapanui (alt. 150 m.) the highest and lowest shade temperatures recorded during 15 years were respectively 36^0 and $- 10^0$, but about $- 5^0$ was much more common. The average rainfall for Dunedin (about sea-level), Tapanui (150 m.), Queenstown (L. Wakatipu, about 300 m.) and Invercargil (about sea-level) is respectively: 100 cm., 90 cm., 76 cm. and 146 cm. and the average number of rainy days 159. 155, 91 and 189. Apparently the distribution of the rainfall is highest in spring and autumn, but least in summer and winter, but the differences are not great.

The *flora* consists of about 775 species of which the following are locally endemic: — *Poa pygmaea, Agrostis Petriei* (extends into NO. for a short distance), *Triodia australis, Carex Hectori, C. pterocarpa, Centrolepis strigosa, Luzula micrantha, L. crenulata, L. triandra* (the three last perhaps epharmonic forms of one species), *L. leptophylla, Gastrodia minor* (may be stunted *G. Cunninghamii), Ranunculus Berggreni, R. novae-zelandiae, R. Scott-Thomsonii, Nasturtium Wallii, Corallospartium racemosum, Carmichaelia virgata* and probably an undescribed species of that genus allied to *C. robusta. Gunnera mixta, Aciphylla simplex, A. Spedeni, A. Scott-Thomsonii, Anisotome lanuginosa, A. imbricata, Tetrachondra Hamiltonii, Hebe Biggarii, H. Poppelwellii, H. annulata, Lagenophora purpurea* (probably *L. petiolata), Celmisia Lindsayi, Helichrysum Selago* var. *tumida* (probably better treated as a species near to *H. coralloides), Cotula Willcoxii, C. obscura, C. sericea,* and *Senecio southlandicus.*

The following species occur also in the Fiord district: — *Deschampsia pusilla, Poa exigua, Ranunculus Buchanani, R. pachyrhizus, Schizeilema exigua, Aciphylla multisecta, A. pinnatifida, Anisotome intermedia* (the type), *A. capil'ifolia, Myosotis pulvinaris, Hebe dasyphylla, H. Petrieii, Celmisia Bonplandii, C. rerbascifolia, C. lanceolata, C. Petriei, C. ramulosa* and *Senecio rerolutus.*

The eastern part of the district was occupied originally by dicotylous-podocarp forest composed of few species as compared with the North Island part of the formation, but the western part is still covered with a wide area of *Nothofagus* forest in the south made up of *N. Menziesii* and *N. cliffortioides* and of these together with *N. fusca* and × *N. cliffusca* in the north, such forest being far from continuous. A good deal of the lowlands and the montane belt is still clothed with tall tussock-grassland (*Danthonia Raoulii* var. *rubra* dominant, or *D. flavescens* — in a wide sense — in places). On the mountains, there are rich herb-fields and, where flat, herb-moors are characteristic. A good many high-mountain species occur at sea-level. Coastal moor is present at certain places on the shore of Foveaux Strait. *Sphagnum* bog originally was quite common.

On ploughable land the *agriculture* of the district consists of rotation farming, in which spring-sown oats and Aberdeen turnips are conspicuous crops. Dairy-farming is a most important industry. Other features of the district are the ridging of turnips and the general application of carbonate of lime, indeed more lime is used than in all the other districts taken together.

The Fiord District. This district occupies all the area pierced by the Otago fiords and lakes south of L. Wakatipu subject to the excessively heavy north-westerly downpour. Its eastern and northern boundaries, as shown on the South Island map, are merely provisional.

The whole of the district is occupied by lofty, precipitous mountains, which, in the northern part carry large glaciers and many years must go by before the vegetation and flora are fairly well known, for not only is the area difficult to explore, but the botanists available are few in number and with other avocations. Probably it will eventually be shown that there are two subdistricts — northern and southern — distinguished by a number of species peculiar to each.

The district has the maximum rainfall and the greatest number of rainy days in the New Zealand Region. Frost and snow, though frequent in the mountains and occurring at all seasons are probably of little moment at sea-level. Taking the few isolated stations for which records are available there is a mean of 413 cm. distributed as follows: — spring 24.8%; summer 22.4%; autumn 29.5%; winter 23.3%.

The *flora* consists of about 700 species of which the following are locally endemic: — *Agrostis magellanica* (also in the Subantarctic province), *Danthonia ovata, D. planifolia, Poa oraria, Poa* sp. of dripping rock in the

Clinton Valley (Fig. 78), *Heleocharis acicularis, Uncinia longifructus, Ranunculus Simpsonii, Pimelea Crosby-Smithiana, Epilobium Matthewsii, E. purpuratum* (only known from the type at Kew), *Aciphylla Cuthbertiana, A. congesta, A. Crosby-Smithii, Anisotome Lyallii, Dracophyllum fiordense* (referred by CHEESEMAN to *D. Townsoni*), *Gentiana flaccida, Myosotis Lyallii. Veronica catarractae, Ourisia macrocarpa* var. *cordata, O. Macphersonii, Euphrasia integrifolia, Olearia operina, O. Crosby-Smithiana* (perhaps identical with *Senecio bifistulosus), Celmisia holosericea, Raoulia Buchanani* (but goes for a short distance in SO.) and *Senecio bifistulosus.*

The chief characteristics of the *vegetation* are the dense forests — *Nothofagus* - podocarp on the west and mostly pure *Nothofagus* on the east; the well defined coastal-scrub of *Olearia operina, Senecio rotundifolius. Hebe elliptica, H. salicifolia* and × *H. ellipsala;* the extensive herb-fields with abundance of *Ranunculus Lyallii* (absent in some localities), *Celmisia Petriei, C. verbascifolia, C. holosericea, Ourisia macrocarpa* var. *cordata* and *Dracophyllum Menziesii* (also forming scrubs) and *Ranunculus Buchanani* and *R. Simpsonii* on stony ground at high levels; and the replacement of forest by sphagnum bogs on which grow *Olearia divaricata* and various herb-field species.

The district is uninhabited except along two tourist routes, and at Preservation Inlet where there is a lighthouse. Nearly all the district is reserved as a National Park.

The Stewart District. This comprises Stewart Island, together with all the islands in Foveaux Strait, those adjacent to Stewart Island itself, and the Solanders. Stewart Island is extremely hilly, the highest portion slightly exceeding 900 m. altitude. It is pierced by two arms of the sea which extend far inland. Much of the coast is rocky but, on the west, there is a fine sandy beach (Mason Bay) backed by very high dunes. Much of the surface of the island, even to the high summits, is wet and peaty.

The *climate* over the greater part of the district is semi-subantarctic in character and this is especially so in the south, south-west, and on the open mountains. The number of rainy days is excessive, even Halfmoon Bay (the driest part of Stewart Island) having a yearly average of 241 days with a maximum of 283 and a minimum of 223. The average annual rainfall (Halfmoon Bay) is 160 cm. and its distribution as follows: — spring 26.7%, summer 22.5%, autumn 27.9% and winter 22.9%. The climate is extremely mild, frost being almost absent on the east coast and the small islands, while snow is generally confined to the subalpine belt. The south-west wind, generally accompanied by rain or, in the mountains, sleet, frequently sweeps over the district in all its fury, except in the sheltered valleys.

The *flora* consists of about 500 species of which the following are locally-endemic: — *Poa Guthrie-Smithiana, Danthonia pungens, Uncinia pedicellata* (according to CHEESEMAN abundant throughout North and South

Island, but he had never seen *living* Stewart Island material), *Chrysobactron Gibbsii, Aciphylla Traillii, Anisotome flabellata, A. intermedia* var. *oblongifolia, Schizeilema Cockaynei, Dracophyllum Pearsoni, Hebe Laingii, Abrotanella muscosa* and *Raoulia Goyeni*. The following are extremely rare elsewhere in the Region: — *Ourisia modesta, Stilbocarpa Lyallii* and *Olearia angustifolia*. Certain species of the Subantarctic province occur which are absent in either North Island or South Island, e. g. *Polypodium Billardieri* var. *rigidum, Asplenium scleroprium, Poa foliosa, Astelia subulata, Urtica australis, Hebe odora, Senecio Stewartiae* and the Stewart *Olearia Colensoi* may be *O. Lyallii*.

The most characteristic features of the vegetation are the two types of forest (rimu-kamahi and *Dacrydium intermedium* swamp-forest); coastal scrubs of *Senecio rotundifolius-Olearia angustifolia;* lowland *Gleichenia-Hypolaena* bog; cushion herb-moor; and associations of high-mountain plants at almost sea-level.

Most of the island is uninhabited. *Farming* is of but little moment, there being only a few hectares of pasture all of which are in the vicinity of Port William, Halfmoon Bay and the Neck.

Certain other Districts. In addition to the districts defined above are the following: The *Chatham*, the *Snares*, the *Lord Auckland*, the *Campbell*, the *Antipodes* and the *Macquarie*. None of these need detailed description, for their endemic plants and plant-communities have been already dealt with in Part II, Section IV.

Chapter II.
The Families, Genera and Elements of the Flora.
1. The General Statistics.

The total number of species of vascular plants. together with such varieties as are of equal rank to many admitted species, is 1843 of which 166 are pteridophytes (*Filices* 147), 20 gymnosperms and 1657 angiosperms (monocotyledons 428 and dicotyledons 1229), and they belong to 109 families and 383 genera. The average number of species to each family is 16.9 and 4.8 to each genus.

The largest families and genera are as follows: — (families) *Compositae* 256 (compound 55), *Filices* 147 (comp. 13), *Cyperaceae* 133 (comp. 15), *Gramineae* 131 (comp. 21), *Umbelliferae* 89 (comp. 7), *Orchidaceae* 71 (comp. 1), *Ranunculaceae* 61 (comp. 13). *Rubiaceae* 55 (comp. 12), *Onagraceae* 45 (comp. 8), *Epacridaceae* 44 (comp. 7), *Leguminosae* 38 (comp. 5), *Borraginaceae* 33 (comp. 4), *Rosaceae* 29 (comp. 9). *Cruciferae* 27 (comp. 2), *Gentianaceae* 24 (comp. 5), *Pittosporaceae* 22 (comp. 4), *Halorrhagaceae* 22 (comp. 3), *Caryophyllaceae* 21 (comp. 3), *Myrtaceae* 20 (comp. 2), *Thymelaeaceae* 20 (comp. 6), *Araliaceae* 20 (comp. 4); (genera) *Hebe* 66, *Carex* 59, *Celmisia* 54, *Coprosma* 48, *Ranunculus* 47, *Epilobium* 41, *Olearia* 35, *Senecio* 35, *Poa* 33, *Myosotis* 32, *Aciphylla* 29, *Dracophyllum* 28, *Carmichaelia* 25, *Cotula* 25, *Pittosporum* 22, *Anisotome* 22, *Gentiana* 22, *Raoulia* 22, *Uncinia* 20, *Pimelea* 20, *Hymenophyllum* 19, *Danthonia* 17, *Asplenium* 15, *Juncus* 15, *Acaena* 15, *Blechnum* 14, *Scirpus* 14, *Thelymitra* 14, *Veronica* 14, *Pterostylis* 13 and *Euphrasia* 13.

Although the above figures show the relative importance of the larger families and genera, so far as the flora is concerned, they exaggerate the part that some play in the vegetation. In this regard many small genera are of more moment although they may be of restricted distribution; the following to which the number of species are appended are examples: — *Cyathea* 5, *Hemitelia* 1, *Dicksonia* 3, *Pteridium* 1, *Gleichenia* 6 (Filices); *Agathis* 1, *Libocedrus* 2 (Cupressac.); *Typha* 1 (Typhac.); *Freycinetia* 1 (Pandanac.); *Arundo* 2 (Gramin.); *Mariscus* 1, *Schoenus* 7; *Gahnia* 8 (Cyperac.); *Rhopalostylis* 2 (Palmae); *Leptocarpus* 1, *Hypolaena* 1 (Restionac.); *Rhipogonum* 1, *Cordyline* 4, *Astelia* 12, *Phormium* 2, *Chrysobactron* 3 (Liliac.); *Nothofagus* 5 (Fagac.); *Elatostema* 1 (Urticac.); *Elytranthe* 4 (Loranthac.), *Muehlenbeckia* 6 (Polygonac.), *Carpodetus* 1 (Saxifrag.), *Weinmannia* 2 (Cunoniac.), *Rubus* 5 (Rosac.), *Edwardsia* 4 (Legum.), *Coriaria* 4 (Coriariac.), *Pomaderris* 5 and *Discaria* 1 (Rhamnac.), *Aristotelia* 2 and *Elaeocarpus* 2 (Elaeocarp.), *Plagianthus* 2 and *Hoheria* 5 (Malvac.), *Melicytus* 5 (Violac.), *Leptospermum* 4 and *Myrtus* 3 (Myrtac.), *Fuchsia* 3 or 4 (Onagrac.), *Nothopanax* 7, *Pseudopanax* 7 and *Schefflera* 1 *(Araliac.);* *Griselinia* 2 (Cornac.); *Gaultheria* 5 (Ericac.); *Cyathodes* 5, *Leucopogon* 3 and *Epacris* 3 (Epacrid.); *Suttonia* 8 (Myrsinac.); *Olea* 4 (Oleac.); *Geniostoma* 1 (Loganiac.); *Parsonsia* 2 (Apocynac.); *Vitex* 1, *Avicennia* 1 (Verbenac.); *Myoporum* 1 (Myoporac.); *Nertera* 6 (Rubiac.); *Alseuosmia* 4 (Caprifoliac.); *Selliera* 1 (Goodeniac.); *Pleurophyllum* 4, *Haastia* 4, *Leucogenes* 2, *Cassinia* 6 and *Brachyglottis* 2 (Compositae).

In contradistinction to the above, the following genera are so rare that their absence from the vegetation would not be noticed: — *Simplicia* 1, *Amphibromus* 1 (Gramin.); *Hydatella* 1 (Centrolepidac.); *Iphigenia* 1 (Liliac.); *Hypoxis* 1 (Amaryll.), *Bagnisia* 1 (Burmanniac.), *Caleana* 1 (Orchid.), *Phrygilanthus* 2 (Loranth.), *Logania* 1 (Loganiac.) and *Tetrachondra* 1 (Tetrachondrac.).

Hybrids. Up to the year 1912 5 hybrids had been recorded for the flora though with some diffidence (COCKAYNE, L. 1912a : 30—31), but in this paper it was suggested that variation in the highly-variable genera *Hebe, Celmisia* and *Acaena* was possibly due in large measure to hybridism. Five years later, the study of wild hybrids began in earnest, so that, at the present time, no less than 290 groups of such have been noted, a large majority consisting not merely of a few individuals more or less similar but of great polymorphic swarms.

The 290 groups of hybrids belong to 42 families ($40^0/_0$ of the families) and 92 genera ($24^0/_0$ of the genera), the largest of which are as follows: — *Compositae* 64 groups of hybrids, *Scrophulariaceae* 36, *Filices* 21, *Gramineae* 17, *Acaena* and *Coprosma* 14 each, *Cyperaceae* 12, *Hebe* 31, *Celmisia* 18, *Coprosma* 14, *Olearia* 13, *Acaena* 10, *Asplenium* 9 and *Ranunculus* 7. In nearly all the above genera additional hybrid groups are certainly expected. In addition, very large polymorphic swarms occur in the following genera: — *Danthonia, Uncinia, Luzula, Phormium, Nothofagus, Mida, Pittosporum, Melicope, Coriaria, Aristotelia, Myrtus, Epilobium, Nothopanax, Fuchsia, Gaultheria, Apium, Corokia, Dracophyllum, Parsonsia, Veronica, Alseuosmia, Gnaphalium, Helichrysum, Cassinia, Craspedia, Cotula* and *Senecio.*

Occasionally, hybrids occur in which 3 species are concerned, e. g. in *Cassinia, Coprosma, Coriaria* and *Alseuosmia.* Also intergeneric hybrids are not unknown, e. g. *Ewartia* × *Helichrysum, Nothopanax* × *Pseudopanax, Helichrysum* × *Gnaphalium, Leucogenes* × *Raoulia* and possibly by *Helichrysum* (subgen. *Ozothamnus*).

Generally, the species which cross are of the same life-form, but this is not always the case. Thus the tree-form crosses with the divaricating-shrub form, e. g. *Plagianthus betulinus* × *divaricatus*; the tall bushy-shrub with the divaricating-shrub, e. g. *Corokia buddleoides* × *Cotoneaster*: the low canopy-tree with the liane, e. g. *Fuchsia excorticata* × *perscandens*; the shrub by the slender slightly woody mat-plant, e. g. *Helichrysum depressum* × *bellidioides*: the liane by the mat-shrub, e. g. *Muehlenbeckia complexa* × *ephedroides*; the tree by the summergreen semi-woody plant, e. g. *Coriaria arborea* × *sarmentosa;* the cupressoid-form by the leafy shrub, e. g. *Hebe Astoni* × *buxifolia*; the dense cushion-plant by the erect low semi-woody plant, e. g. *Raoulia bryoides* × *Leucogenes grandiceps.*

With regard to the validity of the 290 groups of hybrids there is no doubt as to the hybrid nature of 230 of them, most of these having been studied more or less intensively in the field and a good many have been brought into cultivation, especially by W. A. THOMSON, J. S. THOMSON and G. SIMPSON of Dunedin, H. H. ALLAN and myself. Moreover ALLAN has already synthesised two of the hybrids, and seed sown from others has produced polymorphic progeny. The hybrid groups range from those whose parents usually grow in close proximity to those the parents of which very rarely meet. In certain cases, the unpremeditated action of man with his burning &c. has led to great increase in the number of hybrid individuals, e. g. in *Acaena, Myrtus, Hebe, Celmisia* and *Cassinia.* In the case of *Acaena*, hybrids occur between the exotic *A. ovina* and the endemic *A. inermis, A. Sanguisorbae* var. *pusilla* and *A. novae-zelandiae.*

So far as observation and experiment go, it seems safe to assume that most of the hybrids produce viable seed, in fact, so far the only more or less sterile hybrid is one I described as a species many years ago under the name *Rubus Barkeri*. This plant — only the one was found wild — is readily reproduced vegetatively so that it is quite common in cultivation, yet only twice has it been known to bloom since its discovery some 30 years ago. *Rubus parvus* — a creeping and rooting mat-plant — is one of the parents and probably *R. australis* (liane) the other parent. Recently Dr. W. Mc. KAY of Greymouth (NW.) has discovered a most closely-related plant which produced a fair amount of fruit on its shaded shoots, but none where exposed to bright light. Most likely both plants are of the F 1. generation and those of the F 2. &c. generation must be exceedingly rare.

From what has gone before it will be expected that from time to time hybrids have been described as species. This expectation is realised in the fact that no less than 42 hybrids masquerade as good species in the *Manual of the New Zealand Flora*. Also the "variability" of no few "variable species" is due entirely to crossing constantly taking place between the jordanons of which they are composed. This has apparently happened to such an extent in certain species that the jordanons have been entirely swamped out by the hybrids and such alone remain. Over wide areas this seems to be the case with *Leptospermum scoparium* and *Phormium tenax*. Near Queenstown (SO.) swamping can be seen in progress in a small piece of *Nothofagus* forest where hybrids of the × *N. cliffusca* group are now in greater numbers than the parent species; also in indigenous-induced montane *Cassinia* shrubland it is often difficult, if not impossible, to pick out the species, though these are constant when one alone is present. In the case of the endemic genus *Alseuosmia*, polymorphy reaches its highest degree. Theoretically, the genus embraces 4 (HOOKER, KIRK, CHEESEMAN) to 8 species (A. CUNNINGHAM), but where all these grow together a polymorphy unspeakable occurs, perhaps impossible to disentangle. Taking the numerous polymorphic groups ("variable species") it should be now clear enough whence comes the reputed, "extreme variability" of the New Zealand flora, yet, *in the ordinary taxonomic sense, if the epharmones be excluded, there is no variability whatsoever,* but merely a polymorphy due to the grouping together as a species two or more jordanons (closely allied or otherwise) and the hybrids between them.

2. The Elements of the Flora.

The endemic element. The endemic species of *Pteridophyta, Gymnospermae* and *Angiospermae* (monocotyledons 294, dicotyledons 1077) number 1451 and constitute 78.6 per cent of the vascular flora. Taking each group separately, 36 per cent of the pteridophytes are endemic, 67 per cent of the monocotyledons and 88 per cent of the gymnosperms (20) and dicotyledons taken together.

The endemic species are by no means all of one class but range from those extremely distinct from any other species either of the New Zealand or any flora (e. g. *Simplicia laxa, Hectorella caespitosa*) to those almost identical with species of other regions (e. g. *Carex Gaudichaudiana, Oxalis lactea*).

There are 40 endemic genera which like the endemic species show strongly different degrees of endemism, but here for convenience they are arranged in their systematic sequence. The following is a list such genera together with a brief account of the affinities of each: — **Loxsoma** (Filices) distantly related to *Loxsomopsis* of Costarica and Ecuador. **Simplicia** (Gramin.) monotypic considered by HACKEL intermediate between *Sporobolus* (tropical and subtropical especially in America but only introduced in New Zealand) and *Agrostis* (see CHEESEMAN 1906 : 861).

Desmoschoenus (Cyperac.), a monotype agreeing — according to HOOKER (1864 : 303) — with *Isolepis* in its floral characters, as A. RICHARD had shown, but transferred by BOECKELER in 1878 to *Scirpus* under BANKS and SOLANDER's unpublished name *S. frondosus*. **Sporodanthus** *(Restionac.)*, a monotype, is allied to the Australian *Lepyrodia* from which it differs in its 1-celled, 1-seeded fruit, but BENTHAM and HOOKER, considering the 1-celled ovary due to abortion, placed the species in *Lepyrodia*. CHEESEMAN, on the contrary, found no trace of a 3-celled ovary in young buds, and considered von MUELLER's genus, *Sporodanthus,* would be restored. **Chrysobactron** (Liliac.), with 3 species, and, at least, one well-marked variety of one of them, was established by HOOKER in 1844 but in 1864 was transferred by him to *Anthericum* and, later, by BENTHAM and HOOKER *(Genera Plantarum)* and ENGLER *(Pflanzenfamilien)* it was merged into *Bulbinella*, a South African genus. **Pentalochilus** (Orchid.), with 2 species, founded by R. S. ROGERS in 1924, is related to *Caladenia* and "approaches very closely to *Glossodia*" — a small Australian genus. **Mida** (Santalac.), with 2 species and many hybrids between them, is closely related to the Australian genus *Santalum*, and it would not be endemic but for the fact that the only other known species of the genus — *M. fernandezianum* (Phil.) Sprague et Summerh. — is extinct "the last known tree having died between 1908 and 1916" (SPRAGUE, T. A. and SUMMERHAYES, V. S., Kew Bull. Misc. Inform.. No. 5, 1927 : 198). **Tupeia** (Loranthac.), a distinct monotype related both to the *Viscoideae* and the *Loranthoideae*. **Dactylanthus** (Balanophor.), is a very distinct monotype, and according to HOOKER "a most remarkable" genus. **Hectorella** (Caryophyll.), is a monotype of somewhat doubtful affinity possibly related to the Kerguelan *Lyallia* (see EWART, Journ. Linn. Soc. 38 : 1—3.) but placed by HOOKER and others in the *Portulacaceae*. **Wintera** (Winterac.), with 3 species, has been removed from the closely-related *Drimys* by HUTCHINSON, the latter name to be applied to the other members of the family which range from Central and South America to eastern Australia, New Caledonia and the Malay Archipelago. W. R. B. OLIVER (1925 : 3) considers the name *Drimys* should be applied to the New Zealand species and *Wintera* to the South American, and I presume the other species. **Pachycladon** (Crucif.), a monotype, according to HOOKER is intermediate between the *Sisymbrieae* and the *Lepidineae, Notothlaspi* (Crucif.), 2 species, is placed by PRANTL in the *Thelypodieae Stanleyinae* next to *Pringlea* of Kerguelen Land. **Ixerba** (Saxifragac.), monotypic, belongs to the section *Escallonioideae* and is nearest to *Brexia* of Madagascar and the Seychelles. **Carpodetus** (Saxifragac.), monotypic, also belongs to the *Escallonioideae* and is placed by ENGLER between the Australian-New Caledonian *Argophyllum* and the Lord Howe Island *Colmeiroa*. **Corallospartium** (Legum.), 2 species, is closely related to the almost endemic *Carmichaelia*. **Chordospartium** (Legum.), monotypic, confined so far as known to two valleys of the North-eastern district, comes nearest to *Corallospartium* but also possesses characters of *Carmichaelia* and the following genus. **Notospartium** (Legum.), 2 species, differs from the above *Leguminosae* in its rather long, linear many-jointed pod and would be accorded generic rank by any systematist. **Entelea** (Tiliac.), monotypic, is related to the South African *Sparmannia*. **Hoheria** (Malvac.), at least 6 species, belongs to the *Malvineae-Sidinae* and is related to the indigenous and Australian *Plagianthus* and to *Sida*, a genus of wide range including Australia. **Tetrapathaea** (Passiflor.), monotypic, is by some united with *Passiflora*, from which it differs in its dioecious and tetramerous flowers. **Stilbocarpa** (Araliac.), 3 species (2 confined to the Subantarctic Province and 1 to the Stewart and Fiord districts), is somewhat closely allied to *Aralia* (Malayan, North American, east Asian and Australian). **Anisotome** (Umbell.), 21 species, is closely related to almost endemic *Aciphylla* (2 Australian) and is frequently placed in *Ligusticum* from which it differs according to CHEESEMAN "in the narrower fruit with the ribs often unequal in height and in the fewer vittae". **Coxella**, monotypic and confined to the Chathams, has been referred to *Aciphylla, Gingidium, Ligusticum* and *Angelica*, with none of which does it properly agree. **Corokia** (Cornac.), 3 species and a well-marked variety of one of them, is placed by HARMS between the Himalayan-east Asian *Helwingia* and *Cornus*, a genus wide-spread in the Northern-Hemisphere. **Myosotidium** (Borrag.), monotypic, and confined to the Chatham Islands, is a distinct genus belonging to the *Borraginoideae-Cynoglosseae*, its nearest ally appearing to be *Paracaryum* of Central Asia and the Mediterranean. **Teucridium** (Verbenac.), monotypic, is related but not closely to the Himalayan-African *Holmskioldia*. **Pygmaea** (Scroph.), 3 species, is related to *Hebe* but all are herbs of cushion-form, the leaves minute and not quadrifarious, the flowers solitary and terminal and the calyx 5—6-lobed. **Siphonidium** (Scroph.), monotypic, is closely related to *Euphrasia*, a common genus in New Zealand, but amply distinct. **Rhabdothamnus** (Gesneriac.), monotypic, forms with the New Caledonian *Coronanthera* and the Lord Howe *Negria* the small group *Cyrtandroideae-Coronanthereae-Coronantherinae*. **Alseuosmia** (Caprifoliac.), number of species doubtful through overshadowing by their hybrids, is perhaps more closely related to *Lonicera* than to any other genus but differing from all members of the *Caprifoliaceae* in its alternate leaves. **Colensoa** (Campanulac.), monotypic. is very closely related to *Pratia* from which it differs in its tall erect habit, large leaves, racemose inflorescence, large flowers and large stigmatic lobes, but by Schonland it was united to this genus — a course followed by Cheeseman in *Illustrations of the New Zealand Flora*, 1914, but reversed in his *Manual* ed. 2, 1925. **Oreostylidium** (Stylidiac.), monotypic, is related to the Australian *Stylidium*, in which it was placed by Hooker, but differs in its equal corolla-lobes, erect short column and indehiscent capsule. **Pachystegia** (Compos.), a monotype of which there are probably several varieties, for a long time kept in *Olearia*, differs from that genus in the great multibracteolate, ovoid involucre with the bracts in many series and the uniseriate pappus with the hairs equal and clavate. **Shawia** (Compos.), 2 species and many hybrids between them, is closely related to *Olearia* with which it is generally united but it differs in the head consisting of only 1 florest which is tubular and hermaphrodite. **Pleurophyllum** (Compos.), 3 species, is confined to the Subantarctic province and differs only from the almost endemic *Celmisia* in habit and its racemose inflorescence. **Haastia** (Compos.), 3 species and some varieties, is related to no indigenous genus and apparently comes nearest to the Madagascan *Psiadia*. **Raoulia**, 22 species, is related closely to *Helichrysum* and *Gnaphalium* and "is founded more upon habit[1]) than upon fully good and distinctive characters" (Cheeseman 1925 : 968) and it embraces the two subgenera dealt with below. **Leucogenes** (Compos.). 2 species[2]). is more closely related to *Psychrophyton* than to any other of the Gnaphaloid-*Compositae*. **Brachyglottis** (Compos.), 2 species, which are closely allied to the shrubby section of the genus *Senecio*. **Traversia** (Compos.), monotypic, is very closely relatad to *Senecio* and by Bentham and Hooker united to that genus, but it differs in the rigid pappus, coriaceous involucral scales, and leaf-venation, as Hooker pointed out, and is allied to the Juan Fernandez genera *Balbisia* and *Robinsonia*.

Endemism, usually of a weaker character, is shown in the following endemic divisions of 14 genera: — **Deschampsia** (Gramin.) excluding *D. caespitosa*, contains 6 species, which differ from all members of the genus proper in the awn being terminal, very small or wanting. According to

---

1) This is hardly true, for it contains species of two essentially different life-forms — the mat and the dense coushion — nevertheless both develop from the beginning upon exactly the same plan and the cushion-form arises from the mat-form through the coming in of filling material, either inorganic (blown silt, sand &c.), as in the case of *R. lutescens*, or organic (dead rotting shoots &c.) as for *R. Haastii* of stony river-bed and the hard, massive "vegetable-sheep". Cheeseman defines all the species in his diagnosis of the genus (1925 : 967) as "perennial herbs", but all are more or less semi-woody and most of the vegetable-sheep true shrubs. though quite unlike shrubs in general.

2) *Leucogenes Grahami* Petrie. admitted as a species in the *Manual*, is most likely *Helichrysum Selago* × *Leucogenes grandiceps*.

CHEESEMAN (1925 : 164) "they may ultimately form a separate genus". **Luzula** (Juncac.) contains 6 species distinguished by their low stature and cushion-form. BUCHENAU in *Das Pflanzenreich* gives no name to this distinct group but a rather long diagnosis. **Cordyline** section *Dracaenopsis* (Liliac.) includes all the New Zealand species (4) of the genus. **Knightia** (Proteac.) consists of 3 species, the type, *K. excelsa*, being endemic but the other 2 form a sub-genus, **Eucarpa** which is confined to New Caledonia. **Beilschmiedia** (Laurac.), about 40 tropical and Australian species, is represented in New Zealand by 2 species for which HOOKER established the genus **Nesodaphne.** **Acaena** section **Microphyllae** (Rosac.) is distinguished from other species of the genus by the small leaves and spines — present or wanting — without barbs. **Carmichaelia** (Legum.) but for *C. exsul* of Lord Howe Island would be endemic, but in **Huttonella** it has a small endemic subgenus distinguished by its indehiscent pod. **Hoheria** (Malvac.), a distinct endemic genus is sharply divided into two subgenera, **Euhoheria** — the *Hoheria* of the Manual, 3 species or more, distinguished by the 5—7 carpels, the capitate stigma and the fruit conspicuously winged; and **Apterocarpa** — *Gaya* of the *Manual*, 3 species, distinguished by the 10—15 carpels and the fruit either incon-spicuously winged or wingless. **Epilobium** (Onagrac.) possesses the endemic section, **Dermatophyllae,** 12 species, which are supposed to be distinguished by the semi-woody stems, particularly woody at the base, the more or less rigid coriaceous leaves, and the few terminal flowers. The group is not well-marked and its limits badly defined. **Fuchsia** (Onagrac.) is represented by the subgenus **Skinnera**, 3 or 4 species, distinguished by the constriction at the base of the calyx. **Stilbocarpa** (Araliac.), a most distinct endemic genus, falls into 2 well-marked subgenera — **Stilbocarpa proper,** 1 species, and **Kirkophytum,** 2 species — the former distinguished by its 3—4-celled ovary and fruit hollowed at the apex, and the latter by its 2-celled ovary and fruit not hollowed at the apex. **Hebe** (Scroph.) has 3 most distinct endemic groups not connected with any other of the numerous species, namely, the **whipcord group** (*H. tetragona* &c.), the **H. epacridea group** and the **H. macrantha group.** **Olearia** (Compos.) has the group distinct from all its other species which includes **O. angustifolia** and its 3 allies. **Celmisia** (Compos.) possesses the very distinct section, **Jonopsis,** which contains the subantarctic *C. vernicosa* and *C. campbellensis.* **Raoulia** (Compos.) is made up of the subgenera **Eu-Raoulia** and **Psychrophyton,** the former with 5—150 thin, slender pappus-hairs and the latter with 15—25 thickened towards the apex; in this group come the vegetable-sheep with 4 or more species. **Cotula** (Compos.) has 2 Chatham Islands' species — **Cotula Featherstonii** and **C. Renwickii** — quite distinct from any other species of the genus.

The palaeozelandic element. In considering the origin of the New Zealand flora, and indeed that of islands in general, the custom is to assume that the land was gradually peopled by plants from adjacent land-surfaces.

But in the case of New Zealand, since the present dry land dates from mesozoic times; and, that long before the angiosperms had appeared on any part of the earth, its area, then quite extensive, was occupied by similar plants to the world at large, then, if there be anything in evolution, we must conclude that there is no reason why certain genera, *now more or less wide-spread,* should not be of New Zealand origin. In other words, it seems right to consider that *tertiary New Zealand possessed a flora part of which had originated on her own soil,* and that *there exists to-day an ancient New Zealand element* just as there does an Australian, Malayan or South American. Such an element was assumed by Engler, and it formed a portion of his palaeoceanic element. To this assumed autochtonic tertiary element I suggested the name **palaeozelandic.** As to what genera should be included is entirely a matter of conjecture. Certainly, the fact of endemism, at the present time, does not stamp a genus as palaeozelandic any more than does the occurrence of a common New Zealand genus in Australia, Malaya or even South America debar it from being so considered. Thus CHEESEMAN (1909:466) is of opinion that *Edwardsia tetraptera* and *Hebe elliptica* migrated from New Zealand to South America and not *vice versa,* as is generally believed. To go a little further, it seems not unlikely that various New Zealand plants, e. g., — *Chrysobactron, Stilbocarpa, Aciphylla, Celmisia* &c. are quite as much "Antarctic" as even *Nothofagus* and *Donatia.*

The following are possibly **palaeozelandic:** — *Dacrydium* (6 N.Z., 4 New Cal., 1 Tas., 3 Mal.); *Phyllocladus* (3 N.Z., 1 Tas., 2 Mal.), *Carex-Echinochlaenae* (14 N.Z., 2 Tas., 1 Juan Fernandez, 1 Chile); *Deschamspia,* the group already referred to under the last head (6 N.Z.); *Astelia* (12 N.Z, 1 east-Aus.-Tas., 4 Pol., 4 Subant. S. Amer.); *Phormium* (1 N.-Z., 2 N. Z.-Norf. Id.); *Chrysboactron* (3 N. Z.); *Herpolirion* 1 N. Z. east-Aus.-Tas.); *Hectorella* 1 N. Z.; the group referred by CHEESEMAN to *Nasturtium* but probably an undescribed genus (6 N. Z., 1 Tas., 1 N. Z.-Tas.), *Pachycladon* (1 N. Z.); *Carmichaelia* (25 N. Z., 1 Lord Howe), the allied genera are here considered late Tertiary; *Aristotelia* (2 N. Z., 1 Tas., 2 eas Aus.,-1 New Guinea, 1 New Heb., 1 or 2 Chile); *Melicytus* (4 N. Z., 1 N. Z.-Pol.); *Hymenanthera* (6 N. Z., 1 Norf., 1 Aus.); *Gunnera-Milligania* (9 N. Z., 1 Tas.); *Nothopanax* (7 N. Z., 2 Aus., 1 New Cal., 1 Lord Howe); *Pseudopanax* (7 N. Z., 2 Subant. S. Amer.); *Aciphylla* (29 N. Z., 2 east Aus.); *Anisotome* (21 N. Z.); *Coxella* (1 N. Z.); *Angelica,* the endemic group, distinct from the genus in general, and hardly belonging there (5 N. Z.); *Corokia* (3 N. Z.); *Dracophyllum* (28 N. Z., 5 New Cal., 10 Aus. or Tas.); *Myosotidium* (1 N. Z.); *Myosotis-Exarrhena* (13 N. Z., 1 east Aus.-Tas.), *Hebe* (64 and more N. Z., 2 N. Z.-Subant. S. Amer., 2 Aus.-Tas.); *Pygmaea* (3 N. Z.); *Siphonidium* (1 N. Z.); *Coprosma* (48 N. Z., 22 in Aus., Pol., Mal. and Juan Fernandez); *Oreostylidium* (1 N. Z.); *Lagenophora* (8 N. Z., 4 Aus., 2 Pol., 2 Subant. S. Am.), *Pleurophyllum* (3 N. Z.), *Cel-*

misia (53 N. Z., 1 N. Z.-Aus.-Tas., 1 Aus.); *Raoulia* (22 N. Z.); *Haastia* (3 N. Z.); *Abrotanella* (9 N. Z., 2 Tas., 1 east Aus., 3 Subant. S. Amer., 1 Rodriguez Id.) and *Leucogenes* (2 N. Z.).

The Australian element. *Dealing first with the* **genera,** *the follow-ing* 38 genera, or subdivisions of such, are confined to New Zealand and Australia, a few of which have, however been already classed as palaeo-zelandic: — *Phylloglossum* (Lycopod.) 1 N. 1 Z.-Aus.; *Althenia* (Potamoget.) 1 N. Z., 3 Aus.; *Amphibromus* (Gramin.) 1 N. Z., Aus.; *Microlaena* (Gramin.) 5 N. Z., 1 N. Z.-Aus., 2 Aus.; *Echinopogon* (Gramin.) 1 N. Z.-Aus.; *Carex-Inversae* (Cyperac.) 2 N. Z., 1 N. Z.,-Aus. 1 Aus.; *Herpolirion* (Liliac.) 1 N. Z.-Aus.); *Rhipogonum* (Liliac.) 1 N. Z., 4 Aus.; *Orthoceras* (Orchid.) 1 N. Z.-Aus.; *Caleana* (Orchid.) 1 N. Z.-Aus., 3 Aus.; *Calochilus* (Orchid.) 2 N. Z.-Aus., 1 Aus.; *Chiloglottis* (Orchid.) 1 N. Z., 1 N. Z.-Aus., 5 Aus.; *Adenochilus* (Orchid.) 1 N. Z, 1 Aus.; *Townsonia* (Orchid.) 1 N. Z., 1 Aus.; *Persoonia* (Proteac.) 1 N. Z, 60 or more Aus.; *Poranthera* (Euphorb.) 1 N. Z., 1 N. Z.-Aus., 4 Aus.; *Rhagodia* (Chenepod.) 1 N. Z.-Aus., 10 Aus; *Drosera-Bryastrum* (Droserac.) 1 N. Z.-Aus.; *Quintinia* (Saxifrag.) 2 N. Z., 3 Aus.; *Ackama* (Cunoniac.) 1 N. Z., 1 Aus.; *Clianthus* Legum.) 1 N. Z., 1 Aus.; *Swainsona* (Legum.) 1 N. Z., 30 or more Aus.; *Phebalium* (Rutac.) 1 N. Z., 34 Aus., *Plagianthus* (Malvac.) 2 N. Z., 9 Aus.; *Gunnera-Milligania* (Halorrhag.) 9 N. Z., 1 Aus.; *Actinotus* (Umbell.) 1 N. Z.-Aus., 9 Aus.; *Aciphylla* (Umbell.) 29 N. Z., 2 Aus.; *Pernettya-Porandra* (Ericac.) 1 N. Z., 1 Aus.; *Pentachondra* (Epacrid.) 1 N. Z.-Aus., 4 Aus.; *Archeria* (Epacrid.) 1 N. Z., 4 Aus.; *Logania* (Loganiac.) 1 N. Z., 17 Aus.; *Liparophyllum* (Gentian.) 1 N. Z.-Aus. — subantarctic in character; *Myosotis-Exarrhena* (Borraginac.) 13 N. Z., 1 Aus.; *Mentha-Eriodontes* (Labiatae) 1 N. Z., remainder Aus.; *Isotoma* (Campan.) 1 N. Z.-Aus.; 7 Aus.; *Celmisia* (Compos.), 53 N. Z., 1 N. Z.-Aus., 1 Aus.; *Ewartia* (Compos.) 1 N. Z., 4 Aus.; *Craspedia* (Compos.) 4 at least N. Z., several Aus., but genus in confusion owing to hybrids and epharmones.

Almost in the same class as the last are those genera which extend to Norfolk or Lord Howe Islands or even New Caledonia, e. g. *Dichelachne* (Gramin.) 1 N. Z.-Aus., 1 N. Z.-Aus.-Norf.; *Xeronema* (Liliac.) 1 N. Z., 1 New Cal.; *Arthropodium* (Liliac.) 2 N. Z., 6 Aus., 1 New Cal.; *Prasophyllum* (Orchid.) 2 N. Z., 2 N. Z.-Aus., 26 Aus., 2 New Cal.; *Pterostylis* (Orchid.) 11 N. Z., 2 N. Z.-Aus., 2 New Cal., 34 Aus.; *Acianthus* (Orchid.) 1 N. Z., 4 Aus., 2 New Cal.; *Lyperanthus* (Orchid.) 1 N. Z., 4 Aus., 1 New Cal., *Pennantia* (Icacinac.) 1 N. Z., 1 Aus., 1 Norf.; *Pomaderris* (Rhamnac.) 2 N. Z., 3 N. Z.-Aus., 17 Aus., 1 New Cal.; *Hymenanthera* (Violac.) 6 N. Z., 1 Norf., 1 Aus.; *Dracophyllum* (Epacrid.) 28 N. Z., 10 Aus., 5 New Cal.

Certain genera are confined to New Zealand, Australia and subantarctic South America; such are as follows: — *Carpha* (Cyperac.) 1 N. Z.-Aus., also to New Guinea, 1 Sub. S. Am.; *Carex-Echinochlaenae* 14 N. Z., 2 Aus., 2 Sub. S. Am.; *Gaimardia* (Centrolepidac.) 1 N. Z., 1 Sub. S. Am., 1 Aus.; *Libertia*

(Iridac.) 3 N. Z., 1 N. Z.-Aus., 1 Aus., 5 Sub. S. Am., *Nothofagus* (Fagac.) 5 N. Z., 8 Sub. S. Am., 3 Aus. — the evergreen section has 5 N. Z., 2 Aus. and 3 Sub. S. Am.; *Colobanthus* (Caryophyll.) 10 N. Z., 1 N. Z.-Sub. S. Am., 1 Aus., 8 Sub. S. Am., and 1 Kerguelen Land and 1 New Amsterdam; *Caltha-Psycrophila* (Ranun.) 2 N. Z., 2 Aus., 7 Sub. S. Am., but extending northwards along the Andes; *Drosera-Psychophila* (*Droserac.*) 1 N. Z., 1 Aus., 1 Sub. S. Am.; *Acaena-Euancistrum* (Rosac.), 2 N. Z., 1 N. Z.-Aus., 1 N. Z.-Sub. S. Am., 17 or more Sub. S. Am. and also 1 New Amsterdam and 1 Tristan d'Acunha — the above species taken mostly in a very wide sense; *Geranium-Chilensia* (Geran.) 1 N. Z.-Aus., 22 Sub. S. Am.; *Geranium-Andina* (Geran.) 1 N. Z.-Aus.-Sub. S. Am., 16 Sub. S. Am.; *Discaria* (Rhamnac.) 1 N. Z., 1 Aus., 18 Sub. S. Am. and northwards; *Aristotelia* (Elaeocarp.) 2 N. Z., 2 Sub. S. Am., also 1 New Heb. and 1 New Guinea; *Drapetes* (Thymelaeac.) 4 N. Z., 1 Aus., 1 Sub. S. Am., also 1 Mal.; *Epilobium-Sparsaeflora* (Onagrac.) 3 N. Z., 1 Aus., 1 Sub. S. Am.; *Schizeilema* (Umbell.) 11 N. Z., 1 Aus., 1 Sub. S. Am.; *Lilaeopsis* (Umbell.), 2 N. Z., 3 Aus., 4 S. Am. with 1 Sub. S. Am. also 3. North Am. and 2 N. Am.-S. Am.; *Oreomyrrhis* (Umbell.) 3 N. Z., 1 Aus., 7 S. Am. with 1 or more in Mexico; *Pernettya* (Ericac.) 20 Central Am. to the Falklands and represented in N. Z.-Aus. by the section *Pernandra* with 1 N. Z. and 1 Aus.; *Gentiana-Antarctophila* (Gentian.) 1 N. Z., 1 Tas., 1 Aus., 1 Sub. S. Am.; *Hebe* (Scroph.) 64 and more N. Z., 2 N. Z.-Sub. S. Am., 2 Aus., *Ourisia* (Scroph.) 12 N. Z., 2 Aus., 12 Sub. S. Am., 7 Subtrop. Andes; *Plantago-Plantaginella* (Plantag.) 1 N. Z., 2 Aus., 9 Andine, Mexico to Falklands; *Selliera* (Goodeniac.) 1 N. Z.-Aus.-Sub. S. Am., 1 Western Aus.; *Donatia* (Donatiac.) 1 N. Z.-Aus., 1 Sub. S. Am.; *Abrotanella* (Compos.) 9 N. Z., 3 Aus., 3 Sub. S. Am., 1 Rodriguez.

Coming now to the **species,** and excluding the cosmopolitan and subcosmopolitan, the number supposed to be common to New Zealand and Australia (probably few are truly identical) is 236 — pteridophytes 72 (43°/₀ of the New Zealand total), monocotyledons 80 (18.6°/₀) and dicotyledons 84 (only 6.8°/₀) — and 57 families and 139 genera are represented the largest of which are as follows: — (families) *Filices* 63 (26.6°/₀), *Cyperaceae* 28, *Gramineae* 17, *Orchidaceae* 16, *Juncacea* 9, *Compositae* 8 only, *Lycopodiaceae* and *Scrophularinaceae* 7 each; (genera) *Juncus* 8, *Asplenium*, *Lycopodium* and *Cladium* 6 each, *Adiantum*, *Carex* and *Drosera* 5 each.

The New Zealand-Australian species may be arranged in the following group to each of which the number of species is appended: — (1.) **species of Australian origin** (pteridophytes 10, monocotyledons 48, dicotyledons 33); (2.) **species of New Zealand origin** (pterids. 12, monocots. 12, dicots. 19); (3.) **species with equal claim for either Australian or New Zealand origin** (pterids. 10, monocots. 6, dicots. 18); (4.) **species of palaeotropic origin** (pterids. 38, monocots. 9, dicots. 7); (5.) **species of subantarctic origin** (pterids. 1, monocots. 4, dicots. 7); also there is the South African

Todaea barbara and *Carex Gaudichaudiana* with a vicarious representative in the temperate Northern Hemisphere. Taking the 3 systematic groups together the total number of species for each class is: Class (1.) 91, class (2.) 43, class (3.) 34, class (4.) 54 and class (5.) 12.

Leaving the ferns on one side, most of them common forest-plants, about 70 of the remaining 173 species may be considered common and 30 of these extremely abundant, while 103 species are of restricted distribution, many being confined to some small special habitat, and 27 of these are extremely rare. With regard to the vertical distribution of the 236 species under consideration, it is as follows: — coastal 22, lowland 134, lowland-montane 28, lowland-subalpine 39, lowland-alpine 3, montane 1, subalpine 3, montane-subalpine 3 and subalpine-alpine 3.

It may be seen from the above that the importance of the Australian element is easily exaggerated, as also the floristic relation between the two floras. A considerable part of the species common to both occurs only on the mountains of Tasmania and the higher Australian Alps. Many families and genera, characteristically Australian, are either absent or very poorly represented, e. g. *Eucalyptus, Callistemon, Melaleuca* and other *Myrtaceae, Proteaceae* of which there are but 2 species in New Zealand, *Dilleniaceae, Tremandraceae,* genera of *Leguminosae,* e. g. *Acacia* and *Pultenaea,* the important family *Rutaceae* with the genera *Boronia* and *Eriostemon, Casuarina,* various monocotyledonous genera &c. *It is hardly going too far to say that it would be possible for one to have an excellent acquaintance with the botany of Eastern Australia and yet to be acquainted with very few indeed of the species which extend to New Zealand.*

The Subantarctic element. It is the presence of a well-defined element common to New Zealand, Eastern Australia and Tasmania, subantarctic South America, and the extra-New Zealand subantarctic Islands that has given rise to endless speculations as to its origin.

In the first place as to **genera,** many are common to New Zealand and Subantarctic South America &c., as shown in the following list where the mark † denotes that the genus to which it is attached is absent in the Australian flora: — *Hymenophyllum, Trichomanes, Dicksonia, Cheilanthes. Pellaea, Cystopteris, Pteridium, Pteris, Blechnum, Asplenium, Polystichum, Polypodium, Notochlaena, Gymnogramme, Gleichenia, Schizaea, Ophioglossum, Botrychium* (Filices.); *Azolla* (Salviniac.); *Isoetes* (Isoetac.); *Lycopodium* (Lycopod.); *Dacrydium, Podocarpus* (Taxac.); *Libocedrus* (Cupress.); *Triglochin* (Scheuchz.); *Potamogeton* (Potamoget.); *Hierochloe, Agrostis, Deyeuxia. Deschampsia, Trisetum, Danthonia, Koeleria,* sect. *Dorsae-aristatae; Poa. Festuca, Bromus* (Gramin.); *Schoenus, Cladium, (Vincentia†)* 1 species Juan Fernandez, *Oreobolus, Uncinia, Carex* (Cyperac.); *Leptocarpus* (Restion.); *Gaimardia* (Centrolep.); *Rostkovia†, Marsippospermum†, Juncus, Luzula* (Junc.); *Enargea†, Astelia* (Liliac.); *Libertia* (Iridac.); *Spiranthes* (Orchid.); *Notho-*

fagus (Fagac.); *Urtica* (Urticac.); *Phrygilanthus* (Loranth.); *Knightia* (Proteac.); reported as fossil in Seymour Isld. close to Grahamland, *Polygonum, Rumex, Muehlenbeckia* (Polygonac.); *Chenopodium, Atriplex, Suaeda, Salicornia* (Chenopod.); *Mesembryanthemum, Tetragonia,* (Aizoac.); *Montia* (Portulac.); *Stellaria, Colobanthus, Spergularia, Scleranthus* (Caryoph.); *Ranunculus, Myosurus, Caltha* sect. *Psychrophila* (Ranun.); *Laurelia* (Monimiac.); *Cardamine, Sisymbrium, Lepidium* (Crucif.); *Drosera* sect. *Psychophila* (Droserac.); *Tillaea* (Crassulac.); *Rubus, Geum, Acaena,* (Rosac.); *Edwardsia*† (Legum.); *Geranium* (Geran.); *Oxalis* (Oxalidac.); *Linum* (Linac.); *Callitriche* (Callitrich.); *Coriaria*† (Coriariac.); *Discaria* (Rhamnac.); *Aristotelia* (Elaeocarp.-Aristotel.); *Hypericum* (Guttif.); *Drapetes* (Thymel.); *Myrtus, Eugenia* (Myrtac.); *Viola* (Violac.); *Epilobium, Fuchsia*† (Onagrac.); *Halorrhagis, Gunnera, Myriophyllum,* (Halorrhag.); *Pseudopanax*† (Araliac.); *Azorella*†, *Schizeilema, Apium, Lilaeopsis, Oreomyrrhis* (Umbell.); *Griselinia*† (Cornac.); *Gaultheria, Pernettya* (Ericac.); *Samolus* (Primulac.), *Gentiana* (Gentian.); *Calystegia* (Convolv.); *Myosotis* (Borrag.); *Tetrachondra*†, *Scutellaria* (Labiat.); *Jovellana*†, *Limosella, Veronica, Ourisia, Euphrasia* (Scroph.); *Utricularia* (Lentibul.); *Plantago* (Plantag.); *Coprosma* (Juan Fernandez.); *Nertera, Galium* (Rubiac.); *Pratia* (Campan.); *Selliera* (Gooden.); *Phyllachne*†, *Donatia* (Stylid.); *Lagenophora, Gnaphalium, Cotula* sect. *Leptinella, Abrotanella, Senecio, Taraxacum* (Compos.).

In the second place, as to **species** there are at least 58 identical, or almost so, in the New Zealand and Subantarctic floras which have remained unchanged, intense isolation notwithstanding. Moreover there are 44 vicarious species.

The species supposed to be identical are as follows: *Hymenophyllum rarum, H. ferrugineum*† (restricted to Juan Fernandez), *H. tunbridgense, H. peltatum, Pteridium esculentum, Histiopteris incisa, Blechnum penna marina, B. procerum* or a close ally, *Aspelenium obtusatum, Polystichum vestitum, P. adiantiforme, Polypodium Billardieri, Schizaea fistulosa* var. *australis, Botrychium lunaria* (ilic.); *Agrostis magellanica*†, *Trisetum subspicatum, Festuca erecta*† — in New Zealand only on Macquaries (Gramin.); *Scirpus aucklandicus* (Amsterdam Island only), *S. cernuus, S. nodosus* (of wide range), *Uncinia macrolepis*†, *U. compacta* (Kerguelen, Amsterdam Island), *C. Darwinii* var. *urolepis*†, *C. pumila* (of wide range), *C. trifida*†, *C. Oederi* var. *cataractae* (Cyperac.); *Rostkovia magellanica*†, *Juncus planifolius* (Jancac.); *Tetragonia expansa* — of wide range (Aizoac.); *Montia fontana* — cosmopolitan (Portulac.); *Ranunculus acaulis*†, — *R. crassipes*† (Ranun.); *Cardamine glacialis*† — var. *carnosa* is the New Zealand form (Crucif.); *Tillaea moschata*† (Crassulac.); *Geum parviflorum*† — according to SKOTTSBERG there are minor differences between the forms of the two regions, *Acaena adscendens*† — in New Zealand on the Macquaries only (Rosac.); *Edwardsia tetraptera*† in a wide sense — *E. chathamica* is very close to the Chilian plant (Legum.); *Callitriche antarctica*† (Callitrich.); *Halorrhagis erecta*† (Juan Fernandez), *Myriophyllum elatinoides* (Halorrhag.);

Hydrocotyle americana†, *Azorella Selago*† — in New Zealand confined to the Macquaries, *Oreomyrrhis andicola* in a wide sense, probably the forms of the two regions are distinct; *Apium prostratum* (Umbel.); *Samolus repens* (Primulac.); *Myosotis antarctica*† according to SKOTTSBERG occurs in south Patagonia, in New Zealand it is restricted to Campbell Island (Borrag.); *Calystegia tuguriorum*†, *Dichondra repens* — sub-cosmopolitan (Convol.); *Gratiola peruviana*, *Hebe salicifolia* var. *communis*†, *H. elliptica*†, × *H. ellipsala*†, *Limosella tenuifolia* (Scroph.); *Nertera depressa* (Rubiac.); *Selliera radicans* (Goodeniac.); *Cotula plumosa*†, *C. coronopifolia* and *Taraxacum magellanicum*† (Compos.).

Equally important phytogeographically as the identical species are the vicarious, many of which have usually been considered identical. The following is a list of the visarious species, the name given first being the New Zealand species: — *Azolla rubra* — *A. filiculoides* (Salvinac.); *Lycopodium fastigiatum* — *L. magellanicum* (Lycopod.); *Koeleria superba*† — *K. Bergii*; *Poa litorosa*† — *P. novarae* (St. Paul, and New Amsterdam); *Poa foliosa*† — *P. flabellata*; *Hierochloe redolens* — *H. magellanica*; *Festuca novae-zelandiae*† — *F. magellanica* (Gramin.); *Oreobolus pectinatus*† — *O. obtusangulus*; *Carpha alpina* — *C. schoenoides*; *Schoenus pauciflorus*† — *S. antarcticus*; *Carex pseudocyperus* var. *fascicularis* — var. *Haenkeana* (Cyperac.); *Gaimardia setacea*† — *G. australis* (Centrolep.); *Marsippospermum gracile*† — *M. Reichei*; *Luzula crinita*† — *L. alopecurus*; *L. Traversii*† — *L. racemosa*; *Juncus* sp. unnamed of New Zealand Subantarctic Islands† — *J. scheuzerioides* (Juncac.); *Astelia linearis*† — *A. pumila*; *Enargea parviflora*† — *E. marginata* (Liliac.); *Nothofagus Menziesii*† — *N. betuloides* (Fagac.); *Rumex neglectus*† — *R. cuneifolius* (Polygonac.); *Colobanthus muscoides*† — *C. kerguelensis*; *Colobanthus Hookeri*† — *C. subulatus, C. mollis*† — *C. quitensis* (Caryoph.); *Myosurus novae-zelandiae* — *M. aristatus; Caltha novae-zelandiae* — *C. sagittata* (Ranun.); *Laurelia novae-zelandiae* — *L. aromatica* and *L. serrata* (Monimiac.); *Drosera stenopetala* — *D. uniflora* (Droserac.); *Geranium sessiliflorum* var. *glabrum*† — *G. sessiliflorum* (Geran.); *Oxalis lactea* — *O. magellanica* (Oxalidac.); *Coriaria sarmentosa*† — *C. ruscifolia; C. lurida*† — *C. thymifolia* (Coriariac.); *Discaria toumatou*† — *D. discolor* (Rhamnac.); *Drapetes Dieffenbachii*† — *D. muscosus* Thymelaeac.); *Epilobium pedunculare*† — *E. conjugens* (Onagrac.); *Hydrocotyle novae-zelandiae*† — *H. marchantioides; Lilaeopsis novae-zelandiae*† — *L. ovina* (Umbel.); *Tetrachondra Hamiltonii*† — *T. patagonica* (Tetrachondrac.); *Euphrasia zealandica*† — *E. antarctica* (Scroph.); *Plantago Brownii* — *P. barbata* (Plantag.); *Pratia angulata*† — *P. repens* (Canipan.); *Donatia novae-zelandiae* — *D. fascicularis* (Donatiac.); *Lagenophora pumila*† — *L. Commersonii; Abrotanella muscosa*† — *A. emarginata; A. caespitosa*† — *A. linearis.* The above list contains far more plants of a subantarctic character than does the list of identical species and *they are no mere waifs and strays but a definite systematic and especially ecological group,* which appears to point to a much larger group many of the members of which have perished.

The palaeotropic element. Under the term *palaeotropic* come the Malayan, Australian-Malayan, Melanesian and Polynesian elements of the flora. The element stands out ecologically as most distinct from the closely-related palaeozelandic and Subantarctic elements.

The following are the principal families and genera: — (families) *Burmanniaceae, Pandanaceae, Palmae, Chloranthaceae, Moraceae, Loranthaceae, Santalaceae, Balanophoraceae, Nyctaginaceae, Monimiaceae, Lauraceae, Meliaceae, Euphorbiaceae, Icacinaceae, Corynocarpaceae, Elaeocorpaceae, Tiliaceae, Passifloraceae, Myrtaceae, Araliaceae, Myrsinaceae, Loganiaceae, Apocynaceae, Convolvulaceae, Verbenaceae, Gesneriaceae, Rubiaceae, Cucurbitaceae;* (genera) *Hymenophyllum, Trichomanes,. Cyathea, Hemitelia, Alsophila, Dicksonia, Davallia, Leptolepia, Lindsaya, Adiantum, Cheilanthes, Pellaea, Paesia, Pteris, Doodia, Diplazium, Athyrium, Nephrolepis, Gymnogramme, Gleichenia, Lygodium, Marattia* (Filic.); *Tmesipteris, Psilotum* (Lycopod.); *Agathis* (Pinac.); *Freycinetia* (Pandan.); *Imperata, Paspalum, Isachne, Panicum, Oplismenus, Cenchrus, Alopecurus, Eleusine* (Gramin.); *Mariscus, Fimbristylis, Cladium* (Cyperac.); *Rhopalostylis* (Palm.); *Xeronema, Cordyline* (Lil.); *Dendrobium, Bulbophyllum, Earina, Sarcochilus* (Orchid.); *Macropiper, Peperomia* (Piperac.); *Ascarina* (Chloranth.); *Paratrophis* (Morac.); *Elatostema, Boehmeria* (Urtic.); *Elytranthe, Loranthus, Korthalsella* (Loranth.); *Euphorbia, Homalanthus* (Euphorb.); *Pisonia* (Nyctag.); *Hedycarya* (Monimiac.); *Litsaea, Cassytha* (Laurac.); *Weinmannia* (Cunon.); *Pittosporum* (Pittosporac.); *Canavalia* (Legum.); *Melicope* (Rutac.); *Dysoxylum* (Sapind.); *Corynocarpus* (Corynocarp.); *Elaeocarpus* (Elaeocarp.); *Hibiscus* (Malvac.); *Metrosideros, Eugenia* (Myrtac.); *Meryta, Schefflera* (Araliac.); *Sideroxylon* (Sapotac.); *Geniostoma* (Loganiac.); *Parsonsia* (Apocynac.); *Ipomaea* (Convol.); *Myoporum* (Myoporac.); *Vitex, Avicennia* (Verben.); *Solanum* (Solan.); *Sicyos* (Cucurb.); *Scaevola* (Gooden.); *Siegesbeckia, Bidens* (Compos.).

For the most part the species of the above genera are endemic, nevertheless they form, as already explained, a distinct element of the flora which is in large measure restricted to the frostless parts of the Region comparatively few species gaining the subalpine belt or in the south leaving the coast-line.

With regard to the identical species if the numerous palaeotropical pteridophyta, the fairly large widely-distributed tropical and subtropical element and the Norfolk-Howe species be here left out of consideration, the few remaining species shared by New-Zealand with other palaeotropic floras are *Podocarpus ferrugineus* (Podocarp.), New Caledonia only; *Cenchrus calyculatus* (Gramin.), but to Kermadecs only; *Cladium articulatum, C. glomeratum, Gahnia gahniaeformis* (Cyperac.); *Macropiper excelsum* (Piperac.);

Melicytus ramiflorus (Violac.); *Metrosideros villosa* (Myrtac.), but Kermadec only.

The cosmopolitan element. Here the term *Cosmopolitan* is used with a wide significance, and all those species are included that have a considerable range in either temperate or warm climates. Some, especially those of warm countries, are ecologically akin to weeds, to that it is difficult, or impossible to decide, whether their present distribution has come about by *natural means, or whether, also, some of those found in New Zealand are truly indigenous.* Even regarding the well-known floras of European lands, exposed as these have been for long periods to the influence of man, it must be a matter of great uncertainty as to what species are actually indigenous, and, it may well be, that many species, concerning the nativity of which no question has ever been raised, have originated far from what is now the centre of their greatest distribution.

The most important section, from the phytogeographical standpoint, consists of those species supposed to belong without question to the Northern hemisphere, a matter however taken rather for granted than proved. These species fall into the two classes of those in which the northern and southern forms are almost identical and those in which a species or group of species in the north has one or more vicarious southern representatives.

The following are important examples: — (identical species) *Hymenophyllum tunbridgense, H. peltatum, Asplenium Trichomanes, Botrychium lunaria* — also in Australia and Subantarctic America, but only recorded from one locality in New Zealand (Filic.): *Potamogeton natans, P. polygonifolius, Ruppia maritima* (Potam.); *Trisetum subspicatum* (Gramin.): *Carex pyrenaica* — there is *var. cephalotes* but it is known only in two localities, both in South Otago, *C. diandra* — only known in New Zealand in the Southern Hemisphere, *C. stellulata* — only known in Australia and New Zealand in the Southern Hemisphere, *C. lagopina* — in the Southern Hemisphere, only in New Zealand, *C. pseudo-cyperus* (Cyperac.); *Montia fontana* (Portulac.): *Callitriche verna* (Callitrich.); *Calystegia Soldanella* (Convolo.); (vicarious species) *Pteridium esculentum* — *P. aquilinum; Cystopteris novae-zelandiae* — *C. fragilis, Ophioglossum coriaceum* — *O. lusitanicum, O. pedunculosum* — *O. vulgatum, Botrychium australe* — *B. ternatum* (Filic.); *Hierochloe Fraseri* — *H. alpina, Agrostis Dyeri* — *A. canina, Deschampsia caespitosa var. macrantha* — *D. caespitosa, Koeleria novo-zelandica* — *K. cristata, Festuca novae-zelandiae* — *F. rubra* (Gramin.); *Carex appressa* and its allies — *C. paniculata* to which CHEESEMAN, following BENTHAM and VON MUELLER, referred the whole *C. appressa* group in 1884. *C. Gaudichaudiana* — *C. Goodenoughii,* to which under better-known name *C. vulgaris, C. Gaudichaudiana* was referred by CHEESEMAN in 1884; but as a variety, *C. Oederi var. cataractae* — *C. Oederi* (Cyperac.); *Zostera tasmanica* — *Z. marina* (Potam.); *Juncus polyanthemus* — *J. effusus,* which is naturalized only.

J. maritimus var. *australiensis* (might be treated as a species) — *J. maritimus,* the remarkable series of *Luzulae* referred by BUCHENAU to *L. campestris* but as varieties, most of which I consider excellent species and others hybrids between such — *L. campestris* (Juncac.); *Urtica australis* and its near allies — *U. dioica* (Urticac.); *Chenopodium glaucum* var. *ambiguum* — *C. glaucum* (Chenopod.); *Cardamine heterophylla* — *C. hirsuta* (Crucif.); *Potentilla Anserina* var. *anserinoides* — *P. Anserina* (Rosac.); *Geranium pilosum* — *G. dissectum* (Geraniac.); *Elatine gratioloides* — *E. americana* (Elatinac.); *Epilobium Billardierianum* — *E. tetragonum* (Onagrac.); *Apium prostratum* — *A. graveolens* (Umbel.); *Sonchus littoralis* — *S. asper,* only naturalized in New Zealand, *Taraxacum magellanicum* — *T. officinale* (Compos.).

The Lord Howe-Norfolk element. Something has already been said concerning the above when dealing with the Kermadec Islands. Here only a few special details are necessary. The following are the identical and vicarious species which are confined to the two floras: — (identical) *Gahnia xanthocarpa, Muehlenbeckia complexa* var. *microphylla;* (vicarious) *Mariscus ustulatus* — *M. haematodes, Edwardsia microphylla* and its near allies — *E. howinsula* the name given by W. R. B. OLIVER as a variety of the linneon *E. tetraptera, Melicope ternata* — *M. contermina, Hymenanthera novae-zelandiae* either identical, or a closely-related species, or the Norfolk Island *H. latifolia, Coprosma retusa* — *C. prisca* but a matter of opinion whether to treat both as varieties of the compound species, *C. Baueri.* More important than anything in the above list, unless it be the occurrence of *Edwardsia,* is the presence on Lord Howe Island of a species of *Carmichaelia (C. exsul)* — a genus otherwise endemic in New Zealand — for it raises the question as to whether *Carmichaelia* had once a far-wider distribution than at present and that *C. exsul* is the sole survivor.

With regard to the New Zealand-Norfolk Island floristic relationship the following are the identical species not occurring elsewhere and the vicarious species: — (identical) *Phormium tenax, Muehlenbeckia australis, Olea apetala;* (vicarious) *Adiantum obtusatum* — *A. difforme* according to R. M. LAING, *Cyclophorus serpens* — *C. confluens* according to R. M. LAING, *Rhopalostylis sapida* and *R. Cheesemanii* — *R. Baueri, Cordyline australis* — *C. Baueri, Pennantia corymbosa* — *P. Endlicheri, Hymenanthera novae-zelandiae* — *H. latifolia, Suttonia kermadecensis* (Kermadecs only) — *S. crassifolia, Sideroxylon novo-zelandicum* — *S. costatum.*

3. General conclusions.

In order to pave the way for the concluding part of this work it is advisable to state briefly some conclusions derived from the details given in this chapter.

The flora of New Zealand, notwithstanding its strong endemism, possesses two very distinct elements not floristic only but ecological.

The first, and, as I believe, the more primitive, is not one simple floristic entity, but consists of a combination of the palaeozelandic and subantarctic elements of the flora, now difficult to disentangle. They have this one property in common, the power, for the most part to endure a fair amount of cold. In other words, the element is a temperate one.

The second element, also largely endemic, consists of descendents of an ancient palaeotropic stock, so ancient indeed that endemic genera have been developed (*Rhabdothamnus, Ixerba, Alectryon &c.*), as well as many distinct endemic species.

Yet notwithstanding this great age of the members, and their long isolation far from the tropics, but *few have become really fitted to the present average climate of New Zealand, in fact the majority can tolerate very little frost.* For the most part, the species of this class are confined to the lowlands, and in the south some are only found near the coast. This element, in fact, is eminently subtropical; so that the present-day climate is one to which it is not perfectly attuned. Should a change of climate occour, then with increase of temperature the palaeotropic element would advance southwards and the palaeozelandic-subantarctic retreat to the mountains, while, with increase of cold, the contrary would be the case. The isolated colonies of *Nothofagus truncata*, north of lat. 37°, point to such a change of climate, while the presence of the tree-fern (*Hemitelia Smithii*) in the Lord Auckland district suggests a warmer period. But apart from speculations, the non-toleration of frost by so many New Zealand species is good evidence that there has either been a considerable northern land-extension during the glacial period or else that such did not owe its origin even, in part, to increase of cold.

It has been shown that while there is a considerable Australian element, it is made up largely of Subantarctic and Palaeotropic species, while the true Australian element does not play a conspicuous part in the vegetation. Especially is the absence of characteristic Australian genera noteworthy, e. g. *Eucalyptus, Acacia* &c., although virtually all the Tasmanian species are not only quite hardy in the warmer parts of New Zealand, but some can spread spontaneously. Bearing these facts in mind, the possibility of direct land-connection with Eastern Australia except at a very remote period cannot be entertained.

Part V.

The History of the Flora.

General. The origin and subsequent history of the New Zealand flora is, in great measure, a matter of speculation merely, for the material on which conclusions are to be based is in no small degree unsatisfactory and insufficient. In the first place, the all-important statistics as to floristic elements are, of necessity, drawn from existing floras, although the composition of these must be very different from those of the same areas in Tertiary times. It is also assumed, in phytogeographical writings generally, that the absence of a species from any area means that the species in question was never there, a supposition quite at variance with what is happening at present, let alone fossil records. In New Zealand, species can be seen in process of extinction, as in the case of *Podocarpus spicatus* in Stewart Island and the many relict species of the main islands. The family *Podocarpaceae* was formerly in the Chathams, as evidenced by pollen found in the peat (ERDTMAN, O., G., E. 1924: 679:70), the pollen being both of *Dacrydium* and *Podocarpus* type, yet there are now no conifers in the flora. *Mida,* until a few years ago, occurred on Juan Fernandez but the trees have all died and now it is a New Zealand endemic genus. Can it be possible that *Xeronema* has *always* been confined to New Caledonia and one tiny New Zealand islet'. But, it is not necessary to stress the ever-present dying out and coming-in of the members of a flora from families to jordanons. Again, although a genus has its richest development in some particular area, it by no means follows that such is the original centre of its distribution. For instance, in the large genera *Hebe* and *Celmisia* most of the species show signs of youth, and no phytogeographer would look for the origin of the closely-related *Veronica* in the former or for that of the equally closely-related *Aster* in the latter.

The matter of ancient land-connections, where there are now profound ocean-depths, is the burning question in New Zealand biogeography. But, here again, the ground is most insecure, and one can only say that the question of great changes in the relations of land and water is one on which there is about equal evidence for and against. Geology can make no definite pronouncement, and the matter, at present, rests solely on the facts of organic distribution and whether certain critical examples can be clearly explained without the assumption of "land-bridges". To some this biological evidence is all-conclusive, especially from the zoological stand-point; HEDLEY for instance (1899: 393) going so far as to ignore the testimony of ocean-depths and construct a hypothetical land-area on biological considerations alone. Others, again will not allow land-connections at any price. With regard to plants, it is generally assumed, that there is a possibility of their being conveyed by wind, or even birds, across wide stretches of ocean, especially in the case of sporiferous species. But even with these latter, the spores of a plant, inhabiting only a windless forest-interior, could never be the sport of the wind. The case of *Hymenophyllum ferrugineum* of Juan Fernandez, Chile and New Zealand is hard to explain on the supposition of wind-carriage and equally difficult is that of *H. Malingii*, a pseudo-epiphyte, of quite local occurrence, on the dead parts of trunks of certain *Podocarpaceae* or *Cupressaceae* in New Zealand and Tasmania, but absent on neighbouring dicotylous trees. Were spores as readily carried by the wind as is supposed, there should be no special fern-floras, which is not the case; nor should the endemic *Polypodium novae-zelandiae* be confined to one portion of North Island, since it ascends to the subalpine belt. Regarding the seed-plants, the important evidence already given concerning the distribution of alien species in New Zealand, equipped in every way for travel and ecesis, and their relation to the primeval vegetation, shows how exceedingly difficult it is for a plant to gain entrance into a virgin plant-formation, also it has been seen of how slight advantage for long-distance travel is the possession of flying apparatus or of fruits palatable for birds, and how it is *not* the *species* which move but the *associations* to which they belong. Even absolutely bare ground, perfectly suitable as a seed-bed, is only occupied by species from the immediate vicinity, as in the case of the new ground after the Tarawera eruption, river-beds with their seed-catching mat-plants, ground left bare by retreating glaciers and many ideal places for seed-germination made by the operations of man.

Ecesis, rather than the possibility of bird-carriage &c. during long periods of time, is the great stumbling-block. Transoceanic dissemination may rightly be evoked as an explanation of the presence of Australian, Polynesian and even Northern species in the fact of migratory birds from the first two regions and Siberia[1]). The carriage of seeds and even crytozoic

---

1) Several species of birds belonging to the *Limnicolae* migrate from Siberia to New Zealand returning to their northern home to breed. It seems to me quite as likely that seeds should be conveyed by them from north to south as that the path for such seeds must be the Andes as usually suggested. F. W. HUTTON (T. N. Z. I. (1901): 262) considered that "The only possible explanation of oversea migration seems to be that the birds are following old land-lines. The shore-birds follow the old shore-line: the land-birds follow the old land. Migration must have commenced when the two lands were contiguous, or nearly so, so that in no part of the course was an island so far off as to be invisible from those next to it. Gradually the land sank but the force of habit kept up the migration."

animals, on trees brought to the sea by flooded rivers, seems the most feasible method of travel for many species. The indehiscent seed-pod of *Edwardsia* might long preserve the seeds within undamaged by sea-water. Logs are carried from the main islands to the Chatham group; these islands and the Lord Auckland have been colonized by various European birds blown from New Zealand, smoke from forest-fires on the mainland frequently gains the Chathams and, according to MARSHALL[1]), a storm of dust has reached New Zealand from Australia. All these facts, and others of a like kind could be cited, show how seeds could be *rapidly* conveyed over great distances, but, *between the arrival of seed or spore and its becoming a mature plant in a situation favourable, not only for its well-being, but for its increase, is altogether another matter.*

Granting that plants of all kinds can be transported over thousands of kilometres of ocean, a supposition taxing one's judgement to no small degree, there comes in the carriage of animals. Now with regard to various classes of such there is a striking subantarctic affinity and it is the question of how their carriage has come about which has aroused the chief bio-geographical discussion, the matter of a land-bridge with the antarctic or subantarctic being the main point of contention. As for land-connection in the north, almost all who have considered the subject, as will be seen further on, are in its favour. This question of the incoming of the sub-antarctic element of the New Zealand fauna is gone into at considerable detail by HUTTON, CHILTON, BENHAM and others[2]); here only a few cases are noted. But first it must be pointed out, that although it might be possible, though extremely difficult, to suggest a plausible explanation for every case on the supposition of ocean-transit, yet that such could apply to the organisms as a whole is a totally different matter.

Galaxias attenuata, as fresh-water fish, occurs in New Zealand, Tasmania, South-east Australia, the Falklands and Subantarctic South America. *Notiodrilus,* a genus of earth-worms, is found in New Zealand including the Subantarctic province, Kerguelen, Marion Island, the Crozets, South Georgia, the Falklands, Subantarctic South America and South Africa. *Phraeodrilus,* another genus, is represented, according to BENHAM (1909: 254), by 3 or 4 New Zealand species and 6 others distributed on Kerguelen, the Crozets, the Falklands and Fuegia. A species of terrestial crustaceans, genus *Trichoniscus,* according to CHILTON (1909: 799), occurring in the Subantarctic province, is identical with a species of Fuegia, the Falklands and possibly with one of the Crozets. *Idotea lacustris,* a fresh-water Isopod,

---

1) Dust-storms in New-Zealand. Nature LVIII (1903): 223.
2) See General Bibliography to the Subantarctic Islands of New Zealand, II (1909): 808 et seq., where many important publications dealing with southern biogeography are cited. Also the paper of W. R. B. OLIVER (1925: 99—139) should be consulted, since it is not only full of information but presents the case on purely orthodox lines.

occurs in New Zealand, Campbell Island and Subantarctic America. A spider, *Pacificana Cockayni*, from Bounty Island (Subantarctic province) is related to genera from Tasmania and Cape Horn. A group of fresh-water *Crustacea*, including the New Zealand genus *Boeckella*, is represented in extra-New Zealand Subantarctic lands by closely allied genera, while the genus itself occurs in subalpine lakes of Tasmania. The beetle *Loxomerus* is purely Subantarctic[1]).

Just as the botanical evidence of the last chapter and the zoological, of which the above is altogether incomplete, make out a strong case for a Subantarctic or Antarctic "land-bridge", so does the great depth of the ocean to the south and east of the New Zealand continental shelf shake one belief in the possibility of such connection. It is true that Captain DAVIES in Mawson's ship, the Aurora, discovered a small area of comperatively shallow water to the South of Tasmania, but he likewise demonstrated the presence of very deep sea between Macquarie and the Lord Auckland Islands. Further, as seen from the geographical chapter, New Zealand is surrounded by a fairly shallow sea, which to east, west and south suddenly sinks to a profound depth. Obviously, this shallow sea denotes an ancient land-surface, but the sudden drop affords strong evidence that deep water has existed, as at present, for an extremely long period. On the other hand, there may have been a long-continued earth-movement to which the present ocean-depth is due. However, the matter is one of mere speculation, and in the light of our present knowledge only, a belief or disbelief in land-connection rests solely on the belief in the possibility or impossibility of the plants and animals having been able to cross the vast stretch of ocean by means of wind- or bird-carriage alone. The difficulty of believing in this lengthy transoceanic transit is so great that I must declare for the problematical "bridge", but this must have existed at a time antecedent to the advent of mammals on the connected area.

Finally, comes the question of multiple origins. This is rarely seriously advocated at the present time, but, strange to say, in pre-Darwinian days, is was a common belief, though there was not a shred of evidence to show how such a phenomenon could take place. But in the present state of knowledge, there seems to me no reason, according to any accepted theory of evolution, and especially to the doctrine of mutation, why polygenesis should no take place occasionally. It would, of course, occur most frequently with regard to families and genera, and specific polygenesis might be a rare occurrence. But, in biogeographical discussions the idea of polygenesis will never be popular, for were it accepted such discussions would be futile.

---

1) From a study of the distribution of brachiopod faunas in Antarctic and subantarctic lands, J. A. THOMSON postulates land-connection or a relatively shallow sea between Australia, New Zealand and Kerguelen Land, Antarctica and South America in the early Tertiary (Austral. Antarc. Exped., Scientif. Reps., Zool. and Bot., 4 (1918): 59).

The problematical history of the flora. It seems fairly certain that since early Mesozoic times New Zealand has never been completely submerged. From the fossils which have been collected ARBER (The earlier Mesozoic floras of New Zealand. *Pal. Bull. No. 6.* N. Z. Geolog. Surv., 1917) has shown that there was a Triassic-Jurassic flora (more than one flora really) consisting of *Equisetales,* Fern-like plants (some probably seed-bearing), *Cycadophyta, Podozamiteae, Ginkgoales* and *Coniferales.* But of more interest is the record of two Dicotyledons of the Lower Cretaceous which were associated with *Cladophlebis australis* — a fern-like Mesozoic plant of wide distribution the world over and extremely common in Triassic New Zealand.

In the Cretaceous the Mesozoic rocks were folded and by degrees the land rose and extending north and south and south-east became virtually a continent. In the north, basing the statement on the present ocean depths and on the magnitude of the palaeotropic element, **Great New Zealand** included Norfolk Island, the Kermadecs, Lord Howe Island, the New Hebrides and New Caledonia (i. e. their present sites formed a small part of the area) and extended to New Guinea and northern Queensland. In the south, mainly for zoological reasons as already explained and partly through disbelief in long-distance transit over the ocean of seeds &c., my rather unwilling opinion is that the land extended to the Antarctic Continent; and, in the east what are now the Chatham Islands would undoubtedly be included.

During the *Great New Zealand period* the palaeotropic element would people the north while from the fairly warm Antarctica would come the subantarctic element. Synchronously with the invasion of northern and southern plants the palaeozelandic element, which in part had its beginnings in the Jurassic, would advance north and south. On the wide area of Great New Zealand, genera and perhaps families would come into being — these also palaeozelandic.

The main difficulty which now comes to the front is the question of the subantarctic and New Zealand high-mountain elements in the flora of Tasmania. There can have been no *direct* "land-bridge" from New Zealand to Tasmania, for had there been one the *true* Australian element would have been fully represented. It is also nearly impossible to believe that seeds of high-mountain species quite unfitted for wind or bird carriage can have passed from New Zealand to the mountains of Tasmania. The only alternative is to postulate a great extension of Tasmania southwards. With SKOTTSBERG[1]) in contradistinction to W. R. B. OLIVER[2]) I consider *Nothofagus* —

---

1) SKOTTSBERG (195: 138—139) defines his "Old Antarctic element" which consists of "genera or even orders which are virtually bicentric" and, to cite a few, includes Oreobolus, *Rostkovia, Libertia, Nothofagus, Laurelia* and *Donatia* — there are 22 in all. Nor does that eminent, much-travelled botanist favour long-distance carriage over oceans.

2) For OLIVER's arguments regarding *Nothofagus,* see 1925: 120. He considers that "It is probable that *Fagus* and *Nothofagus* originated in North America and spread thence

a critical genus from the biogeographical standpoint — originated in the far south.

Remnants of the rich upper Cretaceous and early Tertiary Flora occur in many localities, especially in Otago and Southland. The fossils have been studied by ETTINGSHAUSEN, who refers some to various Northern genera which one would not expect in the Southern Hemisphere, especially Australasia, e. g. *Myrica, Alnus, Quercus, Ulmus* and *Acer*. He also records the Australian *Casuarina* and *Eucalyptus*. Mixed with these were plants referred to existing New Zealand genera or their representatives. Judging from the figures accompanying the descriptions, ETTINGSHAUSEN's identifications, if accepted at all, must be received with the greatest doubt, indeed I do not think they are of any value. But were these genera present, then there must have been a universal temperate flora, a matter hardly conceivable in the fact of the tropical climate as a barrier. This tertiary fossil flora, however, is a fact, and it teaches us that many species and genera have passed away, just as the present species would have gone, in the future, or changed by natural means, had New Zealand remained a virgin land.

The great elevation and extension of the land was succeeded by an equally great depression which is *the most critical occurrence in regard to any conclusions concerning the origin of the flora, or its distribution in New Zealand itself,* for if the great area was reduced to a few small, flat islands what would happen to the high-mountain species? With this question in my wind I wrote to Professor R. SPEIGHT regarding the area &c. of New Zealand at the maximum period of depression. His reply, which gives an excellent concise account of much of the geological history of New Zealand is here quoted almost in its entirety.

"New Zealand was raised into a mountain region at the close of the Jurassic or the beginning of the Cretaceous, that is, the post-Hokonuian revolution.

By the close of the Cretaceous it was reduced to a peneplain, with probabale elevations of no great height standing above its surface and then this peneplain was slowly depressed beneath sea-level so that it was fairly completely covered with a layer of Sediments. There were no doubt islands of relatively small extent and comparatively low elevation which were not covered with this veneer. It is not postulated that sinking took place all over the land at the same time and it is certain that the sagging of the crust in some areas was posterior to the sagging in adjoining areas. For example the sagging in the North Canterbury-Kaikoura area reached a maximum in the late Eocene, whereas it commenced later in the South Canterbury-North Otago area and reached a maximum in the Miocene.

---

east, south and west. The western moiety passed, *via* Japan, round the Pacific, reaching Australia and New Zealand".

Probably the depression went on later in the North Island. This has, however, little to do with your problem. We may take it therefore that by the Early-Middle Tertiary the reduction of the land area of New Zealand had reached its limiting value, but there may have been adjacent high land of whose precise location we can say nothing.

From the period of maximum submergence onward, the land rose and there must have been land of decided relief in the near neighbourhood in order to furnish the thick deposits of ancient gravels which occur over wide areas in the South Island. You get such gravels in North Canterbury. In the Castle Hill area, in South Canterbury, in the Mackenzie Country, &c. These date from the late Pliocene. This elevation continued till there was the onset of the Pleistocene Glaciation, when the land probably stood higher than at present. There were probably two periods of glacier advance, but it is not thought that the interglacial period was one of very mild climate or of reduced height of the land. I look on the two glacial periods as merely *marked* fluctuations in the ice front.

From then on there has been a reduction in level, probably amounting to 1500 or 2000 feet, but these figures are mere conjectures. The evidence from our artesians certainly show a lowering in level of 700 feet and if bores are put down further this figure will no doubt be substantially increased.

I do not think that the land area during the Oligocene-Miocene times could have been very large and we have little evidence as to height. I think MORGAN considered the presence of glaciers on mountains over the present site of Western Nelson though I fancy that he came to the conclusion that the beds were not glacial. In any case they were coarse in texture and were probably shed from an area of pronounced relief, but of course one cannot say how high it was in so many feet. Coarse material may come from low hills if the slopes are abrupt. Some of the coarse Tertiary conglomerates of Central Otago, if not morainic as postulated by PARK and others, imply the action of rivers with considerable transporting power and this may indicate high land. A consideration of these cases certainly does suggest that there was land with considerable relief but no indication of its height can be arrived at.

The only point which I can see which presents any difficulty is the date of arrival of the high-mountain flora. If it came late in the Tertiary there is no difficulty as far as I can see but if it is a survival of a Cretaceous flora then there are difficulties in providing a refuge for it or a location for it to be established on. There is always the possibility that the refuge may have been lands which have sunk out of sight wether beneath the Tasman Sea or to the east of the present New Zealand. My own investigation of Cretaceous Coal conglomerates leads me to think that the present land features are in no way related to those of that period, and that mountain ranges then existent, which may have persisted fairly far on int the Tertiary, may have entirely disappeared from adjacent areas."

The great Oligocene-Miocene reduction of the land must have brought about the extinction of many species especially among the high-mountain plants and plastic species, perhaps the sole survivors of large genera, would find a haven of refuge on rock-faces and other habitats, perhaps fairly high, unsuited for forest-plants. It follows then that the primitive Subantarctic element would be decimated and that the present species or genera are a mere fraction of the original extensive company. But had there been no high land and the climate warmer than now, which was most likely the case, hardly a true high-mountain species could have survived and it seems impossible that a new subantarctic flora could have arrived during the Pliocene-Pleistocene without another great extension of land southwards, but this appears most unlikely. As for the lowland forest species they would not suffer nearly so much, for it is surprising how great a plant-population can thrive on a small island, such as the Little Barrier, where there are, at the present time most of the forest-plants of Northern Auckland.

During the Pliocene, elevation again took place and extended far into the Pleistocene. The alpine plants could return to the mountains, and under the new stimuli, ancient genera would be revivified and new forms appear. Again the land extended to the north, south and east and to some extent to the west. Exchanges would be possible with Australia by transoceanic methods and with northern palaeotropic floras. Towards the end of the period of elevation in the Pleistocene, when the mountains were at their highest, came the great extension of the glaciers. Then would the palae-otropic element be driven northwards and perhaps eastwards, especially if as is more than probable, the glacial advance was, in part, the result of a colder climate. East of the Southern Alps there would be a steppe-climate on the plateau. Then by *hybridism* and perhaps by *epharmonic change* would arise the intense xerophytes, descendants of mesophytes it may be, such as *Carmichaelia Petriei* from a leafy forest-species, or *Edwardsia prostrata* from *E. chathamica*. Then, too, would appear that semi-stable xeromorphy seen in the palaeotropic *Hoheria* and *Pennantia*.

Towards the end of the Pleistocene, depression of the land set in once more, the glaciers retreated far into the mountains and the re-peopling of the glaciated land, as already described, began. Once more the descendants of the ancient species, some perhaps themselves of high antiquity, com-menced to make new forms, acted on by the novel and diverse environ-ments and probably by secular changes of climate first wet, then drier. Then would *Celmisia, Hebe, Epilobium* and many other genera, no longer held in check by uniform conditions, burst forth into their multiplicity of forms, many of them hybrids (species usually isolated from one another coming together) or epharmones and others true species. Even yet, species-making is in progress.

The land having receded beyond its present limits, elevation once more took place, and the New Zealand of to-day came into being, peopled by its heterogeneous gathering of plants, children of north and south and east and of the New Zealand soil itself, moulded by great earth-movements and climates of extreme variety.[1])   In one thing they differed from the plants of other regions; no grazing mammals had ever been present to molest them, they possessed no structures that could claim to be defensive.

Finally came man; first the Maori, or it may be his predecessor, but their influence on the vegetation was but slight. Then arrived the European. It is more than 100 years since he began to occupy the land, but how great the change his operations have wrought, has been already told. We, who now live in this wonderful country, and love its marvellous vegetation, have set aside sanctuary after sanctuary where the palaeotropic, Australian and palaeozelandic plants, the survivors of that bitter strife with Nature, that commenced millions of years ago, can still pursue their destinies if unmolested by their human enemies and the horde of foreign plants and animals he has let loose.

Will our descendants prize this unique heritage from the dim past and preserve these sanctuaries intact?

1) Nothing shows more clearly how greatly climate must have changed in the temperate Southern Hemisphere than do the angiospermous fossils of Seymour Island and the Antarctic coal discovered by SHACKELTON's expedition. A considerable rise in temperature, accompanied by depression of the land, would undoubtedly be detrimental to the well-being of alpine plants, but, as seen from their present distribution in New Zealand, many thrive under warm lowland conditions, so that a fair number could probably tolerate a considerable rise in temperature. The fossil plants mentioned above clearly show that Antarctica possessed a Tertiary flora distinguished by a small but most characteristic New Zealand element, as evidenced by the following species: *Laurelia insularis, Knightia Andreae, Drimys antarctica* and two species of *Nothofagus*. With these species in one's mind it is easy to agree with SKOTTSBERG that "the Antarctic Continent may have been a centre of evolution from which animals and plants wandered north".

Index of Plant Names.

---

[1]) Correction: Page 416 line 12 from bottom for *Adiantum obtusatum — A. difforme* read *Asplenium obtusatum — A. difforme*.

Arundo conspicua Forst. f. *62*, 78, 91, 93, 94, 97, 98, 99, 102, *128*, 144, 192, 197, 199, 208, 233, 234, 367, 368, Plate VIII, fig. 9; Plate XIII, fig. 16.

Ascarina lanceolata Hook. f. *325*, *326*, 328.

— lucida Hook. f. 63, *122*, 130, 166, 168, 326, 395.

Asperula perpusilla Hook. f. 87, 88, 200, 310.

Asplenium adiantoides (L.) C. Chr. 108, 136, 159, 162, 176, 177.

— anomodon Col. 224.

— bulbiferum Forst. f. 110, 148, 160, 162, 170, 174, 176, 233, 270, 347.

— difforme R. Br. 416.

— flabellifolium Cav. 108.

— flaccidum Forst. f. 77, 99, 113, 136, 159, 162, 170, 220, 221, 223, 267, 270, 334.

— Hookerianum Col. 112, 224.

— lucidum Forst. f. 98, 100, 108, 110, 111, 114, 162, 224.

— obtusatum Forst. f. *67*, *68*, 88, 97, 100, 101, 104, 105, 327, 338, 343, 346, 412, 416.

— Richardi Hook. f. 282.

— scleropodium Homb. et Jacq. 400.

— Shuttleworthianum Kunze 327.

— Trichomanes L. 259, 383, 415.

Astelia 125.

— Banksii A. Cunn. 73, 98, 107, 110, 391.

— Cockaynei Cheesem. 137, 181, 219, 228, 248, 250, 258 to 262, 265, 266, 271, 272, 294, 300, 313, 315, 322, 369. Plate LXXII, fig. 89.

— Cunninghamii Hook. f. 115, *125*, 136, 144, 166, 169, 177.

— — var. Hookeriana T. Kirk 107, 221.

— linearis Hook. f. 204, 228, 316, 318, 320, 321, 323, 349, 413.

— monticola Ckn. 315.

— nervosa Banks et Sol. 270, 334.

— — var. grandis (Hook. f. ex T. Kirk) Ckn. et Allan 105, *125*, 162, 177, 179, 183, 251, 261, 269.

— Petriei Ckn. 315.

— pumila R. Br. 413.

— Solandri A. Cunn. 99, 100, 107, *125*, 136, *137*, 144, 159, 165, 175, 176, 184, 221, 388, Plate XXIV, fig. 27; Plate XXXVII, fig. 41.

— subulata (Hook. f.) Cheesem. 349, 400.

— trinervia T. Kirk *125*, 144, 157 to 159, 260. Plate XXX, fig. 34.

Athyrium australe (R. Br.) Presl 391.

Atriplex crystallina Hook. f. 88 to 90.

— patula L. 86.

Atropis antipoda Petrie 336.

-- stricta (Hook. f.) Hack. 75, 86.

Atropis Walkeri (T. Kirk) Cheesem. 86.

Aulacopilum 25.

Australina pusilla Gaud. 166.

Avicennia 82.

— officinalis L. *65*, 69, *82*. Plate III, fig. 3.

Azolla rubra R. Br. 133, 196, 413.

— filiculoides Lam. 413.

Azorella Selago Hook f. 338, 351, 413.

Bagnisia Hillii Cheesem. *132*, 384.

Balbisia 406.

Banana 325.

Barley 374, 390, 396.

Bean 374.

Beilschmiedia 407.

— taraire (A. Cunn.) Benth. et Hook. 110, *120*, 150, 155, 156, *158*, 159, 160, 161, 171, 176, 381. Plate XXXI, fig. 35; Plate XXXII, fig. 36.

— tawa (A. Cunn.) Benth. et Hook. *121*, 129, 144, 149, 150, 153, 155, 156, *159*, 165, *167*, 169, 170, 173, 183, 186, 222, 381, 384, 386, 388. Plate XXIV, fig. 27.

Black pine 122.

Black southern-beech 122, 234.

Black tree-fern 126.

Blechnum Banksii (Hook. f.) Mett. *67*, *68*, 100.

— discolor (Forst. f.) Keys *126*, 148, 162, 169. 173, 182, 183, 265. Plate XXVIII, fig. 32.

— durum (Moore) C. Chr. *67*, *68*, 87, 88, 99, 100, 104, 333, 338, 343, 345, 346, 391. Plate XXXI, fig. 99.

— filiforme (A. Cunn.) Ettingh. 110, 112, 114, 134, 135, 150, 159, 176.

— fluviatile (R. Br.) Lowe 145, 162, 174. 195, 210, 261, 270.

— Fraseri (A. Cunn.) Luerss. 131, 156 to 158, 161, 391.

— lanceolatum (R. Br.) Sturm 145, 162, 170, 224, 233, 270.

— nigrum (Col.) Mett. 169.

— norfolkianum (Hew.) C. Chr. 325.

— Patersoni (Spreng.) Mett. var. elongata (Bl.) Hook. et Bak. 149.

— penna marina (Poir.) Kuhn 95, 149, 195, 218, 219, 257, 259, 267, 274, 275, 293, 294, 305, 316, 320, 322, 347, 382, 412.

— procerum (Forst. f.) J. G. Anders 77, *126*, 148, 161, 162, 169 to 171, 174, 176, 177, 179, 181, 182, 194, 197, 199, 200, 203, 210, 219, 221, 222, 254, 261, 265, 266, 270 to 272, 286, 328, 334, 346, 347, 350, 367, 386, 412.

— vulcanicum (Bl.) Kuhn 149, 195, 282.

Blue-tussock 305.

Festuca erecta D'Urv. 337, 351, 412.
— gigantea Vill. 367.
— littoralis Labill. 88, 89, 91, 92, 93, 101.
— magellanica Lam. 413.
— multinodis Petrie 99.
— myurus L. 365.
— novae-zelandiae J. B. Armstg. 78, 120, *128*, 193, 211, 212, 216, 217, 220, *241*, 282, 284, 294, *304*, 309, 357, 362, 369, 413, 415.
— rubra L. 415.
— — var. fallax 213, 357.
Ficus macrophylla Desv. 375.
Filices 61, 118, *126*, 220.
Fimbristylis dichotoma Hook. f. 201.
Forstera 119, 126.
— Bidwillii Hook. f. 249, 284, 295, 299, 316.
— var. densifolia Mildb. 314.
— sedifolia L. f. 248, 287, 299, 316.
— — var. oculata Cheesem. 249, 287.
— tenella Hook. f. 287.
Freycinetia 106.
— Banksii A. Cunn. *62*, 63, 103, *106*, 114, *126*, *134*, 144, 150, 157, 159, 166, 174, 177, 395.
Frullania 156.
Fuchsia 407.
— excorticata (J. R. et G. Forst.) L. f. 4, 105, 106, 108, 109, 111, 112, 115, *123*, *142*, 144, 146, 152, 162, 165, 171, 174, 208, 210, 233, 268 to 270, 286, 341.
— — × perscandens 177. 402.
— perscandens Ckn. et Allan 108, *133*, 142, 144, 152, 171.
— procumbens R. Cunn. 72, 73, 144.
Fungi 16, 25.
— parasitic on Rubus 25.
Fusanus 25.

Gahnia 126.
— gahniaeformis (Gaud.) Heller 187, 191, 192, 221, 414.
— pauciflora T. Kirk 152, 179, 183, 192, 194, 259, 260, 261, 268, 271, 272, 281.
— procera J. R. et G. Forst. 260, 269, 271, 272, 278, 280, 281.
— rigida T. Kirk 204.
— robusta T. Kirk 387.
— setifolia (A. Rich.) Hook. f. 159, 192.
— xanthocarpa Hook. f. 126, 144, 157, 159, 161, 174, 176, 368, 416.
Gaimardia 225, 321.
— australis 413.
— ciliata Hook. f. 204, 219, 228, 322, 349.
— minima (T. Kirk) Cheesem. 391.
— setacea Hook. f. 322, 413.
Garden-thyme 360.
Gastrodia 139.

Gastrodia Cunninghamii Hook. f. 397.
— minor Petrie 397.
Gaultheria 226.
— antipoda Forst. f. 188, 192, 194, 221.
— — × oppositifolia 188.
— — var. erecta Cheesem. 221.
— depressa Hook. f. 218, 219, 305, 307, 313, 314, 316, 318, 322, 369.
— oppositifolia Hook. f. 144, 188, 191, 192, 221, 384.
— perplexa T. Kirk 57, 95, 204, 219, 229, 307, 320.
— rupestris (Forst. f.) R. Br. 232, 249, 278, 280, 281, 282, 283, 285, 287, 291, 295, 297, 300, 316, 322, 382.
Gaya 24.
Geniostoma 118, 176.
— ligustrifolium A. Cunn. 78, 108, 109, 114, 139, 144, 152, 157, 159, 161, 174, 184.
Gentiana 226.
— antarctica T. Kirk 337.
— antipoda T. Kirk 337.
— Astoni Petrie 389.
— bellidifolia Hook. f. 249, 291, 297, 313, 316, 323. Plate LXV, fig. 80.
— cerina Hook. f. 337, 342, 348, 349.
— chathamica Cheesem. 325.
— concinna Hook. f. 337.
— corymbifera T. Kirk 218, *240*, 249, 295, 305, 309.
— divisa Cheesem. 287, 316.
— filipes Cheesem. 391.
— flaccida Petrie 399.
— gracilifolia Cheesem. 391.
— Grisebachii Hook. f. 219, 367.
— lineata T. Kirk 320.
— montana Forst. f. 316, 391.
— patula Cheesem. 249, 299, 316.
— saxosa Forst. f. 75, 87, 100, 391.
— serotina Ckn. 393, 310.
— Spenceri T. Kirk 271, 391.
— Townsoni Cheesem. 205, 228, 303, 316, 391.
— vernicosa Cheesem. 391.
Geranium 226.
— dissectum L. 416.
— microphyllum Hook. f. 200, 219, 291, 293, 299, 316, 318, 322.
— molle L. 213.
— pilosum Forst. f. 416.
— sessiliflorum Cav. 88, 90, 91, 93, 94, 212, 214, 413.
— — var. glabrum Knuth 208, 220, 248, 294, 299, 364, 365, 413.
— Traversii Hook. f. 329, 333.
Geum 225.
— albiflorum (Hook. f.) Cheesem. 337, 349, 350.

Hebe Treadwellii Ckn. et Allan 395.
— vernicosa (Hook. f.) Ckn. et Allan 249, 316.
— — var. canterburiensis Ckn. et Allan 266.
Hectorella 404.
— caespitosa Hook. f. 225, 282, 287, 318, 403.
Hedycarya arborea J. R. et G. Forst. 63, 109, 111, 113, *123*, 139.
Helenium quadridentatum Lab. 381.
Helice crassa 82.
Heleocharis acicularis R. Br. 399.
— Cunninghamii Boeck. 94, 324, 335.
— neo-zelandica C. B. Clarke 91, 94, 391.
— sphacelata R. Br. *198*, 201.
Helichrysum alpinum Ckn. et Allan 249, 283, 284, 314.
— bellidioides (Forst. f.) Willd. 95, 209, 212, 218, 219, 248, 282, 283, 284, 292, 295, 299, 301, 305, 309, 316, 318, 347, 348, 369.
— coralloides (Hook. f.) Benth. et Hook. f. 141, 252, 282, *285*, 389, 397. Plate LIV. figs. 64, 65.
— depressum (Hook. f.) Benth. et Hook. f. 292.
— — ✕ bellidioides 402.
— dimorphum Ckn. 133, 134, 243, 275.
— filicaule Hook. f. 209, 212, 214, 219, 336.
— glomeratum (Raoul) Benth. et Hook. f. 152, 220.
— ✕ Gnaphalium 402.
— grandiceps Hook. f. 287.
— microphyllum (Hook. f.) Benth. et Hook. f. 280, 282, 285, 369.
— pauciflorum T. Kirk 284.
— prostratum Hook. f. 283, 284, 337, 347, 348.
— Purdiei Petrie 22.
— Selago (Hook. f.) Benth. et Hook. f. 282, 285.
— — ✕ Leucogenes grandiceps Beauv. 406.
— — var. tumida Cheesem. 397.
Helwingia 405.
Hemitelia 118.
— Smithii (Hook. f.) Hook. *126*, 162, 169, 170, 177, 184, 210, *233*, 270, 341, 417.
Hennedia 22.
Hennediella 22.
Hepaticae 15, 16, 156, 269, 286.
Herpolirion 220, 226.
— novae-zelandiae Hook. f. 219, 249, 307, 367.
Hibiscus 61, 375.
— diversifolius Jacq. 381.
— trionum L. 391.
Hierochloe 118.

Hierochloe alpina Roem. et Schult. 415.
— Brunonis Hook. f. 336, 348.
— Fraseri Hook. f. 316, 318, 319, 415.
— magellanica Hook. f. 413.
— redolens (Forst. f.) R. Br. 197, 200, 248, 307, 318, 335, 413.
Hinau 121.
Histiopteris incisa (Thunb.) J. Sm. 105, 106, 152, 162, 170, 201, 233, 270, 328, 412.
Hoheria 405, 407.
— Allanii Ckn. 393.
— angustifolia Raoul 114, *123*, 141, 144, 368.
— dentata var. angustifolia Hook. f. 172.
— glabrata Sprague et Summerh. 144, 170, 185, 228, 235, 243, 249, 250, 259, 262, 265, 267, 270, 276, 278, 279, 286. Plate LIX, fig. 71.
— Lyallii Hook. f. 138, 144, *235*, 244, 248, 258, 267.
— populnea A. Cunn. 110, 123, 144.
— sexstylosa Col. *123*, 140, 141, 144, 150, 165, 172, 385.
Holcus lanatus L. 90, 192, 213, 355, 369.
Holmskioldia 405.
Honeysuckle 121.
Hordeum murinum (L.) Huds. 90, 361, 364, 365.
Hormosira Banksii Harv. 81.
Horoeka 124.
Horopito 125.
Houhere 123.
Hupiro 123.
Huttonella (sect. of Carmichaelia) 407.
Hydatella 401.
— inconspicua Cheesem. 380.
Hydrocotyle 226.
— americana L. 162, 174, 413.
— marchantioides Clos 413.
— moschata Forst. f. 327.
— novae-zelandiae DC. 162, 209, 212, 214, 305, 413.
— — var. montana T. Kirk 200, 218, 295, 299.
— pterocarpa F. v. Muell. 199, 200, 368.
Hymenanthera 61, 226.
— alpina (T. Kirk) Ckn. 146, 224, 243, 252, 274, 275, 279, 282, 284, 285, 287, 291, 293, 294, 299, 301, 364. Plate LXVII. fig. 82.
— chathamica T. Kirk 334.
— crassifolia Hook. f. *70*, 75, 97, 101.
— dentata R. Br. var. angustifolia Benth. 171.
— latifolia Endl. 416.
— novae-zelandiae (A. Cunn.) Hemsl. 416.
— obovata T. Kirk emen. Ckn. 97, 100.
Hymenodon 156.

442

Index of Plant Names.

Hymenophyllaceae 22, 136, 137, 156, 165, 169, 170, 171, 270.
Hymenophyllum 118.
— Armstrongii T. Kirk 136, 261.
— bivalve (Forst. f.) Sw. 162.
— demissum (Forst. f.) Sw. 159, 160, 162.
— dilatatum (Forst. f.) Sw. 159, 162, 261.
— ferrugineum Colla 412, 419.
— flabellatum Lab. 162, 169, 183, 261, 270.
— Malingii (Hook.) Mett. 244, 261, 273, 382, 419.
— multifidum (Forst. f.) Sw. 162, 183, 232, 254, 257 to 259, 261, 266, 270, 271, 272, 277, 282, 286, 287, 314, 316, 347, 349, 350.
— peltatum Desv. 149, 382, 412. 415.
— pulcherrimum Col. 149, 382.
— rarum R. Br. 162, 261, 412.
— rufescens T. Kirk 261.
— sanguinolentum Swartz 105, 136, 162, 210.
— scabrum A. Rich. 162, 183, 210, 261.
— tunbridgense Sm. 162, 261, 412, 415.
— villosum Col. 136, 259, 261, 265, 266, 269, 270, 271.
Hypericum Androsaemum L. 195, 358.
— gramineum Forst. f. 364.
— perforatum L. 195.
Hypnum hispidum 350.
Hypochaeris radicata L. 192, 213, 355. 358, 369, 373.
Hypolaena 323.
— lateriflora Benth. 120, 127, 202 to 205, 219, 303, 321. 322.
Hypolepis Millefolium Hook. 150 218, 258, 262, 271. 274, 275, 278, 293, 305, 343. 347. 369.
— rugosula (Lab.) Sm. 106, 162, 367.
— tenuifolia (Forst. f.) Bernh. 170. 270, 327.
Hypopterygium 156.
— novae-seelandiae C. Müll. 181.
Hypoxis pusilla Hook. f. 217.

Imperata Cheesemanii Hack. 327.
Ionopsis (sulgen. of Celmisia) 337, 407.
Ipomoea palmata Forsk. 72.
— pes-caprae (L.) Roth 327.
Isachne australis R. Br. 198.
Isoëtes alpinus T. Kirk 246, 324.
Isolepis 404.
Isotoma fluviatilis (R. Br.) F. von Muell. 324.
Ivy-tree 124.
Ixerba 404, 417.
— brexioides A. Cunn. 144, 150, 167, 183, 260, 262.

ovellana 118.
— repens (Hook. f.) Kränzl. 222.
— Sinclairii (Hook.) Kränzl. 150, 222, 385.
Juncus 83, 84.
— antarcticus Hook. f. 323.
— effusus L. 415.
— holoschoenus R. Br. 197.
— maritimus Lam. 65, 416.
— — var. australiensis Buchen. 83, 85, 86, 202, 334, 416. Plate VII, fig. 7.
— novae-zelandiae Hook. f. 324.
— planifolius R. Br. 412.
— polyanthemus Buchen. 200, 328, 415.
— prismatocarpus R. Br. 197.
— scheuchzerioides Gaud. 338, 413.

Kahikatea 122, 163, 164, 174, 177, 178.
Kaikomako 124.
Kamahi 122, 168, 173, 269. 400.
Kanono 123.
Kanuka 127.
Karaka 66, 111.
Karamu 123.
Kauri 5. 120, 153, 156, 157, 161, 162, 190, 191, 201, 271, 381 to 383. Plate XXIV, fig. 27: Plate XXX, fig. 34; Plate XXXI, fig. 35: Plate XXXII, fig. 36; Plate XXXIII, fig. 37.
Kauri-grass 125.
Kirkophytum (sect. of Stilbocarpa) 407.
Knightia 95, 110, 111, 118, 160, 191, 407.
— Andreae Düs 426.
— excelsa R. Br. 107, 121, 129, 130, 141, 144, 150, 152, 159, 161, 165, 170, 406.
Koeleria 119, 282.
— cristata Hook. f. 415.
— Bergii Hieron. 413.
— novo-zelandica Domin 305, 415.
— superba Domin 413.
Kohekohe 114.
Kohuhu 124.
Kopi 66.
Korthalsella 138.
— clavata (T. Kirk) Cheesem. 139, 393.
— Lindsayi (Oliv.) Engl. 139, 244, 393.
— salicornioides (A. Cunn.) Van Tiegh. 139.
Kotukutuku 123.
Kyllinga brevifolia Rottb. 381.

Laburnum 138.
Lacebark 123.
Lagenophora 226.
— Commersonii Cass. 413.
— cuneata Jetrie 212, 214.
— lanata A. Cunn. 190.
— petiolata Hook. f. 212, 219, 257, 262, 397.

Pleurophyllum speciosum Hook. f. 337, *339*, 342, 343, 347, 348.
Plum 138, 375.
Poa 211.
— acicularifolia Buch. 193, 290, 300, *309*, 388.
— anceps Forst. f. 99, 100, 221, 222, 223, 307, 313.
— — var. condensata Cheesem. 97, 98.
— annua L. 213.
— Astoni Petrie 100. Plate XI, fig. 14.
— aucklandica Petrie 336.
— caespitosa Spreng. 78, 93, 102, 120, *128*, 191, 211, 212, *216*, 220, 282, 284, 304, 307, 316, 362, 364.
— chathamica Petrie 335.
— Cockayniana Cheesem. 232, 278, 301, 316.
— Colensoi Hook. f. 120, 194, 212, 214, 282, 283, 291, 294, 301, *305, 306*, 319, 369, 396.
— exigua Petrie 398.
— flabellata Hook. f. 413.
— foliosa Hook. f. *339*, 343, 346, 348, 349, 351, 400, 413.
— — var. condensata 99.
— Guthrie-Smithiana Petrie 399.
— Hamiltoni T. Kirk 336.
— incrassata Petrie 336.
— — var. breviglumis (Hook. f.) Cheesem. 336.
— intermedia Buch. 120, 211, 212, 217, 218, 224, 248, 282, *305*, 309, 357, 365, 396.
— Kirkii Buch. 305.
— Lindsayi Hook. f. 248, 294, 305, 364.
— litorosa Cheesem. *339*, 346, 347, 350, 413.
— maniototo Petrie 6, 357, 364.
— novae-zealandiae Hack. 232, 282, 286, 287.
— oraria Petrie 398.
— polyphylla Hack. 327.
— pratensis L. 213, 355, 363, 371.
— pusilla Berggr. 99, 310, 316, 367.
— pygmaea Buch. 397.
— ramosissima Hook. f. 336, 344. Plate LXXXI, fig. 99.
— sclerophylla Berggr. 288, 290.
Podocarpaceae 418.
Podocarpus 161, 163, 279.
— acutifolius T. Kirk 195, 180, 210, 271, 273.
— alpinus R. Br. 150.
— dacrydioides A. Rich. 95, 108, 115, *122*, 128, 129, 132, 140, 142, 162, 163, 166, 175, 176, *177*, 178, 179, 210, 368. Plate XX, fig. 23; Plate XXI, fig. 24; Plate XXXVII, fig. 41.
— ferrugineus Don 122, 158, 159, 162, 163, 168, 233, 268, 414.

Podocarpus Hallii T. Kirk *122*, 162, 163, *164*, 179, 182, 186, 195, 210, 233, 244, 260, 261, 265, 266, 267, *268* to 270, 273, 277, 278, 392.
— nivalis Hook. 120, 163, 180, 228, *251*, 259, 261, 265, 267, 272, 277, 278 to 280, 282, 285, 291, 294, 297 to 299, 382.
— spicatus R. Br. 95, 107, 108, *122*, 163, 165, *166*, 174, 177, 263, 392, 418.
— totara A. Cunn. 95, 105, 107, 108, *122*, 129, 130, 138, 159, 161, 163, *164*, 165, 166, 171, 177, 353, 384, 392. Plate XXII, fig. 25.
Podozamiteae 422.
Pohutukawa 67, *97*, 110.
Pokaka 121.
Polygala virgata Thunb. 381.
Polygonum serrulatum Lag. *198, 334*.
Polypodium Billardieri (Willd.) C. Christen. 136, 162, 170, 210, 270, 412.
— — var. rigidum (Homb. et Jacq.) Ckn. 400.
— dictyopteris (J. Sm.) Mett. 392.
— diversifolium Willd. 77, 97, 98, 108, 110, 134, 136, 162, 170, 210, 220, 221, 223, 224, 270, 327, 334, 343.
— grammitidis R. Br. 136, 162, 210.
— novae-zelandiae Bak. 134, 150, 261, 383, 384, 419.
— pumilum (J. B. Armstg.) Ckn. 282, 287, 349, 350.
— pustulatum Forst. f. 134, 176.
Polystichum adiantiforme (Forst. f.) J. Sm. 162, 412.
— aristatum (Forst. f.) Presl *325*, 328.
— cystostegia (Hook.) Diels 301, 343, 350.
— hispidum (Sw.) J. Sm. 162.
— Richardii (Hook.) J. Sm. 99, 100, 110, 112, 114, 166, 220, 223.
— vestitum (Forst. f.) Pr. 120, 149, 170, 177, 184, 195, 210, 233, 254, 255, 258, 259, 260, 262, 265 to 267, 270, 273, 275, 279, *338*, 345 to 350, 382, 412.
Pomaderris Edgerleyi Hook. f. 139, 188, 190, 191. Plate XLI, fig. 45.
— elliptica Lab. 188, 190, 191.
— phylicaefolia Lodd. 94, 188, 190, 191.
— rugosa Cheesem. 382.
Poplar 138.
Poranthera alpina Cheesem. 391.
— microphylla Brong. 167, 388.
Porokaiwhiri 123.
Porphyra columbina Mont. 81.
— subtumens J. Ag. 80.
Portulacaceae 119, 404.
Potamogeton 133, 199.
— Cheesemanii A. Benn. *196*, 324, 334.

Buch- und Kunstdruckerei E. Haberland, Leipzig C 1

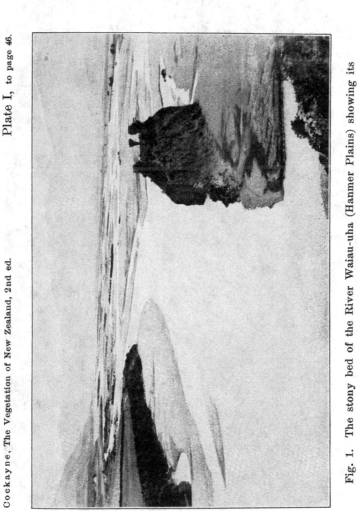

Fig. 1. The stony bed of the River Waiau-uha (Hanmer Plains) showing its
anastomosing streams.

Photo. C. E. Christensen.

Cockayne, The Vegetation of New Zealand, 2nd ed.

Fig. 2. Coastal scrub of Stewart Island. In the foreground wind-swept **Leptospermum scoparium** occupies the place of the usually dominant **Senecio rotundifolius**, which here is in the more sheltered background. — Photo. L. Cockayne.

Cockayne, The Vegetation of New Zealand, 2nd ed.

Plate III, to pages 66, 82.

Fig. 3. Mangrove swamp, showing an adult tree, pneumatophores and young plants of Avicennia officinalis.
Photo. L. Cockayne.

Fig. 4. Aereal roots of **Metrosideros tomentosa** given off from high up the trunk, descending to and entering the ground.

Photo. L. Cockayne.

Fig. 5. **Stilbocarpa Lyallii** growing amongst rocks, Ruapuke Island (Stewart district). In the foreground the inflorescence hidden by the leaves can be dimly seen.

Photo. L. Cockayne.

Cockayne, The Vegetation of New Zealand, 2nd ed.

Fig. 6.   The bull-kelp (**Durvillea antarctica**) as seen at low water.   Dog Island, Foveaux Strait.

Photo. L. Cockayne.

Cockayne, The Vegetation of New Zealand, 2nd ed.

Plate VII, to page 88.

Fig. 7. View of a portion of the extensive **Salicornia** formation at Wakapuaka (Nelson Harbour—SN.), showing various stages of the incoming of **Juncus mariti= mus** var. **australiensis** salt-swamp.

Photo. W. C. Davies (Cawthron Institute, Nelson).

Fig. 8. Natural foredune formed by **Spinifex hirsutus** on the west coast of the Ruahine-Cook district near Waikanae.

Photo. W. H. Field.

Fig. 9. Active dune in process of occupation by **Phormium tenax, Arundo conspicua** and **Cassinia leptophylla.** West of Ruahine-Cook district.

Photo. L. Cockayne.

Fig. 10. **Leptospermum scoparium** heath of a dune-hollow of the Eastern district with **Discaria toumatou** in the foreground.

Photo. L. Cockayne.

Fig. 11. **Mesembryanthemum australe** growing on coastal rock exposed to abundant sea-spray at Lyall Bay, Wellington (Ruahine-Cook district).

Photo. L. Cockayne.

Fig. 12. Sand-worn stones lying on the exposed rock where dunes have
been blown away. Here and there is small **Coprosma acerosa**.
Coast of Egmont-Wanganui district.

Photo. L. Cockayne.

Fig. 13. Where dunes have been blown away small bushes of **Coprosma
acerosa** are growing on the shallow layer of sand with a tongue
of sand in their lee. Coast of Egmont-Wanganui district.

Photo. L. Cockayne.

Fig. 14. **Celmisia Lindsayi** association with mats of that plant in centre; in left-hand corner is a tussock of **Poa Astoni** and to right a scrub of **Hebe elliptica, Phormium tenax** &c. Nugget Point, South Otago district.

Photo. L. Cockayne.

Fig. 15. Mat of **Celmisia Lindsayi** growing on coastal cliff at Nugget Point (South Otago district). In left-hand corner is **Anisotome intermedia**. — Photo. L. Cockayne.

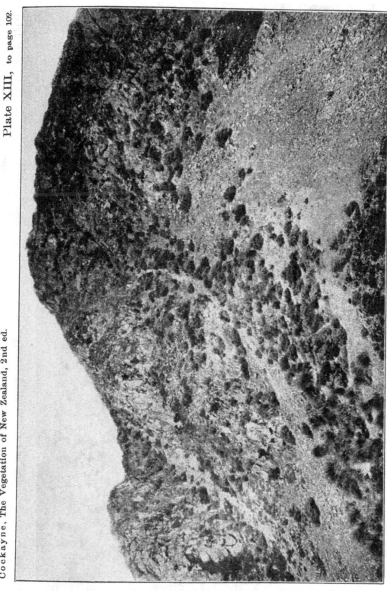

Fig. 16. Coastal shingle-slip on Kapiti Island with an open association consisting chiefly of **Cassinia leptophylla** and **Arundo conspicua**.

Photo. L. Cockayne.

Plate XIV, to pages 103, 104.

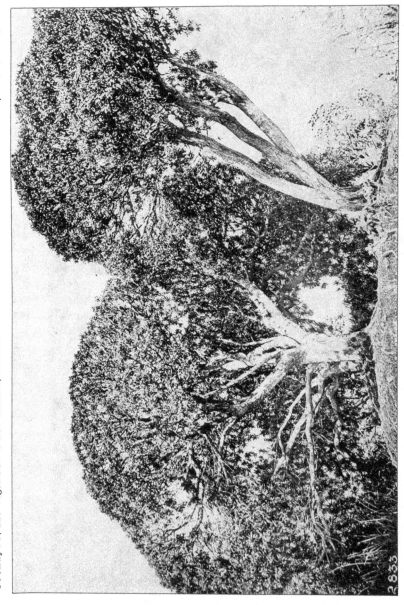

Fig. 17. Isolated trees of **Olearia angustifolia** growing on an exposed part of the coast of Stewart Island. The tree on the left is about 3 m. high, its trunk 37.5 cm. diam. and the rounded crown 12 m. through.

Photo. L. Cockayne.

Fig. 18. **Senecio rotundifolius** scrub on the shore of Patterson Inlet, Stewart Island, showing the dense roof and the branches jutting out over the shore for about 3.9 m.

Photo. L. Cockayne.

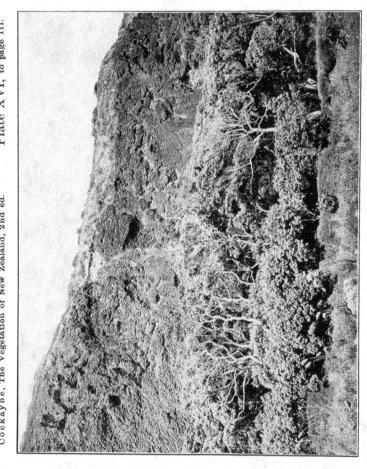

Fig. 19. **Metrosideros tomentosa** scrub at foot of cliffs facing the shore. Base of the Waitakerei Hills, Kaipara subdistrict.

Photo. L. Cockayne.

Fig. 20. Coastal scrub of Rangitoto Island in the Hauraki Gulf, South Auckland district. In centre **Griselinia lucida** which is always epiphytic in forest. — Photo. L. Cockayne.

Cockayne, The Vegetation of New Zealand, 2nd ed.

Fig. 21. Girdle of **Metrosideros tomentosa** near Mangonui, just above high water-mark, North Auckland district.

Photo. L. Cockayne.

Fig. 22. **Dysoxylum** forest of Stephen Island — there truly coastal — showing its close roof (Sounds-Nelson district). — Photo. L. Cockayne.

Fig. 23. Outskirts of dicotylous-podocarp forest near Te Whaiti (East Cape district)
with **Podocarpus dacrydioides** (tree on right) dominant and towards centre
on left a tree-fern **(Dicksonia fibrosa)** and on right the smaller tree-fern,
D. squarrosa.

Photo. L. Cockayne.

Fig. 24. The massive trunk of **Podocarpus dacrydioides,** but more irregular ("fluted")
than usual. Dicotylous- podocarp forest of Sounds-Nelson district.
Photo. W. C. Davies (Cawthron Institute, Nelson).

Fig. 25. Base of a massive tree of **Podocarpus totara** in the dicotylous-podocarp forest of the Mount Peel National Park (Eastern district).
Photo. L. Cockayne.

Fig. 26. Roots resembling liane stems descending along trunk of
Dacrydium cupressinum from an epiphytic epharmone of
Griselinia littoralis in forest, Stewart Island.

Photo. L. Cockayne.

Fig. 27. Epiphytes on branches of **Beilschmiedia tawa** in kauri forest (North Auckland district). Masses of **Astelia Solandri** 60 cm. high and on left a pendent bunch of **Lycopodium Billardieri**, 1.2 m. long.
Photo. L. Cockayne.

Fig. 28. Base of **Metrosideros robusta** showing its irregular habit suitable for occupation by epiphytes.
Forest of Kapiti Island (Cook Strait).
Photo. L. Cockayne.

Plate XXVI, to pages 137, 145, 166, 168.

Fig. 29. Base of trunk of **Weinmannia racemosa** formed out of aereal roots. Forest of Stewart Island.
Photo. L. Cockayne.

Fig. 30. Winter aspect of the deciduous **Plagianthus betulinus,** its branches completely
covered by some of the smaller epiphytes (9 species) with **Cyclophorus serpens** and
Earina mucronata dominant.

Photo. H. H. Allan.

Fig. 31. **Pittosporum divaricatum** — a typical example of the divaricating life-form growing in subalpine scrub on Arthur's Pass (900 m. altitude), Western district. — Photo. L. Cockayne.

Fig. 32. Late stage in the life-history of the lowland vegetation near the Franz Josef glacier (Western district) with **Blechnum discolor** (fronds erect, pale) and **Leptopteris superba** in the foreground. — Photo. L. Cockayne.

Cockayne, The Vegetation of New Zealand, 2nd ed.

Fig. 33. **Metrosideros perforata,** a high-climbing root-climber, growing as a shrub in the open.
Photo. L. Cockayne.

Cockayne, The Vegetation of New Zealand, 2nd ed.

Fig. 34. A portion of the kauri subassociation showing the characteristic undergrowth of Astelia trinervia. Waipoua forest, North Auckland district in 1907.

Photo. L. Cockayne.

Fig. 35. Interior of an extensive area of the kauri subassociation, The large trunks are those of the kauri (Agathis australis) and the slender ones mostly those of Beilschmiedia taraire. Waipoua Kauri forest, North Auckland district in 1907.

Photo. L. Cockayne.

Fig. 36. Interior of an open kauri subassociation showing the slender trunks of **Beilschmiedia taraire** ready to replace the kauri trees when such die; the massive trunks are those of the kauri. Waipoua kauri forest. North Auckland district in 1907. — Photo. L. Cockayne.

Fig. 37. View of roof of kauri forest, the kauris rising high above the other trees. In foreground on right **Cyathea medullaris**, its crown above the foliage of the forest margin. Waipoua kauri forest in 1907.

Photo. L. Cockayne.

Fig. 38.   Interior of rimu **(Dacrydium cupressinum)** forest of the Western district with **Quintinia acutifolia** as undergrowth.

Photo. C. E. Foweraker.

Fig. 39. **Nothofagus cliffortioides** forest at about 560 m. altitude (Waimakariri Valley, junction of Eastern and Western districts) marking the average line reached by the westerly rain. Between the river-bed and forest margin its junction with the tussock-grassland can be seen.

Photo. C. E. Foweraker.

Fig. 40. Colony of **Gleichenia Cunninghamii** on floor of **Weinmannia racemosa** forest near Lake Brunner, North-western district in 1897.
Photo. L. Cockayne.

Fig. 41. Semi-swamp forest near Dargaville (Kaipara subdistrict) in 1911, showing the long straight trunks of **Podocarpus dacrydioides** and the abundance of epiphytic **Astelia Solandri**. — Photo. L. Cockayne.

Fig. 42. Liverwort cushion of **Plagiochila gigantea** nearly 90 cm. high by 75 cm. through at the base and **Lycopodium volubile** growing on it; on left, trunk of **Weinmannia** covered with **Aneura eriocaulon.** **Dacrydium intermedium** association, Stewart Island. — Photo. L. C o c k a y n e.

Fig. 43. Colony of **Nothofagus fusca** in lowland
Nothofagus Menziesii-cliffortioides forest at 270 m.
altitude near the head of Lake Te Anau, Fiord district.
Photo. E. M. Barker.

Fig. 44. Portion of the Mineral Belt — the substratum a magnesian soil — showing the sudden transition from forest (on limestone) to the mostly open vegetation but with some tall tussock-grassland in foreground and some shrubland to left and in centre.

Photo. W. C. Davies (Cawthron Institute, Nelson).

Fig. 45. **Pomaderris Edgerleyi** as a member of Auckland manuka shrubland on the North Cape Promontory, North Auckland district. — Photo. L. Cockayne.

Fig. 46. **Raoulia tenuicaulis** (juvenile form in background) growing on the stony bed of the River Otira (Western district) at 300 m. altitude. — Photo. L. Cockayne.

Fig. 47. Low river-bed forest in the Otira Valley (Western district) at 376 m. altitude with **Pittosporum Colensoi** on the left and **Suttonia divaricata** on the right. — Photo. L. Cockayne.

Fig. 48. Another piece of the low forest of Fig. 47. In front a tree of the "araliad-form" **(Pseudopanax crassifolium var. unifoliolatum)** and various divaricating shrubs in the undergrowth. — Photo. L. Cockayne.

Fig. 49. **Nothofagus cliffortioides** forest giving place to low tussock-grassland where exposed to the full blast of the frequent north-west wind,
near source of River Poulter at about 800 m. Western district just
within the area of high rainfall. — Photo. L. Cockayne.

Fig. 50. Snow avalanche in a gully on Mount Tarndale (North-eastern district)
which has cut a path through the **Nothofagus cliffortioides** forest. — Photo. L. Cockayne.

Fig. 51. New vegetation on ice-worn rock alongside the Franz Josef glacier (Western district) with **Metrosideros lucida** in centre and fully-developed scrub in background. — Photo. L. Cockayne.

Fig. 52. Interior of scrub on old moraine near the terminal face of the Franz Josef glacier. The stems are mostly those of **Coprosma rugosa**. — Photo. L. Cockayne.

Fig. 53. Hard, massive cushion of **Dracophyllum rosmarinifolium** about 60 cm. high, surrounded by low **Olearia Colensoi** scrub on Mount Anglem, Stewart Island, at about 850 m. altitude.

Photo. L. Cockayne.

Plate XLVI, to pages 238, 280.

Fig. 54. Low subalpine scrub, Stewart Island, the tuft-shrub, **Dracophyllum Menziesii** rising above the **Olearia Colensoi.** Mount Anglem at about 850 m.

Photo. L. Cockayne.

Fig. 55. **Celmisia coriacea** var. **stricta** showing the close-growing erect rosettes of this class of the genus. Indigenous-induced (after fire) association at 900 m. altitude on the Takitimu Mountains, South Otago district.

Photo. L. Cockayne.

Fig. 56. **Aciphylla Scott-Thomsonii,** showing the Yucca-form common in the genus,
growing on Flagstaff Hill, South Otago district.

Photo. G. Simpson.

Fig. 57. Subalpine scrub on the Tooth Peaks, South Otago district near its junction
with the Fiord district at 900 m. altitude, composed of **Senecio cassinioides** and
Aciphylla maxima with peduncles 4.5 m. high.

Photo. W. D. Reid.

Fig. 58. Great cushions of **Haastia pulvinaris**—one of the largest vegetable-sheep—growing at 1500 m. altitude on shingle-slip on Mount Tarndale, North-eastern district near its junction with the North-western district. — Photo. L. Cockayne.

Fig. 59. Replacement of tussock-grassland by cushions of **Raoulia lutescens** — the cushion-form being due to the original mat being filled with sand — near Tarras (North Otago district) owing to denudation by drifting sand. — Photo. L. Cockayne.

Fig. 60. Horizontal trunk of **Olearia ilicifolia** in the subalpine **Podo-carpus-Libocedrus** forest near source of the River Rakaia (Southern Alps, Western district) at 920 m. altitude showing the deciduous outer bark. — Photo. L. Cockayne.

Fig. 61. Close view of cushion of **Phyllachne clavigera.** — Photo. S. Page.

Cockayne, The Vegetation of New Zealand, 2nd ed.

Fig. 62. **Veronica spathulata** growing in the scoria desert at base of Mount Ngauruhoe at 1500 m. altitude (Volcanic Plateau district). — Photo. L. Cockayne.

Fig. 63. Mat of **Leucogenes Leontopodium** as a member of the
fell-field association of the Tararua Mountains at about 1300 m.
altitude (Ruahine-Cook district).

Photo. E. Bruce Levy.

Fig. 64. The cupressoid **Helichrysum coralloides** of fairly open habit, growing on a dry rock-face at about 1250 m. altitude on Shingly Range, North-eastern district. — Photo. L. Cockayne.

Fig. 65. **Helichrysum coralloides** growing on dry rock, as in Fig. 64, but exposed to strong wind and forming a close cushion. — Photo. L. Cockayne.

Plate LV, to page 255.

Cockayne, The Vegetation of New Zealand, 2nd ed.

Fig. 66. To right and centre, outskirts of **Nothofagus cliffortioides** forest; to left in foreground tall tussock-grassland of **Danthonia Raoulii** var. **rubra**; on right much **Phormium Colensoi**; in distance, the active volcano Ngauruhoe. Waimarino Plain at about 1000 m. altitude. — Photo. P. Keller.

Fig. 67. Upper limit of **Nothofagus cliffortioides** forest at about 1200 m. altitude on Shingly Range, North-Eastern district. — Photo. L. Cockayne.

Fig. 68. Regeneration of **Nothofagus cliffortioides** forest — the old trees having died naturally — by the same species. Wairau Valley at 900 m. altitude (North-western district). — Photo. L. Cockayne.

Cockayne, The Vegetation of New Zealand, 2nd ed.

Fig. 69. Tall shrubland on outskirts of **Nothofagus cliffortioides** forest at base of **Mount Hahungatahi** (Volcanic Plateau district) at 900 m. altitude with **Cordyline indivisa** (tuft-tree), **Hebe salicifolia** and **Phormium tenax.** — Photo. L. Cockayne.

Fig. 70. Epharmonic cushion of the straggling or turf-making **Dacrydium laxifolium** and growing on it **Hebe tetragona** and **Celmisia longifolia.** Fell-field at 1200 m. altitude near base of Mount Ruapehu (Volcanic Plateau district.) — Photo. L. Cockayne.

Fig. 71. The epiphytic moss, **Weymouthia Billardieri** hanging from twigs of **Hoheria glabrata** in the mountain-ribbonwood low forest of Clinton Valley (Fiord district) at 600 m. altitude. — Photo. L. Cockayne.

Fig. 72. Interior of low forest on Price's Peak, Stewart Island, at an altitude of about 270 m., showing the numerous bryophyte cushions of **Plagiochila gigantea** and **Dicranoloma Billardieri**. — Photo. L. Cockayne.

Fig. 73. Exterior view of shrub-composite subalpine-scrub of Mount Anglem, Stewart Island, at about 700 m. altitude. The main mass is **Olearia Colensoi** but in foreground is some **Dacrydium Bidwillii** and jutting through the roof shoots of **Dracophyllum longifolium**. — Photo. F. G. Gibbs.

Fig. 74. Subalpine scrub of Mount Greenland (Western district) at about 800 m.
altitude; on extreme left is **Quintinia acutifolia**, then **Metrosideros lucida** (both
trees of the forest below), in centre **Pittosporum divaricatum** and on right
Nothopanax simplex. — Photo. L. Cockayne.

Fig. 75. Manuka **(Leptospermum scoparium)** scrub on Frazer Peaks
(South of Stewart Island); on right, where extremely dense the
scrub has regenerated after being burned. — Photo. L. Cockayne.

Cockayne, The Vegetation of New Zealand, 2nd ed.

Fig. 76. Terminal face of old lava-flow from the Red Crater of Mount Tongariro, Volcanic Plateau.

Photo, L. Cockayne.

Fig. 77. **Ewartia Sinclairii** growing on dry rock in the lower subalpine belt at 900 m. altitude of the Awatere Valley, North-eastern district. — Photo. L. Cockayne.

Fig. 78. Vegetation of dripping rock in Clinton Valley, Fiord district with an undescribed grass (not yet collected in flower) and drooping **Celmisia verbascifolia** on either side and to right **Phormium Colensoi**. — Photo. L. Cockayne.

Fig. 79. **Notothlaspi rosulatum** growing on single-shlip of river-terrace at about 750 m. altitude at Castle Hill, Eastern district. — Photo. L. Cockayne.

Fig. 80. **Gentiana bellidifolia** growing on scoria at about 1500 m. altitude on Mount Tongariro, Volcanic Plateau district. — Photo. L. Cockayne.

Plate LXVI, to page 298.

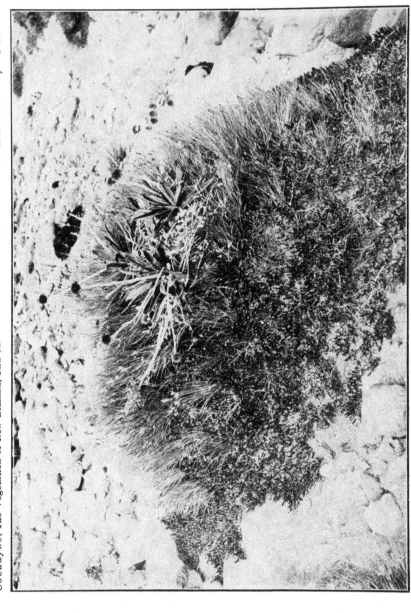

Fig. 81. Open cushion of **Carmichaelia orbiculata** growing as a member of the pumice fell-field at 1140 m. altitude (Volcanic Plateau district) with **Celmisia spectabilis** and **Danthonia setifolia** growing through the cushion. — Photo, L. Cockayne.

Fig. 82. **Hymenanthera alpina** growing on rock at 270 m. altitude surrounded by
the depleted ground, originally tussock-grassland, on the Dunstan Mountains,
North Otago district.

Photo. L. Cockayne.

Fig. 83. Portion of thick mat of **Celmisia hieracifolia** in fell-field of Mount Hector, Tararua Mountains, at about 1400 m. altitude. (Ruahine-Cook district). — Photo. E. Bruce Levy.

Fig. 84. Low cushion of **Raoulia grandiflora** in fell-field of Mount Hector, Tararua Mountains, at about 1400 m. altitude (Ruahine-Cook district). — Photo. E. Bruce Levy.

Fig. 85. **Ranunculus Lyallii** as member of subalpine wet-mountain fell-field on Mount Murray at about 1100 m. altitude near source of River Rakaia (east of Western district), with subalpine-scrub in background.

Photo. L. Cockayne.

Fig. 86. Tall tussock-grassland of almost pure **Danthonia Raoulii** var. **flavescens** in the lower subalpine belt of Mount Dick, South Otago district.

Photo. W. D. Reid.

Fig. 87. Portion of herb-field, Arthur's Pass, at 920 m. altitude with **Celmisia coriacea** in bloom mixed with **Phormium Colensoi** (Western district). — Photo. L. Cockayne.

Fig. 88.   Herb-field of Baird Range near Franz Josef glacier (Western district) at about 1200 m. altitude with a mat of **Celmisia Walkeri** overlying a rock and **C. petiolata** in the background. — Photo. L. Cockayne.

Cockayne, The Vegetation of New Zealand, 2nd ed.

Fig. 89. **Pimelea Gnidia** — of Hebe-form — growing with **Astelia Cockaynei** (foreground), **Phormium Colensoi** (right hand corner of background) and **Dracophyllum filifolium** (left of centre in background) as part of shrubby herb-field of Tararua Mountains (Ruahine-Cook district).

Photo. E. Bruce Levy.

Fig. 90. Low cushion of **Celmisia argentea** in herb-moor of Table Hill, Stewart Island at 570 m. altitude with **Danthonia pungens** in background. — Photo. L. Cockayne.

Fig. 91. **Ranunculus Lyallii** in herb-field near source of River Rakaia (Western district) at 1200 m. altitude. — Photo. M. C. Gudex.

Fig. 92. **Donatia novae=zelandiae,** 1.5 m. long, and growing on it **Celmisia glandulosa** var. **vera, Pentachondra pumila,** and above its upper margin is a line of **Danthonia crassiuscula.** Hanging valley at source of River Routeburn (Fiord district) at about 1200 m. altitude. — Photo. W. D. Reid.

Fig. 93. Cushions of **Montia fontana** in shallow running water at about 1200 m. altitude in the Volcanic Plateau district. — Photo. L. Cockayne.

Fig. 94. Colony of **Rhopalostylis Cheesemanii** in the dry forest of
Sunday Island, Kermadec Islands.

Photo. W. R. B. Oliver.

Fig. 95. Vegetation of sand-covered coastal rock of Chatham Island with **Hebe chathamica** on left, **Sonchus grandifolius** on right and **Festuca Coxii** in centre. — Photo. W. R. B. Oliver.

Cockayne, The Vegetation of New Zealand, 2nd ed.

Fig. 96. Penguin rookery on Snares killing the vegetation; in centre tussock-moor and in background low forest of **Olearia Lyallii**. — Photo. L. Cockayne.

Cockayne, The Vegetation of New Zealand, 2nd ed.

Fig. 97. Interior of **Metrosideros lucida,** forest of Lord Auckland Island showing the prostrate and semi-prostrate trunks.

Photo. S. Page.

Fig. 98. Portion of herb-moor of Adams Island (Lord Auckland group)
at about sea-level with **Pleurophyllum criniferum** in foreground in
blossom and bud.

Photo. S. Page.

Fig. 99. **Poa ramosissima** growing on coastal cliff, Masked Island, Carnley Harbour (Lord Auckland group) and above is a colony of **Blechnum durum.**

Photo. S. Page.

Cockayne, The Vegetation of New Zealand, 2nd ed.

Fig. 100. Interior of southern-rata forest of the Lord Auckland Islands showing the tangle of branches.
Here the usual undergrowth is absent and only a few ferns are present.

Photo. S. Page.

Fig. 101. Roof of **Olearia Lyallii** forest of Ewing Island, Lord Auckland group.
Photo. S. Page.

Cockayne, The Vegetation of New Zealand, 2nd ed.

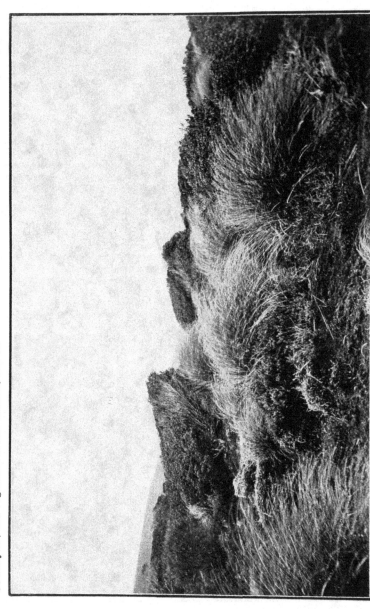

Fig. 102. In foreground is tussock-moor with tussocks of **Danthonia antarctica** and behind is the wind-shorn shrub-form of **Metrosideros lucida** at about 300 m. altitude, Lord Auckland Island.

Photo. S. Page.

Fig. 103. Colony of **Stilbocarpa polaris** in "Fairchild's Garden", Adams
Island (Lord Auckland group) and **Dracophyllum longifolium**
in background.

Photo. S. Page.

Fig. 104. Piece of induced steppe of Central Otago caused by repeated burning and overstocking (sheep and rabbits) but approaching the desert stage; in foreground is a cushion of **Raoulia lutescens.** — Photo. L. Cockayne.

Fig. 105. **Lycopodium fastigiatum** making "fairy rings" in artificial grassland in valley at head of Lake Wanaka, east of Western district. — Photo. W. D. Reid.

Fig. 106. View of a forest-area after a successful bush-burn.
Photo. A. H. Cockayne.

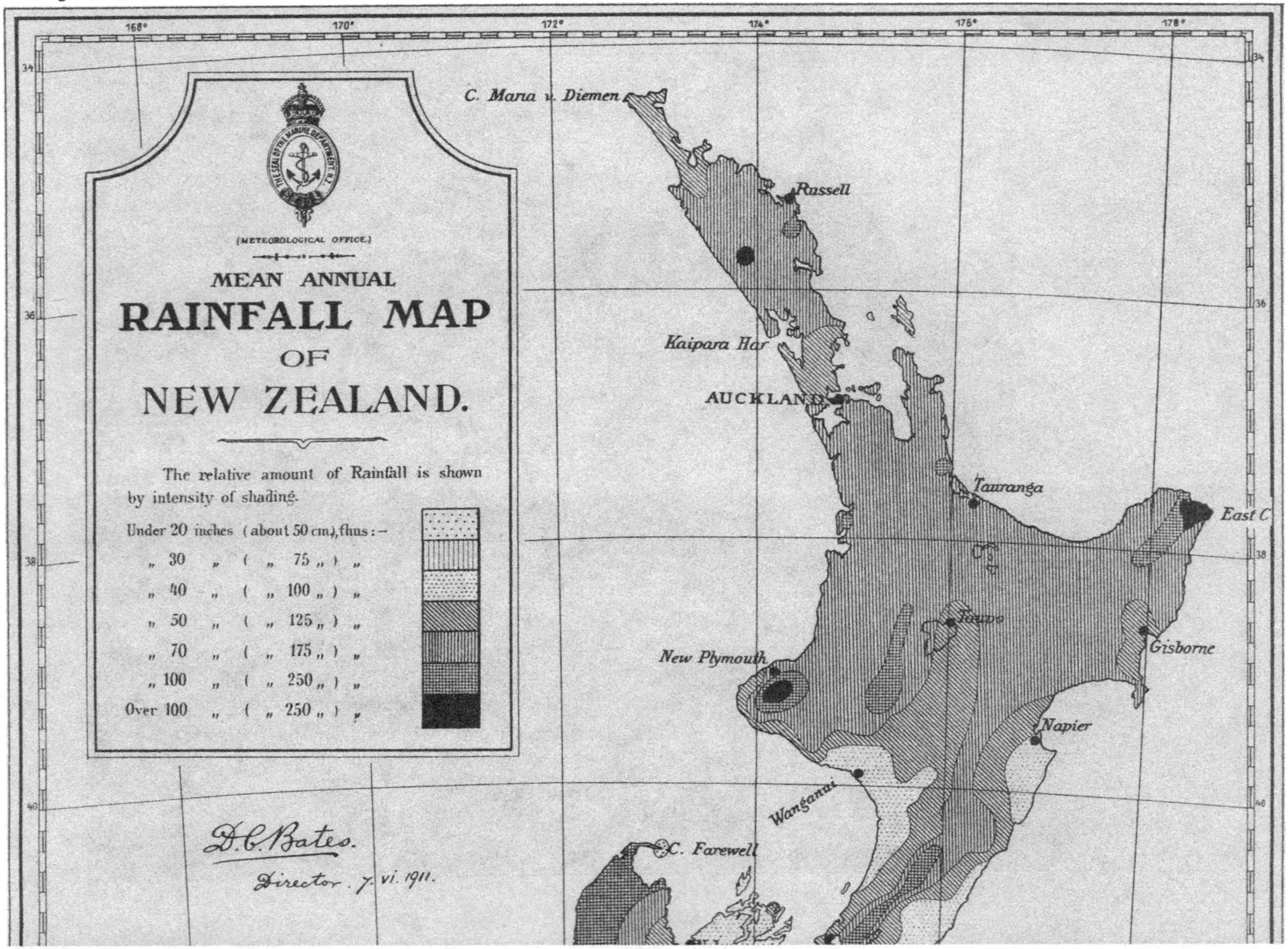

Map I.

Cockayne, New Zealand, 2 nd ed.

(METEOROLOGICAL OFFICE.)

MEAN ANNUAL

RAINFALL MAP

OF

NEW ZEALAND.

The relative amount of Rainfall is shown
by intensity of shading.

Under 20 inches (about 50 cm), thus :—

" 30	"	(	"	75	"	)	"
" 40	"	(	"	100	"	)	"
" 50	"	(	"	125	"	)	"
" 70	"	(	"	175	"	)	"
" 100	"	(	"	250	"	)	"
Over 100	"	(	"	250	"	)	"

L. C. Bates.
Director. 7. vi. 1911.

C. Maria v. Diemen.

Russell

Kaipara Har.

AUCKLAND

Tauranga

East C.

Gisborne

Napier

Wanganui

New Plymouth

C. Farewell

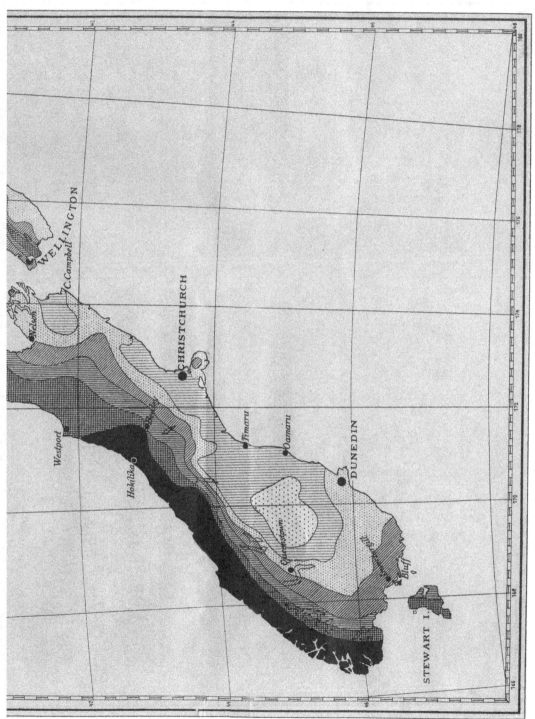

Verlag v. Wilhelm Engelmann in Leipzig.

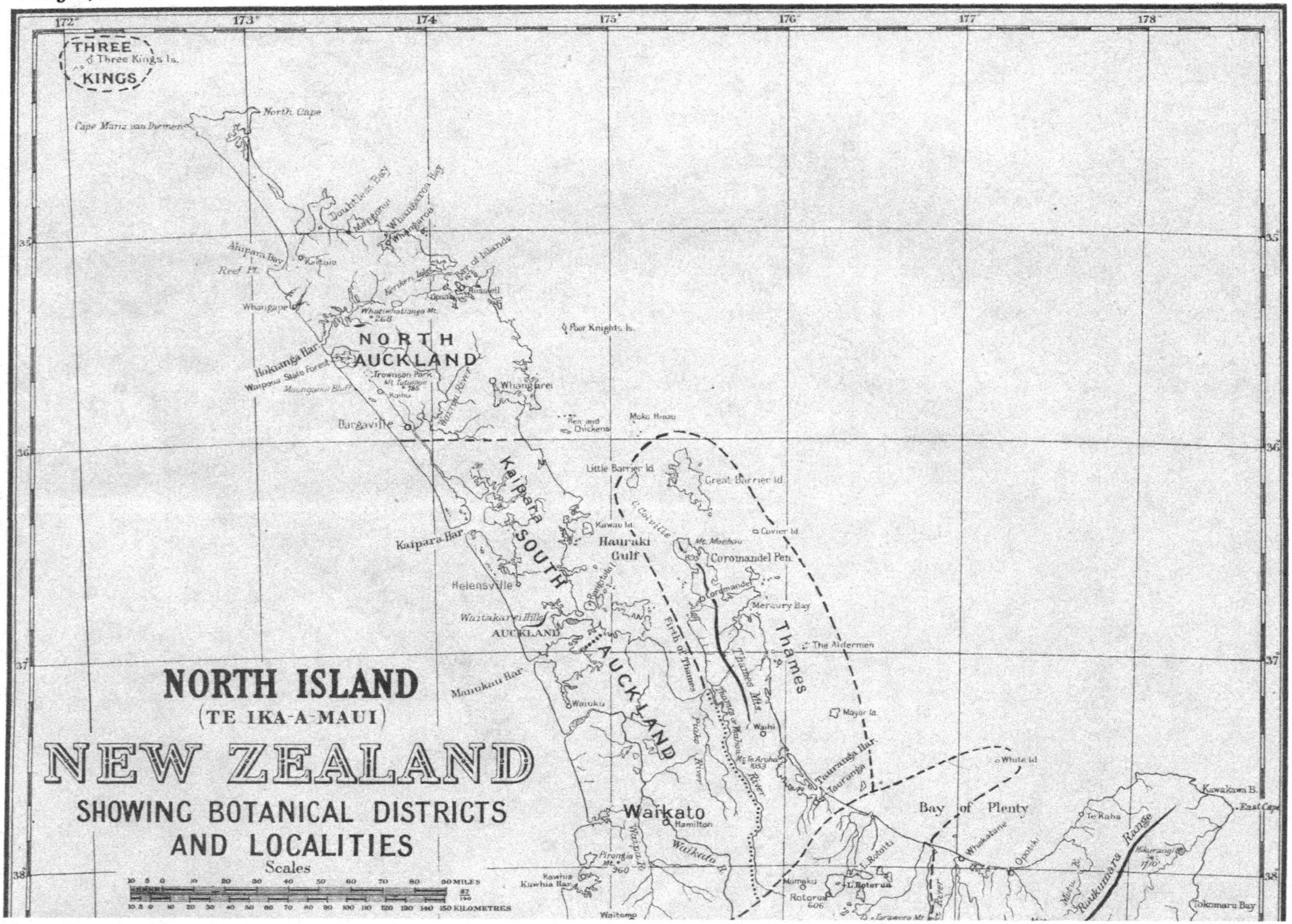

Map II.

Cockayne, New Zealand, 2nd ed.

NORTH ISLAND
(TE IKA-A-MAUI)

NEW ZEALAND

SHOWING BOTANICAL DISTRICTS AND LOCALITIES

Scales

THREE KINGS
(Three Kings Is.)

NORTH AUCKLAND

SOUTH AUCKLAND

Kaipara

Thames

Auckland

Waikato

Bay of Plenty

Hauraki Gulf

Coromandel Pen.

Firth of Thames

Thames Mts.

Great Barrier Id.

Little Barrier Id.

Cape Maria van Diemen

North Cape

Whangaroa

Kaitaia

Hokianga Har.

Kaipara Har.

Helensville

Dargaville

Manukau Har.

Waitakerei Hills

Waipa R.

Thames or Waihou River

Piako River

Waikato R.

Whangarei

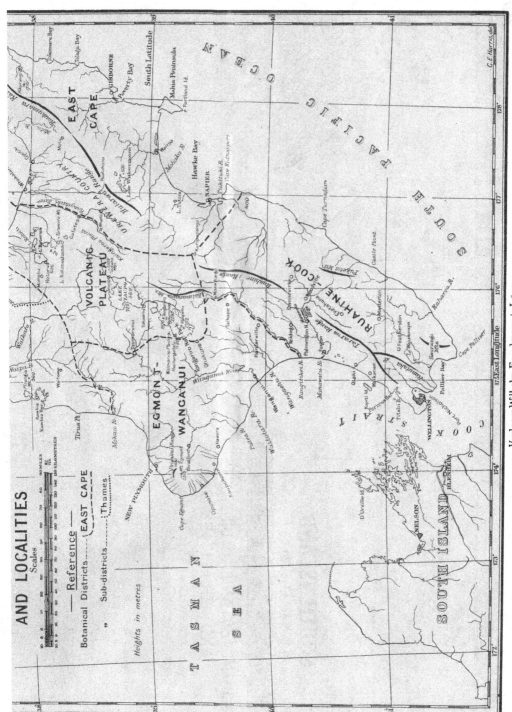

AND LOCALITIES
Scales

——— Reference ———

Botanical Districts ── { EAST CAPE

                         { Thames

     "   Sub-districts ⋯⋯⋯

Heights in metres

SOUTH PACIFIC OCEAN

TASMAN SEA

COOK STRAIT

SOUTH ISLAND

Verlag v. Wilhelm Engelmann in Leipzig.

G.E.Harris, del.

Map III.

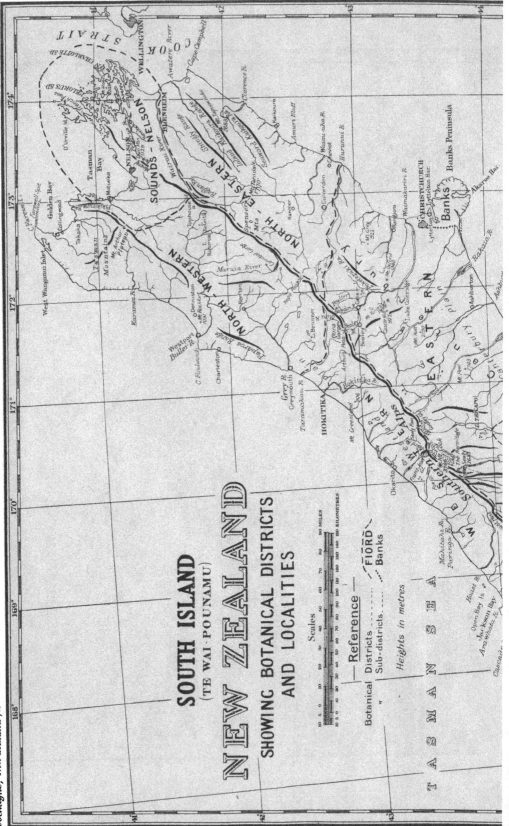

SOUTH ISLAND
(TE WAI-POUNAMU)

NEW ZEALAND

SHOWING BOTANICAL DISTRICTS
AND LOCALITIES

Scales

— Reference —

Botanical Districts
  Sub-districts
  Fiord
  Banks

Heights in metres

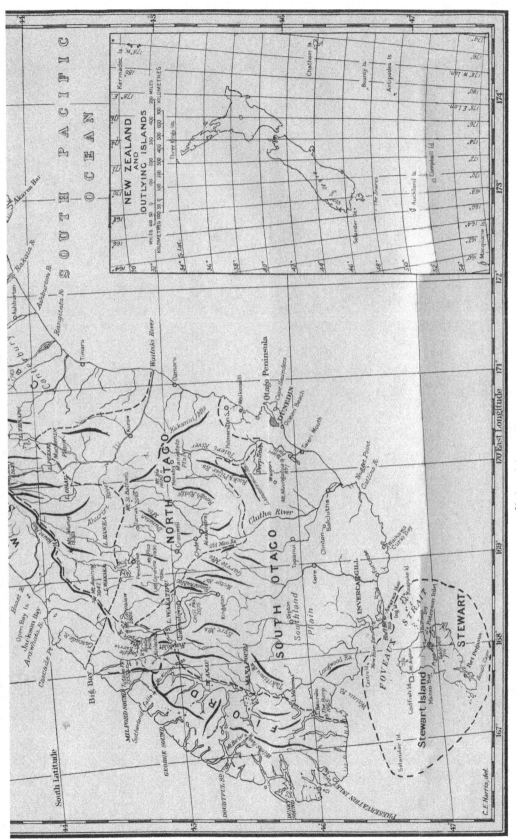

SOUTH PACIFIC OCEAN

Akaroa Har.

Rakaia R.

Ashburton

Rangitata R.

Waitaki River

Timaru

Oamaru

Otago Peninsula

Cape Saunders

DUNEDIN

Ocean Beach

Kakanui Mts.

Kurow

Waikouaiti

NORTH OTAGO

Wanamata Plain

Waipori

Taieri Mouth

Taieri River

Deep Stream

Rock & Pillar Ra.

Palmerston S.O.

Clutha River

Nugget Point

Catlins R.

Clinton

Balclutha

Alexandra

Clyde

Cromwell

Old Man Ra.

Tapanui

SOUTH OTAGO

Gore

Waikawa Curio Bay

Mataura

Garvie Mts.

Hector Ra.

INVERCARGILL

Southland

Plain

Dipton

Knapdale

FOVEAUX STRAIT

Raapuke Id.

Ruapuke Id.

Centre Id.

New River

Bluff

Awarua Bay

Paterson Inlet

STRAIT

Stewart Island

Codfish Id.

Yaran Bay

STEWART

Port Pegasus

South Cape

Solander Id.

PRESERVATION INLET

DUSKY SOUND

DOUBTFUL S'D.

FIORD

L.

MANAPOURI

TEANAU

L. TE ANAU

Takitimu Mts.

Longwood Ra.

Waiau R.

The Hump 1065

Centre I.

Lumsden

Kingston

Eyre Mts.

L. WAKATIPU

Queenstown

Humboldt Mts.

Kinloch

McKinnon

GEORGE SOUND

MILFORD SOUND

Sutherland Falls

Tutoko Pk.

Mt. Earnslaw

Ben Lomond

Remarkable Mts.

WAKATIPU

Cardrona

Pembroke

Wanaka

L. WANAKA

Mt. Aspiring

L. HAWEA

Ahuriri River

Dunstan Mts.

Rough Ridge

Mt. St. Bathans

Naseby

Maniototo

Plain

NORTH OTAGO

Lindis Pass

Lake Ohau

Mackenzie Plain

Cascade Pt.

Big Bay

Haast R.

Jackson Bay

Open Bay Is.

Arawhata R.

Cascade R.

W

South Latitude

Verlag v. Wilhelm Engelmann in Leipzig.

C. E. Harris, del.

Inset map

NEW ZEALAND
AND
OUTLYING ISLANDS

MILES 140 50 0 100 200 300 400 500 MILES

KILOMETRES 100 50 0 100 200 300 400 500 600 800 KILOMETRES

Kermadec Is.

Three Kings Is.

Chatham Is.

Bounty Is.

Antipodes Is.

Auckland Is.

Campbell Id.

Solander Id.

The Snares

Macquarie Id.

Printed in the United States
By Bookmasters